Information Technology

An Introduction for Today's Digital World

Information Technology

An Introduction for Today's Digital World

Richard Fox

CRC Press is an imprint of the
Taylor & Francis Group, an **informa** business
A CHAPMAN & HALL BOOK

CRC Press
Taylor & Francis Group
6000 Broken Sound Parkway NW, Suite 300
Boca Raton, FL 33487-2742

Printed in the United States of America on acid-free paper
Version Date: 20121023

International Standard Book Number: 978-1-4665-6828-0 (Paperback)

Library of Congress Cataloging-in-Publication Data

Fox, Richard, 1964-
Information technology : an introduction for today's digital world / Richard Fox.
p. cm.
Includes bibliographical references and index.
ISBN 978-1-4665-6828-0 (pbk. : alk. paper)
1. Information technology. I. Title.

T58.5.F695 2013
004--dc23 2012036916

Visit the Taylor & Francis Web site at
http://www.taylorandfrancis.com

and the CRC Press Web site at
http://www.crcpress.com

This book is dedicated in loving memory to Barrie and Bernice Fox and Brian Garbarini. I would also like to dedicate this with much love to Cheri Klink, Sherre Kozloff, and Laura Smith and their families.

Contents

Preface

WHY DOES THIS TEXTBOOK EXIST? At Northern Kentucky University, we have offered a 4-year bachelor's degree in Computer Information Technology since 2004. This program, like the field, continues to evolve. In 2011, we added a breadth-first introductory course to our program. The course's role is to introduce students to many of the concepts that they would see throughout their IT studies and into their career: computer organization and hardware, Windows and Linux operating systems, system administration duties, scripting, computer networks, regular expressions, binary numbers, the Bash shell in Linux and DOS, managing processes and services, computer security, and careers in IT.

As I was asked to develop this course, I began looking for a textbook. I realized quickly that this wide range of topics would not be found in any single book. My only options seemed to be to use several books or to utilize some type of computer literacy text. The problem with computer literacy text is that no such book covers the material in the detail we would require, nor would such a text offer a satisfactory introduction to Linux. The problem with requiring several texts is that the students would need at least three or four, and that was an expense that I did not want to burden our students with. As I already had some detailed notes on many of the topics, I took it upon myself to write the text. I did not realize the level of commitment that I was making, but once I had completed the text, I wanted to try to get it published. Fortunately for me, the wonderful folks at CRC Press/Taylor & Francis Group wanted to publish it as well. I am very grateful that they have worked with me to turn my rough draft text into a presentable product.

In creating the publishable text, I reworked many of the chapters and figures (multiple times). Based on reviewer comments, I also added several additional sections and one new chapter. This allowed me to introduce many of the topics recommended by SIGITE (ACM Special Interest Group on IT Education). This book, however, is incomplete and will never be complete for two reasons. First, as the field grows and changes, any text will become quickly outdated. For instance, as I prepare this for press, Microsoft is preparing Windows 8 for release. At that point, will all of my references to Windows 7 be obsolete? Possibly, but hopefully not. Second, the field contains so many topics that I could literally double the length of this text and still not cover all of the material worth mentioning. However, I have to stop at some point, and I hope that I have hit on a wide variety of topics.

This book is suitable for any introductory IT course. It offers students a far more detailed examination of the computer than computer literacy texts, which are commonly used in such courses. Of particular note are the emphasis provided to Windows/DOS and Linux with numerous examples of issuing commands and controlling the operating systems. Topics in hardware, programming, and computer networks further flush out the details. Finally, the text covers concepts that any IT person will need to know from operating system and hardware to information security and computer ethics.

How to Use This Textbook

THE ORDER OF THE 16 CHAPTERS is largely based on the order that the topics have been covered in the first two semesters of offering our CIT 130 course at Northern Kentucky University. This ordering is partially based on the structure of the course, which meets one day per week for a lecture and one day per week for a laboratory session. So, for instance, the placement of computer history in the middle of the text was caused by the fact that the laboratory sessions that pertain to Chapters 5 and 6 carry on for 3 weeks. Similarly, the computer assembly laboratory lasts 2 weeks, and the lectures that we cover over those weeks are from Chapters 2 and 3. As a teacher, you might desire to rearrange the chapters based on your own needs. In some cases, the order of the chapters is important. This is explained below.

- Chapter 1 should be covered first.
- Chapter 2 Section 2.7 may be skipped if you are not covering PC assembly, although it is recommended that students still read the section.
- Chapter 2 should be covered before Chapter 3.
- Chapter 3 should be covered before Chapters 12 and 15. In Chapter 3, Section 3.3 (negative numbers, floating point and fractional values) can be skipped.
- Chapter 4 should be covered before Chapters 8, 9, and 11.
- Chapter 5 should be covered before Chapter 6 (permissions in Chapter 6 to some extent require that you have already covered the file system in Chapter 5).
- Chapter 7 should be covered before Chapter 8.
- Chapter 12 should be covered before Chapter 15.
- Chapter 16 should be covered last.

Acknowledgments and Contributions

I WOULD LIKE TO THANK the following individuals for their assistance in helping me put together this book whether by providing proofreading help, input into the material, or other forms of advice and assistance: Cheri Klink, Scot Cunningham, Justin Smith, Jim Hughes, Wei Hao, Randi Cohen, Stan Wakefield, Renee Human, Jim Leone, Barry Lunt, and Laurie Schlags. I would also like to thank my students in CIT 130 who helped hammer out the content of this book, especially: Kimberly Campbell, Kaysi Cook, David Fitzer, Jason Guilkey, William Hemmerle, James Lloyd, Dustin Mack, Frank Mansfield, Wyatt Nolen, Mark Oliverio, Ronak Patel, Christine Pulsifer, Ishwar Ramnath, Jonathan Richardson, Nathaniel Ridner, Matthew Riley, Jennifer Ross, William Scanlon, Glenn Sparkes, and Adam Thompson. I would like to thank everyone who contributes to the Open Source Community, particularly those who have supported GIMP and Inkscape, two drawing tools that I used extensively to create or manipulate all of the figures in this text.

I would also like to thank many friends and colleagues for all of their support throughout the years, particularly Jim Mardis, Rick Karbowski, Stephanie Stewart, Julie Hartigan, Vicki Uti, Jim Giantomasi, Gary Newell, Russ Proctor, John Josephson, B. Chandrasekaran, Pearl Brazier, John Abraham, and Xiannong Meng.

Author

Richard Fox, PhD, is a professor of computer science at Northern Kentucky University (NKU). He regularly teaches courses in both computer science (Artificial Intelligence, Computer Systems, Data Structures, Computer Architecture, Concepts of Programming Languages) and computer information technology (IT Fundamentals, Unix/Linux, Web Server Administration). Dr. Fox, who has been at NKU since 2001, is the current chair of NKU's University Curriculum Committee, and, in 2012, he received the Faculty Excellence Award for sustained excellence in teaching. Before coming to NKU, Dr. Fox taught for 9 years at the University of Texas–Pan American, where he received the Outstanding Faculty Award for Teaching Excellence (1999). Dr. Fox received a PhD in Computer and Information Sciences from the Ohio State University in 1992. He also has an MS in Computer and Information Sciences from Ohio State (1988) and a BS in Computer Science from the University of Missouri Rolla (now Missouri University of Science and Technology) in 1986. Aside from this textbook, Dr. Fox has written or coauthored several computer literacy laboratory manuals and one introductory programming course laboratory booklet. He is also author or coauthor of more than 45 peer-reviewed research articles primarily in the area of Artificial Intelligence. Richard Fox grew up in St. Louis, Missouri, and now lives in Cincinnati, Ohio. He is a big science fiction fan and a progressive rock fan. As you will see in reading this text, his favorite composer is Frank Zappa.

CHAPTER 1

Introduction to Information Technology

This textbook is an introduction to information technology (IT) intended for students in IT-related fields. This chapter introduces the different career roles of an IT person, with emphasis on system administration, and the types of skills required of an IT professional. In this chapter, the elements that make up the IT infrastructure—the computer, software, users—are introduced.

The learning objectives of this chapter are to

- Describe and differentiate between types of IT careers.
- Describe the set of skills required to succeed in IT.
- Introduce the types of hardware found in a computer system.
- Describe and differentiate between the components of a computer system: hardware, software, and users.

WHAT IS INFORMATION TECHNOLOGY?

So, what is information technology (IT) anyway? IT is a term used to describe several things, the task of gathering data and processing it into information, the ability to disseminate information using technology, the technology itself that permits these tasks, and the collection of people who are in charge of maintaining the IT infrastructure (the computers, the networks, the operating system). Generically, we will consider IT to be the technology used in creating, maintaining, and making information accessible. In other words, IT combines people with computing resources, software, data, and computer networks.

IT personnel, sometimes referred to collectively as "IT," are those people whose job it is to supply and support IT. These include computer engineers who design and build computer chips, computer scientists who write software for computers, and administrators who provide the IT infrastructure for organizations.

What will your role be in IT? There are many varied duties of IT personnel. In some cases, a single individual might be the entire IT staff for an organization, but in many cases, there will be several, perhaps dozens or hundreds of personnel involved, each with separate roles to play. Most IT personnel, however, have two general roles: administration and support. An administrator is someone who is in charge of some portion of the IT infrastructure. There are a variety of administrator roles, as shown in Table 1.1.

Let us examine some of the administrator roles in Table 1.1 in more detail. The most common role in IT is the system administrator. System administration is the process of maintaining the operating system of a computer system. On a stand-alone computer, system administration is minimal and usually left up to the individual(s) using the computer. However, for a network of computers or computers that share files or other resources, system administration becomes more significant and more challenging. The system administrator is the person (or people) who perform system administration.

Maintenance of a computer system (computers, resources, network) will include an understanding of software, hardware, and programming. From a software point of view, administration requires installing software, making it available, troubleshooting problems that arise during usage, and making sure that the software is running efficiently. Additionally, the administrator(s) must understand the operating system well enough to configure the software appropriately for the given organization, create accounts, and safeguard the system from outside attack.

From a hardware point of view, administration requires installing new hardware and troubleshooting existing hardware. This may or may not include low-level tasks such as repairing components and laying network cable. It may also require installing device driver software whenever new hardware is added.

From a programming point of view, operating systems require "fine-tuning," and thus administrators will often have to write their own shell scripts to accomplish both simple

TABLE 1.1 Administrator Roles in IT

Role	Job/Tasks
System Administrator	Administer the computers in an organization; install software; modify/update operating system; create accounts; train users; secure system; troubleshoot system; add hardware
Network Administrator	Purchase, configure, and connect computer network; maintain computer network; troubleshoot network; secure network from intrusion
Database Administrator	Install, configure, and maintain database and database management system; back up database; create accounts; train users
Web Administrator	Install, configure, and maintain website through web server; secure website; work with developers
Web Developer	Design and create web pages and scripts for web pages; maintain websites
Security Administrator	Install, configure, and administer firewall; create security policies; troubleshoot computer system (including network); work proactively against intrusions

and complex tasks. In Linux, for instance, many components of the operating system rely on configuration files. These are often shell scripts. An administrator may have to identify a configuration file and edit it to tailor how that component works within the organization. The goal of writing shell scripts is to automate processes so that, once written, the administrator can call upon the scripts to perform tasks that otherwise would be tedious. A simple example might be to write a script that would take a text file of user names and create a new account for each user name.

System administration may be limited to the administration of the computers, printers, and file servers. However, system administration may extend to network administration and possibly even web server, ftp server, mail server, and database server administration depending on the needs and size of the company and abilities of the system administrator(s). Finally, a system administrator may also be required to train users on the system. Therefore, the skills needed for system administration can vary greatly. Specific common tasks of a system administrator include:

- Account management: creating new user accounts and deleting obsolete user accounts.
- Password management: making sure that all users have passwords that agree with the security policy (e.g., passwords must be changed every month, passwords must include at least one non-alphabetic character)—you might be surprised, but in systems without adequate password management, many users use "" as their password (i.e., their password is just hitting the enter key). Most organizations today require the use of *strong passwords*: passwords that contain at least eight characters of which at least one is non-alphabetic and/or a combination of upper- and lower-case letters, and are changed at least once a month without reusing passwords for several months at a time.
- File protection management: making sure that files are appropriately protected (for instance, making sure that important documents are not writable by the outside world) and performing timely backups of the file system.
- Installing and configuring new hardware and troubleshooting hardware including the network.
- Installing and configuring new software including updating new operating system (OS) patches, and troubleshooting software.
- Providing documentation, support, and training for computer users.
- Performing system-level programming as necessary (usually through scripting languages rather than writing large-scale applications or systems software).
- Security: installing and maintaining a firewall, examining log files to see if there are any odd patterns of attempted logins, and examining suspicious processes that perhaps should not be running.

In many cases, the network administrator is separate from the system administrator. It is the network administrator who is in charge of all aspects of the computer network. The network administrator's duties will include physically laying down cable, making connections, and working with the network hardware (for instance, routers and switches). The network administrator will also have to configure the individual machines to be able to communicate via the network. Thus, like the system administrator, the network administrator will edit configuration files, install software (related to the network), and so forth. Troubleshooting the network will also be a role for the network administrator where, in this case, troubleshooting may combine physical troubleshooting (e.g., is a cable bad?) and software troubleshooting. There is also a security aspect to the computer network. Both the system administrator and network administrator may work on system firewalls. Editing configuration files, writing shell scripts, and installing software and patches will all be part of a network administrators tasks.

Aside from administrative tasks, IT personnel provide support. Support usually comes in two forms: training and help desk. By training, the IT person is responsible for teaching new and current users how to use the IT infrastructure. This may include such simple things as logging into the computer system, setting up printers, accessing shared files, and perhaps training employees in how to use work-related software. The person might create documentation, helpful websites (including wiki pages), and even audiovisual demos, or lead group or individualized training sessions. Because training occurs only as needed (new software, new employees), most support comes in the form of a help desk. In essence, this requires that someone be available to respond to problems that arise at random times. Many large organizations offer 24/7 help desks. The help desk person might simply act as a switchboard, routing the problem to the proper IT person. In other cases, the help desk person can solve the problem directly, often over the phone but sometimes by e-mail or in person.

Where is IT used? IT is ubiquitous today. Nearly everyone on the planet uses some form of computing technology through cellular phones and tablets or home computers, or through school and work. However, most IT personnel are hired to work in IT departments for organizations. These organizations can be just a few people or corporations of tens of thousands. Table 1.2 provides a look at the larger users of IT and how they use IT.

TABLE 1.2 Large-Scale IT Users

Type of Organization	Typical Usage
Business	E-commerce, customer records
Education	Scholastic record keeping, support of teaching
Entertainment	Digital editing, special effects, music composition, advertising
Government	Record keeping, intelligence analysis, dissemination of information
Health/hospitals	Record keeping, medical devices, insurance
Law enforcement	Record keeping, information gathering, and dissemination
Manufacturing	Design, automation/robotics
Research	Computation, dissemination of information

WHO STUDIES IT?

IT personnel in the past were often drafted into the position. Consider the following scenario. Joe received his bachelor's degree in Computer Science from the University of Illinois. He was immediately hired by a software firm in Chicago where he went to work as a COBOL programmer. However, within 3 months, he was asked by the boss, being the new guy, "surely you know something about this Linux operating system stuff, don't you?" Joe, of course, learned Unix as part of his undergraduate degree and answered "Sure." So the boss told Joe "From now on, I want you to spend 10 hours of your week putting together this new network of computers using Linux. Make sure it can connect to our file servers and make it secure." Joe spent 10 hours a week reading manuals, installing the Linux operating system, playing around with the operating system, and eventually getting the system up and running.

After some initial growing pains in using the system, more and more employees switched to the Linux platform. Now, 9 months later, half of the company has moved to Linux, but the system does not necessarily run smoothly. Whenever a problem arises, Joe is usually the person who has to respond and fix it. The boss returns to Joe and says "Fine work you did on the network. I want to move you full time to support the system." Joe did not go to school for this, but because he had some of the skills, and because he is an intelligent, hardworking individual (he would have to be to graduate from University of Illinois's Computer Science program!), he has been successful at this endeavor. Rather than hiring someone to maintain the system, the easier solution is to move Joe to the position permanently. Poor Joe, he wanted to write code (although perhaps not COBOL). But now, the only code he writes are Linux shell scripts!

Sound unlikely? Actually, it was a very common tale in the 1980s and 1990s and even into the 2000s. It was only in the mid 2000s that an IT curriculum was developed to match the roles of IT personnel. Otherwise, such jobs were often filled by computer scientists or by people who just happened to be computer hobbyists. The few "qualified" personnel were those who had associates degrees from 2-year technical colleges, but those colleges are geared more toward covering concepts such as PC repair and troubleshooting rather than system and network administration. Today, we expect to see IT people who have not only been trained on the current technology, but understand all aspects of IT infrastructure including theoretical issues, the mathematics of computers (binary), the roles of the various components that make up a computer system, programming techniques, the operations of databases, networks, the Internet, and perhaps specialized knowledge such as computer forensics.

Common IT curricula include introductions to operating system platforms, programming languages, and computing concepts. We would expect a student to have experience in both Windows and Linux (or Unix). Programming languages might include both scripting languages such as Linux/Unix shell scripting, Ruby or Python, and JavaScript, and compiled languages such as C, C++, Java, or Visual Basic. Concepts will include operating systems and networks but may go beyond these to include web infrastructure, computer architectures, software applications (e.g., business software), digital media and storage, and e-commerce.

TYPES OF IT PROGRAMS

Although the 4-year IT degree is relatively new, it is also not standardized. Different universities that offer such an IT program come at the degree from different perspectives. Here, we look at the more common approaches.

First are the programs that are offshoots of computer science degrees. It seems natural to couple the degrees together because there is a good deal of overlap in what the two disciplines must cover: hardware technology, programming, database design, computer ethics, networking. However, the computer science degree has always heavily revolved around programming, and the IT degree may require less of it. Additionally, math plays a significant role in computer science, but it is unclear whether that amount of math is required for IT.

Next, there are the management information systems variations. The idea is that IT should be taught from a usage perspective—more on the applications, the data storage, the database, and less on the technology underlying the business applications. E-commerce, database design, data mining, computer ethics, and law are promoted here. Furthermore, the course work may include concepts related to managing IT.

Then there is the engineering technology approach that concentrates on hardware—circuit boards, disk drives, PC construction and troubleshooting, physical aspects of networking. There is less emphasis on programming, although there is still a programming component.

Another school of thought is to provide the foundations of computer systems themselves. This textbook follows this idea by presenting first the hardware of the computer system and then the operating systems. We also look at computer networks, programming, and computer storage to have a well-rounded understanding of the technology side to IT. The IT graduate should be able to not only work on IT, say as an administrator, but also design IT systems architecturally from the hardware to the network to the software.

SIGITE, the ACM Special Interest Group on IT Education, provides useful guidelines to build a model IT curriculum.

Who should study IT? To be an IT person, you do not necessarily have to have the rigorous mathematical or engineering background of computer scientists and computer engineers; there are many overlapping talents. Perhaps the most important talent is to have *troubleshooting* skills. Much of being an IT person is figuring out what is going wrong in your system. These diagnostic skills cannot be purely taught. You must have experience, background knowledge, and instinct. Above all, you have to know how the system works whether the system is a Linux operating system, a computer network, a web server, or other. Another talent is the ability to write program code—in all likelihood, you would write small programs, or scripts, as opposed to the software engineer who will be involved in large-scale projects.

You should also be able to communicate with others so that you can understand the problems reported by your colleagues or clients, and in turn describe solutions to them. This interaction might take place over the phone rather than in person. You should also be able to write technically. You may often be asked to produce documentation and reports. Finally, you will need the ability to learn on your own as technology is ever-changing. What you have learned in school or through training may be obsolete within a year or two. Yet, what you learn should form a foundation from which you can continue to learn. See Table 1.3, which highlights the skills expected or desired from IT personnel.

TABLE 1.3 IT Skills

Skill	Description	Example(s)
Troubleshooting, problem solving	Detect a problem Diagnose its cause Find a solution (means of fixing it)	Poor processor performance Disk space full Virus or Trojan horse infection
Knowledge of operating systems	Operating system installation Application software installation User account creation System monitoring	Versions of Linux Versions of Unix Windows Mac OS
System level programming	Shell scripts to automate processes Manipulating configuration files for system services	Bash, Csh scripts DOS scripts Ruby scripts C/C++ programs
System security	Ensuring proper system security is in place Following or drafting policies for users Monitoring for threats	Configuring a system firewall Installing antiviral/antimalware software Examining log files for evidence of intrusion and system security holes Keeping up with the latest security patches
Hardware	Installing and configuring new hardware Troubleshooting, replacing or repairing defective hardware	Replacing CPUs and disk drives Connecting network cables to network hubs, switches, routers

There probably is not a prototypical IT student. But an IT student should:

1. Enjoy playing around with the computer—not just using it, but learning how it works, learning how to do things on it at the system level
2. Enjoy learning on your own—liking the challenge of figuring things out, especially new things
3. Think that technology is cool—to not be afraid of technology but to embrace it in all of its forms
4. Enjoy troubleshooting

It is not necessarily the case that the IT student will enjoy programming. In fact, many students who select IT as a career make this choice because they first start with computer science but soon tire of the heavy programming requirements of that discipline. This is not to say that the IT student does not program, but that the programming is less intensive, requiring mostly writing small shell scripts. As an example, a student of computer science might, in the course of his or her studies, write a word processor, a database management system, or a language translator, whereas an IT student might write scripts to automate user account creation, or write client-side scripts to ensure that web forms have been filled in correctly, or write server-side scripts to process web forms.

There are many facets of the system administration position not covered above that are worth noting. Students may think that by studying for an IT career, they will get a job where they get to "play around" with technology. It is fun, but it is also a challenge—it is almost something that they already do as a hobby. And yet the student, when hired, might be responsible for maintaining the IT infrastructure in an organization of dozens or hundreds of employees. The equipment may cost hundreds of thousands of dollars, but the business itself might make millions of dollars. Therefore, the IT specialist must take their job seriously—downtime, system errors, intrusions, and so forth could cost the organization greatly. The IT specialist has duties that go beyond just being a system administrator. Some of these expectations are elaborated upon below.

To start, the system administrator must be aware of developments in the field. At a minimum, the system administrator has to know the security problems that arise and how to protect against them. These might include securing the system from virus, network intrusions, denial of service attacks, and SQL injection attacks. In addition, the system administrator should keep up on new releases of the operating system and/or server software that he/she maintains. However, a system administrator may have to go well beyond by reading up on new hardware, new software, and other such developments in the field.

In order for the system administrator to keep up with the new technology, new trends, and new security fixes, continuing education is essential. The system administrator should be a life-long learner and a self-starter. The system administrator might look toward Internet forums but should also regularly read technology news and be willing to follow up on articles through their own research. The system administrator should also be willing to dive into new software and experiment with it to determine its potential use within the organization.

A system administrator will often be "on call" during off hours. When disaster strikes, the system administrator must be accessible. An emergency call at 3 A.M. or while you are on vacation is quite possible. Although every employee deserves their own down time, a system administrator's contract may include clauses about being reachable 24/7. Without such assurance, an organization may find themselves with inaccessible data files or the inability to perform transactions for several hours, which could result in millions of dollars of damage. Some companies' reputations have been harmed by denial of service attacks and the inability to recover quickly.

The system administrator must also behave ethically. However, it is often a surprise to students that ethics is even an issue. Yet, what would you do if you are faced with some moral dilemma? For instance, your employer is worried that too many employees are using company e-mail for personal things, and so the boss asks you to search through everyone's e-mail. How would you feel? Now, imagine there is a policy in the company that states that employees can use company e-mail for personal purposes as long as e-mail does not divulge any company secrets. In this case, if you are asked to search through employee e-mail, would this change how you feel about it?

Unethical behavior might include:

- Spying on others (e-mail, web browsing habits, examining files)
- Setting up backdoor accounts to illegally access computer systems
- Illegally downloading software or files, or encouraging/permitting others to do so
- Performing theft or sabotage because of your system administration access

IT INFRASTRUCTURE

IT revolves around the computer. Have you used a computer today? Even if you have not touched your desktop (or laptop) computer to check your e-mail, chances are that you have used a computer. Your cell phone is a computer as is your Kindle. These are far less powerful than desktop units, but they are computers nonetheless. There are computer components in your car and on the city streets that you drive. The building you work or study in might use computers to control the lighting and air conditioning. Yes, computers are all around us even if we do not recognize them.

We will define a computer to be a piece of electronic equipment that is capable of running programs, interacting with a user (via input–output devices), and storing data. These tasks are often referred to as the IPOS (input, processing, output, storage) cycle. A general-purpose computer is one that can run any program. Many devices today are computers but may not be as general purpose as others. For instance, your iPod is capable of playing music; it has a user interface, and may have a small number of applications loaded into it to handle a calendar, show you the time of day, and offer a few games. Your cell phone has an even greater number of applications, but it is not capable of running most software. The degree to which a computer is general purpose is largely based on its storage capacity and whether programs have been specifically compiled for the processor.

Computers range in size and capability—from supercomputers that can fill a room, to desktop units that are not very heavy but are not intended to be portable, to laptop units that are as light as perhaps a heavy textbook, to handheld devices such as cell phones and mp3 players. The general difference between a handheld unit and a desktop or laptop unit is the types of peripheral devices available (full-sized keyboard and mouse versus touch screen, 20-in. monitor versus 2-in. screen), the amount of memory and hard disk storage space, and whether external storage is available such as flash drives via USB ports or optical disks via an optical disk drive.

Computers

We will study what makes up a computer in more detail in the next chapter. For now, we will look at the computer in more general terms. A computer is an electronic, programmable device. To run a program, the device needs a processor [Central Processing Unit (CPU)], memory to store the program and data, input and output capabilities, and possibly long-term storage and network capabilities (these last two are optional). Based on this definition, computers not only encompass desktop and laptop units, servers, mainframe computers, and supercomputers, but also netbooks, cell phones, computer game consoles, mp3 players, and book readers (e.g., Kindles). In the latter two cases, the devices are special-purpose—they run only a few select programs. The notion of the historical computer is gone. Today, we live with computers everywhere.

Figure 1.1 illustrates some of the range in computers. Desktop units with large monitors and system units are common as are laptop computers today with large monitors. Even more popular are handheld devices including personal digital assistants (PDAs), cell phones, and e-book readers. Monitors are flat screens. We no longer expect to find

FIGURE 1.1 Types of computers. (Adapted from Shutterstock/tele52.)

bulky monitors on our desktop computers. Even so, the system unit, which allows us to have numerous disk drive devices and other components, is bulky. We sacrifice some of the peripheral devices when we use laptop computers. We sacrifice a greater amount of accessibility when we move on to handheld devices. In the case of the PDA, laptop, and notebook, the chips and motherboard, and whatever other forms of storage, must be placed inside a very small area. For the PDA, there is probably just a wireless card to permit access to the cell phone network (and possibly wi-fi). For the laptop and notebook computers, there is probably a hard disk drive. The laptop will also probably have an optical disk drive.

The main component of a computer is the *processor*. The processor's role is to process—that is, it executes the programs we run on the computer. To run a program on a given computer, the program has to be *compiled* for that computer. Compilation is a language translation process that converts a program from a more readable form (say Python or Java) into a computer's machine language. A computer can only run programs that are written in that machine's language. We discuss this concept in more detail, and the specific role of the CPU, in Chapter 2. We examine programming languages later in the textbook.

Aside from the processor, computers need storage. There are two types of storage—short-term storage and long-term storage. Short-term storage is most commonly random access memory (RAM). Unfortunately, RAM can describe several different types of memories. Our modern computers typically have three forms of RAM, dynamic RAM (what we typically call main memory), static RAM (cache memory and registers), and ROM (read-only memory). We differentiate between these types of memory in Chapter 2. For now, just consider all three to be "memory".

Main memory (dynamic RAM) is composed of chips. Dynamic RAM offers fast access and often large storage capacity. However, some handheld devices do not have room for much dynamic RAM storage. Instead, they use flash memory storage, which is more limited in capacity. Long-term storage most commonly uses hard disk drives but can also comprise optical disk, flash memory, and magnetic tape. Long-term storage is far greater in capacity than the short-term storage of main memory, and because additional storage space can always be purchased, we might view long-term storage capacity as unlimited. Typical desktop and laptop computer short-term storage is in the 1–8 GB range. 1 GB means 1 gigabyte, which is roughly 1 billion bytes. Think of a byte as 1 character (letter) such that 1 GB will store 1 billion characters. A typical book is probably on the order of 250,000 to 1 million characters. 1 GB would store at least 1000 books (without pictures). Hard disks can now store 1 TB, or 1 terabyte, which is 1 trillion bytes. Obviously, long-term storage is far greater in capacity than short-term storage. Some common storage sizes are shown in Table 1.4. We will study storage sizes in more detail in Chapters 2 and 3.

TABLE 1.4 Storage Sizes

Size	Meaning	Example
1 bit	A single 0 or 1	Smallest unit of storage, might store 1 black-and-white pixel or 1 true/false value, usually we have to combine many bits to create anything meaningful
1 byte (1B)	8 bits	We might store a number from 0 to 255 or −128 to 127, or a single character (letter of the alphabet, digit, punctuation mark)
1 word	32 or 64 bits	One piece of data such as a number or a program instruction
1 KB	1024 bytes	We might store a block of memory in this size
1 MB	~1 million bytes	A small image or a large text file, an mp3 file of a song might take between 3 and 10 MB, a 50-min TV show highly compressed might take 350 MB
1 GB	~1 billion bytes	A library of songs or images, dozens of books, a DVD requires several gigabytes of storage (4–8 GB)
1 TB	~1 trillion bytes	A library of movies

Do we need both forms of storage? It depends on the type of device and your intended usage, but in general, yes we need them both. Why? To run a program, we need to load that program into a storage space that responds quickly. Long-term storage is far slower than RAM, so unless you are willing to have a very slow computer, you need short-term storage. On the other hand, short-term storage is far more limited in capacity and the programs we run tend to be very large. We also often have very large data files (music files, movies, etc.) such that we cannot rely solely on short-term storage. Handheld devices offer a compromise—they often use flash memory instead of RAM, which results in a slower access time when compared to desktop/laptop computers, and they have a limited long-term storage space (if any), requiring that the user move files between the handheld devices and a permanent storage space (say on a desktop computer) fairly often.

Aside from the difference in speed and storage capacity between memory and long-term storage, another differentiating factor is their volatility. The term *volatile*, when describing memory, indicates whether the type of memory can retain its contents when the power supply has been shut off. Main memory (DRAM) and cache/register memory (SRAM) are volatile forms of memory. Once you turn the power off, the contents are lost. This is why, when you turn on your computer, memory is initially empty and the device must go through a "boot" process. Nonvolatile memories include ROM, flash drives, hard disk, and optical disk. The nonvolatile memory retains its contents indefinitely. In the case of ROM, the contents are never lost. In the case of flash drives and disks, the contents are retained until you decide to erase them.

Computers also require *peripheral* devices (although require is not the right word; perhaps we should say that for the user's convenience, we add peripheral devices). The word peripheral means "on the outskirts" but in the case of a computer, we usually refer to peripherals as devices that are outside the computer, or more specifically, outside of the system unit. The system unit is the box that contains the motherboard (which houses the CPU and memory) and the disk drive units. The peripherals are devices that are either too large to fit inside the system unit, or devices that must be accessible by the human users (Figure 1.2). These devices are our input and output devices—keyboard, mouse, track point, track ball or joystick, monitor, printer, speakers, pen and tablet (writing area) or light pen, etc. Without these input and output devices (known as I/O devices collectively), humans would not be able to interact with the computer. If all input data comes from a disk file and all output data similarly will be stored to disk file, there may be no need for the computer to interact with the human. But the human will eventually want to know what the program did.

Among the peripheral devices are the communication device(s). A communication device is one that lets a computer communicate with other computers. These devices are typically MODEMs, which can either require connection to a telephone line (or perhaps a cable TV coaxial line) or be wireless. Nearly all laptop computers today come with wireless MODEMs, whereas desktop units may come with a wired or wireless MODEM. However, in cases where computers are connected to a local area network (LAN), the computer requires a network connection instead of or in addition to the MODEM. The LAN connection is by means of a network card, often an Ethernet card. For high-speed networks,

FIGURE 1.2 Computer peripherals. (Courtesy of Shutterstock/Nevena.)

the network card offers a much higher bandwidth (transmission rate) than a MODEM. We will study wired and wireless MODEMs and network cards later in the textbook when we look at computer networks.

Let us now summarize our computer. A computer is in essence a collection of different devices, each of which performs a different type of task. The typical computer will comprise the following:

1. System unit, which houses
   a. The motherboard, which contains
      i. The CPU
      ii. A cooling unit for the CPU
      iii. Possibly extra processors (for instance, for graphics)
      iv. Memory chips for RAM, ROM
      v. Connectors for peripherals (sometimes known as ports)

vi. Expansion slots for other peripheral device cards

vii. The ROM BIOS for booting and basic input and output instructions

viii. Power supply connector

b. Disk drives

c. Fan units

d. Power supply

2. A monitor and keyboard

3. Typically some form of pointing device (mouse, track point, track ball)

4. Speakers (optional)

5. MODEM or network card (these are typically located inside the system unit, plugged into one of the expansion slots)

6. Printer (optional)

7. External storage devices such as external hard disk and tape drive

Chapter 2 has pictures to illustrate many of the above components.

Now we have defined a computer. But the computer is only a part of the story. Without software, the computer would have nothing to do. And without people, the computer would not know what program to run, nor on what data. So, our computer system includes these components.

Software

What is the point of a computer? To run programs. Without programs, the computer has nothing to do. A program, also known as *software* (to differentiate it from the physical components of the computer, the *hardware*), is a list of instructions that detail to the computer what to do. These instructions are written in a programming language, such as Java or Python. Programming language instructions must be very descriptive. For instance, if you want the computer to input two numbers from the user and output which one is larger, you could not just say "input two numbers and output the larger of the two." Instead, you must describe the actions to take place as an algorithm, broken into step-by-step instructions. The instructions must be written in a programming language. For instance, the problem described in this paragraph might be broken into four steps:

Input number1

Input number2

Compare the two numbers and if the first is greater than the second, output number1

Otherwise output number2

In a language like C, this would look like this:

```
scanf("%d", &number1);
scanf("%d", &number2);
if(number1 > number2) printf("%d is greater", number1);
else printf("%d is greater", number2);
```

The scanf instruction inputs a value, the printf instruction, outputs a value or message. The if instruction is used to compare two values and make a decision. Some of the syntax in C is peculiar, for instance the & before "number1" and "number2" in the scanf statements, the use of the semicolon to end instructions, and the use of %d. Every language will have its own syntax and in many cases, the syntax can appear very odd to someone who is not a programmer or has not learned that language.

Programs are not just a list of executable statements. Programs also require various definitions. These might include variable declarations, functions or methods, and classes. In C, for instance, we would have to define number1 and number2 as being variables to be used in the above code. In this example, they would be declared as integer numbers.

There are many forms of software, but we generally divide them into two categories: system software (the operating system) and application software (programs that we run to accomplish our tasks such as a word processor, an Internet browser or a computer game). Usually, our software is written by professionals—software engineers. However, once you learn to program, you can write your own software if you desire. As an IT student, you will learn to write short pieces of code, scripts. Scripts can be used to support either the operating system or an application. For instance, you might write a Bash shell script to support an activity such as automatically creating user accounts for a new group of users. Or you might write a server-side script in Perl to test a URL for security threats in a web server.

Users

Without the human, the computer would not have anything to do. It is the user who initiates the processes on the computer. "Do this now, do that later." We may want to interact with the programs while they run. This interactivity is done through the I/O devices. Today, we are so used to interactivity that we probably cannot imagine using computers without it. But in earlier days (1940s–1970s), most—if not all—processing was done without human interaction at all. The user specified the program, the source of input, the location of output, and sent the program off to run. The user would see the results once the computer ran the program, which might have been immediately, or many hours later!

Users have progressed over time, just as the technology has progressed. The earliest computer users were the engineers who built and programmed them. Computers were so complicated and expensive that no one else would have access. As computer costs permitted organizations to purchase them (for millions of dollars), computer users were those employees who had received specialized training to use them. Things began to change with the advent of personal computers, first released in the 1970s. But it was not until windowing operating systems came about that computer users could learn to use the computers with little to no training. And so today, it is common that anyone and everyone can use a computer. In fact, computers are so commonplace that people may not realize that they are using a computer when they program their GPS or run an application on their smart phone.

Our View Today

Computers used to be easily identifiable. They were monstrously expensive devices that would weigh tons, filled up a room or more, required clean room environments and special air conditioning. People would not actually touch the computer; they would interface with the computer through terminals and networks. With personal computers, computers for individuals became affordable and many people began to have computers in their own homes. Telecommunication, over LANs or over telephone networks, allowed people to connect their computers together, to communicate to each other and share e-mail messages, files, programs, etc. The Internet, which was first turned on in its earliest form in 1969, became commercially available to home computer users in the mid-1990s. Early in this period, people connected to the Internet via slow MODEM access over their telephones. But over the past 15 years, telecommunications has changed completely. Now, we have wireless access, high-speed Internet connections, cell phones, and more.

Today, computers are not easily identifiable. They are no longer limited to mainframe computers or desktop units. You can have a network computer or a laptop, a notebook computer, a tablet computer, a handheld computer. We even have devices smaller than handheld units that use processors and memory. And our connectivity has changed equally. Your access to telecommunications is no longer limited by the telephone port in your home. With wireless, you can gain access anywhere in your household or anywhere in a remote location that has hot spots. Want a coffee break? No problem, go to Starbucks and you can still access the Internet through your laptop. Or, taking a drive? You can still access the Internet over your cell phone (as long as you are in reasonable proximity to a cell phone tower). We are a world united through nearly instantaneous communication no matter where we are. And we are a world of billions upon billions of processors. We used to count computers by the millions, but today, there are tens of billions of processors and most of these can communicate with each other.

This gentle introduction to IT will serve as our starting point in this text. Over the chapters to come, we will study many IT-related concepts. We first look at computer components, gain an understanding of what they do, how they work, and how we connect them to a computer. We also study a related topic, binary numbers and how we use binary. These next two chapters on computer organization and binary are often material covered in computer science curricula. They are included here so that, as an IT person, you understand more than what a processor and a motherboard are when it comes to the computer hardware. By having a firm foundation of what the computer components do and how they work, you should be able to understand the necessity of when to increase RAM, or how to evaluate a processor. The inclusion of binary in this text is largely to support concepts found in computer networks.

The focus shifts to system software, that is, operating systems. We examine two of the most common operating system platforms: Windows (Windows 7) and Unix (Red Hat

Linux). We will compare and contrast what they look like, how we use them, and how we configure them. Operating system topics include file systems, users, accounts and permissions, processes and process management, and services. We also examine two Linux-specific topics: the Bash shell and the use of regular expressions in Linux.

The text examines several different histories. The evolution of computer hardware, the evolution of operating systems, the evolution of computer programming, the history of both Linux and Windows, and the history of the Internet are all covered (although not in the same chapter). Although perhaps not necessary for an IT person, it does help set a backdrop to how technology has changed so that you will have an appreciation of the rapidity behind the changes in IT. Additionally, by understanding the past, it might help you understand where IT might lead.

The final collection of chapters covers other IT topics. Computer networks are considered from several different perspectives. The logical structure of a network, the physical nature of a network, the network protocols that proscribe how transmitted data are to be treated, and some of the more common network software are all examined in one chapter. Software management describes the types of software available and provides details for how to install software in a computer system, particularly in Linux with open source software. Another chapter concentrates on programming, offering examples of writing scripts in both the Linux shell and DOS. The penultimate chapter of the text covers the information side of IT. In this chapter, we examine such ideas as information management and information assurance and security. A final chapter wraps up the text by considering careers in IT and various topics related to IT professionals.

FURTHER READING

There are a number of websites that provide information on IT careers, some of which are listed below.

- http://www.wetfeet.com/careers-industries/careers/information-technology
- http://www.cio.com/article/101314/The_Hottest_Jobs_In_Information_Technology
- http://www.careeroverview.com/technology-careers.html
- http://www.techcareers.com/
- http://information-technology.careerbuilder.com/

The best source for IT education can be found through the special interest group on IT education (SIGITE) at http://www.sigite.org/it-model-curriculum.

General introductions to computer hardware, software, and users can be found in any number of computer literacy texts such as these:

- Beekman, G. and Beekman, B. *Tomorrow's Technology and You*. Upper Saddle River, NJ: Prentice Hall, 2008.

- Fuller, F. and Larson, B. *Computers: Understanding Technology.* St. Paul, MN: ECM Paradigm Publishing, 2010.
- Meyer, M., Baber, R., and Pfaffenberger, B. *Computers in Your Future.* Upper Saddle River, NJ: Prentice Hall, 2007.
- Laberta, C. *Computers Are Your Future.* Upper Saddle River, NJ: Prentice Hall, 2011.
- Williams, B. and Sawyer, S. *Using Information Technology.* New York: McGraw-Hill, 2010.
- Snyder, L. *Fluency with Information Technology: Skills, Concepts and Capabilities.* Upper Saddle River, NJ: Prentice Hall, 2010.

However, as someone who wishes to make a career of IT, you would be better served with more detailed material. Such texts will be listed in later chapters as we cover material in greater depth. See the further readings in Chapter 2 for more information on computer hardware, Chapter 4 for more information on operating systems, Chapter 14 for more information on programming, and Chapter 16 for more information on IT careers.

REVIEW TERMS

The following terms were introduced in this chapter:

Administrator	Peripheral
Computer	Processor
Hardware	Network administrator
Help desk	Software
Information Technology	Storage capacity
IT specialist	System administrator
MODEM	User

REVIEW QUESTIONS

1. What are the skills expected of an IT specialist?
2. What does administration mean in reference to IT?
3. What does training mean in reference to IT?
4. How does the study of IT differ from a 2-year technical degree in computers?
5. To what extent should an IT specialist be able to write computer programs?

6. What is a system administrator? What is a network administrator? How do the two jobs differ?
7. Define a computer.
8. What is the IPOS cycle?
9. Should a cell phone be considered a computer?
10. How does a computer system differ from a computer?
11. How do short-term and long-term storage differ?
12. What is software? What are the two general forms of software?

DISCUSSION QUESTIONS

1. As a student of IT, what brought about your interests in studying IT? Having read this chapter, are you as interested in IT as you were before, more interested or less interested?
2. Organize the IT skills listed in Table 1.2 in order of most important to least important for a system administrator. Defend your listing based on the types of tasks that a system administrator will be required to undertake.
3. Table 1.2 did not include "soft skills" such as the ability to communicate with others, the ability to work in groups, and the ability to manage projects. Are these types of skills taught or are they learned in other ways? Should a 4-year IT program include courses that cover such skills?
4. What are the differences between computer information technology and computer science? Should a program in computer science include computer information technology courses, or should they be separate programs?
5. How does a 4-year IT degree differ from a 2-year IT degree or a degree earned at an IT technical school?
6. Compare computers of today to those that existed in the 1950s.
7. In your lifetime, what changes have you seen in computers and other information technology (particularly handheld devices)? What changes do you expect to see in the next 10–15 years?
8. Many people are surprised to learn that smart phones should be considered computers. In what ways are smart phones similar to desktop and laptop computers? In what ways are they different? Should ordinary cell phones be considered computers?

CHAPTER 2

Computer Organization and Hardware

This chapter examines the hardware of a computer: the central processing unit (CPU), memory, the I/O subsystem, and the bus. Each of these components is described in detail to provide an understanding of its role. In examining the CPU, the chapter emphasizes the fetch–execute cycle. An example program is used to illustrate the fetch–execute cycle. The CPU's components are themselves examined: the control unit, the arithmetic logic unit, and registers. In examining memory, the memory hierarchy is introduced and the different forms of memory are discussed: RAM, SRAM, DRAM, ROM, virtual memory. In the discussion of input and output devices, topics of human–computer interaction (HCI) and ergonomics are emphasized. The chapter includes a discussion on how to assemble the various computer components. The intent of this chapter is to provide the IT student with a solid foundation in computer hardware.

The learning objectives of this chapter are to

- Identify the components of a computer and their roles.
- Describe the fetch–execute cycle.
- Discuss characteristics that impact a processor's performance.
- Describe the different levels of the memory hierarchy.
- Describe the role of the various types of input and output devices.
- Discuss the impact that I/O devices have on the human body and the importance of HCI.
- Illustrate how a computer can be assembled from component parts.

Computer organization is the study of the components of a computer, their function, their structure (how they work), and how they are connected together. This topic is common

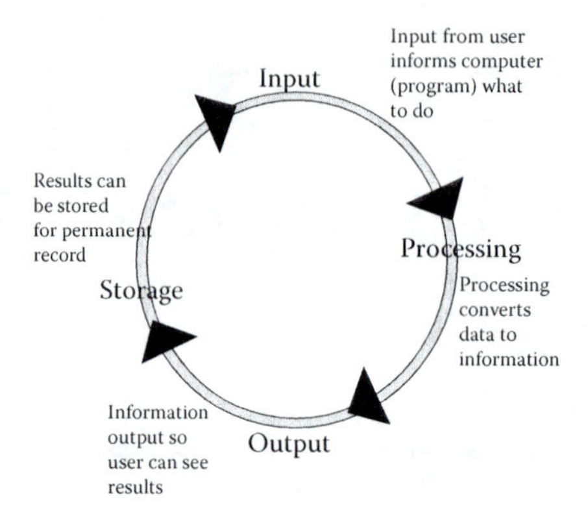

FIGURE 2.1 IPOS cycle.

in computer science programs as a sophomore or junior level class. Here, while we will look at these components, we will look more at their function and how they are connected together. By studying computing organization, you will gain an understanding of how computers work and the importance of each of the primary components, as introduced in Chapter 1. This chapter concludes with a section that discusses computer hardware and the process of assembling (building) a computer from component parts.

A computer consists of a central processing unit (CPU), memory and storage, and peripheral devices (including some form of network connection). A computer performs four operations:

1. Input
2. Processing
3. Output
4. Storage

The IPOS cycle—input–processing–output–storage—describes roughly how we use a computer. We input data, process it, output the results, and store any information that we want to keep permanently. Figure 2.1 illustrates the IPOS cycle. Describing the computer at this level does not really tell us what is going on—for instance, how does storage differ from memory, and how does the processing take place? So we will take a closer look here.

STRUCTURE OF A COMPUTER

Figure 2.2 shows the overall structure of most computer systems. As you can see, there are four components: the CPU (or processor), memory, input/output (I/O) subsystem, and the bus. The CPU is the device that not only executes your programs' instructions, but also commands the various components in the computer. Memory stores the program(s) being executed and the data that each program is using. The I/O subsystem includes all peripheral devices (input, output, storage, network) where storage consists of the long-term

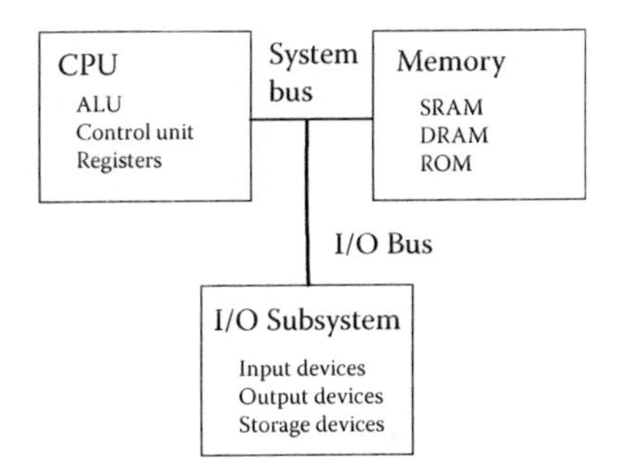

FIGURE 2.2 Structure of modern computers.

storage devices (disk drives, tape). The bus is the device that permits information to move between each component.

Before we continue with our look at the computer's components, we have to understand that the devices in the computer are *digital* devices. Inside the computer, digital data are represented as electrical current being in one of two states: high or low. High means that there is a noticeable current flowing through the component, and low means that there is no, or nearly no, current flowing. We will assign the number 1 to be high current and 0 to be low (or no) current. The bus then is perhaps an easy device to understand. The bus consists of a number of wires, each wire allows 1 bit (a single 1 or 0, high or low current) to flow over it. We discussed storage capacity previously in Chapter 1 (see Table 1.4) but for now, we will define three terms: a bit (a single 1 or 0), a byte (8 bits, using eight wires on the bus, usually the smallest unit of data transfer), and a word (today, computers have either 32-bit or 64-bit words; the word size is the typical size of a datum).

The bus actually consists of three parts: the address bus, the control bus, and the data bus. These three parts of the bus perform the following operations:

- The address bus is used by the CPU to send addresses to either memory or the I/O subsystem. An address is the location that the CPU either wants to retrieve a datum from, or the location that the CPU wants to store a datum to. The CPU more commonly addresses memory than I/O. Note that a datum might either be a value to be used in a program, or a program instruction.

- The control bus is used by the CPU to send out commands. Commands might be "read" or "write" to memory, or "input", "output", or "are you available" to an I/O device. The control bus is also used by the various devices in the computer to send signals back to the CPU. The primary signal is known as an interrupt, which is a request by a device (e.g., disk drive, printer) to interrupt the CPU and ask for its attention.

- The data bus is used to send data (including program instructions) from memory to the CPU, or data from the CPU to memory, or between I/O devices and the CPU or memory.

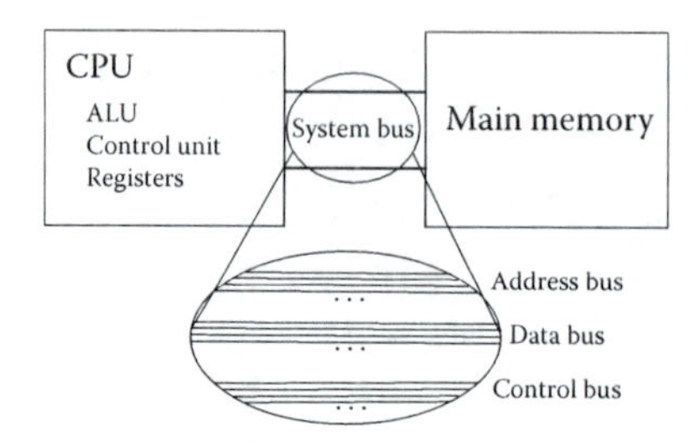

FIGURE 2.3 System bus connecting CPU and memory.

The size of the data bus is typically the size of the computer's word. The size of address bus is based on the size of addressable memory. The size of the control bus is based on the number of different commands that the CPU might send out. See Figure 2.3, where the data bus would probably be 32 bits in size, and the address and control bus size would depend on a number of different issues. The bus shown in this figure is the system bus. Another bus, the local bus, connects the components of the CPU [ALU (arithmetic/logic unit), control unit, registers] and yet another bus connects the I/O devices together.

WHAT ARE PINS?

Take a look at a CPU. You will see a number of metal pins that come out of the bottom. The pins are used to connect the CPU into a socket. The socket connects to the system bus. Therefore, each pin connects to one line of the bus. Earlier CPUs had few pins, whereas today CPUs can have more than a thousand pins! Thus, pins have to be very small to fit. Also, every pin requires a certain amount of power so that current could potentially flow over any or all pins at any time. So our modern CPUs require greater power consumption. This, in turn, gives off more heat, so we need more powerful cooling of the CPU. Below are two Intel CPUs, an early CPU from the 1970s with just a few dozen pins, and a modern Pentium processor with almost 500 pins.

(Adapted from Kimmo Palossaari, http://commons.wikimedia.org/wiki/File:IC_DIP_chips .JPG, and Stonda, http://commons.wikimedia.org/wiki/File:Pentium-60-back.jpg.)

A PROGRAM

In order to understand the CPU, we have to understand a computer program. This is because the CPU executes programs. So, we will first examine a simple program. The following program is written in C although it could be written in nearly any language. The only thing that would change is the syntax. The // marks indicate comments. Comments will help you understand the program.

```
#include <stdio.h>              //input/output library
void main( )                    //start of the program
{
    int a, b, c;                //use 3 integer variables
    scanf("%d", &a);            //input a
    scanf("%d", &b);            //input b
    if(a < b)                   //compare a to b, if a is less then b
        c = a + b;              //then set c to be their sum
        else c = a-b;           //otherwise set c to be their difference
    printf("%d", c);            //output the result, c
}
```

This program inputs two integer values from the user, and if the first is less than the second, it computes c = a + b (c is the sum of the two) otherwise, it computes c = a – b (c is a minus b). It then outputs c. If, for instance, the user inputs 5 and 10, it computes c = 15 and outputs 15. If instead the user inputs 5 and 3, it computes c = 2 and outputs 2. This is a fairly pointless program, but it will serve for our example.

Once written, what happens to the C program? Can I run this program? No. No computer understands the C programming language. Instead, I have to run a special program called a *compiler*. The compiler translates a program into a simpler language called *machine language*. The machine language version of the program can then be executed. Machine language is archaic and very difficult to read, so we will look at an intermediate form of the program, the *assembly language* version. Again, comments will occur after // marks.

```
Input 33        //assume 33 is the keyboard, input a value
                //from keyboard
Store a         //and store the value in the variable a
Input 33        //repeat the input for b
Store b
Load a          //move a from memory to CPU, a location
                //called the accumulator
Subt b          //subtract b from the accumulator
                //(accumulator = a - b)
Jge else        //if the result is greater than or equal
                //to 0, go to location "else"
Load a          //otherwise, here we do the then clause,
                //load a into accumulator
Add b           //add b (accumulator is now a + b)
Store c         //store the result (a + b) in c
```

```
        Jump next       //go to the location called next
else:   Load a          //here is the else clause, load a into the
                        //accumulator
        Subt b          //subtract b (accumulator is now a - b)
        Store c         //store the result (a - b) into c
next:   Load c          //load c into the accumulator
        Output 2049     //send the accumulator value to the output
                        //device 2049, assume this is the monitor
        Halt            //end the program
```

The assembly language version of the program consists of more primitive instructions than the C version. For instance, the single scanf statement in C, which can actually be used to input multiple values at a time, is broken into input and store pairs for each input. The C if statement is broken into numerous lesser statements. Now, recall that an assembly language program is supposed to be simpler than a high level language program. It is simpler in that each assembly language instruction essentially does one primitive thing. However, for our programs to accomplish anything useful, we need far more assembly language instructions than high level language instructions. You might think of the two types of languages in this way: an assembly language instruction is like the nuts and bolts of construction, whereas the high level language instruction is like prefabricated components such as hinges and locks. The computer cannot do something like scanf, or an if–else statement directly in one instruction, but it can do a load, add, store, or jump in single operations.

As with the C program, a computer cannot execute an assembly language program either. So, our program must next be converted from assembly language into machine language. The reason why we showed the assembly language program is that it is easier to understand than a machine language program.

A compiler is a program that translates a source program written in a high-level language, such as C, into machine language. An assembler is a program that translates an assembly language program into machine language. So imagine that we have translated the previous assembly program into machine language. A portion of the final program might look something like this:

```
1000100 0000000000000000000100001 —input (from keyboard)
1000111 0010011000100101101010001 —store the datum in a
1000100 0000000000000000000100001 —input (from keyboard)
1000111 0010011000100101101010010 —store the datum in b
```

The remainder of the machine language program is omitted because it would make no more sense than the above listing of 1s and 0s. If you look at the 1s and 0s, or binary numbers, you might notice that they are formatted into two parts. These two parts represent each instruction's operation (the first 7 bits) and the operand, or datum. Here, the operation denotes one of perhaps 100 different instructions. 1000100 represents "input" and 1000111 represents "store". There would similarly be operations for "add", "subt", "jge", "load", and "jump". The operand denotes a memory location written in binary (for instance, datum a is stored at location 0010011000100101101010001, which is address 5,000,017, and the value

00000000000000000000100001 denotes a device number, 33, for the keyboard). Although this machine language code is made up for the example, it is not too different from what we might find if we studied specific machine languages.

EXECUTING THE PROGRAM

Okay, back to computer organization. We want to run this simple program. The first thing that happens when you want to run a program is that the operating system loads the program from where it is stored on the hard disk into a free section of memory. We will assume that the operating system has placed it at memory location 5,000,000. Because there are 17 instructions, the program will be stored consecutively from memory location 5,000,000 to 5,000,016. We will also assume that the variables, a, b, and c, are stored immediately after the program, in memory locations 5,000,017 through 5,000,019, respectively, for a, b, and c. Once loaded, the operating system transfers control to the processor (CPU) to begin running this program.

Now we have to understand in more detail what the CPU and memory do. The CPU consists of the control unit, the ALU, and registers. One of those registers is called the program counter, or PC (not to be confused with the generic name of a home computer). The PC gets the address of the first program instruction in memory, 5,000,000. Another register is called the instruction register (IR). It stores the current instruction. Another register is called the status flags (SFs); it stores the result of the most recent ALU computation in terms of whether the result was positive, negative, zero, caused an overflow, caused an interrupt, had a value of even parity, and so forth. Each of these items is stored in 1 bit, so the SF will store multiple results, although most of the bits will be 0. Other registers are data registers—they store data that we are currently using. One special data register, called the accumulator (AC), is used for storing the most recent value computed or used.

We are ready to look at how the CPU runs the program. The CPU performs what is known as the *fetch–execute cycle*. The idea behind this cycle is that the CPU first fetches an instruction from memory, and then executes it. In fact, there is more to it than that. The typical fetch–execute cycle (the cycle differs depending on the processor) will have four or five, or maybe more, stages (Figure 2.4).

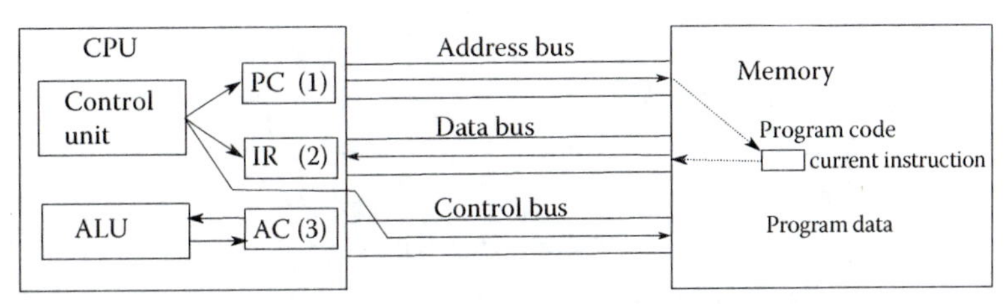

1. Control unit moves PC to address bus and signals memory "read" command over control bus, memory returns instruction over data bus to be stored in IR
2. Control unit decodes instruction in IR
3. Execute instruction in the ALU using datum in AC, putting result back in the AC

FIGURE 2.4 Three-part fetch–execute cycle.

For our example, we will assume a five-part cycle* with the following stages:

1. Fetch instruction from memory.
2. Decode the instruction from machine language into microcode.
3. Fetch the operands from registers.
4. Execute the instruction.
5. Store the result back into a register.

In Figure 2.4, steps 3–5 are all indicated as step 3, executing the instruction using the data register. However, in many CPUs, there are several, even dozens of registers. This requires that steps 3 and 5 be separated from step 4.

Step 4 from our five-part cycle is where the instruction is executed. However, as you see, this is not the only step required to execute an assembly language operations. Without all of the stages, the program does not run correctly. Microcode, mentioned in step 2, will be discussed later in the chapter.

So let us see how the program discussed in the last section will be executed on this five-part cycle. The first thing that happens is that the CPU fetches the first program instruction. This occurs by sending the value stored in the PC (the memory location of the first instruction) to memory over the address bus, and sending a "read" signal over the control bus. Memory receives both the address and the "read" command, and performs the read access at the given memory location. Whatever it finds is then sent back to the CPU over the data bus. In this case, what is sent back to the CPU is not a datum but a program instruction, and in the case of our program, it is the instruction "Input", written in binary as:

```
1000100 0000000000000000000100001
```

Once the instruction is received by the CPU, it is stored in the IR. To end this stage, the CPU increments the PC to 5,000,001 so that it now indicates the location of the next instruction to be fetched.

The next stage is to decode the instruction. The control unit breaks the instruction into two or more parts—the operation (in this case, the first 7 bits) and the operand(s). In essence, the control unit consults a table of operations to find 1000100, the instruction "Input". This informs the CPU that to execute the instruction, it needs to perform an input operation from the input device given by the address in the operand. The operand is the binary number for 33, which (in our fictitious computer) is the keyboard.

As there are no operands to fetch, stage 3 is skipped. Stage 4 is the execution of the instruction. The input instruction requires that the CPU communicate with the input device (the keyboard) and retrieve the next datum entered. The execution of the input instruction is not typical in that the CPU does not proceed until the user has entered something. At that point, the CPU retrieves the datum over the data bus and brings it into

* Different processors have different lengths for their fetch–execute cycles, from just a few to dozens.

the CPU. The fifth and final stage requires moving the value from the data bus into the AC register.

We have now seen the full execution of our program's first instruction. What happens next? The entire fetch–execute cycle repeats. In fact, it will continue repeating until the program terminates.

For the second instruction, the first step is to fetch the instruction from memory. Now, the PC has the address 5,000,001, so the CPU fetches the instruction at that location from memory. The PC value is placed on the address bus, the control unit signals a memory read across the control bus, memory looks up that address and returns the contents over the data bus, and the item is stored in the IR. The last step of the fetch phase is to increment the PC to now point at 5,000,002. The instruction in the IR is

```
1000111 0010011000100101101010001
```

which is "store a".

The decode stage determines that the operation is a store operation, which requires moving a datum from the AC into main memory. The address that will receive the datum from the AC is stored as the second portion of the instruction, 5,000,017 (the address of a).

To execute the instruction, the latter part of the IR is moved to the address bus, the value in the AC is moved to the data bus, and the control unit signals a memory "write" over the control bus. The execution of the instruction is now in the hands of main memory, which takes the value from the data bus and stores it at the memory location received over the address bus (5,000,017). This instruction does not require a fifth step as the CPU itself does not need to store anything.

The next two instructions occur in an almost identical manner except that these instructions are at 5,000,002 and 5,000,003, respectively, and the second datum is stored at memory location 5,000,018 (b). By the time these two instructions have been executed, the PC will be pointing at location 5,000,004, and memory locations 5,000,017 and 5,000,018 will store the first two input values, a and b, respectively.

The fifth instruction is fetched and stored in the IR. This instruction is "load a". The execution of this instruction sends the address, 5,000,017, across the address bus and a memory read signal over the control bus. Memory looks up the datum stored at this location and sends it back over the data bus. The final stage of this instruction is to store the resulting datum in the AC. Notice that unlike the previous "memory read" operations discussed in this section, this is a memory read of a true datum, not of an instruction. Figure 2.5 illustrates the difference between the "load" (upper portion of the figure) and "store" (lower portion of the figure) instructions, that is, between a memory read and a memory write.

The sixth instruction, "subt b", starts off similar to the fifth, "load a". The instruction is fetched as before, with the "subt b" placed in the IR and the PC incremented. The control unit then decodes this instruction. However, the execution stage for subtraction differs from load because this instruction requires two separate execution steps. First, memory is read, similar to "load a" but in this case, the address is 5,000,018 (the variable b). The datum that is returned from memory is not stored in the AC though. Instead, the contents

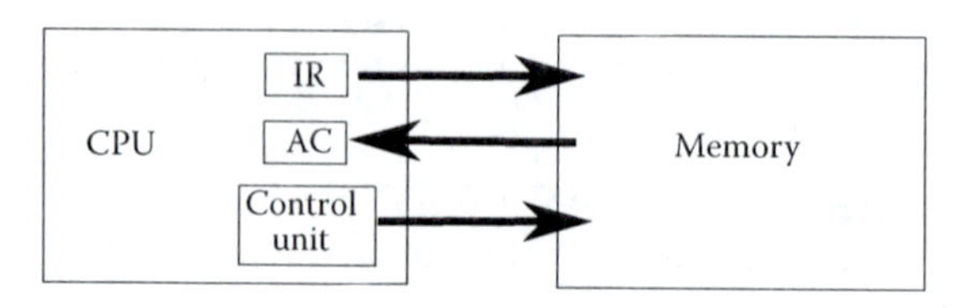

1. Address from IR to address bus
   control unit signals memory read
   over control bus
2. Memory accesses address
   returns datum over data bus
3. Datum stored in AC

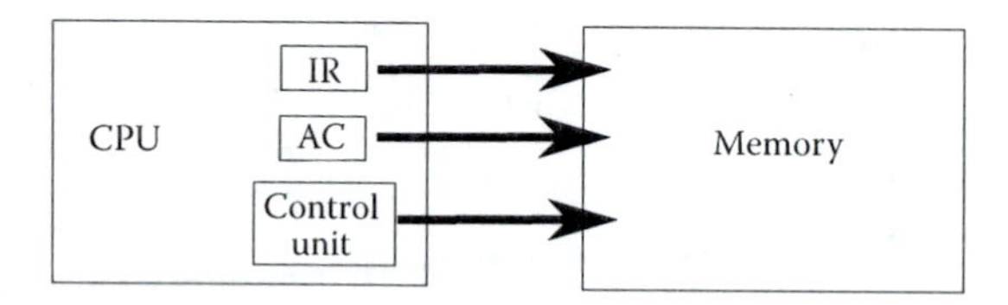

1. Address from IR to address bus
   datum from AC over data bus
   control unit signals memory write
   over control bus
2. Memory accesses address
   stores datum from data bus to
   memory location

FIGURE 2.5 Memory read (top) versus memory write (bottom).

of the AC and the datum returned from memory are both sent to the ALU. The control unit signals the ALU to perform a subtraction. The subtraction circuit operates on the two data and the result, coming out of the ALU, is the value from the AC minus the value from memory (i.e., a – b). The store result step of this instruction is to take that result from the ALU and move it to the AC.

The ALU also sets a status flag based on the result of the subtraction. The SFs are used to indicate the results of ALU operations, such as whether the result was positive, zero, or negative, or whether the result caused an overflow or a carry out or an interrupt. The flags are usually denoted as PF (positive flag), ZF (zero flag), NF (negative flag), OF (overflow flag), and so forth. Any ALU operation will result in at least one flag being set (changing from 0 to 1). We will see how these flags are used in the next instruction.

The seventh instruction (at 5,000,006) is the first instruction that you might not understand by reading the assembly language code. It says "jge else". This means "jump on greater than or equal to the location else". The location "else" is another memory location. Because the first instruction in the program is at memory location 5,000,000, the "else" location is at 5,000,011 (it is the 12th instruction in the program). Refer back to the program in the last section. As with all previous instructions, this instruction ("jge else") is fetched from memory into the IR and decoded. The execution stage works as follows:

> If either the PF or ZF is set (the subtraction resulted in a positive or zero result), then reset the PC to the location of "else". If the NF is set, then do nothing.

This means that, if the previous instruction (the subtraction) resulted in a positive or zero result, branch to location "else". We use the PF and ZF here because the instruction specifies "greater than or equal to". Had the instruction been "less than", we would only use the NF. Other comparisons include "equal to", "not equal to", "greater than", and "less than or equal to". At the end of this instruction, the PC will either be 5,000,007 (the next sequential instruction) or 5,000,011 (the location of else).

The eighth instruction then depends on what happened in the seventh. Either the CPU fetches the instruction at 5,000,007 or 5,000,011. Whichever is the case, the next three instructions are nearly identical. They are to "load a", "add/subt b", and "store c". That is, the group of three instructions either perform c = a + b or c = a – b depending on which of the two sets of code is executed. Assuming that we have executed the instructions starting at 5,000,007, then the instruction at 5,000,010 is "jump next". This instruction will change the PC value to be 5,000,012 (the location of next). Whichever path led us here, the last three instructions are "load c", "output", "halt". The output instruction takes whatever is in the AC and moves it to the output device listed (2049, presumably the monitor). When the CPU executes the "halt" instruction, the program ends and control reverts to the operating system.

ROLE OF CPU

The CPU processes our programs. It does so using two different pieces of hardware. The first is the ALU, which executes all arithmetic and logic operations. The ALU has individual circuits for performing the various operations. An adder is used to perform both addition and subtraction. A multiplier is used for multiplication. A divider is used for division. A comparator is used to compare two values (for instance, to determine if a > b). Other circuits perform shifting, rotating, and parity computation (see Chapter 3).

The second piece of hardware is the control unit. The control unit is responsible for controlling (commanding) all of the components in the computer. As we saw in the example in Executing the Program, the control unit sends out such signals as a memory read or a memory write. It also sends signals to the ALU such as to perform a subtraction or a comparison. The control unit controls the fetch–execute cycle. First, it accomplishes the fetch stage. Then it decodes the fetched instruction into *microcode*. This, in turn, instructs the control unit on how to execute the instruction. Therefore, the control unit will have separate sets of commands for the instruction fetch stage, and for every machine instruction. Some instructions require an operand, and so the control unit handles how to acquire the operand (stage 3 from the fetch–execute cycle discussed in Executing the Program). Some instructions require the ALU, so the control unit informs the proper ALU circuit to operate. Some instructions require that a result be stored (stage 5 of the fetch–execute cycle from Executing the Program), so the control unit moves the datum from its source to the AC (or other register).

Microcode is a confusing topic; however, we will briefly describe it here. Recall that the control unit sends out control signals to all of the components in the computer. This might be a signal to memory to perform a read, or a signal to the ALU to perform an add, or a signal to move a datum from one location (say the output of the adder) to a register. Microcode is the specific operations that should take place within the given clock cycle. For instance, at the beginning of a fetch, the first step is to move the value from the PC to the address bus and signal memory to perform a read. Those two commands are the only two actions to take place during that clock cycle. The microcode for that step will be a binary listing of which actual components should receive a command. The binary listing is almost entirely made up of zeroes because, as we see with the instruction fetch, only two components out of the entire computer do anything. So the microcode for this step has two 1s and the rest are 0s. In this way, microcode looks much like machine language. However, although our example machine language in A Program consisted of 32-bit instructions, our microcode instructions are as long as there are components in the computer. For instance, if there are 50 different components to command, the microcode instruction would be 50 bits long. Here, let us imagine that the computer has 50 components to command, and that the command to move the PC value to the address bus is control signal 0 and the signal to memory to perform a read is control signal 15. Microcode for this step would be:

```
10000000000000100000000000000000000000000000000000
```

A single microcode instruction is sometimes called a micro-operation. There is a different micro-operation for each step of the fetch–execute cycle, plus one or more micro-operations for every machine language instruction. Recall that step 2 of the fetch–execute cycle was to convert the machine instruction (such as "load a") into microcode. Once this is done, executing the step is merely a matter of the control unit sending out the signals, where each bit in the micro-operation denotes a bit to be sent out over a different control bus line. This topic is very advanced and if you do not understand it, do not worry, because it is not important for the material in this text.

We have already talked about various registers. There are two different classes of registers: those used by the control unit and those used to store data for the ALU. The control unit uses the PC (program counter), IR (instruction register), SFs, Stack Pointer, and possibly others. Although our example in A Program and Executing the Program referenced a single data register, the AC (accumulator), modern computers have several, possibly hundreds of, registers. The Intel 8088 processor (used in early IBM PC computers) and later the Pentium processors use four integer data registers given the names EAX, EBX, ECX, and EDX. Other computers might name their registers as R0, R1, R2, and so forth. These registers store data to be used during computations in the ALU.

The speed of a computer is usually provided in terms of clock speed. Modern computers have clock speeds of several GHz (gigahertz). What does this term actually mean? 1 GHz means 1 billion clock cycles per second, or that the clock operates at a speed of 1 billionth of a second. This makes it sound like a 1-GHz CPU executes 1 billion instructions per second. This is not true. Recall that a fetch–execute cycle might consist of five stages or more. It turns out that each stage of the fetch–execute cycle requires at least 1 clock cycle for that stage to be accomplished.

Consider the fetch stage as discussed in Executing the Program. It required at least three clock cycles. In the first cycle, the PC value is sent across the address bus to memory and the control unit signals a memory read. In the second cycle, memory returns the instruction across the data bus. In the third cycle, the instruction is stored in the IR, and the PC is incremented. If we assume that decoding takes one cycle, if operand fetching takes one cycle, that execution of the instruction takes anywhere from one to three cycles, and storing the result takes one cycle, then a single fetch–execute cycle, equivalent to one machine language instruction, will take six to eight clock cycles. Notice though that different processors will have different length fetch–execute cycles. A 1-GHz processor might have a five-stage cycle requiring eight clock cycles per instruction. A 2.5-GHz processor might have a 12-stage cycle requiring some 20 clock cycles per instruction. Which is faster? The 2.5-GHz processor has a faster clock but takes more cycles to execute an instruction. In fact, in this case, the two would have equal performance (20/2.5 billion = 8/1 billion).

Another way to gauge the speed of a processor is to count how many instructions can be executed within 1 second. This value is often expressed in MIPS (millions of instructions per second). Since computer graphics and many computations require the use of floating point values (numbers with decimal points), another term is Megaflops (millions of floating point operations per second).

None of these, however, tells the full story of our processor speed. Processor speed is also impacted by the following:

- Word size, which limits the size of data moved at any one time (the word size is typically the size of the data bus and the size of the registers; smaller word sizes usually mean more data transfers over the bus and lengthier execution times to compute large values)
- Cache performance (see the next section) and memory speed and size
- The program itself (some programs require resources that are slower or more time consuming than others)
- Whether the computer is running multiple programs at a time versus running on an unloaded system
- The impact of the operating system on performance
- The impact of virtual memory on performance
- Many other issues

To obtain a true picture of the processor's performance, computer architects will test processors against *benchmark* programs. The performance on any single program can be misleading, so the benchmarks that are used to test out processors are suites of programs that test different aspects of the processor. Only by looking at the overall benchmark performance can we gain a good understanding of how a processor performs. But the important point here is not to be fooled by the GHz rating of a processor—it tells you something about the processor, but not as much as you might think. If you had to order computers for your organization, you would be best served by not just reading the packaging and seeing the GHz rating, but by reading about each processor's performance in publications (e.g., *Consumer Reports*, *PC World*) and websites (www.cpubenchmark.net, www.geek.com).

ROLE OF MEMORY

Early in the history of computers, it was thought that memory would only store the data being processed. The computer program being executed would be input one instruction at a time from punch card. However, with computers such as the ENIAC (Electronic Numerical Integrator and Computer) being able to execute 5000 instructions per second, inputting program instructions from punch cards would reduce the ENIAC's performance to that of the punch card reader. John von Neumann was the first to conceive of the stored program computer. The idea is that a computer's memory would store both program code and program data. In this way, the CPU would not be burdened by the slow input offered by the punch card reader. However, there is a significant problem with relying on memory (RAM) and that is that CPU speeds continue to improve substantially every year but RAM access speed improves only very gradually over the years. The result is that the CPU now has to wait on memory to respond. Figure 2.6 illustrates the enormous improvement in processor speed versus the modest improvement in memory access time. Notice that the performance increase shown on the Y axis is in an exponential scale, increasing by units of 2. For instance, between 1982 and 1986, memory access time barely increased, whereas CPU performance quadrupled. As the CPU's performance roughly doubles every couple of years, whereas main memory access time barely increases over the years, main memory access speed lags behind CPU speed more and more with each passing year.

The CPU relies on memory at least once per instruction—to fetch the instruction from memory. Some instructions require either reading from memory (load instructions) or

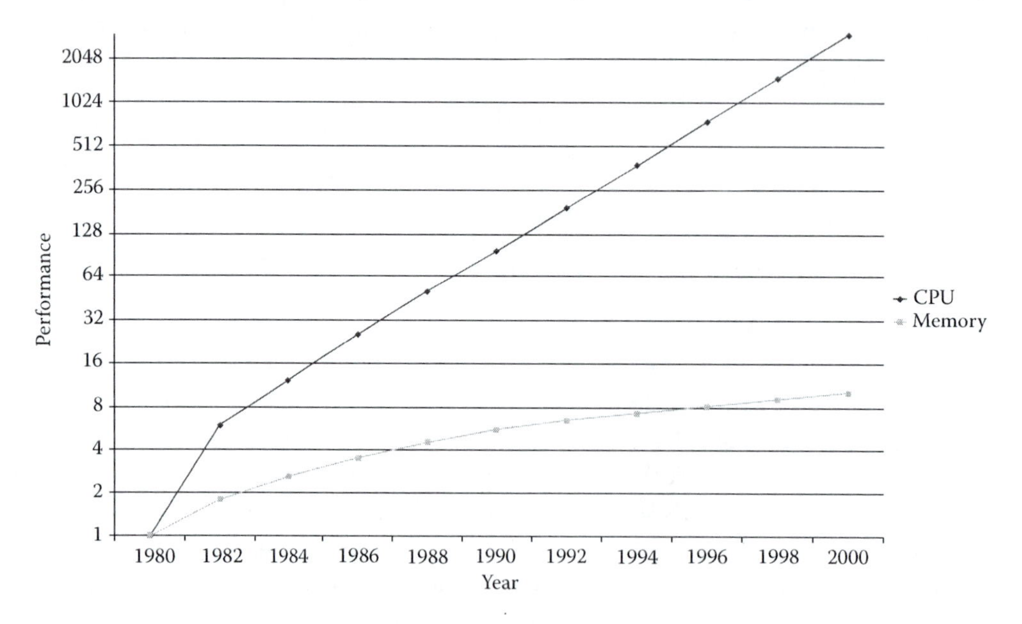

FIGURE 2.6 Improvement in CPU speed over memory access time.

writing results to memory (store instructions). In such an instruction, there will be two memory accesses per instruction, one instruction fetch and one data load or store. In fact, some processors allow multiple reads and writes to memory with each instruction. If the CPU is much faster than memory, what happens? Does this ruin the processor's performance because it is always waiting on memory?

Before continuing, we need to define RAM. RAM stands for *random access memory.* The idea is that we send memory an address and a read command. Memory looks up the item and returns it. Alternatively, we send memory an address, a datum, and a write command. Memory then stores the item at that location. The term "random" is meant to convey that addresses will be sent in a seemingly random pattern. The term is also meant to express that access to any address should take the same amount of time no matter which address is sent. This differentiates RAM from forms of storage that take different amounts of time depending on the location of the item being sought (tape, for instance, might require a good deal of fast-forwarding or rewinding, and disk takes time for the read/write head to be moved accordingly). Unfortunately, the term RAM is somewhat ambiguous—there are three forms of memory that all qualify as RAM.

We differentiate these forms of RAM as dynamic RAM (DRAM), static RAM (SRAM), and ROM. ROM is *read-only memory.* It is memory where the information being stored is permanently fused into place, so it can only be read, not written to. Most computers store the boot program in ROM along with the basic input/output system (BIOS) but little else. We will not worry about ROM in this text. Of the other two forms of RAM, DRAM is the older. DRAM consists of capacitors, which can be miniaturized to such an extent that we can literally have billions on a single chip. DRAM is also very inexpensive. DRAM, however, is relatively slow when compared to the CPU (refer back to Figure 2.6). So, on the one hand, we have an inexpensive means of providing a great deal of memory storage, but on the other hand, the access time is much slower than the CPU (by a factor of between 25 and 100!).

SRAM is built using several transistors, units known as flip-flops. We use SRAM to build both registers in the CPU and *cache* (pronounced "cash") memory. Cache is a newer technology in computing than DRAM but very valuable. SRAM is far more expensive than DRAM though, so we tend to use a good deal less of it in our computers. However, SRAM is roughly as fast as the CPU, and therefore when the CPU accesses SRAM for instructions or data, the CPU is not forced to wait like it does when accessing DRAM.

This leaves us with an interesting dilemma. Since we have a small amount of SRAM, how can we ensure that when we need something from memory, it has been moved (copied) into SRAM from DRAM? For that, we design the memory of our computer into what we call the *memory hierarchy.* This comprises several layers where each layer going down will be slower, less expensive memory, but also of a far larger storage capacity (Figure 2.7).

In the memory hierarchy, registers are part of the CPU. There are anywhere from a few to a few dozen registers, each able to store a word,* so a CPU might have a few dozen to a

* Some processors have double registers, capable of storing two Words.

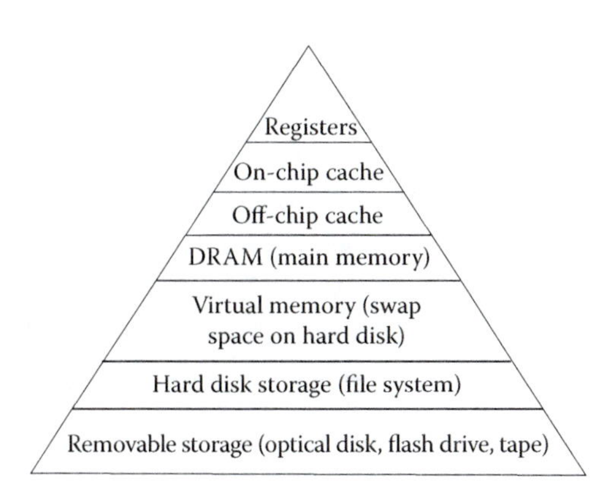

FIGURE 2.7 Memory hierarchy.

few hundred bytes of register storage. They operate at roughly the same speed as the system clock. There are often two different levels of cache: on-chip and off-chip. The on-chip cache is located on the CPU although is not in the same area of the CPU as the registers. Although on-chip cache is the same technology as registers, additional addressing hardware is needed that causes on-chip cache to be slightly slower, perhaps as much as twice as slow as register access. The off-chip cache, located separately on the motherboard, uses the same technology as on-chip cache. However, because the off-chip cache is a greater distance away from the CPU, it requires a longer time to access it, perhaps 3 to 10 times slower than the on-chip cache. Because the on-chip cache is located on the CPU itself, it must share space with the rest of the CPU. The off-chip cache can be far larger and therefore a far greater storage capacity. An on-chip cache might consist only of 32 or 64 KB, whereas an off-chip cache might be anywhere from 1 to 8 MB in size. Even so, the off-chip cache is far smaller than DRAM capacities, which today are at least 4 GB. Yet, the DRAM speed is far slower still, perhaps 25 to 100 times slower than the CPU itself.

Notice that the memory hierarchy does not end with DRAM. The programs that we run today are often larger than the size of memory itself! How then can we squeeze them into memory? Also consider that we tend to run several programs at a time. In order to handle the need for greater amount of memory than we have, we extend main memory using a concept called *virtual memory*. In virtual memory, whatever is not currently being used is shipped off to an area known as *swap space*. Swap space is stored on the hard disk in a separate area known as the swap partition. This partition is set up differently from the rest of the file system disk space—it is partitioned to be faster even though the partitioning is not as efficient in terms of space utilization. But by having it operate quickly, we can swap pieces of memory in and out as needed.

Each level of the memory hierarchy acts as a "backstop" for the next higher level. We start by loading the entire program and space for data from hard disk (or removable storage) into virtual memory and copy only the parts that are needed immediately (e.g., the start of the program) into DRAM. Then, as we begin to access instructions and data, they are copied into the on-chip and off-chip caches. Once the on-chip cache fills up, any additional items loaded into the cache first require that something be discarded.

As you move up the hierarchy, the space restrictions are greater and so discarding happens more often. Selecting wisely results in items being in the cache when the CPU needs them. Selecting poorly causes poorer performance. We refer to the efficiency of cache by the hit rate (how often what we want is found in the cache). Surprisingly, even for small on-chip caches, hit rates can be as high as 98% to 99%. As you move down the hierarchy, hit rates improve. Variations of the hierarchy have been tried including a third level of cache and including a cache with the hard disk. These variations, although more expensive, tend to improve performance.

ROLE OF INPUT AND OUTPUT

The I/O subsystem consists of input and output devices (including storage devices), a bus to connect the devices together and to the system bus, and interface devices such as expansion cards and ports. Strictly speaking, a computer does not need input or output devices to function. The computer program is stored in memory along with program data, and results are stored back into memory. However, without input and output devices, we are unable to interact with the computer. We cannot tell the computer what we want done, nor can we view the results.

The earliest forms of input and output were restricted to magnetic tape, punch cards, and computer printout. There was no direct interaction; instead, the user would prepare both the program code and data using a teletype device, whose results would appear on punch cards (one card per instruction or datum). The stack of punch cards would be input through a punch card reader and stored on magnetic tape. The tape would be mounted, the program and data input and executed, and the output stored back to magnetic tape. The tape would be removed and mounted onto a printer, which would print the output. This was not a very satisfying way to use a computer!

Today, of course, there are a great number of devices available to make the computer accessible in any number of situations. The key word behind I/O today is ease of use. We refer to the design philosophy of this interaction as *human–computer interaction* (HCI). In HCI, we view computer usage as human-centric rather than machine-centric. Table 2.1 lists some of the more common I/O devices in use today and the primary reason for their use. Storage devices are omitted from the table but would include internal and external hard disk, optical disk, flash memory, magnetic tape, and network storage devices. Network storage is commonly hard disk, accessible over a network.

In HCI, the emphasis is on promoting more natural ways of communicating with a computer. This area of study, which brings in branches of computer science, psychology, design, and health, among others, provides guidelines for more accessible computer usage. For instance, Braille output devices are available for users with visual impairments. Larger monitors and operating systems that can easily change screen resolution (the size of the objects on the screen) can also aid those with visual impairments. Microphones are often used by people who cannot use a keyboard and/or mouse.

HCI also focuses on devices that will reduce the strain placed on the human body through excessive computer interaction. *Ergonomics* is the study of designing systems and objects to better fit human usage. For computers, these include improved keyboards that put less strain on the human wrist, better pointing devices such as an improved mouse or a

TABLE 2.1 Input and Output Devices

Bar code reader	Input packaging information, used primarily in stores
Camera	Input video image (still or moving)
Goggles	Output video image, used primarily in virtual reality
Joystick	Input motion/pointing information, used primarily for computer games
Keyboard	Input text information, primary means of input for most users
Microphone	Input voice information, used in cases where either the user is unable to use hands or has a large amount of information to input; sometimes inaccurate
Monitor	Primary output device
Mouse	Input user interactions with GUI windowing system
MIDI device	Input of musical instrument data
Pen tablet	Input written information when keyboard is undesirable; sometimes inaccurate
Printer	Output text and/or images to paper
Scanner	Input text and/or images from paper
Speakers	Output music and sound
Touch pad/point	Alternate pointing input device when mouse is not desired (because of portability issues or health issues)
Touch screen	Alternate pointing input device when mouse and touch pad/point is not desired, primary input device for handheld devices
Trackball	Alternate pointing input device when mouse is not desired (because of portability issues or health issues), sometimes used for computer games

touch point, as well as improved furniture. Both keyboard and mouse usage have led many people to developing carpal tunnel syndrome, one of the many forms of *repetitive stress injuries*. A repetitive stress injury arises from doing similar actions over and over in such a way that it causes wear and tear on the body. For instance, using a keyboard incorrectly can strain the muscles in the wrist, leading to carpal tunnel syndrome. Poor posture when sitting in front of a computer for hours at a time can lead to other forms of stress.

The Rehabilitation Act of 1973, which authorizes grants to states that promote services for citizens with severe handicaps, has been amended (section 508) to promote accessibility in all forms of IT. The standards set forth in section 508 include guidelines for IT products including websites, software, operating systems, and hardware. For instance, all desktop and laptop computers are now required to come with expansion slots that support the various HCI devices (e.g., microphone, trackball). Operating systems are required to permit change in screen resolution to accommodate visual impairments. Government-supported websites are required to be easily traversable, for instance, by permitting the use of the tab key (or other hot keys) rather than the mouse to move from one area to another.

Aside from the Rehabilitation Act section 508, there are many other accessibility guidelines. These include the national instructional materials accessibility standards, promoted by the American Foundation for the Blind, UDL Guidelines 2.0 from the National Center on Universal Design for Learning, and the Web Content Accessibility Guidelines as recommended by the World Wide Web Consortium. Each of these focuses on different technologies, but they all promote the ideology that accessibility is important. Through HCI, information technology has become more usable to a greater population.

Although both accessibility and reduced injury are important goals of HCI, studies today primarily explore how to put the human at the center of the computer's interaction. From a data perspective, how can we improve human ability to input into the computer? For instance, would a microphone coupled with speech recognition technology allow a human to input more accurately and quickly than a keyboard? For the processed information (output), is the monitor the best output device? Would 3-D goggles provide a more precise way of conveying information?

Issues in HCI include not only improved devices, but improvements to existing devices. Two examples are to provide higher resolution displays to make output more legible and to provide redundancy in output to ensure that the output signals are understood. Other concepts include matching mental models of the human mind, providing related information in close proximity to other related information, using visual output rather than text to better match human memory capabilities, and so forth.

One last issue with HCI is permitting the human to move away from the stationary computer and take computing with him or her. This concept, sometimes referred to as ubiquitous computing, is now available through the use of handheld devices. However, small screens do not seem to support the concepts listed above. Therefore, new I/O devices are being developed that allow us to take our computing with us. These devices combine wireless and cell phone-based technologies with some of the more advanced input and output devices such as goggles for displays, microphones for input, and headphones for audio output. Sometimes referred to as *wearables*, these devices will probably become commonplace over the next 5 to 10 years.

As an example of what a wearable could do, imagine that you are riding the bus. The bus provides wireless Internet access. Your wearable devices connect to your home computer through the wireless, so all processing and storage is on your home computer. Your goggles are presenting to you visual information such as text-based news or a video. Your headphones not only serve to provide you audio output, but also attempt to block sounds coming from the outside world. Your microphone allows you to speak to your computer to control what you are doing. Finally, the goggles are not entirely opaque, they are semi-transparent so that you can also see the outside world as necessary (for instance, so that you can see when you arrive at your bus stop). But the goggles do more than just present an image from your computer. Your computer has been programmed to block unwanted images. And so, as you walk off the bus, you do not see the advertisements displayed on the side of the bus, instead those images are replaced with blank spots or images that you find appealing. Does this sound farfetched? Perhaps it is today, but it will not be as HCI continues to improve.

Aside from wearables, there are a number of other forms of emerging technology worth mentioning. Touch screens are obviously a large part of our lives today as they are a part of all smart phones and tablet devices. Touch screens are both input and output devices as they are displays that have sensors to note locations where a person is making contact with the screen. Touch screens have existed since the late 1960s, but were too prohibitively expensive to use in computing devices. However, today, touch screens are used not only for handheld device interfaces, but for computer games, kiosks, and medical devices, and may soon also be found as a standard interface for desktop and laptop computers. Today's touch screens do not require a pen-based (stylus) interface and are capable of responding to multiple touches at the same time.

Virtual reality (VR) is a technology still in its infancy even though it has been around since the late 1980s. In VR, the computer creates an illusionary world and through various types of input and output devices, the human is deposited in that world. The human interacts with the virtual world based on the motion and orientation of the head, via a headset with a motion sensor, and data gloves. The user sees and hears the virtual world through goggles and headphones (Figure 2.8). In addition, a full body suit can provide the user with other forms of sensation.

With VR, a human can be immersed in a location that normally the human could not reach, such as walking on Mars or walking into an active volcano, or swimming deep in the ocean. Practical uses of VR include exploration, design and analysis (imagine inspecting an aircraft that only exists in the design stage), education, and entertainment. However, in spite of the potential that VR offers, it is not commonplace because there are many obstacles to overcome. Primarily, the limitations in I/O devices to portray a realistic illusion leave a lot to be desired. Additionally, VR requires an excessive amount of computation to properly model and present an illusionary world. Thus, VR is still fairly expensive with data gloves and headsets costing between $500 and $5000 each.

FIGURE 2.8 Virtual reality headset and data glove in use. (Courtesy of NASA, http://gimp-savvy.com/cgi-bin/img.cgi?ailsxmzVhD8OjEo694.)

One last form of emerging I/O technology worth noting is the *sensor network*. Sensors are devices that can sense some aspect of the environment such as temperature, water pressure, or sound vibration. Most devices that contain sensors usually have limited sensors, positioned in specific locations on the device. A sensor network is a distributed collection of sensors so that the device is able to obtain a more realistic "view" of the environment. Today, sensors are cheap enough that sensor networks are becoming more commonplace. We combine sensors with wireless communication to create a wireless sensor network, consisting of as many as a few thousand individual sensors, each of which can communicate with each other and a base computer to process the sense data. Sensor networks are found in a variety of consumer devices, vehicles, health monitoring devices, and military hardware. An entire sensor network may cost as little as a few hundred dollars. Applications for sensor networks include surveillance, monitoring atmospheric conditions (weather, greenhouse gases, air pollution), health monitoring, seismic activity, and exploration.

To wrap up this discussion of I/O devices, we also need to address two other technologies that have improved users' abilities to connect the devices to their computers. These are plug-and-play and USB. The idea behind *plug-and-play* is that you are able to connect a new device to your computer at any time and that the device, once recognized by the operating system, is ready for use. Before plug-and-play, you would have to reboot your computer once you connected the new device so that the device could be recognized during the boot process. Plug-and-play was first pioneered for the Windows 95 operating system but has since become standard in most operating systems.

The most common (although not only) means of attaching a new I/O device to your computer is through the USB port. USB stands for *Universal Serial Bus*, a standard interface for computers since the 1990s. The USB standard defines the connectivity, which dictates the type of cable used, the method of communication between device and computer, and the power supply that will exist between the computer and device. Today, most devices can be attached via USB including keyboards, pointing devices, printers, external disk drives, smart phones, and other handheld devices. USB drives, or flash memory, give a user a decent amount of external storage (1–8 GB) that is portable. The USB port has replaced serial and parallel ports used by such devices. In addition, the USB port can supply power to the device.

COMPUTER HARDWARE AND COMPUTER ASSEMBLY (INSTALLATION)

In this section, we examine the more common computer components and examine a basic assembly (installation) for windows-based desktop computers.

CPU

The central processing unit, or processor, is stored on a single chip, known as the *microprocessor*. The CPU has a number of small pins on one side of it. It is best to never touch these pins—bending them will most likely ruin the CPU. You insert the CPU into a special socket on the motherboard. You might notice that one corner of the CPU has a triangle on it. This will help you position the CPU correctly when you insert it into its socket.

(Adapted from Gyga, http://commons.wikimedia.org/wiki/File:Sockel7-cpus.JPG.)

Memory

Most computer memory is stored in RAM. We usually refer to this as memory or main memory, although there are other types of memory in the computer (ROM, cache). Memory chips today come already placed on a circuit board. You slip the circuit board into a special slot on the motherboard. Again, you will have to position the circuit board correctly into the slot. If you notice in the figure, there are pins that run along the bottom of this circuit. One set of pins is longer than the other; this will help you decide the proper orientation in the memory slot.

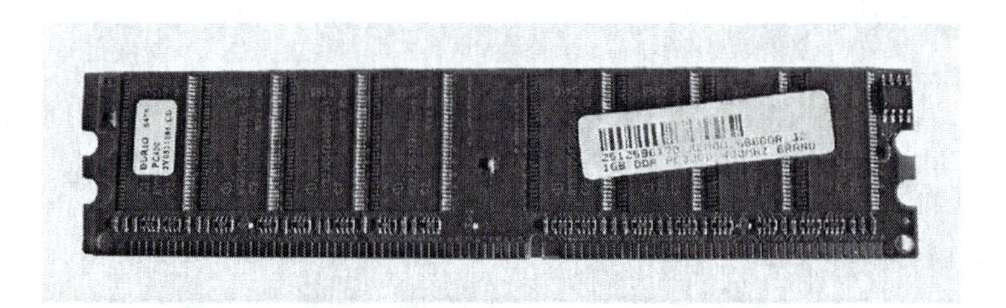

(Courtesy of Laszlo Szalai, http://commons.wikimedia.org/wiki/File:DDR_RAM-3.jpg.)

System Unit

This is a case into which you will place several key components of the computer. The most important component is the motherboard. The motherboard will sit on top of "standoffs" to elevate it above the surface. It will actually be placed along one side of the system unit, not on the bottom. Looking at the system unit, you can see many screw holes; some of these will have the standoffs. Also inserted into the system unit are a power supply, a fan, and some storage devices such as an optical disk drive and a hard disk drive. There is room for other devices in the slots along the front of the system unit.

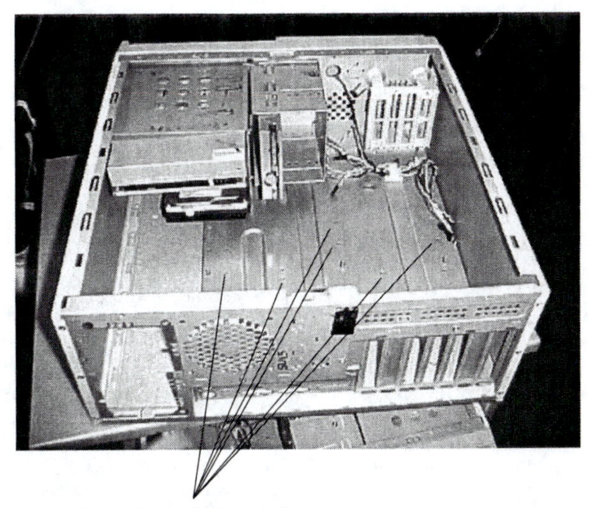

(Courtesy of Norman Rogers, http://commons.wikimedia.org/wiki/File:Stripped-computer-case.JPG.)

Motherboard

The motherboard provides the connections between the CPU, memory, and other components. These devices will connect to the motherboard via expansion slots and rear panel ports. The motherboard is a piece of fiberglass or plastic on which sockets and bus lines are attached. The sockets include one (or more) socket for the CPU, sockets for DRAM memory, expansion slots for cards that are interfaces with peripheral devices, connectors for storage devices, power supply connections, and ports that will stick through the rear panel of the system unit. ROM memory chips are already attached to the motherboard.

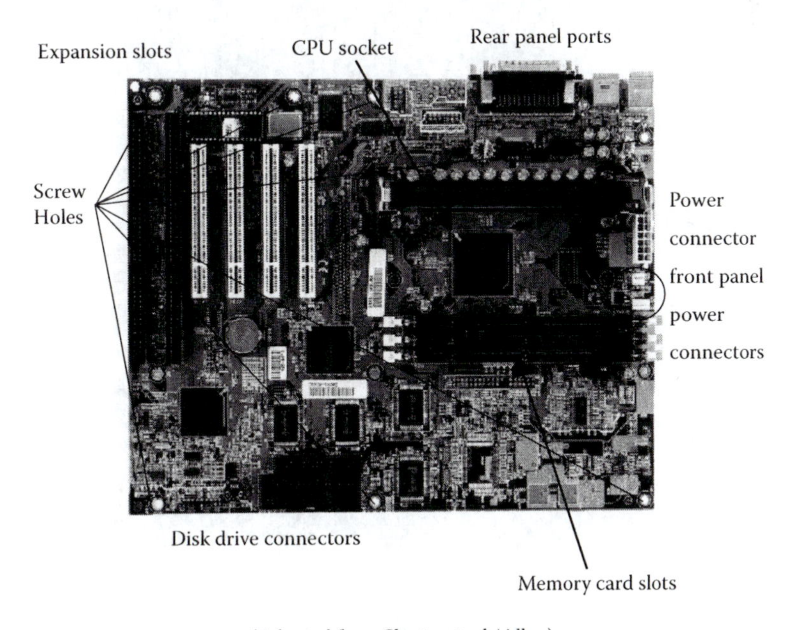

(Adapted from Shutterstock/Albo.)

The motherboard has screw holes so that you can attach it to the system unit. As the system unit will usually stand vertically, the motherboard will actually be positioned vertically as well, so it is important that the motherboard be attached firmly. Additionally, the standoffs ensure that the motherboard does not physically touch the inner surface of the system unit. Since electrical current flows across the motherboard, if it were to touch the metallic inner surface of the system unit, this could short out the current flow resulting in a lack of current making it to the appropriate chips.

The underside of the motherboard contains "wires"—soldered lines that make up a portion of the system bus. Current flows over these lines. Therefore, it is important that these lines do not touch anything else that is metal. This is one reason why we will mount the motherboard on top of "standoffs". This will elevate the motherboard off of the side of the system unit. Notice that the underside of the motherboard is rough (bumpy) with little pieces of hardware and solder sticking out. When you work on the motherboard, it is best to place it on top of a soft, yielding surface, but also one that will not conduct any electricity. The box that the motherboard came in would be one possibility, and the plastic bag that contained the motherboard would be another.

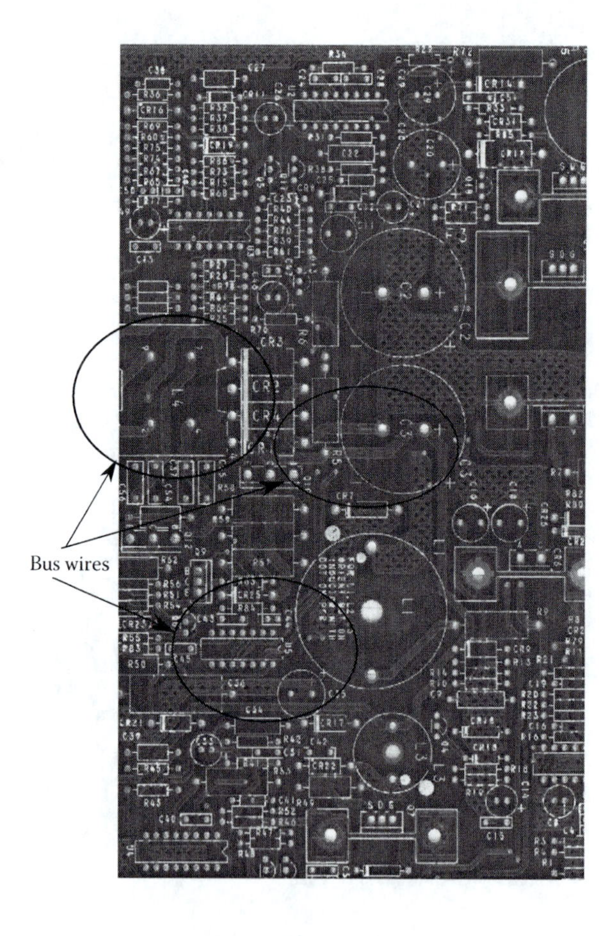

(Courtesy of Shutterstock/jakelv7500.)

Hard Disk Drive

The hard disk is the standard unit of long-term storage. In the earliest days of computers, magnetic tape was used (reel-to-reel then magnetic tape cassettes and cartridges).

(Courtesy of Hubert Berberich, http://commons.wikimedia.org/wiki/File:Seagate-ST4702N-03.jpg.)

Today, the hard disk drive stores a great quantity (typical storage sizes are about 1/2 TB to a full TB). Our computers also contain optical disk drives more to allow our computers to serve as entertainment consoles (to play music CDs and movie DVDs) than for storage purposes. The hard disk drive contains several hard disk platters onto which information is stored as magnetic charges. Read/write heads move in unison over the various disk surfaces. The hard disk drive also contains the logic and mechanisms to spin the disks and move the read/write head arm.

There are generally two types of drives that we use today: IDE (Integrated Drive Electronics) and SATA (Serial Advanced Technology Attachment). Shown below are the back of a SATA drive where you can see two connections, one for the power supply and one for data. To the right of the drive is a power cable for the SATA drive. Notice how the SATA connector has an "L-shape" to it, which helps you orient the connector when you plug it in.

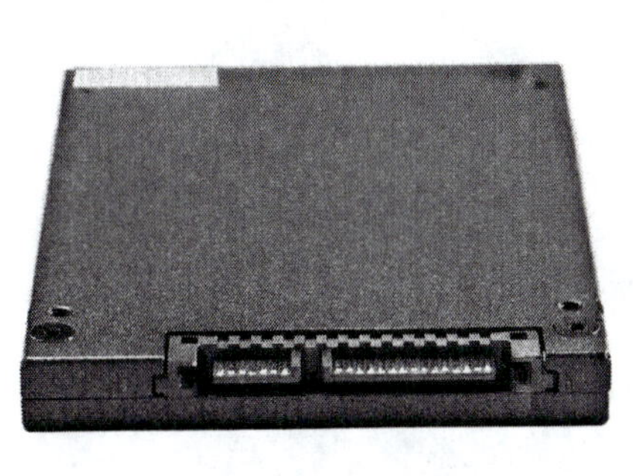

(Courtesy of Shutterstock/Matee Nuserm.)

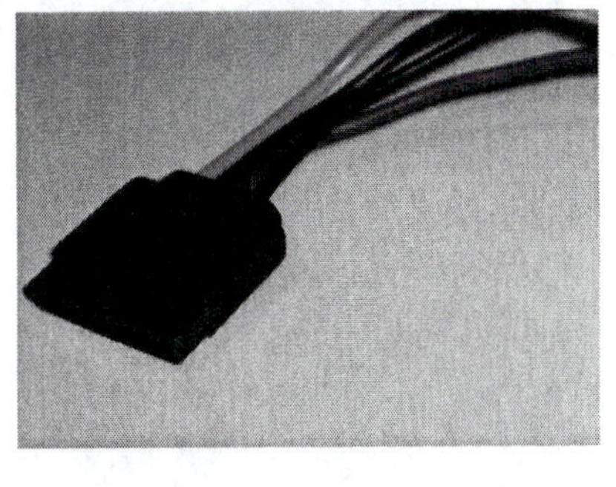

(Adapted from Martixer, http://commons.wikimedia.org/wiki/File:Molex-to-SATA-power_009.jpg.)

An example of the back of an IDE drive is shown below. To its right is an IDE data connector. The power connection for the IDE drive consists of four holes to match the four prongs on the left (as shown here, the drive is upside down so in fact the four prongs would be on the right) of the drive.

(Adapted from Zzubnik, http://commons.wikimedia.org/wiki/File:Hard-drive.jpg.)

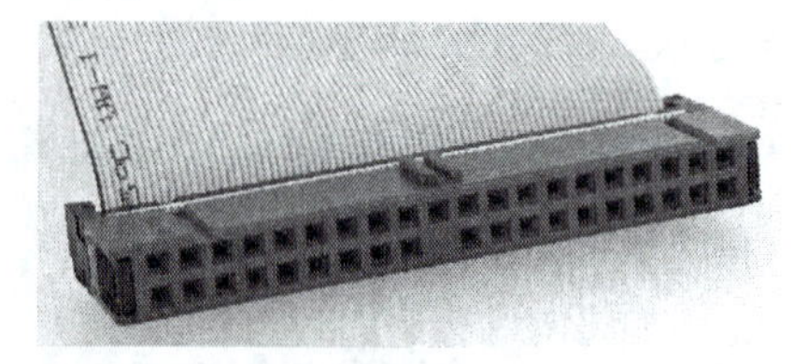

(Courtesy of Jonas Bergsten, http://commons.wikimedia.org/wiki/File:Ata_20070127_001.jpg.)

In addition to the motherboard and drive units, you will also insert a power supply unit into the system unit. The power supply unit connects to a number of cables. These cables are connected to the motherboard to power the motherboard, CPU, the display panel (on the front of the system unit), the fan, and the drives. This figure shows the cables all bundled together. You will have to be careful as you connect the various power cables to their locations on the motherboard or drives as these cables can be awkward to connect correctly such that they do not get in your way as you continue with the assembly process.

BUYING A COMPUTER

So you want to buy a computer. What are your choices? Before answering this question, you must first decide what you will use the computer for. This will help you classify which type you might need:

- Server
- Desktop
- Laptop
- Notebook

Once you have selected the type of computer, you have other choices. If you are looking at anything other than a server, your choices basically boil down to

- Macintosh
- PC running Windows
- PC running Linux
- Other (this is primarily a choice for notebook and server)

And now within these choices, you must select

- Processor type (speed, cache size, 32-bit or 64-bit)
- Amount of main memory (DRAM)
- Size of hard disk
- Optical drive?
- Monitor size
- MODEM speed (or network card)
- Other peripheral devices

Your choices here are predicated on the use of the computer, the software requirements of software you intend to use, and the amount of money you are willing to spend.

(Courtesy of Winhistory, http://commons.wikimedia.org/wiki/File:At-netzteil.jpg.)

You might wonder, what about the connectors for the keyboard, mouse, and display? These connections are already part of the motherboard. You will remove a panel from the back of the system unit so that the various connectors (ports) are accessible from outside.

Now that you have seen the various components, let us look at how they go together. The first thing you should do is identify all of the components. These will include the motherboard, CPU, memory circuit boards (chips), power supply, hard disk drive, optical disk drive, screws and standoffs, a grounding strap, and tools. Although you do not have to wear the grounding strap yet, be aware that it is very important that you put it on before touching any of the electronics (motherboard, chips). To further protect the hardware, do not work on carpeting and wear lose clothes that will not retain a static charge. A static charge could damage any of the chips, most especially the CPU. In addition, keep the work area clean of clutter and debris. Do not plug in the power supply yet, and always plug the power supply in to a power strip, not directly into a wall socket

(Courtesy of http://www.freeimagespot.com/Countries/grounding-strap.html, author unknown.)

After identifying and organizing your components, the first step is to mount the power supply into the power supply bay of the system unit. You will need to remove the side panel off of the system unit. Place the power supply in its proper place and screw it in. You will also need to remove the rear port panel. This should snap right out.

Next, you will insert the DVD and hard drive. You will have to remove the front "bezel" off of the system unit in the location of where you want to insert the DVD drive (you do not need to do this for the hard disk since the hard disk drive's surface will not need to be visible, but the DVD drive needs to open). Depending on the type of system unit shell you have, you will either have to attach plastic rails onto the drives to slide them into their slots, or slide the drives in and attach them using the slot "arms". Once your drives are in place, do not yet connect any cables to them. The cables would interfere with other work, so it is best to save the connections until the last steps.

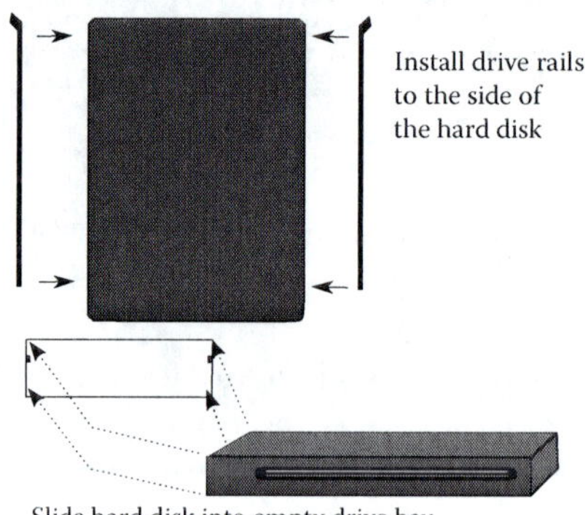
Install drive rails
to the side of
the hard disk
Slide hard disk into empty drive bay
in system init sliding drive rails appropriately

Now, you will begin to assemble the components on the motherboard. For this, you need to wear a grounding strap. You should attach the lose end of the grounding strap to something metal, preferably the metal "cage" of the system unit. Using the motherboard, determine where the standoffs are needed in the system unit. Do not affix the motherboard yet, but screw in the standoffs in their proper position (again, using the motherboard to determine where—you will need to look for the screw holes in the motherboard and match then to screw holes in the side of the system unit).

The first component to insert onto the motherboard is the CPU. Make sure you have the CPU aligned properly (look for the arrow), lift the locking lever bar, flip up the CPU socket lid, slip the CPU into its socket, lower the lid, and then lock the lever into place.

(Courtesy of Vdblquote, http://commons.wikimedia.org/wiki/File:Intel_80486DX4_Pins_and_socket_3.jpeg.)

Next, you will affix the CPU's cooling unit. First, you need to place a dab of heat transfer paste onto the outer surface of the CPU. The amount you place onto the CPU is small, about the size of a small pea. Next, orient the cooling unit so that its four connectors (pegs) are positioned over the four holes surrounding the CPU socket. Make sure that the unit is oriented so that the power cord can easily fit into the power socket on the motherboard. Once it is positioned correctly, push down on each of the four pins one at a time. Once all four pins have been pushed through the motherboard, lift up the motherboard to confirm this. Then, turn each of the pins as indicated on the surface to lock the pins in place.

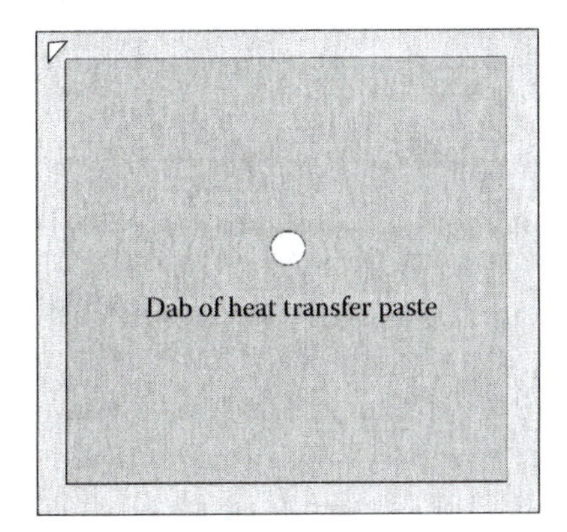

(Courtesy of Shutterstock/Bretislav Horak.)

Next, take a memory circuit board and insert it into one of the memory slots. Make sure that the board is aligned properly by looking for the gap in the pins and match it to the gap in the slot. Make sure the tabs along the side of the slots are "open". Slide the circuit board onto the opening and then using your thumbs only, press down until you see the tabs close automatically. You would repeat this step for each memory circuit that you wish to add to your computer.

(Courtesy of Shutterstock/Jultud.)

With these units in place, you can now screw the motherboard into place on the standoffs inserted earlier. To finish off the installation, you will now need to make all of the connections between the power supply unit, drives, fan, and motherboard. First, find the cable that has four 2-pin connectors (note: depending on the power supply unit, this might be a single eight-pin connector). The attachments will be shown on the motherboard installation document. For instance, the drawing below indicates the connections to power the disk drive LED, the reset (reboot) button, the power LED, and the on/off switch. Failure to connect the connectors to the proper pins will result in these components not functioning. In the diagram below, pins 1/3 are used for the hard disk drive LED. Pins 2/4 power the on/off LED. Pins 5/7 are used for the reboot button. Pins 6/8 are used for the on/off button's light. Finally, pin 9 does nothing.

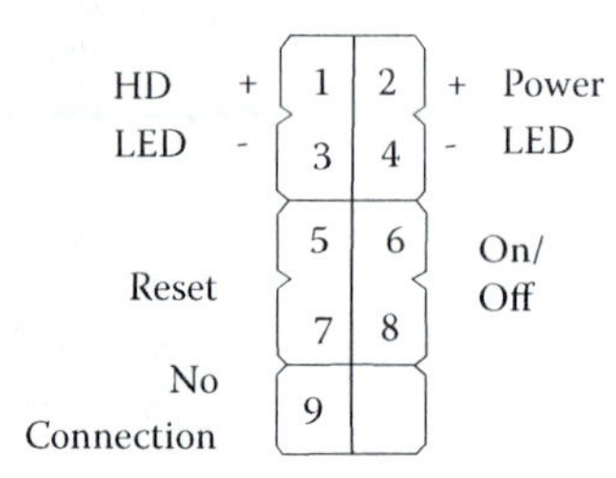

Next, connect your drives. You will probably have two drives to connect, an optical drive and a hard disk drive. Chances are, the hard disk is an IDE as IDE drives are more common in PC-style computers. The optical drive may be either SATA or IDE. Identify the type to determine which plugs to use. First, insert the drives into available bays in the system unit. You will probably want to insert the optical drive on top. This is because the optical drive will interface with the outside world so it is more common to have it appear near the top of the system unit. You will have to remove the cover over that bay so that the optical drive is accessible from outside. Different system units will have different types of bays. Commonly though, all you need to do is slide the drive into the bay and then turn the key on the side of the bay.

Once both (or all) drives are in their bays, you must connect them to the motherboard and the power supply unit. Find the appropriate power cable to attach to the power supply and the appropriate data cable to connect to the motherboard. The last connection is to plug the fan power cable into the power supply cable and to connect this to the motherboard. At this point, you can plug in the power supply unit to both the computer and the surge protector. Make sure the surge protector is plugged into an electrical outlet and turned on. Now you can boot your computer!

If the computer successfully boots, it will boot to the ROM BIOS but because there is no operating system on your hard disk, it will not reach windows. If you have successfully assembled your computer, insert a Windows installation CD into the optical drive, shut down the computer, and reboot. Once the computer has booted to ROM BIOS, you can go through the Windows installation (this is briefly covered in Chapter 4). Congratulations!

FURTHER READING

Computer science and computer engineering programs typically require a course on computer organization and architecture. Texts for such courses go well beyond the introductory level presented in this chapter. Among the topics covered in such classes are Boolean algebra and circuit design, binary representations and arithmetic (which we introduce in Chapter 3), CPU implementation details and performance, cache and DRAM technologies, and the I/O system. The seminal texts are the following:

- Patterson, D. and Hennessy, J. *Computer Organization and Design: The Hardware/Software Interface.* San Francisco: Morgan Kaufmann, 1998.
- Hennesey, J. and Patterson, D. *Computer Architecture: A Quantitative Approach.* San Francisco: Morgan Kaufmann, 2012.

The first text is primarily targeted at juniors in college, whereas the latter is aimed at seniors and graduate students. Both of these texts are challenging even for the brightest computer science majors. Other computer organization and architecture textbooks include the following, the third of which might be one of the more accessible books particularly for non-majors:

- Clements, A. *The Principles of Computer Hardware.* New York: Oxford, 2000.
- Hamacher, C., Vranesci, Z., Zaky, S., and Manjikian, N. *Computer Organization and Embedded Systems.* New York: McGraw Hill, 2012.
- Null, L. and Lobur, J. *The Essentials of Computer Organization and Architecture.* Sudbury, MA: Jones and Bartlett, 2012.
- Stallings, W. *Computer Organization and Architecture: Designing for Performance.* Upper Saddle River, NJ: Prentice Hall, 2003.
- Tanenbaum, A. *Structured Computer Organization.* Upper Saddle River, NJ: Prentice Hall, 1999.

You can find a nice demonstration on the fetch–execute cycle at http://www.eastaughs.fsnet.co.uk/cpu/execution-fetch.htm. Additionally, the website http://courses.cs.vt.edu/csonline/MachineArchitecture/Lessons/CPU/Lesson.html provides a nice diagram of the most important parts of a CPU along with a description of how they are used during the fetch–execute cycle.

One of the best sources for comparing CPU performance on benchmarks is the website www.cpubenchmark.net. You can also find useful pricing comparisons of computers and computer hardware at www.pricewatch.com.

Texts covering just the memory hierarchy tend to cover design and algorithms for efficient memory usage or leading-edge technology with DRAM and SRAM. The following text though will shed more light on the role of the components that make up the memory system:

- Jacob, B. and Wang, D. *Memory Systems: Cache, DRAM, Disk*. San Francisco: Morgan Kaufmann, 2007.

There are a number of texts dealing with input, output, HCI, and related topics, some of which are listed here.

- Dargie, W. and Poellabauer, C. *Fundamentals of Wireless Sensor Networks: Theory and Practice*. Hoboken, NJ: Wiley, 2010.
- Dix, A., Finlay, J., Abowd, G., and Beale, R. *Human–Computer Interaction*. Englewood Cliffs, NJ: Prentice Hall, 2003.
- Heim, S. *The Resonant Interface: HCI Foundations for Interaction Design*. Reading, MA: Addison Wesley, 2007.
- Kortum, P. *HCI Beyond the GUI: Design for Haptic, Speech, Olfactory and Other Nontraditional Interfaces*. San Francisco: Morgan Kaufmann, 2008.
- Lumsden, J. *Human–Computer Interaction and Innovation in Handheld, Mobile and Wearable Technologies*. Hershey, PA: IGI Global, 2011.
- McCann, J. and Bryson, D. (editors). *Smart Clothes and Wearable Technology*. Cambridge: Woodhead Publishing, 2009.
- Salvendy, G. (editor). *Handbook of Human Factors and Ergonomics*. Hoboken, NJ: Wiley, 2006.
- Sherman, W. and Craig, A. *Understanding Virtual Reality: Interface, Application and Design*. San Francisco: Morgan Kaufmann, 2002.

The following government-run website details section 508 of the Rehabilitation Act of 1973, describing accessibility standards: http://www.section508.gov/index.cfm.

This chapter also provided a brief introduction to personal computer components and assembly. There are numerous books on the topic such as

- Chambers, M. *Build Your Own PC Do-It-Yourself for Dummies*. Hoboken, NJ: Wiley, 2009.
- Heaton, J. *Build a Computer From Scratch*. St. Louis, MO: Heaton Research, Inc., 2006.
- Majlak, D. *Building Your Own Computer. No Frills, No Filler, Just Answers* (a Kindle book). Seattle, WA: Amazon Digital Services, 2011.
- Mueller, S. *Upgrading and Repairing PCs*. Indiana, IN: Que, 2011.
- Thompson, B. and Thompson, B. *Building the Perfect PC*. Massachusetts: O'Reilly, 2010.

Although we briefly introduced programming, we will cover that topic in greater detail later in the text.

REVIEW TERMS

Terminology used in this chapter:

Accessibility
Accumulator (AC)
Assembly program
Address bus
ALU
Bit
Bus
Byte
Compiler
Control bus
Control unit
CPU
CPU cooling unit
Data bus
Decode
DRAM
Ergonomics
Fan
Fetch–execute cycle
GHz
Grounding strap
Hard disk drive
HCI
Hit rate
IDE drive
Input
Instruction register (IR)
IPOS cycle
Load
Megaflops
Memory
Memory chips
Memory hierarchy
MIPS
Mother board
Off-chip cache
On-chip cache
Output
Power supply
Processing
Processor (CPU)
Program counter (PC)
Read
Register
SATA drive
Sensor network
SRAM
Standoffs
Status flags (SF)
Storage

Store
Swap space
System unit
Virtual memory
Virtual reality
Wearable
Word
Write

REVIEW QUESTIONS

1. What happens during the fetch stage of the fetch–execute cycle?
2. What happens during the decode stage of the fetch–execute cycle?
3. What happens during the execute stage of the fetch–execute cycle?
4. What is a load?
5. What is a store?
6. What is an operand?
7. What is the ALU? What are some of the circuits in the ALU?
8. What does the control unit do?
9. What does the PC store?
10. What does the IR store?
11. What does the AC store?
12. Is the AC the only data register?
13. What is moved over the data bus?
14. What is moved over the address bus?
15. What is moved over the control bus?
16. Why does a processor's GHz rating not necessarily tell you how fast it is?
17. What is MIPS and how does it differ from GHz?
18. What is Megaflops and how does it differ from MIPS?
19. What is a benchmark?
20. What is the memory hierarchy?
21. Which form of memory is faster, DRAM or SRAM?
22. Why are there both on-chip and off-chip caches?

23. What happens if the CPU looks for something at one level of the memory hierarchy and does not find it?
24. Where is virtual memory stored?
25. What were the forms of input and output found in early computers?
26. What is HCI? What does it study?
27. What is a repetitive stress injury?
28. What types of input and output devices could you use as wearable technology?
29. What are some of the applications for virtual reality?
30. Why is it important to wear a grounding strap?
31. What might happen if you discharge static while working with the motherboard?
32. Why should you install the CPU onto the motherboard before installing the motherboard into the system unit?
33. What does the CPU cooling unit do? Why is it necessary?
34. When installing a storage drive, how does the SATA data connector differ from the IDE data connector?

DISCUSSION QUESTIONS

1. How important is it for an IT person to understand the functions of individual computer components such as the role of the CPU and memory? How important is it for an IT person to understand concepts such as cache memory, pipelining processors, the use of the bus, the fetch–execute cycle, and the use of registers? How important is it for an IT person to understand the differences between SRAM, DRAM, and ROM?
2. As an IT person, do you ever expect to program in assembly language? If so, provide some examples and if not, explain why not.
3. Which is more significant in IT education, understanding the function of the hardware of the computer or understanding how to assemble (build) a computer? Explain.
4. Most people believe that the processor's clock speed is the most important factor in a processor's performance. Discuss all of the factors that can impact the performance of a processor. Rank your factors in the order that you feel will have the most significant impact on performance. Where did the clock speed rank?
5. In HCI, why do they need to study human psychology to design improved I/O devices?

6. For the input and output devices listed in Table 2.1, which ones could cause repetitive stress injuries? Which ones might you identify as replacements to prevent repetitive stress injuries?

7. Provide a list of five ways that you might use virtual reality in your day-to-day life.

8. You work for a large organization. Your employer asks you to put together the specifications for new desktop computers for the employees. As these are work computers, there will not be a high demand on multimedia performance, but instead there is a desire for ease of use, efficient communications, and large storage. Put together a specification in terms of what you would look for, including platform (e.g., Windows, Mac, Linux), processor type and speed, amount of cache, memory, word size, and secondary storage. What other factors will impact your decision?

9. Imagine that you are assembling your own computer as described in "Computer Hardware and Computer Assembly (Installation)". You have completed the assembly, plugged in the computer and turned on the power switch. Nothing happens. What are some of the possible problems that might have occurred during assembly that will prevent the computer from booting?

CHAPTER 3

Binary Numbering System

In Chapter 2, we saw that the digital computer operates by moving electrical current through its component parts. We must then find a way to convert information into a binary representation. We may then examine how to compute using binary. In this chapter, we study the binary numbering system. The chapter covers conversion methods from binary to decimal and decimal to binary for unsigned (positive) integers, negative integers, fractional, and floating point numbers. The chapter also introduces the octal and hexadecimal numbering systems. Character representations are discussed. The chapter then turns to binary (Boolean) operations and how to compute with them. Finally, the chapter introduces three example uses of binary: network addressing, image files, and parity for error detection and correction.

The learning objectives of this chapter are to

- Provide methods for numeric conversion between bases: decimal, binary, octal, and hexadecimal, for positive and negative integers and floating point numbers.
- Describe how characters are stored in computer memory.
- Demonstrate the application of binary (Boolean) operations of AND, OR, NOT, and XOR on binary numbers.
- Illustrate the use of binary in a computer with a focus on IP addresses, image file storage, and parity bits.

The computer processes and stores information in binary form. So we need to understand what binary is, how to represent information in binary, and how to apply binary (Boolean) operators. This chapter looks at the binary numbering system and a couple of related numbering systems. Although it is critical that a computer scientist understand binary, in information technology (IT) you will find that binary comes up occasionally and so is worth understanding.

NUMBERING SYSTEMS

A numbering system (also known as a numeral system) is a writing system for expressing numbers. In mathematics, there are an infinite number of numbering systems although we tend to largely focus on only two, decimal and binary. Table 3.1 presents the commonly used and historically used numbering systems.

Formally, we will define a numbering system as base k (k being a positive integer), where

- Any number in this base will contain digits from 0 to $k - 1$.
- The interpretation of a digit is its value * power.
- The power is based on the column, or position, where power = $\text{base}^{\text{column}}$, and where column is counted from the right to the left with the rightmost column being column 0.

For example, decimal is base 10. This means we can use digits 0 through 9. The power will be 10^{column}. If our number is 130, this represents the value $1 * 10^2 + 3 * 10^1 + 0 * 10^0$, or 1 in the hundreds column, 3 in the tens column, and 0 in the ones column.

Unary, base 1, is an odd base for two reasons. First, following the definition above, base 1 means that our only digits are 0, but this would mean that all of our possible numbers, say 0000, 0000000000, 0, will all be 0 because the value in the formula value * power is always 0. For this reason, in unary, we will use the digit 1 instead of 0. Second, notice that all of the powers will be 1 ($1^0 = 1$, $1^1 = 1$, $1^2 = 1$, $1^3 = 1$). The result is that the number being represented is merely the sum of the number of digits. So, for instance, in unary, 111 is the representation for the value 3, and 111111111 is the representation for the value 9.

Let us consider base 5. For this base, we use only digits 0 to 4 and the "columns" of any number represent from right to left the one's column, the five's column, the twenty-five's column (5^2), the hundred twenty-five's column (5^3), the six hundred twenty-five's column (5^4), and so forth. The number 1234 in base 5 is the value $1 * 125 + 2 * 25 + 3 * 5 + 4 * 1 = 194$ in decimal.

TABLE 3.1 Common Numbering Systems

Base	Name	Reason Used/Popularity
1	Unary	Used in theoretical computer science when dealing with Turing machines, not very common
2	Binary	Digital computers use binary and therefore is studied in computer-related disciplines
8	Octal	Groups of three binary digits create an octal digit, used in computer-related disciplines to make binary more readable
10	Decimal	Primary human numbering system
16	Hexadecimal	Groups of four binary digits create a hexadecimal digit, used in computer-related disciplines to make binary more readable, easier to read than octal (requires fewer digits) but is more complex because of the inclusion of digits A–F
20	Vigesimal	Used by the Pre-Columbian Mayan civilization; dots and lines represented digits, 0–4 are denoted by 0 to 4 dots, 5–9 are denoted by 0 to 4 dots with an underline, 10–14 are denoted by 0 to 4 dots with 2 underlines, 15–19 are denoted by 0 to 4 dots with 3 underlines; not used outside of the Mayan civilization
60	Cuneiform	Used by the ancient Babylonian civilization

Given any number $a_n a_{n-1} \ldots a_1 a_0$ in some base b, the number is interpreted using the following summation $\sum_{i=0}^{n} (a_j * b^i)$. That is, we take the rightmost digit and multiply it by b^0 and add it to the next rightmost digit and multiply it by b^1 and add it to the next rightmost digit and multiply it by b^2, ..., and add it to the second to leftmost digit (in column $n - 1$), then multiply it by b^{n-1} and add it to the leftmost digit (in column n) and multiply it by b^n.

Notice that the rightmost column's power will always be 1 no matter what base. Why is this? The rightmost column will always have a power of $\text{base}^{\text{column}}$ where column is 0, or base^0. In mathematics, any positive integer (1 or greater) raised to the 0th power is always 1.

Since there are an infinite number of numbering systems (any positive k is a legal base), we need to be able to denote what base a number is in. We do this by placing the base as a subscript at the end of the number. So 1234_5 is 1234 in base 5, which as we saw above is the same as 194_{10}. For convenience sake, we are allowed to omit the base if we are dealing with base 10 (decimal) because that is the default base, so we would say $1234_5 = 194$.

Let us consider a base larger than 10, say base 12. We might have the number 246_{12}. This number is actually $2 * 12^2 + 4 * 12 + 6 * 1 = 342$. But recall that in any base k, we are permitted to use digits $0..k - 1$. For base 12, this would be any digits 0–11. Now, we have a problem, because 10 and 11 are not single digits. If I were to write the number 11_{12}, it would look like I meant 1 in the twelve's column and 1 in the one's column instead of 11 in the one's column. This would lead us to have confusion, is $11_{12} = 13$ (a two digit number) or $11_{12} = 11$ (a one-digit number)? So we need to find an alternative way to represent the digits in a base larger than 10. Luckily for us, we tend to only use one base larger than 10, base 16, also known as *hexadecimal*. For this base, it was decided that the digits used to represent 10, 11, 12, 13, 14, and 15 would be A, B, C, D, E, and F, respectively. Thus, the value $3AB2_{16}$ means 3 in the 16^3 column, A (or 10) in the 16^2 column, B (or 11) in the 16^1 column and 2 in the one's column. So what does $3AB2_{16}$ represent? $3 * 16^3 + 10 * 16^2 + 11 * 16^1 + 2 * 1 = 15{,}026$.

BINARY NUMBERING SYSTEM

Why is all this important? It would not be if our computers used base 10 (decimal), but computers store all information, transmit all information, and process all information in binary instead of decimal. And although as an IT specialist, you may never have to worry about looking at or expressing data in binary, the use of non-decimal numbering systems does come up from time to time. The most common occurrence of a non-decimal numbering system is that of network communication. IP addresses are sometimes denoted in binary, and IPv6 addresses are denoted in hexadecimal. Octal is also used for various computer applications, but that is happening far less often. In this section, we look first at binary and then octal and hexadecimal.

Binary is probably the most important numbering system outside of decimal. Because of the digital nature of computers (i.e., everything is stored in one of two ways, current or no current), everything has to be represented as ons and offs. So we use binary, where 1 means on and 0 means off. This means that everything the computer stores will have to first be converted into 1s and 0s. We will call a single 1 or 0 value a *bit* (for *b*inary dig*it*).

Data movement requires one wire for every bit so that the transmission of a datum (say of 8 bits) can be done at one time, 1 bit per wire, rather than 1 bit at a time. The digital circuits in the CPU *operate* on bits, both registers and memory *store* bits. As a computer scientist or computer engineer, you might occasionally have to look at data or program code in binary. If you were to look at a thousand bits, it would be difficult to focus on it. So instead, we break up our data into groups. Eight bits are placed together into a unit called a *byte*. It was common in earlier computers that a datum would be stored in one byte, possibly two. A byte, being 8 bits, can store any combination of 8 zeroes and ones, for instance, 11111111, 10101010, 10001000, 00000000, 01111110, and 01011011. There are 256 different combinations of ones and zeroes that can be placed in 8 bits (1 byte). We get that number by simply computing 2^8 (multiply two together eight times).

The numeric conversion from binary to decimal can be done using the formula from the previous page. However, there is an easier way to think about it. Let us take as an example a byte storing 01010101_2. According to the formula, we would convert this as $0 * 2^7 + 1 * 2^6 + 0 * 2^5 + 1 * 2^4 + 0 * 2^3 + 1 * 2^2 + 0 * 2^1 + 1 * 2^0 = 0 * 128 + 1 * 64 + 0 * 32 + 1 * 16 + 0 * 8 + 1 * 4 + 0 * 2 + 1 * 1 = 85$. Notice that every digit in our binary number is either a 0 or a 1, unlike say the base 5 example in the previous section, which allowed digits 0, 1, 2, 3, and 4. When converting from binary to decimal, you do not need to worry about multiplying the digit times the power. Instead, the digit will be a 0, in which case 0 * power = 0 and can be dropped out of the summation, or the digit is a 1, in which case 1 * power = power. Therefore, our summation is merely the powers of 2 of the columns that contain a 1. Going back to the previous example, we really could write 01010101_2 as $0 + 64 + 0 + 16 + 0 + 4 + 0 + 1 = 85$, as shown in Figure 3.1. If we can learn the powers of each column (the powers of 2), then all we are doing is adding up those column values whose associated digits are 1s.

Table 3.2 shows many of the powers of base 2. You get this by starting at 1 and then doubling the number for each higher power (or each column as you move to the left). Notice that the powers of 2 raised to units of 10 (10, 20, 30) are all abbreviated for convenience. $2^{10} = 1024$ or 1 K or approximately 1000 (if we do not mind rounding off a little). $2^{20} = 1$ M or approximately 1 million (as shown in Table 3.2, the exact value is 1,048,576). $2^{30} = 1$ G or approximately 1 billion. $2^{40} = 1$ T or approximately 1 trillion. 2^{50} (not shown in the table) = 1 P for a peta, which is approximately 1 quadrillion.

Let us try to put the powers of 2 in Table 3.2 to use. Given any binary number, just add up the power of 2 from the chart in Table 3.2 based on the location of the 1s, where the rightmost digit represents the value 2^0. For instance, 1101 would contain a 2^3, a 2^2, but no 2^1, and a 2^0, or $2^3 + 2^2 + 2^0 = 8 + 4 + 1 = 13$. We will stick with byte-long numbers (8 bits) for simplicity.

2^7	2^6	2^5	2^4	2^3	2^2	2^1	2^0	
0	1	0	1	0	1	0	1	
	=		=		=		=	
	64		16		4		1	= 85

FIGURE 3.1 Sample binary to decimal conversion.

TABLE 3.2 Powers of Two

Power of 2	Value
2^0	1
2^1	2
2^2	4
2^3	8
2^4	16
2^5	32
2^6	64
2^7	128
2^8	256
2^9	512
2^{10}	1,024
2^{11}	2,048
2^{12}	4,096
2^{13}	8,192
2^{14}	16,384
2^{15}	32,768
2^{16}	65,536
2^{20}	1,048,576
2^{30}	1,073,741,824
2^{40}	1,099,511,627,776

11000110: $2^7 + 2^6 + 2^2 + 2^1 = 128 + 64 + 4 + 2 = 198$

10101010: $2^7 + 2^5 + 2^3 + 2^1 = 128 + 32 + 8 + 2 = 170$

01000111: $2^6 + 2^2 + 2^1 + 2^0 = 64 + 4 + 2 + 1 = 71$

00001110: $2^3 + 2^2 + 2^1 = 8 + 4 + 2 = 14$

00110011: $2^5 + 2^4 + 2^1 + 2^0 = 32 + 16 + 2 + 1 = 51$

00000000: 0

11111111: $2^7 + 2^6 + 2^5 + 2^4 + 2^3 + 2^2 + 2^1 + 2^0 =$

$$128 + 64 + 32 + 16 + 8 + 4 + 2 + 1 = 255$$

Notice in this last case, we have the largest 8-bit number, which is equal to $2^8 - 1$.

What about converting from decimal to binary? There are two strategies that we can apply. Below is the traditional strategy, but another—possibly easier—strategy will be shown next.

Take the decimal value and divide by 2. Take the quotient and divide by 2. Continue doing this until you are left with 0. Record the remainders of the division (the remainder when dividing by 2 will either be a 1 or 0). The collection of remainders in reverse order (or from bottom up) is the binary value. For example, let us try 89:

89/2 = 44 with a remainder of 1

44/2 = 22 with a remainder of 0

22/2 = 11 with a remainder of 0

11/2 = 5 with a remainder of 1

5/2 = 2 with a remainder of 1

2/2 = 1 with a remainder of 0

1/2 = 0 with a remainder of 1

We are done (we have reached 0). The binary equivalent of 89 is 01011001 (the remainders in reverse order, with an added bit in the front to make it an 8-bit number). We can confirm this by converting this number back to decimal: 01011001 = 64 + 16 + 8 + 1 = 89.

Let us try another one, 251:

251/2 = 125 with a remainder of 1

125/2 = 62 with a remainder of 1

62/2 = 31 with a remainder of 0

31/2 = 15 with a remainder of 1

15/2 = 7 with a remainder of 1

7/2 = 3 with a remainder of 1

3/2 = 1 with a remainder of 1

1/2 = 0 with a remainder of 1

So 251 is 11111011. We convert back to check, 11111011 = 128 + 64 + 32 + 16 + 8 + 2 + 1 = 251.

The division approach is simple although not necessarily intuitive, because you have to reverse the order of the remainders to form the binary number. Another way to convert from decimal to binary is to use subtractions. Given a decimal number, identify the largest power of 2 that can be subtracted from it while still leaving a positive number or 0. Do this until your decimal number is 0. For each power of 2 that is subtracted into the number, write a 1 in that corresponding column.

Let us consider as an example the same value we just converted, 251. Referring back to Table 3.2, the largest power of 2 that can we can subtract from 251 is 128. In subtracting 128 from 251, we have 123. The largest power of 2 that we can subtract from 123 is 64, leaving 59. The largest power of 2 that we can subtract from 59 is 32 leaving 27. The largest power of 2 we can subtract from 27 is 16 leaving 11. The largest power of 2 we can subtract from 11 is 8, leaving 3. The largest power of 2 we can subtract from 3 is 2 leaving 1. The largest power of 2 we can subtract from 1 is 1 leaving 0. We are done. Our conversion process went like this:

251 – 128 = 123

123 – 64 = 59

59 – 32 = 27

27 – 16 = 11

11 – 8 = 3

3 – 2 = 1

1 – 1 = 0

Thus, 251 = 128 + 64 + 32 + 16 + 8 + 2 + 1. So our binary equivalent has 1s in the columns representing 128, 64, 32, 16, 8, 2, and 1 and 0s elsewhere (the only other column represents 4). This gives us 11111011. We can check our work, as shown in Figure 3.2, which converts 11111011 back to decimal, 251.

Let us do another example. Convert 140 into binary. The largest power of 2 that we can subtract into 140 is 128 leaving 12. The largest power of 2 that we can subtract into 12 is 8 leaving 4. The largest power of 2 that we can subtract into 4 is 4 leaving 0.

140 – 128 = 12

12 – 8 = 4

4 – 4 = 0

140 = 128 + 8 + 4. So we have 1s in the columns that represent 128, 8, and 4, 0s elsewhere. This is the value 10001100, so 140 = 10001100.

So far, we have limited our binary numbers to 8 bits. Recall that with 8 bits, we have 256 different combinations of 1s and 0s. Since 00000000 is 0, the 8 bit (1 byte) range of values is 0 to 255. What if we want to store a larger number? Then we need more bits. For instance, in 2 bytes (16 bits), we can store any number from 0 to 65535. Recall that $2^{16} = 65536$. In general, if we have n bits, we can store any number from 0 to $2^n - 1$. Although we will often use the byte as a convenient amount of storage (instead of 1 or a few bits), limiting ourselves to 1 byte is too restrictive. Today's computers typically provide 32 bits (4 bytes) or 64 bits (8 bytes) for the typical storage size. We refer to the typical storage size as the computer's *word* size.

2^7	2^6	2^5	2^4	2^3	2^2	2^1	2^0	
1	1	1	1	1	0	1	1	
‖	‖	‖	‖	‖		‖	‖	
128	64	32	16	8		2	1	= 251

FIGURE 3.2 Converting back to check answer.

How big is 32 bits? It gives us this range of values:

00000000000000000000000000000000 = 0

...

11111111111111111111111111111111 = 4294967296 (nearly 4.3 billion)

For most applications, 32 bits gives us ample space for storage. In 64 bits, the largest number we can store is more than 18 quadrillion (18 followed by 15 zeroes!) These are huge numbers. Are 32- or 64-bit word sizes wasteful of memory? It depends on the application. However, with memory sizes of at least 4 billion bytes and disk storage of at least half a trillion bytes, using 32 or 64 bits to store numbers should not be a cause for concern.

Consider having to look at a series of numbers stored in 32-bit binary values. Because all of the digits are 1 and 0 and because there are a lot of them, the resulting list becomes very difficult to look at, almost like looking at an optical illusion. For convenience, we may translate binary numbers into different representations, commonly octal (base 8) or hexadecimal (base 16). Let us try with the following list. Imagine pages and pages of this stuff!

111011001000010110010101100000111

00101100010100101010110110010001

10001011001001001011011100100111

10001000101001111010001110101011

00001001010100000111100010101100

You might notice that 8 and 16 are multiples of 2. This makes translating from base 2 to base 8 or base 16 simple (as compared to translating from base 2 to base 10 or base 10 to base 2). For base 8, take 3 binary bits and group them together and then convert those 3 bits from binary to octal. This works because $2^3 = 8$. Note that because octal uses only eight digits (0 to 7), converting 3 bits from binary to octal is the same as converting 3 bits from binary to decimal. Table 3.3 shows you how to convert 3 bits between the two bases.

TABLE 3.3 Binary to Octal Conversion Table

Binary	Octal
000	0
001	1
010	2
011	3
100	4
101	5
110	6
111	7

Using the first 32-bit number from the previous page, we will convert it to octal.

11101100100010110010101100000111

First, divide the 32-bit number into 3-bit segments starting from the right end

11 101 100 100 010 110 010 101 100 000 111

Second, since the leftmost segment is not 3 bits, we add leading 0s to make it 3 bits

011 101 100 100 010 110 010 101 100 000 111

Third, using the above table, we convert each group of 3 bits into its equivalent octal value

3 5 4 4 2 6 2 5 4 0 7

Finally, we combine the individual octal digits into a single value: 35442625407.
Therefore, $11101100100010110010101100000111_2 = 35442625407_8$.
The other four numbers are converted as follows:

$00\ 101\ 100\ 010\ 100\ 101\ 010\ 110\ 110\ 010\ 001 = 05424526621_8$

$10\ 001\ 011\ 001\ 001\ 001\ 011\ 011\ 100\ 100\ 111 = 21311133447_8$

$10\ 001\ 000\ 101\ 001\ 111\ 010\ 001\ 110\ 101\ 011 = 21051721653_8$

$00\ 001\ 001\ 010\ 100\ 000\ 111\ 100\ 010\ 101\ 100 = 01124074254_8$

The entire list of numbers is then

35442625407

05424526621

21311133447

21051721653

01124074254

There are still a lot of digits to deal with in our octal listing of numbers. We can reduce the size of the list even more by converting the binary number to hexadecimal. In hexadecimal, we will group 4-bit binary numbers and convert them to the equivalent hexadecimal digit. Table 3.4 shows the binary, hexadecimal, and decimal equivalents for single hexadecimal digits (the first numbers 0–15 in decimal). Notice, as discussed earlier, that for 10–15, we use A–F so that we can store these values as single digits.

TABLE 3.4 Binary and Hexadecimal Conversion

Binary	Hexadecimal	Decimal
0000	0	0
0001	1	1
0010	2	2
0011	3	3
0100	4	4
0101	5	5
0110	6	6
0111	7	7
1000	8	8
1001	9	9
1010	A	10
1011	B	11
1100	C	12
1101	D	13
1110	E	14
1111	F	15

Let us use the same number as before, $11101100100010110010101100000111_2$, and convert it to hexadecimal. We use the same approach that we did when converting from binary to octal except that we use groups of 4 bits. Refer to Table 3.4 to see the hexadecimal digits.

1110 1100 1000 1011 0010 1011 0000 0111

E C 8 B 2 B 0 7

As shown above, $11101100100010110010101100000111_2 = \text{EC8B2B07}_{16}$.
The other numbers in our list will be converted as follows:

0010 1100 0101 0010 1010 1101 1001 0001 = 2C52AD91_{16}

1000 1011 0010 0100 1011 0111 0010 0111 = 8B24B727_{16}

1000 1000 1010 0111 1010 0011 1010 1011 = 88A7A3AB_{16}

0000 1001 0101 0000 0111 1000 1010 1100 = 095078AC_{16}

To convert an octal value back into binary, just replace each octal digit with its three-digit binary equivalent. Make sure you use all 3 digits in binary though. For instance, $723_8 = 111\ 010\ 011_2$, not $111\ 10\ 11_2$. The value $12345670_8 = 001\ 010\ 011\ 100\ 101\ 110\ 111\ 000 = 001010011100101110111000_2$, and if this was to be stored in 32 bits, we would add leading zeroes to extend it to 32 bits, giving us:

$00000000001010011100101110111000_2$

To convert a value from hexadecimal to binary, just replace each hex digit with its equivalent 4-bit binary value. For instance $2468BDF0_{16}$ = 0010 0100 0110 1000 1011 1101 1111 0000_2 = $00100100011010001011110111110000_2$. The advantage of using octal or hexadecimal over binary is that the numbers are easier to read.

NEGATIVE NUMBERS AND FRACTIONS

In Binary Numbering System, we saw how any positive whole number (integer) can be converted from decimal to binary. All we need is enough bits to store the binary number. How do we store negative numbers and numbers with a decimal point? We need to use different representations.

For negative numbers, computers use a representation called *two's complement.* A two's complement number is identical to the normal binary number if the value is positive. If the number is negative, we have to perform a conversion. The conversion requires that you take the positive binary number, flip all of the bits (0s become 1s, 1s become 0s), and add 1. Binary addition is covered in Binary Operations, but we will apply it to some simple examples here.

Imagine that we have the number 61, and we want to store it as an 8-bit two's complement value. Since it is positive, we use the normal decimal to binary conversion: 61 = 32 + 16 + 8 + 4 + 1 = 00111101, so 61 is 00111101.

What about –61? Since 61 = 00111101, to get –61, flip all of the bits and add 1. Flipping the bits of 00111101 gives us 11000010 (just make every 0 a 1 and every 1 a 0). Now add 1 to 11000010 and you get 11000011. So –61 = 11000011. Notice that the leading bit is a 1. In two's complement, a negative number will always start with a 1 and a positive number will always start with a 0.

Let us try another one, 15. This will be 8 + 4 + 2 + 1 = 00001111. So –15 requires that we flip all the bits to get 11110000 and then add 1 to get 11110001. Again, notice that the positive version starts with a 0 and the negative version starts with a 1.

To convert from binary to decimal, if the number is positive, just convert it as usual. For instance, 00111101 is positive, so the number will be 32 + 16 + 8 + 4 + 1 = 61, and 00001111 is positive so the number will be 8 + 4 + 2 + 1 = 15.

What about 11000011? This is a negative number. We use the same conversion technique that we used to convert from decimal to binary: flip all bits and add 1. So 11000011 becomes 00111100 + 1 = 00111101, which is the decimal number 61, so 11000011 must be –61. And 11110001 is negative, so we flip all bits and add 1 giving us 00001110 + 1 = 00001111, which is the decimal number 15, so 11110001 must be –15.

Now let us try to convert –48 to binary. We again start with +48 = 32 + 16 = 00110000. Since we want –48, we have to flip all bits and add one. In flipping all bits, we go from 00110000 to 11001111. Now we have to add 1. What is 11001111 + 1? Well, that requires a little binary arithmetic to solve. It is not that difficult; you just have to perform some carries that would give you 11010000. So –48 is 11010000.

Since binary arithmetic is covered in Binary Operations, for now, we will consider an alternative approach to converting negative numbers to binary. We again start with a negative number in decimal. Convert the positive version of the number into binary, as we did

in the previous approach. We must convert the positive binary number into its negative equivalent in two's complement. Now you can apply this simpler strategy:

1. Start at the rightmost digit, working right to left, copy every 0 until you reach the first 1 bit
2. Copy the first 1 bit
3. Continue working right to left flipping all remaining bits

For instance, 00110000 would be converted as follows:

1. Starting from the right, we copy down all of the 0s

   ...0000

2. When we reach the first 1, copy it

   ...1 0000

3. Flip all remaining bits, in this case 001 becomes 110

   110 1 0000

So converting 00110000 from positive to negative gives us 11010000. What if the number has 1 in the rightmost position? Let us start with the number 11000011 (–61). Step 1 is not used because there are no 0s; so, instead, we write down the rightmost 1 and then flip the rest of the bits, giving us 00111101 (+61).

Fractional numbers are both easier and more complex to understand. They are easier to understand because they are an extension to the formula presented earlier, $\sum_{i=0}^{n}(a_i * b^i)$, except that we extend i to include negative numbers for those that appear on the right side of the decimal point. You might recall from basic algebra that $b^{-i} = 1/b^i$. So, b^{-3} is really $1/b^3$. In binary, b is always 2, so we have to learn some negative powers of 2. Table 3.5 shows several negative powers of 2.

TABLE 3.5 Negative (Fractional) Powers of 2

Exponential Form	Value as a Fraction	Value as a Real Number
$2^{-1} = 1/2^1$	1/2	0.5
$2^{-2} = 1/2^2$	1/4	0.25
$2^{-3} = 1/2^3$	1/8	0.125
$2^{-4} = 1/2^4$	1/16	0.0625
$2^{-5} = 1/2^5$	1/32	0.03125
$2^{-6} = 1/2^6$	1/64	0.015625
$2^{-7} = 1/2^7$	1/128	0.0078125

Let us look at an example. What is the binary number 110111.01? Using our formula from above, 110111.01 = $1 * 2^5 + 1 * 2^4 + 0 * 2^3 + 1 * 2^2 + 1 * 2^1 + 1 * 2^0 + 0 * 2^{-1} + 1 * 2^{-2} =$ 32 + 16 + 4 + 2 + 1 + ¼ = 55.25.

Another example is 101.101 = 4 + 1 + 1/2 + 1/8 = 5.625. Can you work out the next examples?

1111.1001

0.1111

10101.101

Converting from binary to decimal is not very difficult. However, there is a problem. When we want to store a value such as 55.25 in the computer, it must be translated into binary, so we get 110111.01. However, the computer only stores 1s and 0s, how do we specify the decimal point? The answer is that we do not try to store a decimal point, but we record where the decimal point should be.

In order to denote where the decimal point is, we use a more complex strategy based on *scientific notation*. In scientific notation, the value 55.25 is actually represented as $0.5525 * 10^2$. That is, the decimal point is shifted to precede the first non-zero digit. In shifting the decimal point however, we must multiply the value by a power of 10. Thus, 55.25 becomes $0.5525 * 10^2$. Now we have two numbers to store, 0.5525 and 2. We can omit the decimal point as implied, that is, the decimal point will always precede the first non-zero digit. We can also omit the 10 because it is implied to be base 10.

In binary, we do the same basic thing. We start with 110111.01, shift the decimal point to precede non-zero digit (the first 1) by multiplying by a power of 2. We then omit the decimal point and the 2 and store the number as two parts. Using 110111.01 as our example, we first shift the decimal point giving us $.11011101 * 2^6$. We get 2^6 because we shifted the decimal point 6 positions to the left. Since our number got smaller (went from 110111.01 to .11011101), we had to multiply by a large value. Now, we simply store this as two numbers, 11011101 and 110 (the exponent 6). The value 2 used in the multiplication is the base. Since base 2 is implied, we can remove it from our number. We actually add a third number to our storage, a sign bit. We will use 0 for positive and 1 for negative. So our number becomes 0 for the sign bit, 110 for the exponent and 11011101 for the actual number. We refer to this portion as the mantissa.

We organize the three parts of this storage as sign bit, exponent bits, mantissa bits. The sign bit is always a single bit. We must select a size in bits for the exponent and mantissa. The size we select will impact the *precision* of our representation. Let us create a 16-bit *floating point* representation as follows: 1 sign bit, 5 exponent bits, 10 mantissa bits. The exponent will be stored in two's complement so that we can have both positive and negative exponents. The mantissa is a positive number because the sign bit will indicate whether the number is positive or negative.

Taking 55.25 as our original decimal number, we first convert it into binary as 110111.01. Now we shift the decimal point to give us $.11011101 * 2^6$. We convert the exponent into binary using 5 bits, 00110. Finally, we convert the mantissa to 10 bits giving us 1101110100. Notice that when we added 0s to the mantissa, we added *trailing zeroes*. This is because the

number is a fraction, and so adding zeroes at the end does not change the value. We added *leading zeroes* to the exponent because the exponent is not a fraction. It would be like adding 0s to 12.34, we would add leading zeroes to the left of the whole portion and trailing zeroes to the right of the fractional portion, giving us for instance 0012.340000. Now, our floating point value for 55.25 is 0 (sign bit), 00110 (exponent bits), 1101110100 (mantissa). This is stored as a 16-bit value as 0001101101110100.

Let us try to convert another decimal number to binary using this 16-bit floating point representation. We will try –14.75. First, we will disregard the minus sign for now. We convert 14.75 into binary as follows: 14.75 = 8 + 4 + 2 + .5 + .25 or 8 + 4 + 2 + 1/2 + 1/4. This is the binary number 1110.11. Now, we must shift the decimal point to precede the first 1 giving us $.111011 * 2^4$. Thus, our mantissa is 111011 and our exponent is 4, or 100. We extend the mantissa to 10 bits giving us 1110110000 and our exponent to 5 bits giving us 00100. Finally, we started with a negative number, so the sign bit is 1. Thus, –14.75 = 1 00100 1110110000 = 1001001110110000.

We will do one more for conversion from decimal to binary. We will convert 0.1875 to binary. This looks tricky because there is no whole portion in our number. 0.1875 consists of 1/8 + 1/16 or the binary number 0.0011. Recall that we must shift the decimal point to precede the first 1. In this case, we are shifting the decimal point to the right instead of the left. Thus, 0.0011 is becoming larger, not smaller as we saw with the previous two examples. We must therefore multiply by a negative power of 2, not a positive power of 2. $0.0011 = 0.11 * 2^{-2}$. Our exponent is a negative number. We can store positive and negative numbers in two's complement. In 5 bits, –2 = 11110. So our exponent is 11110. Our sign bit is 0. Finally, our mantissa, extended to 10 bits is 1100000000. Therefore, 0.1875 = 0 11110 1100000000 or 0111101100000000.

Let us now convert from our 16-bit floating point representation back into decimal. We will use the number 1001111100010110. We break this number into its three parts, we have 1 00111 1100010110. The initial 1 tells us it is a negative number. The next group of digits is the exponent. 00111 is binary for 7, so our exponent is 7 meaning that we will shift the decimal point 7 positions to the right. Finally, the last group of digits is the mantissa. A decimal point is implied to precede these digits. So 1100010110 is really $.1100010110 * 2^{00111}$. When we multiply by 2^{00111}, we get 1100010.110. Now, we convert this number from binary to decimal. This is the number 64 + 32 + 2 + ½ + ¼ = 98.75, and since the sign bit was 1, this is the number –98.75.

Here is another example, 0111011010000000, which is 0 11101 1010000000. This number is positive. The exponent is 11101, which is a negative number in two's complement. We convert it to its two's complement positive version, which is 00011, or 3. So the exponent is really –3, which means that we will shift the decimal point three positions to the left (instead of right). The mantissa is .10100000000, or .101 for convenience. Thus, our number is $.101 * ^{-3}$ = .000101. Using our binary to decimal conversion algorithm, $.000101 = 1 * 2^{-4} + 1 * 2^{-6} = 1/2^4 + 1/2^6 = 1/16 + 1/64 = 3/64 = 0.046875$.

Our 16-bit representation is not actually used in computers. There are two popular formats, or representations, used by today's computers. They are often referred to as single precision and double precision. In single precision, 32 bits are used for the sign, exponent, and mantissa. In double precision, 64 bits are used.

CHARACTER REPRESENTATIONS

Aside from numbers, we also want to store text in the computer. We will store text by breaking it up into individual characters. Each character will be stored in some distinct amount of memory (for instance, 1 byte per character). Therefore, text, such as this chapter, would require converting each character to its character representation and then storing each character representation in a byte of memory. The character representations would have to include not only the 26 letters of the alphabet, but whether the letters are upper or lower case (52 total letters), the 10 digits (for when we are referencing digits, not numbers, such as someone's zip code or phone number), punctuation marks such as #, $, &, *, +, the blank space (space bar), and the end of line (carriage return or enter key).

We cannot use a representation like we did for numbers because there is no natural mapping from characters to numbers. So, instead, we make up our own. There are two common forms of character representation, ASCII (American Standard Code for Information Interchange) and Unicode. A third form, EBCDIC, was extensively used in the 1960s and 1970s. We will briefly discuss these three forms of representation here.

Each of these forms of representation is manmade—that is, someone invented them. In each case, someone decided that a given character would be stored using a specific bit pattern. Although it is not all that important for us to know these bit patterns, two of the three representations are shown in Figures 3.3 and 3.4. Figure 3.3 demonstrates the ASCII

Dec	Hex	Oct	Char	Dec	Hex	Oct	Char	Dec	Hex	Oct	Char	Dec	Hex	Oct	Char
0	0	0		32	20	40	[space]	64	40	100	@	96	60	140	`
1	1	1		33	21	41	!	65	41	101	A	97	61	141	a
2	2	2		34	22	42	"	66	42	102	B	98	62	142	b
3	3	3		35	23	43	#	67	43	103	C	99	63	143	c
4	4	4		36	24	44	$	68	44	104	D	100	64	144	d
5	5	5		37	25	45	%	69	45	105	E	101	65	145	e
6	6	6		38	26	46	&	70	46	106	F	102	66	146	f
7	7	7		39	27	47	'	71	47	107	G	103	67	147	g
8	8	10		40	28	50	(	72	48	110	H	104	68	150	h
9	9	11		41	29	51	)	73	49	111	I	105	69	151	i
10	A	12		42	2A	52	*	74	4A	112	J	106	6A	152	j
11	B	13		43	2B	53	+	75	4B	113	K	107	6B	153	k
12	C	14		44	2C	54	,	76	4C	114	L	108	6C	154	l
13	D	15		45	2D	55	-	77	4D	115	M	109	6D	155	m
14	E	16		46	2E	56	.	78	4E	116	N	110	6E	156	n
15	F	17		47	2F	57	/	79	4F	117	O	111	6F	157	o
16	10	20		48	30	60	0	80	50	120	P	112	70	160	p
17	11	21		49	31	61	1	81	51	121	Q	113	71	161	q
18	12	22		50	32	62	2	82	52	122	R	114	72	162	r
19	13	23		51	33	63	3	83	53	123	S	115	73	163	s
20	14	24		52	34	64	4	84	54	124	T	116	74	164	t
21	15	25		53	35	65	5	85	55	125	U	117	75	165	u
22	16	26		54	36	66	6	86	56	126	V	118	76	166	v
23	17	27		55	37	67	7	87	57	127	W	119	77	167	w
24	18	30		56	38	70	8	88	58	130	X	120	78	170	x
25	19	31		57	39	71	9	89	59	131	Y	121	79	171	y
26	1A	32		58	3A	72	:	90	5A	132	Z	122	7A	172	z
27	1B	33		59	3B	73	;	91	5B	133	[	123	7B	173	{
28	1C	34		60	3C	74	<	92	5C	134	\	124	7C	174	|
29	1D	35		61	3D	75	=	93	5D	135	]	125	7D	175	}
30	1E	36		62	3E	76	>	94	5E	136	^	126	7E	176	~
31	1F	37		63	3F	77	?	95	5F	137	_	127	7F	177	

FIGURE 3.3 ASCII table. Each character is shown in decimal, hex, and octal. (Courtesy of ZZT32, http://commons.wikimedia.org/wiki/File:Ascii_Table-nocolor.svg.)

Decimal	Character	Decimal	Character	Decimal	Character	Decimal	Character
000	NUL	043	CU2	127	"	210	K
001	SOH	045	ENQ	129	a	211	L
002	STX	046	ACK	130	b	212	M
003	ETX	047	BEL	131	c	213	N
004	PF	050	SYN	132	d	214	O
005	HT	052	PN	133	e	215	P
006	LC	053	RS	134	f	216	Q
007	DEL	054	UC	135	g	217	R
008	GE	055	EOT	136	h	224	\
009	RLF	059	CUB	137	i	226	S
010	SMM	060	DC4	145	j	227	T
011	VT	061	NAK	146	k	228	U
012	FF	063	SUB	147	l	229	V
013	CR	064	BLANK	148	m	230	W
014	SO	074	¢	149	n	231	X
015	SI	075	.	150	o	232	Y
016	DLE	076	<	151	p	233	Z
017	DC1	077	(	152	q	240	0
018	DC2	078	+	153	r	241	1
019	TM	079	|	161	~	242	2
020	RES	090	!	162	s	243	3
021	NL	091	$	163	t	244	4
022	BS	092	*	164	u	245	5
023	IL	093	)	165	v	246	6
024	CAN	094	;	166	w	247	7
025	EM	095	¬	167	x	248	8
026	CC	096	-	168	y	249	9
027	CU1	097	/	169	z		
028	IFS	106	¦	192	{		
029	IGS	107	,	193	A		
030	IRS	108	%	194	B		
031	IUS	109	_	195	C		
032	DS	110	>	196	D		
033	SOS	111	?	197	E		
034	FS	121	`	198	F		
036	BYP	122	:	199	G		
037	LF	123	#	200	H		
038	ETB	124	@	201	I		
039	ESC	125	'	208	}		
042	SM	126	=	209	J		

FIGURE 3.4 EBCDIC table. (From www.LookupTables.com, http://www.lookuptables.com/ebcdic_scancodes.php. With permission.)

representation. Figure 3.4 demonstrates the EBCDIC representation. Note that some of the characters are control or non-printable characters and so do not appear in the table.

EBCDIC was only used in IBM mainframe computers and was discontinued in part because it was not a very logical representation. If you look over the ASCII representation, you will see that for any two letters, the ASCII representation is in the same cardinal order as the letters appear in the alphabet; so, for instance, 'a' immediately precedes 'b', which immediately precedes 'c', and so forth. The representation is organized so that upper case letters appear before lower case letters. This means that 'A' is less than 'a' when we compare them. In EBCDIC, although letters are in order ('a' before 'b' before 'c'), they do not necessarily immediately follow one another, for instance 'j' does not immediately follow 'i'.

The way ASCII works is that in place of the character, we have the binary sequence. For instance, 'a' is stored as the decimal value 97, or the hexadecimal value 61. Using either decimal to binary or hexadecimal to binary, you will see that 'a' is 01100001. Compare this to 'A', which is decimal 65 (hexadecimal 41), which is 01000001. Interestingly, 'a' and 'A' are nearly identical, the only difference is with the third bit from the left. The upper case version of every letter has a third bit of 0 and the lower case version of every letter has a third bit of 1.

The ASCII representation uses 7 bits, which gives us 128 different combinations. With the 52 letters, 10 digits, and various punctuation marks, there is room left over for other characters. These include what are known as escape sequences such as the tab key and the return key (or "end of line"). In spite of using 7 bits for ASCII, each ASCII character is stored in 1 byte, so the last bit becomes a 0. This 0 will be inserted as the leading bit (leftmost). The EBCDIC representation uses 8 bits and so contains 256 different combinations.

Let's consider an example. We want to store the string "Hello". How is this done? In ASCII, each of the characters is represented by a different code. "H" is 01001000, "e" is 01100101, "l" occurs twice, both times 01101101, and "o" is 01101111. So "Hello" is stored as a sequence of five 8-bit sequences, or five bytes: 01001000 01100101 01101101 01101101 01101111. In EBCDIC, it takes as much storage (five bytes) to store "Hello", but the representations differ. In EBCDIC, we have instead: 11001000 10000101 10010111 10010111 10010110.

Neither ASCII nor EBCDIC has any representations for letters outside of the English alphabet. With the international nature of computing today, there is a need to represent other languages' symbols; for instance, the Arabic letters, the Cyrillic letters or Russian, the Chinese and Japanese characters. However, a majority of computer users do speak English. So the decision was made to expand ASCII from 7 bits to 16 bits, which gives us 65,356 different combinations, or the ability to store more than 65,000 characters. This became UNICODE. In UNICODE, the first 128 characters are the same as in ASCII. However, the remainder are used to store the characters found in other languages along with other symbols. These other symbols include the zodiac elements, symbols for chess pieces, the peace sign, the yin/yang symbol, smiley/frowny faces, currency symbols, mathematical symbols, and arrows.

The previous example for "Hello" will look similar in Unicode because the first 128 characters of Unicode are the same as the 128 characters of ASCII. However, in UNICODE, each character is stored in 16 bits rather than 7. So, "Hello" becomes:

0000000001001000 "H"

0000000001100101 "e"

0000000001101101 "l"

0000000001101101 "l"

0000000001101111 "o"

Thus, "Hello" requires twice the storage space in Unicode as it does in either ASCII or EBCDIC. At the end of this chapter, we will find another use for the leading bit in ASCII (the leftmost bit, into which we inserted a 0).

BINARY OPERATIONS

Aside from converting numbers, a computer needs to process values. We typically think of computers performing operations such as addition, subtraction, multiplication, division, and comparison (compare two values to determine if they are equal or which is less than the other). However, the computer operates only on binary values and performs binary operations (these are the same as Boolean operations). The operations are

AND—are the two values both on (1s)?

OR—is either value on (1s)?

NOT—convert the value to its opposite (1 becomes 0, 0 becomes 1)

XOR—exclusive or, are the two values different?

The truth tables for these four basic Boolean operations are shown in Figure 3.5.

Let us consider a few brief examples of applying the binary, or Boolean, operations. First, we will look at simple 1-bit examples.

0 OR 0 = 0	0 OR 1 = 1	1 OR 1 = 1
0 XOR 0 = 0	0 XOR 1 = 1	1 XOR 1 = 0
0 AND 0 = 0	0 AND 1 = 0	1 AND 1 = 0

X	Y	AND
0	0	0
0	1	0
1	0	0
1	1	1

X	Y	OR
0	0	0
0	1	1
1	0	1
1	1	1

X	NOT
0	1
1	0

X	Y	XOR
0	0	0
0	1	1
1	0	1
1	1	1

FIGURE 3.5 Basic logic operation truth tables.

In essence, we take the two bits and look up the row in the appropriate truth table. For 0 XOR 1, we look at the row where X = 0 and Y = 1.

Even though we apply AND, OR, and XOR to just two bits (and NOT to just 1), it is more interesting and useful to look at larger binary operations. For instance, 11110000 AND 10101010. In this case, we apply the binary operation in a bit-wise fashion. That is, we would apply AND to the first bit of each binary number, and then to the second bit of each binary number, and then..., recording in this case an 8-bit answer because each number is 8 bits long. The operation can be easily performed by lining the two numbers up on two lines, and then applying the operation column by column.

Let us compute 11110000 AND 10101010.

	1	1	1	1	0	0	0	0
AND	1	0	1	0	1	0	1	0
	1	0	1	0	0	0	0	0

Starting with the leftmost bit, 1 AND 1 is 1. In the next column moving right, 1 AND 0 is 0. Next, 1 AND 1 is 1. Next, 1 AND 0 is 0. Now, notice that the remainder of the bits in the top number as we move right are all 0. 0 AND anything is 0, so the last 4 bits will all be 0.

Let us try XOR, which is slightly more complicated. 10101111 XOR 00111100.

	1	0	1	0	1	1	1	1
XOR	0	0	1	1	1	1	0	0
	1	0	0	1	0	0	1	1

Here, if the two bits are the same, the result is a 0 and if the two bits differ, the result is 1.

A more complicated series of operations can be applied, for instance, NOT (10110011) AND (11110000 XOR 11001100). First, we will apply the XOR:

	1	1	1	1	0	0	0	0
XOR	1	1	0	0	1	1	0	0
	0	0	1	1	1	1	0	0

Next, we apply NOT (10110011). This merely flips each bit giving us 01001100. Finally, we apply the AND.

	0	1	0	0	1	1	0	0
AND	0	0	1	1	1	1	0	0
	0	0	0	0	1	1	0	0

Therefore, NOT (10110011) AND (11110000 XOR 11001100) = 00001100.

You might wonder why we are bothering with such simple operations. At its most basic level, these four operations are all the computer uses to compute with. So we will have to use these four operations to perform such tasks as addition, subtraction, multiplication, division, and comparison. How?

Before answering that question, let us look at how to perform binary addition using our ordinary addition operator, +. Let us do 0111 + 0110. We would accomplish this addition in a similar way to adding two decimal values together. We line them up, we add them column by column in a right-to-left manner, writing down the sum, and if necessary, carrying a unit over to the next column on the left. However, for binary, the only digits we are allowed are 0 and 1, and our carries will be not in units of 10 but units of 2.

Let us now solve 0111 + 0110. The solution is shown in Figure 3.6. We start with the rightmost column. There is no carry in to that column (or more precisely, the carry in is 0). So we want to add 0 (the carry) + 1 + 0 (the rightmost digits of our two numbers). What is 0 + 1 + 0? It is 1. So we write a sum of 1, with no carry over. In the next column, we have 0 (the carry) + 1 + 1. 0 + 1 + 1 is 2. However, we are not allowed to write 2 down as the only digits allowable are 0 and 1. What is 2 in binary? 10_2. This tells us that we should write down a 0 (the rightmost digit of 10_2), and carry a 1. Recall that the carry is a power of 2, so we are carrying a unit of 2, not 10. Even so, we write the carry as a 1 digit over the next column. Now in the third-from-right column, we have 1 (the carry) + 1 + 1. This is 3. Again, we cannot write down a 3. But 3 in binary is 11_2. So we write a 1 and carry a 1. In the leftmost column, we have 1 (the carry) + 0 + 0 = 1. So we write a 1 and have no carry. The carry coming out of the leftmost column can be recorded in the status flags (see Chapter 2) under the carry bit. This may be useful because, if there is a carry out produced, it may be a situation known as an *overflow*. That is, the two numbers were too large to fit into the destination, so we overflowed the destination (a register or memory).

Notice in Figure 3.6 that, for each column, we are computing two things. We compute the sum of the three digits (including the carry in) and the carry out to be taken to the next column as the carry in. We will always add three numbers: the two digits from the number and the carry in. We will always compute two different results, the sum of that column and the carry out. The carry out for one column becomes the carry in for the next column.

The sum algorithm can be expressed as follows:

1. Line up the two binary numbers.

2. The carry in for the rightmost column is always 0.

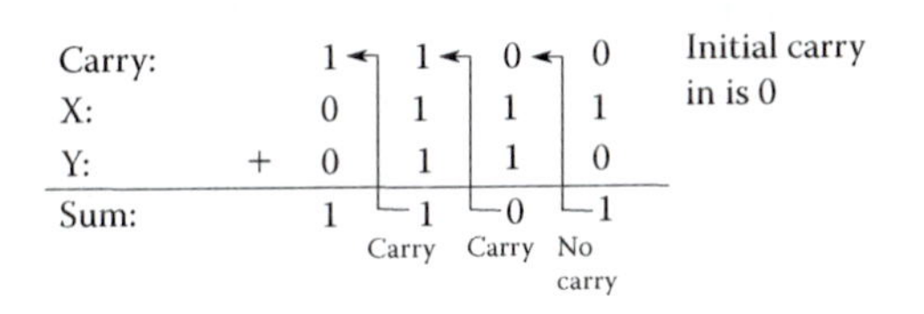

FIGURE 3.6 Binary addition showing sum and carry out/carry in.

3. Working right to left, repeat the following, column by column.

   a. Compute the sum of the two binary digits plus the carry in, as follows:

      All three are 0: 0

      One of the three is a 1: 1

      Two of the three are 1: 0 (1 + 1 + 0 = 2, or 10, sum of 0, carry of 1)

      All three are 1: 1 (1 + 1 + 1 = 3, or 11, sum of 1, carry of 1)

   b. Compute the carry out of the two binary digits plus the carry in, as follows:

      All three are 0: 0

      One of the three is a 1: 0

      Two of the three are 1: 1

      All three are 1: 1

   c. The carry out of this column becomes the carry in of the next column.

Let us give it a try. We will add 1001 + 0011. First, we line up the numbers:

$$\begin{array}{r} 1001 \\ +0001 \\ \hline \end{array}$$

For the rightmost column, the carry in is a 0. This gives us 0, 1, and 1. According to step 3a above, two 1 bits give a sum of 0. According to step 3b, two 1 bits give us a carry of 1. So we have the following partial result after one column.

$$\begin{array}{r} \text{Carry: --}10 \\ 1001 \\ +0011 \\ \hline \text{---}0 \end{array}$$

Now going into the second column (from the right), we have a carry in of 1, and the two digits are 0 and 1, so we have two 1 bits. This gives us a sum of 0 and a carry of 1. Now our partial result is as follows.

$$\begin{array}{r} \text{Carry: -}110 \\ 1001 \\ +0011 \\ \hline \text{--}00 \end{array}$$

For the third column, we have a carry in of 1, and two 0 digits. This gives us a sum of 1 and a carry of 0. Now our partial result is as follows:

$$\begin{array}{r} \text{Carry}: 0110 \\ 1001 \\ +0011 \\ \hline -100 \end{array}$$

Our leftmost column has a carry in of 0 and digits of 1 and 0. This gives us a sum of 1 and a carry of 0. Our resulting sum is then 1100. If we were to convert problem into decimal, we would find that we have just performed 9 + 3 = 12.

Try a few on your own and see how you do.

0001 + 0101 = 0110

1001 + 0011 = 1100

0111 + 0011 = 1010

Now, how do we *implement* this addition algorithm in the computer since we only have AND, OR, NOT, and XOR operations to apply? We will break our addition algorithm into two computations, computing sum and computing carry. We start with carry because it is a little easier to understand. Recall from above that the carry out is a 1 if either two or all three of the digits are 1. Let us call the three values X, Y, and C (for carry in). There are four possible combinations whereby there are two or three 1s. These are X = 1 and Y = 1, X = 1 and C = 1, Y = 1 and C = 1, or X = 1, Y = 1 and C = 1. In fact, we do not care about the fourth possibility because the three other comparisons would all be true. Therefore, we can determine if the carry out should be a 1 by using the following Boolean logic: (X AND Y) OR (X AND C) OR (Y AND C).

To compute the sum, we want to know if there is either a single 1 out of the three bits (X, Y, C), or three 1s out of the three bits. We can determine this by cleverly chaining together two XORs:

(X XOR Y) XOR C

Let us redo our example from Figure 3.6 using our Boolean operations. The two numbers are 0111 + 0110. Starting with the rightmost column, we have X = 1, Y = 0, C = 0. The sum is (X XOR Y) XOR C = (1 XOR 0) XOR 0 = 1 XOR 0 = 1. So we have a sum of 1. The carry is (X AND Y) OR (X AND C) OR (Y AND C) = (1 AND 0) OR (1 AND 0) OR (0 AND 0) = 0 OR 0 OR 0 = 0. So out of the rightmost column, we have a sum of 1 and a carry of 0.

In the second-to-right column, we have X = 1, Y = 1, C = 0. Our sum is (1 XOR 1) XOR 0 = 0 XOR 0 = 0. Our carry is (1 AND 1) OR (1 AND 0) OR (1 AND 0) = 1 OR 0 OR 0 = 1. So we have a sum of 0 and a carry of 1.

In the third-to-right column, we have X = 1, Y = 1, C = 1. Our sum is (1 XOR 1) XOR 1 = 0 XOR 1 = 1. Our carry is (1 AND 1) OR (1 AND 1) OR (1 AND 1) = 1 OR 1 OR 1 = 1. So we have a sum of 1 and a carry of 1. In our leftmost column, we have X = 0, Y = 0, C = 1. Our sum is (0 XOR 0) XOR 1 = 0 XOR 1 = 1, and our carry is (0 AND 0) OR (0 AND 1) OR (0 AND 1) = 0 OR 0 OR 0 = 0. So our leftmost column has a sum of 1 and carry of 0.

Given two binary numbers, we can perform addition using AND, OR, and XOR. What about subtraction? Let us assume we have two numbers, X and Y, and we want to compute X – Y. You might recall from algebra that X – Y = X + (–Y). Assuming X and Y are two's complement numbers, we have to modify Y to be –Y. How do we convert a positive number to negative, or negative number to positive, in two's complement? We flip all of the bits and add 1. How do we flip all of the bits of a number? We use the NOT operation. So in fact, X – Y = X + ((NOT Y) + 1). Thus, subtraction can use the same operations as above for sum and carry, but we add to it a NOT operation first on Y, and then add 1 to NOT Y. We add 1 by again using the sum and carry operations. See Figure 3.7.

Multiplication is a little more complicated, but a multiplication is just a series of additions (x * y requires adding x together y times). Division is a series of subtractions (x/y requires subtracting y from x until we reach 0 or a negative). Therefore, we can apply the addition and subtraction algorithms described above repeatedly to perform multiplication and division. They use a combination of AND, OR, XOR, and NOT.

How about comparison? Actually, this is done very simply. To compare two values X and Y, we do X – Y and see if the result is positive (in which case X > Y), negative (in which case X < Y), or zero (in which case X = Y). To test for negative, look at the leftmost bit of the difference (X – Y); if it is a 1, then the difference is negative (recall in two's complement, the leftmost bit is the sign), which means that X < Y. To test for zero, OR all of the bits together and see if the result is 0, which means that X – Y = 0 or X = Y. For instance, if we have 00000001, the OR of these bits is 1, so the value is not zero. Positive is actually the most difficult to determine as we have to make sure that the result is not zero and that the leftmost bit is 0. However, we can also determine if a number is positive if it is neither negative nor zero! If the difference is positive, then X – Y > 0 or X > Y.

How do we compare non-numbers? Typically, aside from numbers, the only other things we compare are characters and strings of characters. To see which is greater, "Frank" and "Fred", we compare character by character until we either reach the end of the string (in which case the strings are equal) or we have a mismatch. Here, we find a mismatch at the

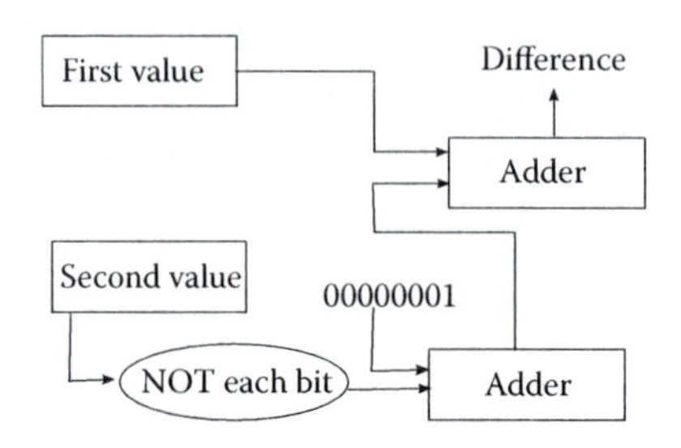

FIGURE 3.7 Subtracting two numbers performed by two additions and NOT.

third character 'a' versus 'e'. How do we compute 'a' – 'e'? Recall that characters are stored in memory using a representation called ASCII. In ASCII, 'a' is 96 and 'e' is 100. Therefore to compare "Frank" to "Fred", we do 'F' – 'F', which is 0, so we do 'r' – 'r', which is 0, so we do 'a' – 'e', which is –4. Since 'a' < 'e', "Frank" < "Fred".

A MESSAGE FROM THE STARS?

Physicist Frank Drake founded the Search for Extraterrestrial Intelligence (SETI). This effort, started around 1961, was to use radio telescopes (and other forms of observation) to listen for messages from the stars. What form might a message from aliens take? Although we have no idea, it has been proposed that any message would be encoded in some binary representation. Why? Because information is easily transmitted in binary, just send a series of on/off pulses. But what might such a message look like? The aliens should pick a representation that could be easily interpreted. For instance, we would not send out a message in ASCII because ASCII is strictly a human invention and would be meaningless to aliens. To demonstrate how difficult deciphering a message might be, Drake provided the following to his colleagues. Can you figure out what its meaning is? Neither could they!

11110000101001000011001000000010000010100
10000011001011001111000001100001101000000
00100000100001000010001010100001000000000
00000000001000100000000001011000000000000
00000001000111011010110101000000000000000
00001001000011101010101000000000101010101
00000000011101010101110101100000001000000
00000000000100000000000001000100111111000
00111010000010110000011100000001000000000
10000000010000000111110000001011000101110
10000000110010111110101111100010011111001
00000000000111110000001011000111111100000
10000011000001100001000011000000011000101
001000111100101111

Another type of Boolean operation that we will need our computers to perform is known as *masks*. A mask compares two binary values together using one of AND, OR, or XOR. One use of the AND mask is covered in Examples of Using Binary Numbers.

EXAMPLES OF USING BINARY NUMBERS

Here, we examine three very different ways that binary numbers are used in computers. In the first, we look at binary numbers as used in network addressing and netmasks. We then look at some common video file representation formats. Finally, we look at the use of binary numbers to perform error detection and correction.

Network Addresses and Netmasks

Computer networks provide a good example of how the binary concepts you have learned so far can be applied. Every computer in a network must have a unique network address. IPv4 (Internet Protocol version 4) addressing uses a dotted decimal format for these addresses.

The IPv4 address consists of four numbers, each called an *octet*. Each number consists of an unsigned 8-bit binary number. Recall that in 8 bits, you can store any decimal number from 0 to 255. So an IP address then is four numbers from 0 to 255, each number separated by a dot (period). For example, your computer's IP address might be 10.251.136.253. A network address is simply represented as a single 32-bit binary number. We break it down into 4 octets to make it simpler to read.

IP addresses are grouped so that those computers in one network share some of the same octets. This is similar to how street addresses are grouped into zip codes. Computers within a single network can communicate with each other without the need of a router. Routers are used to facilitate communication between networks. This is similar to how post offices are used to facilitate mail delivery between different zip codes. You can tell which street addresses belong to a given post office based on the zip code of that address. Computers have a similar concept. Hidden within your IP address is sort of a zip code that is the address of the network. We refer to this as the *network address*.

In order to decipher what network address an IP address belongs to, computers use a *netmask*. Let us look at an example where your IP address is 10.251.136.253, and your netmask is 255.255.255.0. The binary representation (with dots to ease reading) of these numbers is:

IP Address:	00001010.11111011.10001000.11111101
Netmask:	11111111.11111111.11111111.00000000

To compute the network address, a binary AND operation is performed between these two 32-bit binary numbers resulting in a network address of:

Network Address:	00001010.11111011.10001000.00000000 (10.251.136.0)

Recall that a binary AND is 1 only if both binary bits are a 1. This netmask has the result of returning the first three octets in their entirety because they are being ANDed with 11111111, with the final octet being masked, or zeroed out, because that octet is being ANDed with 00000000.

In the example above, we use a netmask of 255.255.255.0. This netmask is selected because all computers on this network share the first three octets. That is, the network address is the first three octets of the four octets. Each device's address on the network is the final octet, thus giving every device a unique address. Now, with a netmask of 255.255.255.0, there are 256 (2^8) possible addresses in that network. So, this organization would be able to provide 256 unique IP addresses. In dotted decimal format, the network address returned from this operation is 10.251.136.0. This provides network addresses in the range from 10.251.136.0 through 10.251.136.255, including our initial device's address of 10.251.136.253.

Two computers that share the same network address can communicate with each other without the need of a router. Two computers with different network addresses will know

that they need to contact their friendly neighborhood router (often referred to as the default route) in order to communicate between those two devices.

This concept is easy to understand when your netmask aligns evenly with the octet boundaries (i.e., the dots) as it did in our example above. However, let us look at another example where this is not the case. We will use the same IP address, 10.251.136.253, but with a netmask of 255.255.240.0. To compute our network address, we perform a binary AND of these two numbers. Figure 3.8 demonstrates the application of the netmask 255.255.240.0, where our network address turns out to be 10.251.128.0 rather than 10.251.136.0.

Netmasks are determined by the number of bits in the address that make up the network portion of the address. Some network addresses only comprise the first octet. For instance, one network may be 95.0.0.0 (or 01011111.00000000.00000000.00000000). This network would use a netmask of 255.0.0.0.

Obviously, understanding binary numbers and their operations is very important to those studying networking. Networking professionals must be fluent in understanding Boolean operations of AND, OR, NOT, and XOR. We cover networks in detail in Chapter 12.

Image Files

Earlier in this chapter, we saw that numbers (positive, negative, integer, fractional) can all be stored in binary either directly or through a representation such as two's complement or floating point. We also saw that characters are represented in binary using a code called ASCII. How are images stored in a computer? The most straightforward format is known as a *bitmap*. We will first consider a black and white image.

Any image is a collection of points of light. Each individual point is known as a *pixel* (short for picture element). For a black and white image, a pixel will take on one of two colors, black or white. We can represent these two colors using 1s and 0s (commonly, white is 1 and black is 0). To store a black and white image is merely a matter of placing into the file each pixel, row after row. The image in Figure 3.9 is a simple bitmap. The figure contains both the image and the storage in binary (to the right of the image).

For a black and white image of X × Y pixels (X rows, Y columns), the storage required is X * Y bits. A 1000 × 1000 pixel image requires 1,000,000 bits. This would be 125,000 bytes, or roughly 125 KB.

A variation of a black and white image is a gray scale image. Here, different shades of gray are added to the strictly black and white image. To represent an image, similar to the black and white image, each pixel is stored as one value, but now the value is in a range from least intensity (black) to most intensity (white). Any value in between is a shade of

IP address:	00001010 . 11111011 . 10001000 . 11111101 (10.251.136.253)
Netmask:	11111111 . 11111111 . 11110000 . 00000000 (255.255.240.0)
Network address:	00001010 . 11111011 . 10000000 . 00000000 (10.251.128.0)

FIGURE 3.8 Applying a netmask.

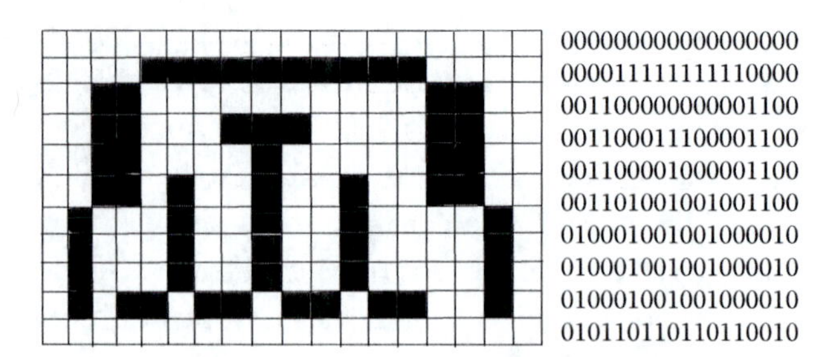

FIGURE 3.9 Black and white bitmap and its corresponding binary file.

grey. Most commonly, the level of intensity will be an integer number from 0 to 255, so that each pixel can be stored in 1 byte. A 1000 × 1000 pixel image would require 1,000,000 bytes (roughly 1 MB). This is 8 times the size of the black and white image, which makes sense because we increased the storage requirement from 1 bit per pixel to 1 byte per pixel.

Colors are represented in the computer by the amount of red, green, and blue that make up that pixel's light. We refer to this strategy as *RGB*. For a color image, we would store each pixel as a trio of numbers, the amount of red, the amount of green, and the amount of blue. As with gray scale images, we will store each of these intensities as an integer in 1 byte, giving us a range of 0–255. The color represented by (0, 0, 0)—that is, no red, no green, no blue—is black. The color represented by (255, 255, 255) is white. The color represented by (255, 0, 255) is purple (full red, no green, full blue). In binary, purple would be indicated as the three numbers (11111111, 00000000, 11111111). For convenience, we often denote RGB values in hexadecimal instead. So purple is FF00FF. This notation is used in designing web pages. Yellow, for instance, is all green and all blue with no red, or (0, 255, 255), or 00FFFF. The value 884400 would be (128, 64, 00), that is, it would be half of red and one fourth of green, a brownish-orange color.

For RGB, a 1000 × 1000 pixel image would require 3,000,000 bytes or roughly 3 MB. Note that using RGB as a representation, every pixel in an image can be one of more than 16 million colors (256 * 256 * 256).

As you can see, going from black and white to color greatly increases the file size. Are there any ways to reduce file sizes of images? Yes, by *file compression*. The two most common forms of compression are GIF and JPG. GIF stands for Graphics Interchange Format. The idea is that humans generally cannot differentiate between the millions of colors available in any kind of precise way. Why use so much storage space when one shade of red might look like another? In GIF, a color palette is provided that consists of 256 of the most popular colors. This permits each pixel to reference a color in the palette using only 1 byte instead of the 3 bytes (required for RGB); therefore, the size of the color image is reduced by a factor of 3. The 1000 × 1000 pixel color image now requires only 1 MB of storage. Unfortunately, if the original image uses colors that are not a part of the standard palette, then the image will look different, and in some cases you will get a very improper replacement color, for instance, blue in place of a pale yellow.

The JPG format is perhaps more common than GIF. JPG stands for Joint Photographic Experts Group (the original abbreviation for a JPG file was .jpeg). In GIF, if the colors are

available on the palette, the resulting GIF file looks identical to the BMP file and is therefore known as *lossless* compression. But JPG uses *lossy* compression. The difference is that after compressing the image, some quality is lost. You sacrifice image quality for storage size. The JPG compression format works by combining some of the pixels together into blockier groups of pixels. In effect, the image loses some resolution and becomes blurrier. In addition, if the image is a color image, it is translated from RGB to an alternative color format known as YUV, where each pixel is represented using brightness (Y) and luminescence (U and V). This format requires either two times or four times less storage space than RGB. Additionally, the lossy compression in the JPG format can reduce the image size even more so.

The newer PNG (Portable Network Graphics) enhances the GIF format. Like GIF, it is a lossless form of compression that uses a bitmap. It can support a 24-bit RGB palette, gray scale, and bitmaps without palettes. And like JPG, it compresses images to reduce space, but using a lossless algorithm so that the compressed image is of equal quality. Unlike JPG, the PNG compression may or may not succeed in truly reducing file size.

Error Detection and Correction

One significant challenge when dealing with computer-stored information is handling errors that arise during transmission of data from one location to another. It is unlikely but possible for errors to occur when data are moved from one component to another in the computer, but it is far more likely and even expected that errors can arise during transmission. Therefore, some form of error detection and correction is desired.

The simplest form of error detection is through the use of *parity*. In general, parity means equality. But when discussing a computer, parity refers to the evenness (or oddness) of a number. Specifically, we will count the number of 1 bits in a sequence of data and see if that count is even or odd. We can then use this to detect simple errors in transmission.

We will introduce a *parity bit* to every byte (8 bits) of storage. The parity bit will be 1 if the number of 1 bits in the byte is odd, and the parity bit will be 0 if the number of 1 bits in the byte is even. That is, the 8 bits plus the parity bit should *always* have an even number of 1 bits. This strategy is known as *even parity* (there is a variation called *odd parity* where the 8 bits plus parity bit must always have an odd number of 1 bits).

We will use even parity as follows. Given a byte, before this datum is moved, we will generate a parity bit. Now we have 9 bits. We transmit/move the 9 bits. The recipient receives 9 bits. If the 9 bits has even parity, then no error occurred. If the 9 bits has odd parity, an error occurred, the datum is discarded, and the recipient requests that the datum be resent.

How can we determine even or odd parity? It is very easy just using the XOR operation (refer back to Figure 3.5). The XOR operation only operates on 2 bits at a time, but we can chain together the XOR operations. So, given 8 bits, we XOR the first two bits and then XOR the result with the third bit and then XOR the result with the fourth bit, and so forth. Let us look at some examples of computing and using the parity bit. We want to move a datum 10010011 to another location in the computer or over a network. The first step is to compute the parity bit:

(((((((1 XOR 0) XOR 0) XOR 1) XOR 0) XOR 0) XOR 1) XOR 1) =

((((((1 XOR 0) XOR 1) XOR 0) XOR 0) XOR 1) XOR 1) =

((((((1 XOR 1) XOR 0) XOR 0) XOR 1) XOR 1) =

((((0 XOR 0) XOR 0) XOR 1) XOR 1) =

(((0 XOR 0) XOR 1) XOR 1) =

((0 XOR 1) XOR 1) =

(1 XOR 1) = 0

Therefore, 10010011 has a parity bit of 0. This makes sense because the total number of 1 bits should be even. 10010011 has four 1 bits, and so to keep the number of 1 bits even, we add a 0. Now, the 9 bits consist of four 1s and five 0s.*

We transmit these 9 bits and the recipient receives 10010011 & 0. The recipient performs a similar operation, XORing all 9 bits together, and winds up with 0. Therefore, it is correct, that is, no error arose.

Now consider if, upon receipt, the 9 bits were 10010011 & 1. In this case, the XOR computes:

((((((((1 XOR 0) XOR 0) XOR 1) XOR 0) XOR 0) XOR 1) XOR 1) XOR 1) = 1

So an error arose. Where did the error arise? By using a single parity bit, we do not know, but because there is an error, the recipient discards the data and asks for it to be re-sent. Notice in this case that the error arose in the parity bit itself. Another example might be 10010010 & 0, where the parity bit has no error but the byte has an error (the last bit should be 1). The recipient computes the XOR of these 9 bits and gets

((((((((1 XOR 0) XOR 0) XOR 1) XOR 0) XOR 0) XOR 1) XOR 0) XOR 0) = 1

Notice that the parity bit not only does not help us identify the error, but would also fail us if two errors arose. Imagine that the recipient gets 10010000 & 0. In this case, the last 2 bits are incorrect. If we XOR these 9 bits, we get 0, implying that no error occurred. Thus, the parity bit can detect 1 error but not 2, and cannot correct the error. If we want to correct the error, we would need more than 1 parity bit.

Also, the parity bit does not necessarily have to be associated with a byte. Recall from earlier in this chapter that ASCII values use 7 bits of a byte. The eighth bit can be used to store parity information. For instance, the letter 'a' is represented as x1100001. The initial x is unused, so we would normally place a 0 there. However, rather than waste the bit, we can

* Computing parity can be done much more simply than applying the XOR operation; just add up the number of 1 bits and make sure that the bits in the datum plus the parity bit is always an even number.

put a parity bit there. So, in fact, 'a' will be stored as 11100001 if we wish to include a parity bit. This saves us from having to add yet another bit.

Another use of parity comes with RAID technology. RAID is a newer form of disk storage device where there are multiple disk drives in one cabinet. The idea is that extra disk drives can provide both redundant storage of data and the ability to access multiple drives simultaneously to speed up disk access. RAID is discussed in more detail in Chapter 15 (Information Assurance and Security). Here, we briefly look at how parity can be used to provide error correction.

Imagine that we have five disk drive units. We will store 4 bytes of data on four different drives. The fifth drive will store parity information of the 4 bytes. For each 4-byte grouping that is distributed across the disks, we will also compute parity information and store it on the fifth drive. Consider one 4-byte group of data as follows:

11001111	01010101	11011011	00000011

We will compute bit-wise XOR across each of the 4 bytes. This means that we will take the first bits (leftmost bits) of the 4 bytes and compute XOR of those 4 bits. This will be the parity bit for the first bit of the four bytes. We will do this for each bit of the 4 bytes.

	11001111	//first byte, stored on the first disk
XOR	01010101	//second byte, stored on the second disk
XOR	11011011	//third byte, stored on the third disk
XOR	00000011	//fourth byte, stored on the fourth disk

To obtain the parity information then, we compute each XOR going down. The first parity bit is (((1 XOR 0) XOR 1) XOR 0) = 0. The second parity bit is computed as (((1 XOR 1) XOR 1) XOR 0) = 1. The third parity bit is computed as (((0 XOR 0) XOR 0) XOR 0) = 0. The fourth parity bit is computed as (((0 XOR 1) XOR 1) XOR 0) = 0. The fifth parity bit is computed as (((1 XOR 0) XOR 1) XOR 0) = 0. The sixth parity bit is computed as (((1 XOR 0) XOR 0) XOR 1) = 0. The seventh parity bit is computed as (((1 XOR 0) XOR 1) XOR 1) = 1. The eight parity bit is computed as (((1 XOR 1) XOR 1) XOR 1) = 0. So, our parity byte is 01000010. This byte is stored on the fifth disk.

Now, let us assume over time that the second disk drive fails—perhaps there is a bad sector, or perhaps the entire disk surface is damaged. Whatever the case, we can restore all of the data by using the three other data drives and the parity drive. For our 4 bytes above, we would have the following data available:

11001111	//first byte, stored on the first disk
xxxxxxxx	//second byte data is damaged
11011011	//third byte, stored on the third disk
00000011	//fourth byte, stored on the fourth disk
01000010	//parity byte, stored on the fifth disk

We XOR the four corresponding bits of the three bytes plus parity byte to obtain the missing data. For instance, using the first bits of each byte, we have (((1 XOR 1) XOR 0) XOR 0) = 0. So the first bit of our missing datum is 0. The second bit would be (((1 XOR 1) XOR 0) XOR 1) = 1. See if you can work through the remainder of this example to obtain the entire missing datum.

Using a single parity bit per byte is sometimes referred to as a horizontal redundancy check because the parity bit is computed across a block of data. Another form of horizontal redundancy check is a checksum (briefly mentioned in Chapter 12). The parity computation used for RAID is sometimes referred to as a vertical redundancy check because the parity computation is performed across a number of bytes, bit by bit.

FURTHER READING

A study of binary representations can be found in nearly every computer organization book (See the section Computer Hardware and Computer Assembly (Installation) in Chapter 2). The Kindle text, Binary Made Simple (R. Barton) available from Amazon Digital Services offers a useful look at how to perform conversions between binary, decimal, and hexadecimal. More thorough books on numbering systems provide background on theory and computation, such as the following:

- Conway, J. *The Book of Numbers*, New York: Springer, 1995.
- Guilberg, J. and Hilton, P. *Mathematics: From the Birth of Numbers*, New York: Norton, 1997.
- Ifrah, G. *The Universal History of Numbers: From Prehistory to the Invention of the Computer.* New York: Wiley, 1999.
- Niven, I., Zuckerman, H., and Montgomery, H. *An Introduction to the Theory of Numbers*, New York: Wiley, 1991.

Boolean (binary) operations and logic are also covered in computer organization texts. You can also find whole texts on the topic. A few choice texts are listed here.

- Brown, F. *Boolean Reasoning: The Logic of Boolean Equations.* New York: Dover, 2012.
- Gregg, J. *Ones and Zeros: Understanding Boolean Algebra, Digital Circuits, and the Logic of Sets.* Piscataway, NJ: Wiley-IEEE Press, 1998.
- Whitesitt, J. *Boolean Algebra and Its Applications.* New York: Dover, 2010.

As Chapter 12 covers computer networks in detail, see that chapter for further readings for information on such things as IP addresses and netmasks.

A comprehensive reference on the major forms of computer graphics files is available:

- Miano, J., *Compressed Image File Formats: JPEG, PNG, GIF, XBM, BMP.* Reading, MA: Addison Wesley, 1999.

These formats are all stored as binary files, which we discuss in Chapter 5.

REVIEW TERMS

Terminology introduced in this chapter:

ASCII
Base
Binary
Bit
Bitmap
Byte
Compression
Decimal
EBCDIC
Floating point
GIF
Hexadecimal
JPG
Lossy compression
Mantissa
Netmask
Numbering systems
Octal
Octet
Parity
Parity bit
Pixel
PNG
RAID
Two's complement
Unicode
Word

SAMPLE PROBLEMS

1. Convert 10111100 from binary to decimal.
2. Convert 10111100 from two's complement binary to decimal.
3. Convert 10100011 in two's complement to its negation (i.e., change its sign).
4. Repeat #3 for the two's complement binary value 00010010.
5. Convert 43 from decimal to 8-bit binary.
6. Convert –43 from decimal to two's complement 8-bit binary.
7. Convert $F3EA1_{16}$ to binary.
8. Convert 31752_8 to binary.
9. Convert 876 from decimal to octal.

10. Convert 6732 from decimal to hexadecimal.

11. Convert –45.625 from decimal to the 16-bit floating point representation of the chapter.

12. Convert 0111001100110000 from the 16-bit floating point representation of the chapter into decimal.

13. What is 10011110 AND 00001111?

14. What is 10011110 OR 00001111?

15. What is 10011110 XOR 00001111?

16. What is NOT 10011110?

17. Perform the 8-bit addition: 00110011 + 01010101

18. Perform the 8-bit addition: 01111011 + 00111110

19. Perform the 8-bit subtraction: 01010101 – 00001111 (hint: convert the second number to its two's complement opposite and then perform addition, the result will produce a carry out of the leftmost column, ignore it)

20. How many bytes would it take to store the following text using ASCII?

    Holy cow! The brown fowl jumped over the moon?

21. Repeat #20 in Unicode.

22. What is the parity bit of 11110010?

23. Repeat #22 for 11110011.

24. Repeat #22 for 00000000.

25. Given the byte 10011101 and the parity bit 1, assuming even parity, is there an error?

26. Repeat 25 for the byte 11110001 and the parity bit 0.

27. Compute the value ((((1 XOR 0) XOR 1) XOR 1).

28. Given the four bytes 00000000, 00000001, 11110000, and 10101101, compute the parity byte.

REVIEW QUESTIONS

1. In your own words, what does NOT do?

2. In your own words, explain when the AND operation will be true.

3. In your own words, explain when the OR operation will be false.

4. In your own words, explain when the XOR operation will be false.
5. How does XOR differ from OR?
6. Why is binary of importance to our society? Why is it important for you to understand binary in information technology?
7. What is a mantissa? What is a base?
8. What is lossy compression?
9. Explain why a 1024 × 1024 pixel color bitmap requires 3 MB of storage space.
10. Explain how a jpg image is able to reduce the storage space required of the same bmp file.
11. Why do you suppose we do not use EBCDIC as our preferred form of character representation?
12. What is an octet?
13. What is the difference between even parity and odd parity?

DISCUSSION QUESTIONS

1. Is it important for an IT person to understand the binary numbering system? Aside from the examples covered in this chapter (ASCII/Unicode, bitmaps, netmasks), attempt to identify five other reasons why knowledge of binary may come in handy in IT.
2. Where might you find hexadecimal notation in use? Provide some examples.
3. Provide several examples of how or why you might need to apply Boolean operators (AND, OR, NOT, XOR).
4. Does the fact that computers use binary to store information make it more challenging to understand how computers work? If so, in what way(s)?
5. Compression has become an increasingly important part of computing because of multimedia files—moving such files across the Internet (e.g., downloading, streaming) and storing files on your computer. Examine the various forms of audio and video compression. What are the issues that might help you choose a form of audio compression to stream a song? To permanently store a song? What are the issues that might help you choose a form of video compression to stream a video? To permanently store a video?
6. Examine Frank Drake's "message from the stars" in the sidebar of this chapter. Attempt to locate meaningful units of information. For instance, is there a binary code that is used for "start" and "stop" tags to indicate word, phrase, or concept boundaries? Can you identify any patterns in the bits?

7. Research typical error rates in telecommunications (i.e., how many errors typically arise when transmitting say 1 million bits). Given this error rate, how likely is it to find two errors in any single byte transmission?

8. Two other forms of error detection and correction are checksums and Hamming distance codes. Briefly research these and describe them. How do they differ from the use of the parity bit? Which would of the three would be the best way to detect errors in telecommunications?

9. Devise your own message from the stars. What would you want to say to another species that indicates that you are intelligent and wish to communicate with them?

CHAPTER 4

Introduction to Operating System Concepts

In Chapter 2, we saw the hardware of the computer. In order to facilitate communication between the user and the hardware, computer systems use an operating system. This chapter introduces numerous operating system concepts, many of which are referenced in later chapters of this book. Specifically, the roles of process management, resource management, user interface, protection, and security are introduced here so that later examinations can concentrate on how to utilize them in Windows and Linux operating systems. Operating system concepts of interrupts, the context switch, the booting process, and the administrator account are also introduced. The chapter concludes with the steps required for operating system installation for Windows 7 and Linux Red Hat.

The learning objectives of this chapter are to

- Describe the roles of the operating system.
- Compare the graphical user interface and the command line interface.
- Differentiate between the forms of process management.
- Discuss the interrupt and the context switch.
- Describe virtual memory.
- Discuss the issues involved in resource management: synchronization and deadlock.
- Describe the boot process and system initialization.
- Provide step-by-step instructions on Windows 7 and Linux Red Hat installation.

WHAT IS AN OPERATING SYSTEM?

At its most simple description, an operating system (OS) is a program. Its primary task is to allow a computer user to easily access the hardware and software of a computer system. Beyond this, we might say that an OS is *required* to maintain the computer's environment.

An OS is about control and convenience. The OS supports control in that it allows the user to control the actions of the software, and through the software, to control (or access) hardware. The OS supports convenience in that it provides access in an easy-to-use manner. Early OSs were not easy to use but today's OSs use a graphical user interface (GUI), and commands can be issued by dragging, pointing, clicking, and double clicking with the mouse. In some OSs such as Linux and Unix, there are two ways to issue commands: through the GUI and by typing in commands at a command line prompt. In actuality, both Windows and Mac OS also have command line prompts available, but most users never bother with them. Linux and Unix users, however, will often prefer the command line over the GUI.

SOME USEFUL TERMS

In earlier chapters, we already introduced the components of any computer system. Let us revisit those definitions, with new details added as needed. We also introduce some OS terms.

Hardware

The physical components of the computer. We can break these components into two categories: internal components and peripheral devices. The internal components are predominantly situated on the motherboard, a piece of plastic or fiberglass that connects the devices together. On the motherboard are a number of sockets for chips and expansion cards. One chip is the CPU, other chips make up memory. The CPU is the processor; it is the component in the computer that executes a program. The CPU performs what is known as the fetch–execute cycle, fetching each program instruction, one at a time from memory, and executing it in the CPU. There are several forms of memory: cache (SRAM), DRAM, and ROM. Cache is fast memory, DRAM is slow memory, but cache is usually far more expensive than DRAM so our computers have a little cache and a lot of DRAM. ROM is read-only memory and is only used to store a few important programs that never change (like the boot program). Connecting the devices together is the bus. The bus allows information to move from one component to another. The expansion slots allow us to plug in expansion cards, which either are peripheral devices themselves (e.g., wireless MODEM card) or connect to peripheral devices through ports at the back of the computer. Some of the more important peripheral devices are the hard disk drive, the keyboard, the mouse, and the monitor. It is likely that you will also have an optical disk drive, a wireless MODEM (or possibly a network card), a sound card, and external speakers, and possibly a printer.

Software

The programs that the computer runs, referred to as software to differentiate them from the physical components of the computer (hardware). Software exists only as electrical current

flowing through the computer (or as magnetic charges stored on disk). Software is typically classified as either applications software (programs that we run to accomplish some task) or systems software (programs that the computer runs to maintain its environment). We will primarily explore systems software in this text. In both Linux and Windows, the OS is composed of many small programs. For instance, in Linux, there are programs called ls, rm, cd, and so forth, which are executed when you issue the commands at the command line. These programs are also called by some of the GUI tools such as the file manager. If you look at the Linux directories /bin, /usr/bin, /sbin, and /usr/sbin, you will find a lot of the OS programs. Many users will rarely interact directly with the OS by calling these programs. If you look at directories such as /usr/local/bin, you will find some of the applications software. Applications software includes word processors, spreadsheets, internet browsers, and computer games. In Windows, you find the application software under the Program Files directory and the OS under the Windows directory.

All software is written in a computer programming language such as Java or C++. The program written by a human is called the source program or *source code*. Java and C++ look something like English combined with math symbols. The computer does not understand source code. Therefore, a programmer must run a translation program called a *compiler*. The compiler takes the source code and translates it into *machine language*. The machine language version is an *executable program*. There is another class of language translators known as an *interpreter*. The interpreter is a little different in that it combines the translation of the code with the execution of the code, so that running an interpreted program takes more time. As this is far less efficient, we tend to use the interpreted approach on smaller programs, such as *shell scripts*, *web browser scripts*, and *webserver scripts*. As an IT student, you will primarily write code that will be interpreted. Scripting languages include the Bash shell scripting language, PHP, Perl, JavaScript, Python, and Ruby.

Computer System

A collection of computer hardware, software, user(s) and network. Our OS is in reality a collection of software components. We might divide those components into four categories, described below.

Kernel

The core of the OS; we differentiate this portion from other parts that are added on by users: shells, device drivers, utilities. The kernel will include the programs that perform the primary OS tasks: process management, resource management, memory management, file management, protection, and security (these are explored in the next section).

Device Drivers

Programs that are specific interfaces between the OS running your computer and a piece of hardware. You install a device driver for each new piece of hardware that you add to your computer system. The drivers are usually packaged with the hardware or available for download over the Internet. Many common drivers are already part of the OS, only requiring installation. Rarer drivers may require loading off of CD or over the Internet.

Shell

An interface for the user, often personalized for that given user. The shell provides access to the kernel. For instance, a GUI shell will translate mouse motions into calls to kernel routines (e.g., open a file, start a program, move a file). Linux and Unix have command-line shells as well that may include line editing commands (such as using control+e to move to the end of the line, or control+b to back up one position). Popular GUI shells in Linux/Unix include CDE, Gnome, and KDE, and popular text-based shells in Linux/Unix include bash (Bourne-again shell), ksh (Korn shell), and C shell (csh). The text-based shells include the command-line interpreter, a history, environment variables, and aliases. We cover the Bash shell in detail in Chapter 9.

Utility Programs

Software that helps manage and fine-tune the hardware, OS, and applications software. Utilities greatly range in function and can include file utilities (such as Window's Windows Explorer program), antiviral software, file compression/uncompression, and disk defragmentation.

VIRTUAL MACHINES

A software emulator is a program that emulates another OS, that is, it allows your computer to act like it is a different computer. This, in turn, allows a user to run software that is not native to or compiled for his/her computer, but for another platform. A Macintosh user, for instance, might use a Windows emulator to run Windows software on a Macintosh.

Today, we tend to use virtual machines (VMs) rather than stand-alone emulators. A VM is software that pretends to be hardware. By installing a different OS within the VM, a user then has the ability to run software for that different OSs platform in their computer that would not otherwise be able to run that software.

However, the use of VMs gives considerably more flexibility than merely providing a platform for software emulation. Consider these uses of a VM.

- Have an environment that is secure in that downloaded software cannot influence or impact your physical computer
- Have an environment to explore different platforms in case you are interested in purchasing different OSs or different types of computer hardware
- Have an environment where you can issue administrator commands and yet have no worry that you may harm a physical computer—if you make a mistake, delete the VM and create a new one!
- Have an environment where multiple users could access it to support collaboration
- Have an environment that could be accessed remotely

There are different types of VMs. The Java Virtual Machine (JVM) exists in web browsers to execute Java code. VM software, such as VirtualBox (from Sun) and vSphere (from VMWare), provide users the ability to extend their computer hardware to support multiple OSs in a safe environment.

OS TASKS

The OS is basically a computer system *manager.* It is in charge of the user interface, process and resource management, memory management, file management, protection, and security.

User Interface

The GUI allows a user to control the computer by using the mouse and pointing and clicking at objects on the screen (icons, menus, buttons, etc.). Each OS offers a different "feel". For instance, Mac OS has a standard set of menus always listed along the top of the desktop and each application software adds to the menu selections. In Windows 7, there is no desktop level menu selection, but instead each software title has its own set of menus. Linux GUIs include Gnome and KDE. Each of these OSs provides desktop icons for shortcuts and each window has minimize, maximize, and close buttons. Additionally, each of these interfaces is based on the idea of "point and click". The mouse is used to position the cursor, and the mouse buttons are used to specify operations through single clicking, double clicking, and dragging. Cell phones and tablets, based on touch screens, have a gesture-based interface where movements include such operations as swipe, tap, pinch, and reverse pinch. Windows 8 is being marketed as following the touch screen approach rather than the point-and-click.

The GUI is a much simpler way to control the computer than by issuing commands via the command line. The command line in most OSs* runs in a shell. A shell is merely a part of the OS that permits users to enter information through which the user can command the OS kernel. The shell contains a text interpreter—it interprets entered text to pass along proper commands to the OS kernel. As an example, a Linux user may type in a command like:

```
find ~ -name 'core*' -exec rm {} \;
```

This instruction executes the Linux *find* command to locate, in this user's home directory, anything with a name that starts with core, and then executes the rm command on any such files found. In other words, this command finds and deletes all files starting with the letters "core" found in the user's home directory. Specifically, the instruction works as follows:

- The command is find which receives several parameters to specify how find should operate.
- ~/ specifies where find should look (the user's home directory).
- -name 'core*' are a pair, indicating that the name of the file sought will contain the letters "core" followed by anything (the * is a wildcard character).
- -exec indicates that any file found should have the instruction that follows executed on it.
- rm is the deletion operation, {} indicates the found file.
- \; ends the instruction.

* What is the plural of OS? There are several options, I chose OSs out of convenience, but some people dislike that because OSS stands for Open Source Software. Other people advocate OSes, OS's, and operating systems. For a discussion, see http://weblogs.mozillazine.org/gerv/archives/007925.html.

Many OS command line interpreters (such as DOS, which is available in the Windows OS) are far more simplistic. Linux is favored by some computer users because of the ability to express very complex instructions. You will study the Bash interpreter in Chapter 9, where you will learn more about how the command line interpreter works.

Process Management

The main task of any computer is to run programs. A program being executed by the computer is called a *process*. The reason to differentiate between program and process is that a program is a *static* entity, whereas a process is an *active* entity. The process has a status. Its status might include "running" (process is currently being executed by the CPU), "waiting" (process is waiting for input or output, or waiting to be loaded into memory), or "ready" (process is loaded into memory but not currently being executed by the CPU). The process also has specific data stored in memory, cache, and registers. These data change over time and from execution to execution. Thus, the process is dynamic.

A computer might run a single process at a time, or multiple processes in some form of overlapped (concurrent) fashion. The OS is in charge of starting a process, watching as it executes, handling interrupting situations (explained below) and input/output (I/O) operations, handling multiple process interaction, and terminating processes.

An interrupt is a situation where the CPU is interrupted from its fetch–execute cycle. As discussed in Chapter 2, the CPU continuously fetches, decodes, and executes instructions from memory. Left to itself, the CPU would do this indefinitely. However, there are times when the CPU's attention needs to be shifted from the current process to another process (including the OS) or to address a piece of hardware. Therefore, at the end of each fetch–execute cycle, the CPU examines the interrupt flag (part of the status flags register) to see if anyone has raised an *interrupt*. An interrupt signal can come from hardware or from software.

If an interrupt arises, the CPU handles the interrupt as follows. First, it saves what it was doing. This requires taking the values of the various registers and saving them to memory. Second, the CPU identifies the type of interrupt raised. This requires the CPU to determine which device raised the interrupt, or if the interrupt was caused by software. Third, the CPU switches to execution of the OS. Specifically, the CPU begins executing an *interrupt handler*. The OS will have an interrupt handler (a set of code) for each type of interrupt. See Table 4.1 for a list of types of interrupts. The CPU then resumes the fetch–execute cycle, but now is executing the OS interrupt handler. Upon completion of executing the interrupt handler, which will have taken care of the interrupting situation, the CPU restores the register values saved to memory. This "refreshes" the interrupted process. The CPU then resumes the fetch–execute cycle, but now it is continuing with the process it has previously been executing, as if it had never stopped. The interruption may have taken just a few machine cycles or seconds to minutes if the interruption involves the user. For instance, if the interruption was caused by a "disk not in drive", then the user has to physically insert a disk. The interruption would only be resolved once the user has acted.

OSs will implement process management in different ways. The simplest approach is to use single tasking where a process starts, executes, and terminates with no other processes

TABLE 4.1 Types of Interrupts

Device	Reason(s) for Interrupt
Disk drive	File not found Disk not in drive Disk not formatted
Keyboard	User enters keystroke User presses ctrl+alt+del
Mouse	Mouse moved Mouse button pressed or depressed
Network/MODEM	Message arrives
Printer	Paper jam Printer out of paper Printout complete
Program	Run time error (e.g., divide by 0, bad user input) Requires communication with another program
Timer	Timer elapses

running. This is not a very satisfactory use of computer resources however, nor do users typically want to be limited to running one program at a time. Various forms of concurrent processing exist. Concurrency means that processes are executed in some overlapped fashion. These include multiprogramming, multitasking (or time sharing as it was originally called), multithreading, and multiprocessing. What each of these have in common is that the OS permits multiple processes to be in some state of execution. With the exception of multiprocessing (which uses multiple CPUs), there is only one CPU so that only one process can be executing at any moment in time. The others wait. How long they wait, why they are waiting, and where they wait differs to some extent based on the type of process management.

We will explore various forms of process management in the next section. However, before proceeding, we must first introduce multitasking as it is referenced later in this section. In multitasking, there are two or more processes running. The CPU only executes one process at any one time, but switches off between all of the running processes quickly. So, for instance, the CPU might execute a few thousand machine cycles each on process 0, and then process 1, and then process 2, and then return to process 0. A timer is used to time how long each process is executed. When the timer elapses, it interrupts the CPU and the OS then is invoked to force a switch between processes. The timer is reset and the CPU continues with the next process.

Scheduling

If we want the computer to do anything other than single processing—that is, running one program until it completes and then moving on to the next user task, the OS will have to perform *scheduling*. There are several types of scheduling. For instance, in multitasking, when the timer elapses, the CPU switches from the current process to another. Which one? Typically, the OS uses a round-robin scheduler. If there are several processes running, they will be placed into a queue. The processes then are given attention by the CPU in the order that they reside in the queue. When the processor has executed some number of instructions on the last process in the queue, it resumes with the first process in the queue. That is, the queue is a "wrap-around" queue, thus the term round-robin. The OS is also in charge of deciding which processes should be loaded into memory at any time. Only processes present in memory will be placed in the queue. Commonly today, all processes are loaded upon demand, but low priority processes may be removed from the ready queue and memory and reside in a waiting queue. Finally, the user can also specify that a process should be run. This can be done by selecting, for instance, the process from the tab at the bottom of the desktop. This forces a process to move from the background to the foreground. We will consider scheduling in more detail in the next section, as well as in Chapter 11.

Memory Management

Because main memory is smaller in size than the size of the software we typically want to run, the OS is in charge of moving chunks of programs and data into and out of memory as needed. Every program (along with its data) is divided into fixed-sized blocks called *pages*. Before a program can be executed, the OS copies the program's "image" into *swap space* and loads some of its pages into memory. The "image" is the executable program code along with the memory space that makes up the data that the program will access. Swap space is a reserved area of the hard disk. Now, as the process runs, only the needed pages are loaded into memory. This keeps each program's memory utilization down so that memory can retain more programs. If each program were to be loaded into memory in its entirety, memory would fill quickly and limit the number of programs that could fit.

The use of swap space to back up memory gives the user an illusion that there is more main memory than there actually is. This is called *virtual memory*. Figure 4.1 demonstrates virtual memory with two processes (A and B), each of which currently have pages loaded into main memory and the remainder of the processes are stored in swap space. The

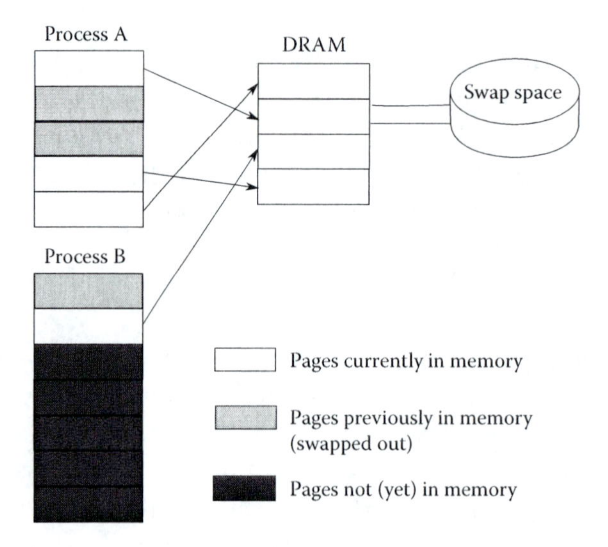

FIGURE 4.1 Virtual memory.

unmapped pages refer to pages that have not yet been loaded into memory. Swapped out pages are pages that had at one time been in memory but were removed in order to accommodate newer pages.

When the CPU generates a memory address (for instance, to fetch an instruction or a datum), that address must be translated from its logical (or virtual) address into a physical address. In order to perform that translation, the OS maintains a *page table* for every process. This table denotes for every page of the process, if it is currently in memory and if so, where. Figure 4.2 shows what the page tables would look like for the two processes in Figure 4.1. Process A has three pages currently loaded into memory at locations 0, 1, and 3 (these locations are referred to as *frames*) and process B has one page currently loaded into memory at location 2. The valid column is a bit that indicates if the page is currently in memory (valid bit set to 1) or not.

If a referenced page is not currently in memory, a *page fault* is generated that causes an interrupt. The OS gets involved at this point to perform *swapping*. First, the OS locates an available frame to use in memory. If there are no available frames, then the OS must select a page to discard. If the page has been modified, then it must first be saved back to swap space. Once a frame is available, the OS then loads the new page from swap space into that frame in memory. The OS modifies the page table to indicate the location of the new page (and if a page has been removed from memory). Finally, the OS allows the CPU to resume the current process.

As swap space is stored on hard disk, any paging (swapping) involves hard disk access. Because the hard disk response time is so much slower than memory response time, any swapping will slow down the execution of a program. Therefore, swapping is to be avoided as much as possible. If the discarded page had to be written back to swap space first, this increases the swap time even more.

Process A page table

Page	Frame	Valid
0	1	T
1	–	F
2	–	F
3	3	T
4	0	T

Process B page table

Page	Frame	Valid
0	–	F
1	2	T
2	–	F
3	–	F
4	–	F
5	–	F
6	–	F

FIGURE 4.2 Example page tables for processes in Figure 4.1.

Another factor requiring memory management is that processes generate addresses for data accesses (loads and stores). What is to prevent a process from generating an address of a section of memory that does not contain data of that process? A situation where a process generates an address of another process' memory space is called a *memory violation*. This can happen through the use of pointers, for instance, in programming languages such as C and C++. A memory violation should result in termination of the process, although in C++, a programmer could also handle this situation through an exception handler, which is an interrupt handler written by the programmer and included with the current program.

Resource Management

Aside from memory and the CPU, there are many other resources available including the file system, access to the network, and the use of other devices. The OS maintains a table of all active processes and the resources that each process is currently using or wants to use. Most resources can only be accessed in a *mutually exclusive* way. That is, once a process starts using a resource, no other process can use the resource until the first process frees it up. Once freed, the OS can decide which process next gets to access the resource.

There are many reasons for mutual exclusion, but consider this situation. A file stores your checking account balance. Let us assume it is currently $1000. Now, imagine that you and your significant other enter the bank, stand in line, and each of you is helped by different tellers at the same time. You both ask to deduct the $1000. At the same time, both tellers access the data and both find there is $1000 available. Sad that you are closing out the account, they both enter the transaction, again at the same time. Both of the tellers' computers access the same shared disk file, the one storing your account information. Simultaneously, both programs read the value ($1000), determine that it is greater than or equal to the amount you are withdrawing, and reset the value to $0. You and your significant other each walk out with $1000, and the balance has been reduced to $0. Although there was only $1000 to begin with, you now have $2000! This is good news for you, but very bad news for the bank. To prevent such a situation from happening, we enforce mutually exclusive access to any shared resource.

Access to any shared datum must be *synchronized* whether the datum is shared via a computer network or is shared between two processes running concurrently on the same computer. If one process starts to access the datum, no other process should be allowed to access it until the first process completes its access and frees up the resource. The OS must handle interprocess synchronization to ensure mutually exclusive access to resources. Returning to our bank teller example, synchronization to the shared checking account datum would work like this. The OS would receive two requests for access to the shared file. The OS would select one of the two processes to grant access to. The other would be forced to wait. Thus, while one teller is allowed to perform the database operation to deduct $1000, the other would have to wait. Once the first teller is done, the resource becomes freed, and the second teller can now access the file. Unfortunately, the second teller would find the balance is $0 rather than $1000 and not hand out any money. Therefore, you and your significant other will only walk away with $1000.

Aside from keeping track of the open resources, the OS also handles *deadlock*. A simple version of deadlock is illustrated in Figure 4.3. The OS is multitasking between processes P0 and P1, and there are (at least) two available resources, R0 and R1.

P0 runs for a while, then requests access to R0.

The OS determines that no other process is using R0, so grants the request.

P0 continues until the timer elapses and the OS switches to P1.

P1 runs for a while, then requests access to R1.

The OS determines that no other process is using R1, so grants the request.

P1 continues until the timer elapses and the OS switches to P0.

P0 continues to use R0 but now also requests R1.

The OS determines that R1 is currently in use by P1 and moves P0 to a waiting queue, and switches to P1.

P1 continues to use R1 but now also requests R0.

The OS determines that R0 is currently in use by P0 and moves P1 to a waiting queue.

The OS is ready to execute a process but both running processes, P0 and P1, are in waiting queues. Furthermore, neither P0 nor P1 can start running again until the resource it is waiting for (R1 and R0, respectively) becomes available. But since each process is holding onto the *resource* that the other needs, and each is waiting, there will never be a time when the resource becomes available to allow either process to start up again, thus deadlock.

To deal with deadlock, some OSs check to see if a deadlock might arise before granting any request. This tends to be overcautious and not used by most OSs. Other OSs spend some time every once in a while to see if a deadlock has arisen. If so, one or more of the deadlocked processes are arbitrarily killed off, freeing up their resources and allowing the other process(es) to continue. The killed-off processes are restarted at some random time

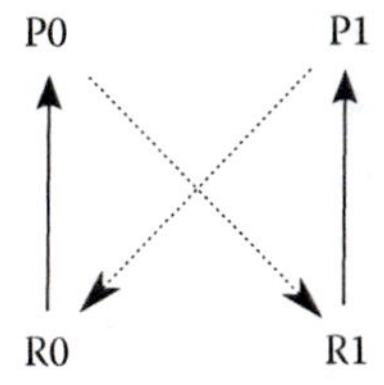

Solid lines indicate granted resources
Dotted lines indicate requested resources

FIGURE 4.3 A deadlock situation.

in the future. Yet other OSs do not deal with deadlock at all. It is up to the user to decide if a deadlock has arisen and take proper action!

File System Management

The primary task of early OSs was to offer users the ability to access and manipulate the file system (the name MS-DOS stands for Microsoft Disk Operating System, and the commands almost entirely dealt with the disk file system). Typical commands in file management are to open a file, move a file, rename a file, delete a file, print a file, and create, move, rename, and delete directories (folders). Today, these capabilities are found in file manager programs such as Windows Explorer. Although these commands can be performed through dragging, clicking, etc., they are also available from the command line. In Unix/Linux, file system commands include cd, ls, mv, cp, rm, mkdir, rmdir, and lp. In DOS, file system commands include cd, dir, move, copy, del, mkdir, rmdir, and print. There are a wide variety of additional Unix/Linux commands that allow a system administrator to manipulate the file system as a whole by mounting new devices or relocating where those devices will be accessed from in the file space, as well as checking out the integrity of files (for instance, after a file is left open and the system is rebooted).

Behind the scenes, the OS manages file access. When a user submits a command such as to move a file or open a file, whether directly from the command line or from some software, the user is in actuality submitting a request. The OS now takes over. First, the OS must ensure that the user has access rights for the requested file. Second, the OS must locate the file. Although users will specify file locations, those locations are the file's logical location in the file system. Such a specification will include the disk partition and the directory of the file (e.g., C:\Users\Foxr\My Documents\CIT 130). However, as will be discussed in Chapter 5, files are broken up into smaller units. The OS must identify the portion of the file desired, and its physical location. The OS must map the logical location into a physical location, which is made up of the disk surface and location on the surface. Finally, the OS must initiate the communication with the disk drive to perform the requested operation.

Protection and Security

Most computers these days are a part of a larger network of computers where there may be shared files, access to the network, or some other shared resource. Additionally, multiple users may share the same computer. The files stored both locally and over the network must be protected so that a user does not accidentally (or maliciously) overwrite, alter, or inappropriately use another user's files. Protection ensures that a user is not abusing the system—not using someone else's files, not misusing system resources, etc. Security extends protection across a network.

A common mechanism for protection and security is to provide user accounts. Each user will have a user name and authentication mechanism (most commonly, a password). Once the OS has established who the user is, the user is then able to access resources and files that the user has the right to access. These will include shared files (files that other users have indicated are accessible) and their own files. Unix and Linux use a fairly simplistic approach by placing every user into a group. Files can be read, written, and executed, and a file's accessibility can be controlled to provide different rights to the file's owner, the file's group, and the rest of the world. For instance, you might create a file that is readable/writable/executable by yourself, readable/executable by anyone in your group, and executable by the world. We discuss user accounts, groups, and access control methods in Chapter 6.

FORMS OF PROCESS MANAGEMENT

The various ways in which OSs will execute programs are

- Single tasking
- Batch
- Multiprogramming
- Multitasking
- Multithreaded
- Multiprocessing

A single tasking system, the oldest and simplest, merely executes one program until it concludes and then switches back to the OS to wait for the user to request another program to execute. MS-DOS was such an OS. All programs were executed in the order that they were requested. If the running process required attention, such as a lengthy input or output operation, the CPU would idle until the user or I/O device responded. This is very inefficient. It can also be frustrating for users who want to do more than one thing at a time. Aside from early PCs, the earliest mainframe computers were also single tasking.

A batch processing system is similar to a single tasking system except that there is a queue (waiting line) for processes. The main distinction here is that a batch processing system has more than a single user. Therefore, users may submit their programs for execution at any time. Upon submitting a program request, the process is added to a queue. When the processor finishes with one process, the OS is invoked to decide which of the waiting processes to bring to the CPU next. Some form of *scheduling* is needed to decide which process gets selected next. Scheduling algorithms include a priority scheme, shortest job first, and first come first serve. For the *priority scheme*, processes are each given a priority depending on the user's status (e.g., administration, faculty, graduate student, undergraduate student). Based on the scheduling algorithm, the load on the machine, and where your process is placed in the queue, you could find yourself waiting minutes, hours, or even days before your program executes! In *shortest job first*, the OS attempts to estimate the amount of CPU time each process will require and selects the process that will take the least amount of time. Statistically, this keeps the average waiting time to a minimum although it might seem unfair to users who have time-consuming processes. *First come first serve* is the traditional scheme for any queue, whether in the computer or as found in a bank. This scheme, also known as FIFO (first-in, first-out), is fair but not necessarily efficient.

Another distinction between a single tasking system and a batch processing system is that the batch processing system has no *interactivity* with the user. In single tasking, if I/O is required, the process pauses while the computer waits for the input or output to complete. But in batch processing, all input must be submitted at the time that the user submits the process. Since batch processing was most commonly used in the first three computer generations, input typically was entered by punch cards or stored on magnetic tape. Similarly, all output would be handled "offline", being sent to magnetic tape or printer. Obviously, without interactivity, many types of processes would function very differently than we are used to. For instance, a computer game would require that all user moves be specified before the game started, and a word processor would require that all edit and formatting changes to be specified before the word processor started. Therefore, batch processing has limitations. Instead, batch processing was commonly used for such tasks as computing payroll or taxes, or doing mathematical computations.

Notice that in batch processing, like in single tasking, only one process executes at a time, including the OS. The OS would not be invoked when a new process was requested by a user. What then would happen when a user submits a new process request? A separate batch queuing system was in charge of receiving new user submissions and adding them to the appropriate queue.

Although batch processing was common on computers in the first through third generation, running on mainframe and minicomputers, you can still find some batch processing today. For instance, in Unix and Linux, there are scheduling commands such as cron and at, and in Windows, the job scheduler program is available. Programs scheduled will run uninterrupted in that they run without user intervention, although they may run in some multitasking mode.

In a batch processing system, the input is made available when the process begins execution through some source such as punch cards or a file on magnetic tape or disk. A single tasking system, on the other hand, might obtain input directly from the user through keyboard or some other device. Introducing the user (human) into the process greatly slows down processing. Why? Because a human is so much slower at entering information than a disk drive, tape drive, or punch card reader (even though the punch card reader is very slow). Similarly, waiting on output could slow down processing even if output is handled offline. The speed of the magnetic tape, disk, or printer is far slower than that of the CPU. A single tasking system then has a significant inefficiency—input and output slows down processing. This is also true of batch processing even though batch processing is not as significantly impacted because there is no human in the processing loop.

A multiprogramming system is similar to a batch system except that, if the current process requires I/O, then that process is moved to another queue (an I/O waiting queue), and the OS selects another process to execute. When the original process finishes with its I/O, it is resumed and the replacement process is moved back into a queue. In this way, lengthy I/O does not cause the CPU to remain idle, and so the system is far more efficient. The idea of surrendering the CPU to perform I/O is referred to as *cooperative multitasking*. We will revisit this idea below.

There are several different uses for queues in multiprogramming, so we need to draw a distinction between them. There is the *ready queue* (the queue of processes waiting for the CPU), *I/O queues* (one for each I/O device), and the *waiting queue* (the queue of processes that have been requested to be run, but have not yet been moved into the ready queue). The reason that there is a waiting queue is that those processes in the ready queue are already loaded into memory. Because memory is limited in size, there may be processes requested by users that cannot fit, and so they sit in the waiting queue until there is room in the ready queue. There will be room in the ready queue if a process ends and exits that queue, or if many processes have been moved from the ready queue to I/O queues.

The multiprogramming system requires an additional mechanism known as a *context switch*. A context switch is merely the CPU switching from one process to another. We examined this briefly earlier. Let us take a closer look. In order to switch processes, the CPU must first save the current process' status and retrieve the next process' status. Figure 4.4 provides a snapshot of a computer to illustrate the context switch. At this point, process P3 is being executed by the CPU. The PC register stores the address in memory of the next instruction of P3, the IR stores the current instruction of P3, the SP stores the top of P3's run-time stack in memory, the flags store the status of the last instruction executed, and the data registers store relevant data for P3. The context switch requires that these values be stored to memory. Then, P3 is moved to the appropriate I/O queue in memory, moving process P7 up to the front of the ready queue. Finally, the CPU must restore P7's status, going to memory and retrieving the stored register values for P7. Now, the PC will store the address in memory of P7's next instruction, the IR will store P7's current instruction, the SP will store the top of P7's run-time stack, the flags will store the status of P7's last executed instruction, and the data registers will store the data last used by P7. The CPU resumes its fetch–execute cycle, but now on P7 rather than P3. This continues until either P3 is ready to resume or P7 requires I/O or terminates. In the former case, a context switch occurs in which P3 is restored and P7 moved back in the queue, and in the latter cases, the CPU switches from P7 to P0.

As the context switch requires saving and restoring the context of two processes, it is not instantaneous, but rather is somewhat time consuming. The CPU idles during

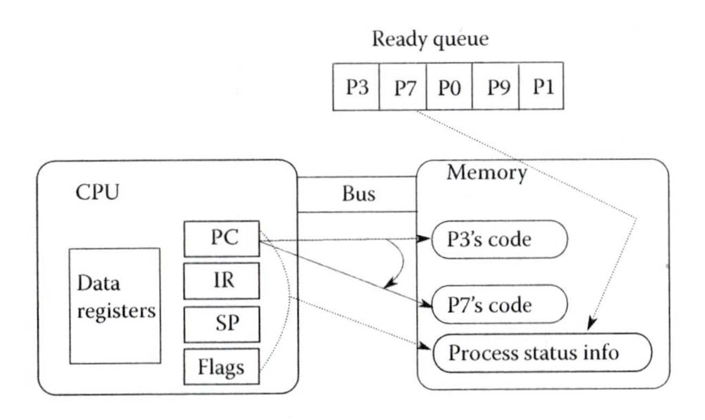

FIGURE 4.4 A context switch requires changing register values.

the switch. Most computers store process status in main memory. Therefore, the context switch requires several, perhaps a dozen to two dozen, memory operations. These are slow compared to CPU speed. If the context switch can store and restore from cache, so much the better. However, some high-performance computers use extra sets of registers for the context switch. This provides the fastest response and thus keeps the CPU idle during the least amount of time. However, in spite of the CPU idle time caused by the context switch, using the context switch in multiprogramming is still far more efficient than letting the CPU idle during an input or output operation.

Multitasking takes the idea of multiprogramming one step further. In multiprogramming, a process is only suspended (forced to wait) when it needs I/O, thus causing the CPU to switch to another process. But in multitasking, a *timer* is used to count the number of clock cycles that have elapsed since the last context switch started this process. The timer is set to some initial value (say 10,000). With each new clock cycle, the timer is decremented. Once it reaches 0, the timer interrupts the CPU to force it to switch to another process. Figure 4.5, which is a variation of Figure 4.4, demonstrates the context switch as being forced by the timer. The suspending process gets moved to the end of the ready queue and must wait its turn. This may sound like a tremendous penalty for a process, but in fact since modern processors are so fast, a suspended process only waits a few milliseconds at most before it is moved back to the CPU. In multitasking systems, humans will not even notice that a process has been suspended and then resumed because human response time is greater than the millisecond range. Therefore, although the computer appears to be executing several programs simultaneously with multitasking, the truth is that the computer is executed the processes in a *concurrent*, or overlapping, fashion.

In multitasking, we see two mechanisms by which the CPU will move from one process to another: because the current process is requesting some form of I/O or because the timer has elapsed on this process. Multitasking is more properly considered to be *cooperative multitasking* when the process voluntarily gives up the CPU. This happens in

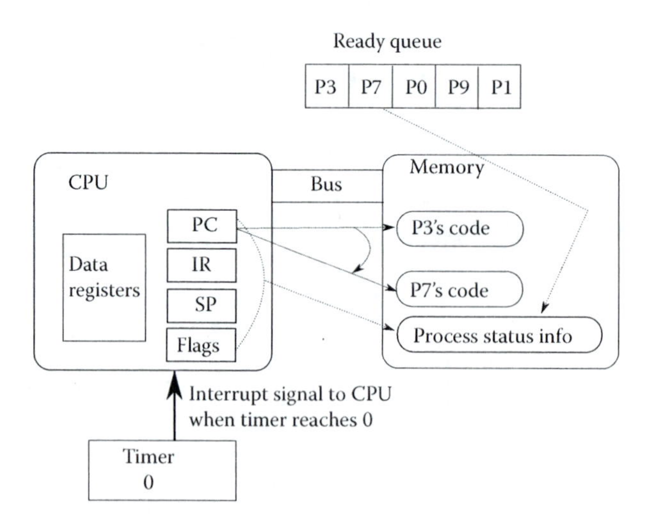

FIGURE 4.5 A context switch in multitasking.

multiprogramming when the process requires I/O. In the case of the timer causing a context switch, this form of multitasking is called *competitive multitasking*, or *preemptive multitasking*. In this case, the process is forced to give up the CPU.

There are other reasons why a process might move itself to another queue or back to the end of the ready queue. One such reason is forced upon a process by the user who moves the process from the foreground to the background. We use the term *foreground* to denote processes that are either immediately available for user input or currently part of the ready queue. *Background* processes "sit in the background". This means that the user is not interacting with them, or is currently waiting until they have another chance at gaining access to the processor. In Windows, for instance, the foreground process is the one whose tab has been selected, the one "on top" of the other windows as they appear in the desktop. In Linux, the same is true of GUI processes. From the command line, a process is in the foreground unless it has been executed using the & command. This is covered in more detail in Process Execution in Chapter 11.

Another reason that a process may voluntarily surrender the CPU has to do with waiting for a *rendezvous*. A rendezvous occurs when one process is waiting on some information (output, shared datum) from another process. Yet another reason for a process to surrender the CPU is that the process has a low priority and other, higher priority processes are waiting.

The original idea behind multitasking was called *time sharing*. In time sharing, the process would give up the CPU on its own. It was only later, when OSs began implementing competitive multitasking, that the term time sharing was dropped. Time sharing was first implemented on mainframes in 1957 but was not regularly used until the third generation. By the fourth generation, it was commonplace in mainframes but not in personal computers until the 1980s. In fact, PC Windows OSs before Windows 95 and Mac OS before OS X performed cooperative, but not competitive, multitasking.

Multitasking uses a *round-robin* scheduling routine. This means that all processes in the ready queue are given their own time with the CPU. The CPU moves on from one process to the next as time elapses. The timer is then reset. Although this is a very fair approach, a user may want one specific process to have more CPU attention than others. To accomplish this, the user can specify a priority for any or every process. Typically, all processes start with the same priority. We will examine setting and changing process priorities in Chapter 11.

Today, processes can contain multiple running parts, called *threads*. A thread shares the same resources as another thread. The shared resources are typical shared program code stored in memory, although they can also share data, register values, and status. What differentiates one thread from another is that they will also have their own data. For instance, you might run a Firefox web browser with several open tabs or windows. You are running a single process (Firefox) where each tab or window is its own thread. The only difference between each tab or window is its data. As the threads are of the same program, the OS is running one process but multiple threads. A multitasking system that can switch off between processes and between threads of the same process is known as a *multithreaded* OS. See Figure 4.6, which illustrates a process of four threads. Each thread makes its own

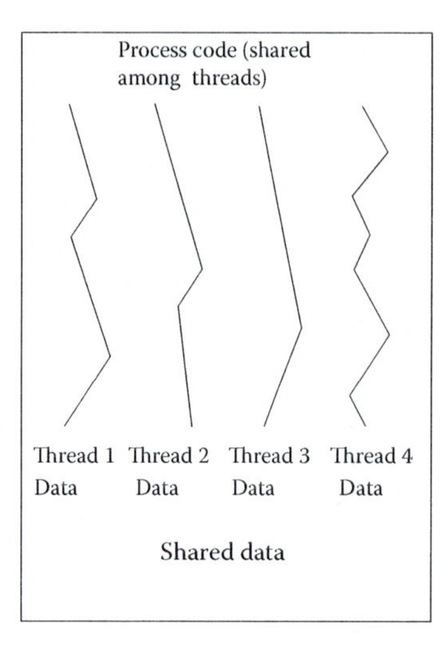

FIGURE 4.6 Threads of a process.

way through the code, and although the four threads share some data, they each have their own separate data as well.

Threads serve another purpose in programming. With threads in a process, the threads can control their and other threads' availabilities. Consider a process that contains two threads. One thread produces values for a memory buffer. The other consumes values from the buffer. In multithreading, the processor switches off between threads. However, the consumer thread cannot proceed if the buffer is currently empty. This is because there is nothing to consume. Therefore, the consumer thread can force itself to wait until the producer thread signifies that a new datum has become available. On the other hand, the producer thread may force itself to wait if the buffer is full, until the consumer signals that it has consumed something. The two threads can "meet up" at a rendezvous. By forcing themselves to wait, the threads in this example use cooperative multitasking (multithreading) in that they voluntarily give up the processor, as opposed to being forced to surrender the processor when the timer elapses.

Going one step further than the rendezvous, it is important that two threads or processes do not try to access a shared datum at the same time. Consider what might happen if thread 1 is attempting to access the shared buffer and the thread is interrupted by a context switch. It has already copied the next buffer item, but has not had a chance to reset the buffer to indicate that the particular item is now "consumed". Now, thread 2 is resumed and it attempts to place a datum in the buffer only to find that the buffer is full. It is not actually full because one item has been consumed. Therefore, access to the buffer by thread 2 causes the thread to wait when it should not. Another situation is when the two threads share the same datum. Recall our banking example from earlier in the chapter motivating the use of mutual exclusive access. For threads, again, access to the shared datum must be mutually exclusive. Now, we consider the same example using threads. Thread 1 reads the checking account balance of $1000. Thread 1 is ready to deduct $1000 from it, but is interrupted by the timer. Thread 2 reads the checking account balance, $1000, deducts $1000 from it and stores the new value back, $0. Switching back to thread 1, it has an out-of-date value, $1000.

It deducts $1000, stores $0 back. The result is that the value is set to $0, but two different threads deducted $1000. This would corrupt the data.

To ensure proper access to any shared data, the OS must maintain proper *synchronization*. Synchronization requires that the process or thread is not interrupted while accessing a datum, or that the datum is not accessible once one process or thread has begun accessing it until it has been freed. Multiprogramming, multitasking, and multithreading require synchronization.

The advantage of the multithreaded OS over the multitasking OS is that context switching between threads is faster and more efficient than switching between processes. The reason for this is that the context switch requires fewer operations to store the current thread and restore the next thread than it does to store and restore processes. Imagine, referring back to Figure 4.5, that process P7's code has been moved out of memory and back to virtual memory (swap space). A context switch between P3 and P7 would be somewhat disastrous as the OS would not only have to store P3's status and restore P7's status, but also would have to load some pages of P7 from disk to memory. This will not happen when switching between threads. Assume that P3 and P7 are threads of the same process. The context switch between them requires far less effort and therefore is much quicker.

Most computers, up until recently, were known as single-processor systems. That is, they had only a single CPU. With one CPU, only one process or thread could be actively executing at any time. In multiprogramming, multitasking, and multithreading, several processes (or threads) could be active, but the CPU would only be executing one at any given time. Thus, concurrent execution was possible using any of these modes of process execution, but not true parallel processing. In multiprocessing, a computer has more than one CPU. If a computer, for instance, had two processors, it could execute two processes or threads simultaneously, one per processor. Today, most computers are being produced with multiple processors.

The more traditional form of multiprocessor system was one that contained more than one CPU chip. Such computers were expensive and few personal computers were multiprocessor systems. However, with the continued miniaturization seen over the past decade, enough space has been made available in a single CPU to accommodate multiple *cores*. Thus, modern processors are referred to as dual or quad core processors. Each core is, in effect, a processor; it has its own ALU (arithmetic/logic unit), control unit, registers, and cache. The only thing it shares with the other processors on the chip are the pins that attach the entire set of cores to the computer's system bus.

Parallel processing presents an opportunity and a challenge. The opportunity is to execute multiple processes simultaneously. For instance, one core might execute a web browser process while another might execute an entertainment program such as a video or mp3 player. The challenge, however, is to ensure that each core is being used as much as possible. If the user is only running one process, could that process somehow be distributed to the multiple cores such that each core executes a separate portion of the process? This requires a different mindset when writing program code. A programmer may try to separate the logic of the program from the computer graphics routines. In this way, one core might execute the logic and the other core the computer graphics routines. The idea of distributing the processing over multiple processors is known as parallel programming, and it remains one of the greater challenges open to programmers today.

We wrap up this section by comparing some of the approaches to process management described in this section. Imagine that you want to run three programs called p1, p2, p3. We will see how they run on a single tasking system, batch system, multiprogramming system, and multitasking system.

In single tasking, p1 is executed until it completes, and then p2 is run, and then p3 is run. This is not very satisfying for the user because you are unable to do multiple things at a time (for instance, you might be editing a paper, searching the Internet for some references, and answering email—you do not do them simultaneously, as you do not have them all running at the same time). You cannot copy and paste from one program to another either, because one program completes before the next starts.

In batch processing, the execution of the three processes is almost identical to single tasking. There are two changes. First, since batch processing foregoes any interaction with the user, all input must be supplied with the program and any output is performed only after execution concludes. Second, a separate processing system might be used to schedule the three processes as they are added to a waiting queue. Scheduling might be first come first serve, priority, shortest job first, or some other form.

In multiprogramming, the computer would run p1 until either p1 terminated or p1 needed to perform I/O. At this point, the OS would move p1 to a waiting queue and start p2. Now p2 will run until either p2 needs I/O, p2 terminates, or p1 is ready to resume. If p2 needs I/O or terminates, then the OS would start p3. If p1 becomes ready, p1 is resumed and p2 is forced to wait. Whereas in multiprogramming, input and output can be done on demand, as opposed to batch, multiprogramming may not give the user the appearance of interaction with the computer as processes are forced to wait their turn with the CPU and I/O devices.

In multitasking, we truly get an overlapped execution so that the user cannot tell that the CPU is switching between processes. For multitasking, the process works like this:

- Load p1, p2, p3 into memory (or as much of each process as needed)
- Repeat
    - Start (or resume) p1 and set the timer to a system preset amount (say 10,000 cycles)
    - Decrement the timer after each machine cycle
    - When the timer reaches 0, invoke the OS
    - The OS performs a context switch between p1 and p2
        - The context switch requires saving p1's status and register values and restoring p2's status and register values, possibly also updating memory as needed
    - Repeat with p2
    - Repeat with p3
- Until all processes terminate

If any of p1, p2, and p3 were threads of the same process, then multithreading would be nearly identical to multitasking except that the context switch would be less time consuming.

BOOTING AND SYSTEM INITIALIZATION

Main memory is *volatile*, that is, it requires a constant power supply to retain its contents. Shut off the power and main memory becomes empty (all of its contents become 0s). After shutting the computer down, memory is empty. The next time you turn on the computer, memory remains empty. This is important because, in order to use the computer, you need to have the OS loaded into memory and running. The OS is the program that will allow us to load and run other programs. This presents a paradoxical situation: how can we get the OS loaded and running when it is the OS that takes care of loading and running programs for us and the OS is not in memory when we first turn the computer on? We need a special, one-time process, called booting.*

The boot process operates as follows:

1. The CPU initializes itself (initializes its registers) and sends out control signals to various components in the computer system.
2. The basic I/O system (BIOS) performs a power-on self-test (POST) where it checks memory and tests devices for responses. It may also perform such tasks as to set the system clock, enable or disable various hardware components, and communicate with disk controllers.
3. The hard disk controllers (SCSI first, then IDE) are signaled to initialize themselves, other devices are tested, network card, USB, etc.
4. BIOS determines where the OS is stored. This is usually done by testing each of these devices in a preset order looking for the OS until it is found: floppy disk, CD-ROM, first disk drive, second disk drive, or network.
5. If your computer has multiple OSs, then a boot loader program is run to determine which OS to load and run. Two programs used to dual boot in Windows and Linux are GRUB (Grand Unified Boot loader) and LILO (Linux Loader).
6. The OS *kernel* is loaded from disk into memory. The kernel represents the portions of the OS that must be in memory to use the computer, or the core components of the OS.
7. At this point, control moves from the boot program to the OS kernel, which then runs initialization scripts to finish the boot process. Initialization scripts are exactly what they sound like, shell scripts that, when run, initialize various aspects of the OS.

* The term booting comes from the word bootstrapping, which describes how one can pull himself up by his own bootstraps, like a fireman when he gets up in the middle of the night in response to a fire alarm. The term itself is attributed to the tales of Baron Munchhausen.

(a) Linux begins by running the program init. Its primary job is to start the OS in a particular run-level, such as text-only, text-with-network, graphics, or graphics-with-network. Once the level has been selected, the program runs other scripts such as the rc.sysinit script. This script, based on the run level, starts up various services. For instance, if the system starts in a level with network availability, then the network services must be started.

(b) In Windows, one of the decisions is whether to start the OS in safe mode or full user mode. Safe mode is a diagnostic mode primarily used so that a system administrator can remove malicious software without that software making copies of itself or moving itself. Another task of initialization is to determine if a secure login is necessary and if so, bring up a log in window. One of the last steps of initialization is to start up a user-specific shell based on who logged in.

The boot process is a program. However, it cannot reside in RAM since, when the computer is turned on, RAM is empty. Since the boot process does not change, we will store it in ROM, which is nonvolatile since its contents are permanently fixed into place. However, since ROM tends to be expensive, and since portions of the boot process need to be flexible, we can store portions of the boot program, such as the boot loader program and parts of the BIOS, on hard disk.

ADMINISTRATOR ACCOUNT

In order to perform system-oriented operations, most OSs have a set of privileged instructions. For instance, to create new user accounts, test for security holes, and manipulate system files you must have administrator access. Most OSs have two types of accounts, user accounts, and administrator accounts.* The administrator account is sometimes referred to as *root* or *superuser.* The administrator account comes with an administrator password that only a few should know to prevent the casual user from logging in as administrator and doing something that should not be done.

In Windows, the account is called Administrator. To switch to Administrator, you must log in as Administrator, entering the proper password. In Linux, the account is called root. To change to the root, you use the su command (switch user). Typically, su is used to change from one user account to another by saying *su username*. The OS then requires the password for username. If you do *su* without the username, then you are requesting to change to root. Because you can run su from the command line prompt, you do not have to log out. In fact, you can open numerous windows, some of which are controlled as you, the user, and some as root.

The administrator account is the owner for all of the system software and some of the applications software. In Linux, an ls –l (long listing) of directories such as /sbin, /bin, /usr/bin, and /usr/sbin will demonstrate that many of the programs and files are owned by root. In many cases, root is the only user that can access them. In other cases, root is the only

* There are some operating systems that have more than two levels of accounts. In the MULTICS operating system for instance, there are eight levels of access, with each higher level gaining more access rights.

user that can write to or execute them but others can view them. For instance, you must be root to run useradd and userdel so that only the system administrator(s) can add and delete user accounts. The root account in Linux is also peculiar in that root's home directory is not in the same partition as the users' home directories. User home directories are typically under /home/username, whereas root is under /root.

As an IT person, you should always be aware of when you are logged in as an administrator and when you are not. In Linux, you can differentiate between the two by looking at the prompt in the command line. The root prompt in Linux is typically # and the user prompt is typically $ (unless you alter it). You can also find out who you are in Linux by using the command whoami. You might wonder why it is important to remember who you are, but you do not want to issue certain commands as root casually. For instance, if you are a user and you want to delete all of the files in a directory, including any subdirectories, you might switch to that directory and issue rm –rf *. This means "remove all files recursively without asking permission". By "recursively" deleting files, it also deletes all subdirectories and their files and subdirectories. If you are in the wrong directory, the OS will probably tell you that it is not able to comply because you do not own the files. But if you are in the wrong directory AND you are root, then the OS performs the deletion and now you have deleted the wrong files by mistake. You may think that you will have no trouble remembering who you are, but in fact there will be situations where you will log into one window as root and another as yourself in order to change OS settings (as root) and test those changes out (as a normal user).

INSTALLING AN OS

When you purchase most computers today, they have a preinstalled OS. Some users may wish to install a different OS, or because of such situations as computer virus infections, deleted files, or obsolete OSs, a user may wish to reinstall or upgrade an OS. In addition, adding an OS does not necessarily mean that the current OS must be replaced or deleted.

A user may install several OSs, each in its own disk partition, and use some boot loading program such as GRUB, LILO, and BOOTMGR. When the computer first boots, the four hardware initialization steps (See the section Booting and System Initialization) are performed. Step 5 is the execution of the bootloader program, which provides a prompt for the user to select the OS to boot into. Once selected, the boot process loads the selected OS, which is then initialized, and now the user is able to use the computer in the selected OS. To change OSs, the user must shut down the current OS and reboot to reach the bootloader program. Loading multiple OSs onto a computer can lead to difficulties especially when upgrading one of the OSs or attempting to repartition the hard drive. Another way to have access to multiple OSs is to use VMs. The VM is in essence a self-contained environment into which you can install an OS. The VM itself is stored as data on the hard disk until it is executed. Therefore, the VM only takes up disk space. If your computer has an older processor, it is possible that executing a VM will greatly slow down your computer, but with modern multicore processors available, VMs can run effectively and efficiently. This also allows you to have two (or more) OSs open and running at the same time, just by moving in and out of the VM's window. With VM software, you can create multiple VMs and run any or all of them at the same time.

The remainder of this section discusses how to install two OSs, Red Hat Linux (specifically, CentOS 5.5), and Windows 7. It is recommended that if you attempt either installation, that you do so from within a VM. There are commercial and free VM software products available such as VMWare's VMWare Client, VMWare Player, VMWare Server, and vSphere, and Sun's VirtualBox (or VBox).

Installing Windows

To install Windows 7, you start by inserting a Windows 7 CD into your optical drive and then booting the computer. As your computer is set up to boot to an OS on hard disk, unless your computer has no prior OS, you will have to interrupt the normal boot process. This is done, when booting, by pressing the F12 function key.* It is best to press the key over and over until you see your computer respond to it. This takes you to the boot options, which is a list of different devices that can be booted from. Your choices are typically hard disk, optical disk, network, USB device, and possibly floppy disk. Select the optical disk. The optical disk will be accessed and the installation process begins.

In a few moments, you will be presented with the first of several Windows 7 installation dialogs. The typical installation requires only selecting the default (or recommended) settings and clicking on Next with each window. Early on, you will be asked to select the language to install (e.g., English), time and currency format, and input/keyboard type (e.g., US). Then, you will be prompted to click on the Install Now button (this will be your only option to proceed). You will be asked to accept the licensing terms.

You are then given options for a custom installation or an upgrade. You would select upgrade if you already had a version of Windows 7 installed and were looking to upgrade the system. This might be the case, for instance, if your Windows 7 were partially damaged or years out of date. The custom installation does not retain any previous files, settings, or programs, whereas upgrade retains them all. The custom installation can also allow you to change disks and partitioning of disks. You are then asked where windows should be installed. If you have only a single hard disk drive, there is no choice to make.

From this point, the installer will run for a while without interruption or need for user input (perhaps 10 to 20 minutes depending on the speed of your optical drive and processor). During this time, Windows will reboot several times. When prompted again, you will be asked to create an initial account and name your computer. The default name for the computer is merely the account name you entered followed by –PC. For instance, if you specify the name Zappa, then the computer would default to Zappa-PC. You can, of course, change the default name. The next window has you finish the account information by providing an initial password for the account along with a "hint" in case you are prone to forgetting the password. This initial account will allow the user to immediately begin using the computer without requiring that an Administrator create an account.

* The actual function key may differ depending on your particular manufacturer. Instructions appear during system booting to tell you which function key(s) to use.

Before proceeding, Windows now requests a product key. This is a code that is probably on the CD packaging. The key will be a combination of letters and numbers and be 25 characters long. This ensures that your version of Windows is authorized and not pirated.

The next step is for Windows to perform automated updates. Although this step is optional, it is highly useful. It allows your installation to obtain the most recent patches of the Windows 7 OS. Without this step, you would be limited to installing the version of the OS as it existed when the CD was manufactured. If you choose to skip this step, Windows would install the updates at a later time, for instance, the first time you attempt to shut down your computer.

After updates are installed, you set the time zone and have the option of adjusting the date and time. Finally, you are asked to specify the computer's current location, which is in essence selecting a network for your computer to attempt to connect to. Your options are Home network, Work network, and Public network. This is an option that you can reset at a later time. Once selected, your computer will try to connect to the network. Finally, your desktop is prepared and the OS initializes into user mode. You are ready to go!

Although you are now ready to use your computer, Windows booted with settings that were established by the Windows programmers. At this point, if you wish to make changes to your desktop, you should do so through the Control Panel. You may, for instance, change the style of windows, the desktop background, the color settings used, the resolution of the screen, and the size of the desktop icons. You can also specify which programs should be pinned to the taskbar that runs along the bottom of the desktop, and those that should appear at the top of the programs menu. You should also ensure that your network firewall is running.

Installing Windows 7 is easy and not very time consuming. From start to finish, the entire installation should take less than an hour, perhaps as little as 30 minutes.

Installing Linux

Similar to Windows 7, installing Red Hat Linux can be done through CD. Here, we assume that you have a CentOS 5.5 installation CD. The installation is more involved than with Windows and requires an understanding of concepts such as disk partitioning (disk partitions are described in Chapter 5).

Upon booting from the installation CD, you will be presented with a screen that has several installation options such as testing the media (not really necessary unless you have created the install CD on your own) and setting the default language.

Now you reach the disk partitioning step. In a new install, it is likely that the hard disk is not partitioned and you will have to specify the partitions. If you are installing Linux on a machine with an existing OS, you will have to be careful to partition a free disk. Note that this does not necessarily require two or more hard disk drives. The "free disk" is a logical designation and may be a partition that is not part of the other OS. If partitions already exist that you want to remove, you must select "Remove Linux partitions on selected drives and create default layout" and select the "Review and modify partitioning layout" checkbox.

Here, you must specify the partitions of your Linux disk. You will want to have different partitions for each of the root partition, the swap space, and the user directories. You may want a finer group of partitions by, for instance, having a partition for /var and for /usr; otherwise, these directories will be placed under root. For each partition, you must select the mount point (the directory) or the file system (for swap), the size of the partition, and whether the partition should be fixed in size, or fill the remaining space available. As an example, you might partition your Linux disk as follows:

- Mount Point: select /(root), size of 4000 (i.e., is 4 GB), fixed size
- Mount Point: select /var, size of 1000, fixed size
- File System Type: swap (do not select a Mount Point), size of 1000, fixed size
- Mount Point: select /home, fill to maximum allowable size

At this next screen, you will be asked for a boot loader. GRUB is the default and should be selected and installed on the main hard disk, which is probably /dev/sda1. For network devices, you need to specify how an IP address is to be generated. The most common technique is to have it assigned by a server through DHCP. The last question is to select your time zone.

You are asked to specify the root password of the system. This password will be used by the system administrator every time a system administration chore is required such as creating an account or installing software. You want to use a password that you will not forget.

At the next screen, you can specify what software should automatically be installed with CentOS. Desktop—Gnome should already be selected. You may select other software at this point, or Customize Later. The installation is ready to begin. This process usually takes 5 to 10 minutes. When done, you will be prompted to reboot the system. Once rebooted, you finalize the installation process. This includes setting up the initial firewall settings (the defaults will probably be sufficient) and whether you want to enforce SELinux (security enhanced). Again, the default (enforcing) is best. You are able to set the date and time. Finally, you are asked to create an initial user account, much like with Windows. This account is required so that, as a user, you can log into the GUI. It is not recommended that you ever log in to the GUI as root. Unlike Windows, you do not have to reboot at this point to start using the system; instead, you are taken to a log in window and able to proceed from there by logging in under the user account just created.

FURTHER READING

As with computer organization, OS is a required topic in computer science. There are a number of texts, primarily senior-level or graduate reading. Such texts often discuss OS tasks such as process management, resource management, and memory management. In addition, some highlight various OSs. The following texts target computer science students, but the IT student could also benefit from any of these texts to better understand the implementation issues involved in designing OSs.

- Elmasri, R., Carrick, A., and Levine, D. *Operating Systems: A Spiral Approach.* New York: McGraw Hill, 2009.
- Garrido, J. and Schlesinger, R. *Principles of Modern Operating Systems.* Sudbury, MA: Jones and Bartlett, 2007.
- Silberschatz, A., Galvin, P., and Gagne, G. *Operating System Concepts.* Hoboken, NJ: Wiley & Sons, 2008.
- Stallings, W. *Operating Systems: Internals and Design Principles.* Upper Saddle River, NJ: Prentice Hall, 2011.
- Tanenbaum, A. *Modern Operating Systems.* Upper Saddle River, NJ: Prentice Hall, 2007.

For the IT student, texts specific to Linux, Windows, Unix, and Mac OS will be essential. These texts typically cover how to use or administer the OS rather than the theory and concepts underlying OSs. Texts range from "for dummies" introductory level texts to advanced texts for system administrators and programmers. A few select texts are listed here for each OS. Many of these texts also describe how to install the given OS.

- Adelstein, T. and Lubanovic, B. *Linux System Administration.* Sebastopol, CA: O'Reilly Media, 2007.
- Bott, E., Siechert, C., and Stinson, C. *Windows 7 Inside Out.* Redmond, WA: Microsoft Press, 2009.
- Elboth, D. *The Linux Book.* Upper Saddle River, NJ: Prentice Hall, 2001.
- Frisch, E. *Essential System Administration.* Cambridge, MA: O'Reilly, 2002.
- Fox, T. *Red Hat Enterprise Linux 5 Administration Unleashed.* Indianapolis, IN: Sams, 2007.
- Helmke, M. *Ubuntu Unleashed.* Indianapolis, IN: Sams, 2012.
- Hill, B., Burger, C., Jesse, J., and Bacon, J. *The Official Ubuntu Book.* Upper Saddle River, NJ: Prentice Hall, 2008.
- Kelby, S. *The Mac OS X Leopard Book.* Berkeley, CA: Peachpit, 2008.
- Nemeth, E., Snyder, G., Hein, T., and Whaley, B. *Unix and Linux System Administration Handbook.* Upper Saddle River, NJ: Prentice Hall, 2010.
- Russinovich, M., Solomon, D., and Ionescu, A. *Windows Internal.* Redmond, WA: Microsoft Press, 2009.
- Sarwar, S. and Koretsky, R. *Unix: The Textbook.* Boston, MA: Addison Wesley, 2004.
- Sobell, M. *A Practical Guide to Linux Commands, Editors, and Shell Programming.* Upper Saddle River, NJ: Prentice Hall, 2009.

- Wells, N. *The Complete Guide to Linux System Administration*. Boston, MA: Thomson Course Technology, 2005.
- Wrightson, K. and Merino, J., *Introduction to Unix*. California: Richard D. Irwin, 2003.

There are also thousands of websites set up by users and developers, and many are worth exploring.

Virtualization and VMs are becoming a very hot topic although the topic is too advanced for this text. Books again range from "for dummies" books to texts on cloud computing. Three references are listed here:

- Golden, B. *Virtualization for Dummies*. Hoboken, NJ: Wiley and Sons, 2007.
- Hess, K., and Newman, *A. Practical Virtualization Solutions*. Upper Saddle River, NJ: Prentice Hall, 2010.
- Kusnetzky, D. *Virtualization: A Manager's Guide*. Massachusetts: O'Reilly, 2011.

REVIEW TERMS

Terminology from this chapter

Background
Batch
BIOS
Booting
Boot Loader
Command line
Competitive multitasking
Concurrent processing
Context switch
Cooperative multitasking
Deadlock
Device driver
File system management
Foreground
Initialization
Initialization script
Interactivity
Interrupt
Interrupt handler
I/O queue
Kernel
Memory management
Memory violation
Multiprocessing
Multiprogramming
Multitasking
Multithreading
Mutually exclusive

Nonvolatile memory

Page

Page fault

Page table

Process

Process management

Process status

Program

Protection

Queue

Ready queue

Rendezvous

Resource management

Root

Root account

Round-robin scheduling

Security

Shell

Single tasking

Swap space

Swapping

Synchronization

Thread

Timer

User account

User interface

Utility program

Virtual machine

Virtual memory

Volatile memory

Waiting queue

REVIEW QUESTIONS

1. In what way does the OS support convenient access for a user?
2. What does the OS help a user control?
3. What components of an operating system are always in memory?
4. Why might a user prefer to use a command line for input over a GUI?
5. What types of status can a process have?
6. What is an interrupt and when might one arise?
7. What types of situations might cause a disk drive to raise an interrupt? The keyboard?
8. Why might a USB flash drive raise an interrupt?
9. Aside from hardware, programs can raise interrupts. Provide some examples of why a program might raise an interrupt. Note: if you are familiar with a programming language such as C++ or Java, an interrupt can trigger an exception handler. This might help you more fully answer this question.

10. When does the CPU check for an interrupt? What does the CPU do immediately before handling an interrupt?

11. In virtual memory, what is swapped into and out of memory? Where is virtual memory stored?

12. Given the following page table for some process, X, which of X's pages are currently in memory and which are currently not in memory? Where would you look for page 4? For page 5?

Process X Page Table

Page	Memory Frame	Valid Bit
0	12	1
1	3	1
2	–	0
3	–	0
4	9	1
5	–	0
6	–	0
7	2	1

13. What happens if a memory request is of a page that is not currently in memory?

14. Why is swapping a slow process?

15. If a resource's access is not synchronized, what could happen to the resource?

16. If a resource does not require mutually exclusive access, can it be involved in a deadlock?

17. Assume a deadlock has arisen between processes P0 and P1. What has to happen before either process can continue?

18. If a deadlock arises, what are the choices for the user?

19. What is the difference between the OS roles of protection and security?

20. What is the difference between single tasking and batch processing?

21. What is the difference between batch processing and multiprogramming?

22. What is the difference between multiprogramming and multitasking?

23. What is the difference between multitasking and multithreading?

24. What is the difference between competitive and cooperative multitasking?

25. Why might you use shortest job first as a scheduling algorithm?

26. What is the difference between placing a process in the waiting queue versus the ready queue?

27. What does the CPU do during a context switch?

28. What changes when a context switch occurs?

29. Why are context switches more efficient between threads than processes?

30. What is the difference between a process in the foreground and a process in the background?

31. What are the steps of a boot process?

32. Why is the boot program (at least in part) stored in ROM instead of RAM or the hard disk?

33. What is BIOS and what does it do?

34. What is an initialization script? Provide an example of what an initialization script does.

35. Why should a system administrator log in to a computer running a GUI environment as his/herself and then switch to the administrator account instead of logging in directly as the administrator?

DISCUSSION QUESTIONS

1. Are you a fan of entering OS commands via the command line instead of using a GUI? Explain why or why not.

2. What are the advantages and disadvantages of using the command line interface for an ordinary end user? What are the advantages and disadvantages of using the command line interface for a system administrator?

3. Are you more likely to use the command line in Windows or Linux, both, or neither? Why?

4. As an end user, how can an understanding of concepts such as virtual memory, process management, deadlock, protection, and security help you?

5. Identify specific situations in which understand operating system concepts will be essential for a system administrator.

6. Consider a computer without an interrupt system. In such a system, if a program is caught doing something that you do not want it to (e.g., an infinite loop), you would not be able to stop it by closing the window or typing control-c or control-alt-delete. What other things would your computer not do if it did not have an interrupt system?

7. As a user, do you need to understand concepts such as a context switch, cooperative multitasking, competitive multitasking, and multithreading? Explain.

8. As a system administrator, do you need to understand concepts such as a context switch, cooperative multitasking, competitive multitasking, and multithreading? Explain.

9. As a system administrator, do you need to understand concepts of threads, synchronization, and rendezvous? Explain.

10. Rebooting a computer starts the operating system in a fresh environment. However, booting can be time consuming. If you are a Windows user, how often do you reboot? How often would you prefer to reboot? Under what circumstances do you reboot your computer?

11. You are planning on purchasing a new computer. In general, you have two choices of platform, Mac or Windows. You can also install other operating systems. For instance, it is common to install Linux so that your computer can dual boot to either Windows or Linux. Alternatively, you could wipe the disk and install Linux. How would you decide which of these to do? (Mac, Windows, dual boot, Linux, other?)

CHAPTER 5

Files, Directories, and the File System

The logical and physical implementations of file systems are covered in this chapter. The logical view of a file system is the organizational units of files, directories, and partitions. The physical implementation is the translation from the logical view to the physical location on disk storage of the file components. File systems of both DOS/Windows and Linux are presented including the top level directories and the implementation of block indexing. The chapter concludes with a section that explores how to navigate through the file systems using command line instructions.

The learning objectives of this chapter are to

- Describe the elements of the file system from both the logical and physical viewpoint.
- Describe how files are physically stored and accessed.
- Present the use of the top level directories in both Windows and Linux.
- Illustrate through example Linux and DOS commands to navigate around a file system.

In this chapter, we will examine the components of a file system and look specifically at both the Windows file system and the Unix/Linux file system.

FILES AND DIRECTORIES

A file system exists at two levels of abstraction. There is the physical nature of the file system—the data storage devices, the file index, and the mapping process of taking a name and identifying its physical location within the hard disks (and other storage media). There is also the logical nature of the file system—how the physical system is partitioned into independently named entities. As both users and system administrators, we are mostly interested in the logical file system. The details for how the files and directories are distributed only becomes of interest when we must troubleshoot our file system (e.g., corrupt files, the need to defragment the file system). We will take a brief look at the physical nature of file systems later in this chapter, but for now, we will concentrate on the logical side.

In a file system, you have three types of entities:

- Partitions
- Directories
- Files

A partition is a logical division within the file system. In Windows, a partition is denoted using a letter, for instance C:\ versus D:\ might both refer to parts of the hard disk, but to two separate partitions. In Linux, a partition refers to a separate logical disk drive although, in fact, several partitions may reside on a single disk. We place certain directories in a partition, thus keeping some of the directories separated. For instance, the swap space will be placed on one partition, the users' home directories on a second, and quite likely the Linux operating system (OS) on a third. You usually set up a partition when you first install the OS. Once established, it is awkward to change the size or number of partitions because that might cause an existing partition's space to be diminished and thereby require that some of the files in that partition be moved or deleted. Partitions are not that interesting, and it is likely that once you set up your OS, you will not have to worry about the individual partitions again.

Directories, also known as folders, allow the user to coordinate where files are placed. Unlike partitions, which are few in number, you can create as many directories as desired. Directories can be placed inside of directories giving the user the ability to create a hierarchical file space. Such a directory is usually referred to as a subdirectory. The hierarchical file space is often presented in a "tree" shape. At the top, or root, of the tree are the partitions. Underneath are the top level directories. Underneath those are files and subdirectories. The tree branches out so that the leaves are at the bottom. File manager programs often illustrate the tree in a left to right manner (the left side has the root of the tree); as you move toward the right, you have subdirectories branching out, and the leaves (which are individual files) are at the right end of any branch. Figure 5.1 illustrates typical file spaces as viewed in file manager software: Windows on the left-hand side and Linux on the right. In both OSs, the topmost level is the computer itself, which has underneath it the various file spaces or drives. In modern computers, it is common to have the hard disk drive (C:) and an optical drive (D:). There may be additional drives listed, such as an external hard disk, floppy disk, USB flash drive, file server available by network, or partitions within the internal hard disk itself. In this latter case, in Linux, they may all still appear underneath Filesystem, as shown in Figure 5.1. The top level of the Linux file system contains a number of preestablished directories. In Windows, the file system is typically limited to initial directories of Program Files (two directories in Windows 7 to denote the 32-bit programs from the 64-bit programs), Users (the user home directories), and Windows (the OS). In Linux, the directory structure is more defined. We will explore the use of the various directories in detail later in this chapter.

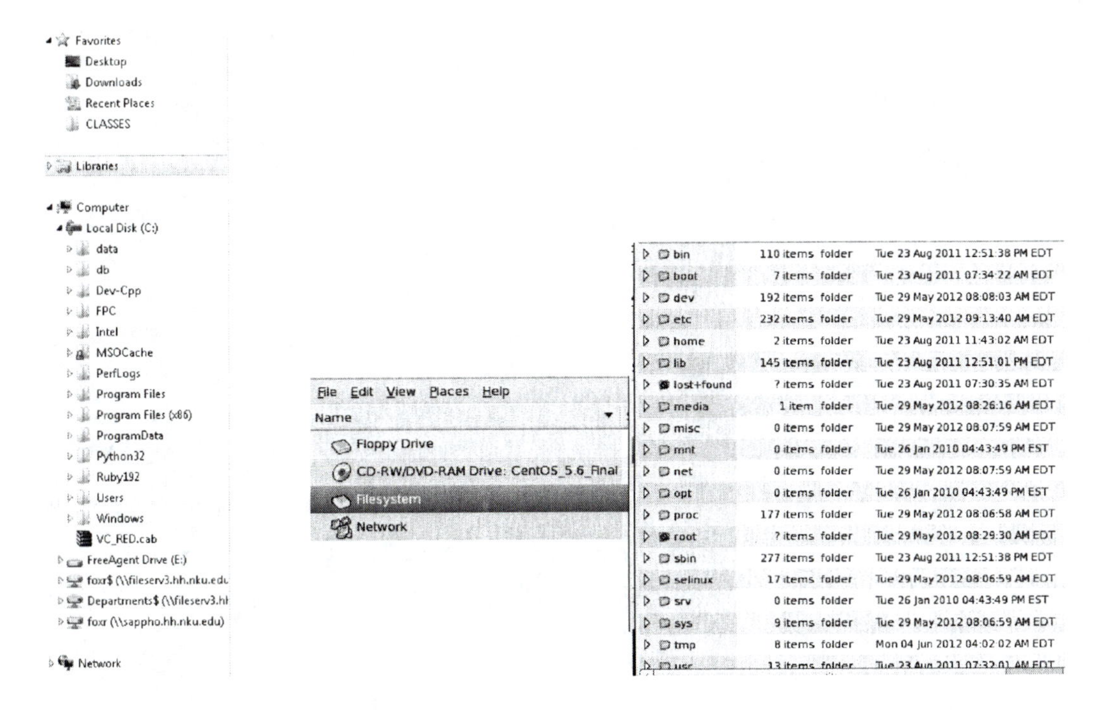

FIGURE 5.1 Typical file spaces.

Files are the meaningful units of storage in a file system. They comprise data and programs, although we can think of programs as being special types of data files. Data files come in all sizes and types. Data files might be text files, formatted text files using special escape or control characters, or binary files. In Linux, text files are very common. In Windows, data files are more commonly those produced from software such as Word documents (docx) and Excel files (xlsx). Data files may also be multimedia files such as music or sound files (wav, mp3), image files (gif, jpg, bmp), and movie files (mpg, avi, mov).

Files have names. In older OSs, file names were limited to up to eight characters followed by an optional extension that could be up to three characters. Separating the name from the extension is a period. In much older OSs, the characters used in file names were limited to alphabetical, numeric, underscore, and hyphen characters only. Today, file names can include spaces and some other forms of punctuation although not every character is allowable. Longer names are often useful because the name should be very descriptive. Some users ignore extensions but extensions are important because they tell the user what type of file it is (i.e., what software created it). File extensions are often used by the OS to determine how a file should be opened. For instance, if you were to double click on a data file's icon, the OS could automatically open that file in the proper software if the file has an extension, and the extension is mapped to the proper software.

Aside from file types and names, files have other properties of note. Their logical location in the file system is denoted by a *pointer*. That is, the descriptor of the file will also contain its location within the file system space. For instance C:\Users\foxr\My Documents\cit130\notes\ch7.docx describes the file's logical position in the file system but not its physical location. An additional process must convert from the logical to physical location. The file's size, last modification time and date, and owner are also useful properties. Depending on the OS, the creation time and date and the group that owns the file may also be recorded. You can find these properties in Windows by right clicking on any file name from the Windows Explorer and selecting Properties. In Linux, you can see these properties when you do an ls –l command.

Files are not necessarily stored in one contiguous block of disk space. Instead, files are broken into fixed sized units known as *blocks*. A file's blocks may not be distributed on a disk such that they are in the same track or sector or even on the same surface. Instead, the blocks could be distributed across multiple disk surfaces. This will be covered in more detail later in the chapter.

Many older OSs' commands were oriented toward managing the file system. The DOS operating system for instance stands for Disk Operating System. There were few commands other than disk operations that a user would need. This is not to say that the OS only performed operations on the file system, but that the user commands permitted few other types of operations. Common commands in both MS-DOS and Linux are listed below. Similarly, Windows mouse operations are provided. We will explore the Linux and DOS commands in more detail in Moving around the File System.

Linux:

- ls—list the files and subdirectories of the given directory
- mv—move or rename a file or directory
- cp—copy a file or directory to a new location
- rm—remove (delete) a file
- cat, more, less—display the contents of a file to the screen
- mkdir—create a new (sub)directory
- rmdir—remove a directory; the directory must be empty for this to work

DOS:

- dir—list the files and subdirectories of the given directory
- move—move a file to a new location, or rename the file
- copy—copy a file to a new location
- del—remove (delete) a file
- type—display the contents of a file to the screen
- mkdir—create a new (sub)directory
- rmdir—remove a directory; the directory must be empty for this to work

Windows (using Windows Explorer; see Figure 5.2):

- To view a directory's contents, click on the directory's name in the left-hand pane; its contents are shown in the right-hand pane—if the directory is not at the top level, you will have to expand its ancestor directories from the top of the file system until you reach the sought directory.
- To move a file, drag the file's icon from the right-hand pane in its current directory to a directory in the left-hand pane.
- To copy a file, right click on the file's icon, select copy, move to the new directory by clicking on that directory in the left-hand pane, and then in the right-hand pane, right click and select paste.
- To rename a file, right click on the file's icon, select rename (or click in the file's name in the icon twice slowly), and then type the new name.
- To delete a file (or folder), drag the icon into the recycle bin (or right click on the icon and select delete or left click on the icon and press the delete key) —you will have to separately empty the recycle bin to permanently delete the file.
- To create a new directory (folder), click on the button New Folder. Or, right click in the right-hand pane and select new and then folder, then name the folder once it appears.

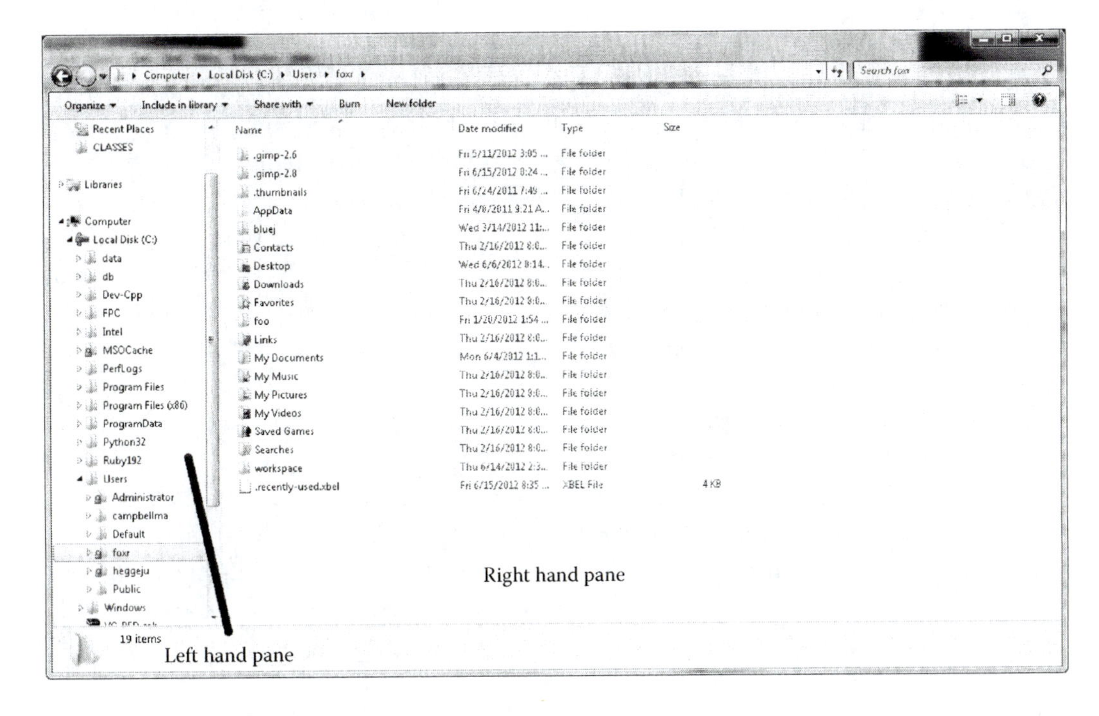

FIGURE 5.2 Windows Explorer.

- To display the contents of a file, you can double click on the file icon and it should launch in whatever software is associated with the file's extensions.

Note that in Windows, file extensions may not appear in the file manager, so you would have to set it up so that file extensions appear (if you want to see them).

In order to reduce the impact that large files have on the available space in a file system, the user can perform *file compression*. There are two forms of file compression. In *lossless* file compression, a file is reduced in size such that, when it is uncompressed to its original form, no data are lost. There are many algorithms for performing lossless file compression, and they will work especially well on text files. With lossless file compression, you usually have to uncompress the file before you can access it (there are exceptions such as the FLAC audio compression algorithm). The advantage of lossless file compression is that files that are not currently in use can be stored in greatly reduced sizes. The primary disadvantage is that compressing and uncompressing the file is time consuming.

Lossy file compression actually discards some of the data in order to reduce the file's size. Most forms of streaming audio and video files are lossy. These include .avi, .mpg, .mov (video) and .wav and .mp3 (audio). Most forms of image storage also use lossy file compression, such as .jpg. If you were to compare a bitmap (.bmp) file to a .jpg, you would see that the .bmp has greater (perhaps much greater) clarity. The .jpg file saves space by "blurring" neighboring pixels. The .gif format instead discards some of the colors so that if the image uses only the standard palette of colors, you would not see any loss in clarity. However, if the .gif uses different colors, the image, while having the same clarity, would not have the correct colors.

When you have a collection of files, another useful action is to bundle them together into an *archive*. This permits easy movement of the files as a group whether you are placing them in backup storage or uploading/downloading them over the Internet. File archiving is common when a programmer wants to share open source programs. The source code probably includes many files, perhaps dozens or hundreds. By archiving the collection, the entire set of source code can be downloaded with one action rather than requiring that a user download files in a piecemeal fashion. One very common form of archiving is through the zip file format, which performs both file compression and archiving. To retrieve from a zipped file, one must unzip the file. There are numerous popular software packages for performing zip and unzip including winzip and winrar in Windows, and gzip and gunzip in Linux. Linux also has an older means for archiving called tar. The tar program (tape archive) was originally used to collect files together into a bundle and save them to tape storage. Today, tar is used to archive files primarily for transmission over the Internet. The tar program does not compress, so one must compress the tar file to reduce its size.

Lossy compression is primarily used on image, sound, and movie files, whereas lossless compression is used on text files and software source files. The amount of compression, that is, the reduction in file size, depends on the compression algorithm applied and the data itself. Text files often can be greatly reduced through compression, by as much as 88%. Lossy audio compression typically can compress data files down to 10% or 20% of their original version, whereas lossless audio compression only reduces files to as much as 50% of their original size.

FILE SYSTEMS AND DISKS

The file system is the hierarchical structure that comprises the physical and logical file space. The file system includes an index to map from a logical file name to a physical location. A file system also includes the capability of growing or shrinking the file space by adding and deleting media. This is sometimes referred to as *mounting*. A mounted file space is one that can later be removed (unmounted). The file system is physically composed of secondary storage devices, most predominantly an internal hard disk drive, an internal optical drive, and possibly external hard disk drives. Additionally, users may mount flash drives and tape drives. For a networked computer, the file space may include remote drive units (often called file servers).

The OS maintains the file space at two levels: logical and physical. Our view of the file space is the logical view—partitions, directories, and files. In order to access a file, the OS must map from the logical location to the physical location. This is done by recording the physical location of a file in its directory information. The location is commonly referenced by means of a pointer. Let us assume that we are dealing with files stored only on hard disk. The pointer indicates a particular file system *block*. This must itself be translated from a single integer number into a drive unit, a surface (if we are dealing with a multiplatter disk drive), a sector, and a track on the disk (Figure 5.3). The term cylinder is used to express the same track and sector numbers across all platters of a spinning disk. We might reference a cylinder when we want to access the same tracks and sectors on each surface simultaneously.

As an example, imagine file f1.txt is located at block 381551. Our file system consists of one hard disk with four platters or eight surfaces. Furthermore, each surface has 64 sectors and 2048 tracks. We would locate this block on surface 381551/(64 * 2048) = surface 2 (starting our count at surface 0). Surface 2 would in fact contain blocks 262144 through 393216. We would find block 381551 in sector 58 (again, starting our count at sector 0) since

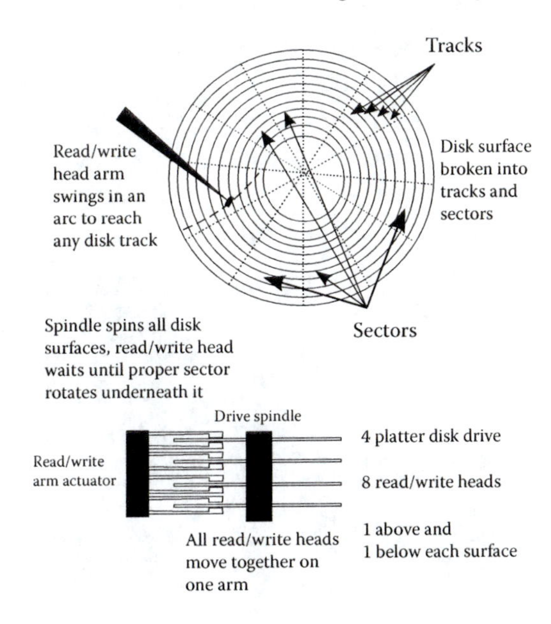

FIGURE 5.3 Hard disk layout. (Adapted in part from the public domain image by LionKimbro at http://commons.wikimedia.org/wiki/File:Cylinder_Head_Sector.svg.)

(381551 – 262144)/2048 = 58.3. Finally, we would find the block on track 623 (381551 – 262144) – 2048 * 58 = 623. So, fl.txt, located at block 381551, is found on surface 2, sector 58, track 623.

As shown in Figure 5.3, hard disk drives contain multiple platters. The platters rotate in unison as a motorized spindle reaches up through the holes in each platter. The rotation rate of the disks depends on the type of disk, but hard disk drives rotate at a rate of between 5400 and 15,000 revolutions/min. Each platter will have two read/write heads assigned to it (so that both surfaces of the platter can be used). All of the read/write heads of the hard disk drive move in unison. The arms of the read/write heads are controlled by an actuator that moves them across the surface of the disks, whereas the disks are spun by the spindle. The read/write heads do not actually touch the surface of the disk, but hover slightly above/below the surface on a cushion of air created by the rapid rotation rate of the disks. It is important that the read/write heads do not touch the surface of the disk because they could damage the surface (scratch or dent it) and destroy data stored there. For this reason, it is important to "park" the read/write head arm before moving a computer so that you do not risk damaging the drive. The read/write head arm is parked when you shut down the computer. Many modern laptop computers use motion sensors to detect rapid movements so that the read/write head arm can be parked quickly in case of a jerking motion or if the entire laptop is dropped.

The data stored on hard disk is in the form of positive and negative magnetic charges. The read/write head is able to read a charge from the surface or write a new charge (load/open and store/save operations, respectively). An example of a read/write head arm for eight platters (sixteen heads, only the top head is visible) is shown in Figure 5.4.

Files are not stored consecutively across the surface of the disk. Instead, a file is broken into fixed size units called *blocks*, and the blocks of a file are distributed across the disk's surface, possibly across multiple surfaces. Figure 5.5 illustrates this idea, where a file of 6 blocks (numbered 0 through 5) are located in various disk block locations, scattered throughout the hard disk. Note that since every block is the same size, the last block may not be filled to capacity. The dotted line in disk block 683 indicates a fragment—that is, the end of the file does not fill up the entire disk block, so a portion of it goes unused.

There are several reasons for scattering the file blocks across the disk surface(s). First, because of the speed by which a disk is spun, if the read/write head had to access multiple

FIGURE 5.4 Read/write head arm. (Courtesy of Hubert Berberich, http://commons.wikimedia.org/wiki/File:Seagate-ST4702N-03.jpg.)

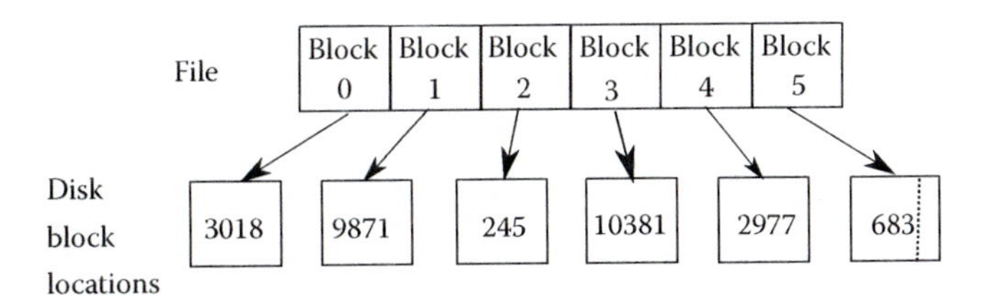

FIGURE 5.5 File blocks mapped to disk blocks.

consecutive disk blocks, it is likely that the time it takes to transfer a block's worth of content to memory would be longer than the time it would take for the disk to move the next block under the read/write head. The result would be that the read/write head would be out of position for the next access and therefore the access would have to wait until the proper block spun underneath the read/write head, that is, another rotation of the disk. One way to combat this problem is to space disk blocks out, in the same general location but not adjacent. For instance, disk blocks might be saved on every other sector of a given track and then onto the next adjacent track (e.g., sector 0/track 53, sector 2/track 53, sector 4/track 53, sector 6/track 53, sector 8/track 53, sector 0/track 54, sector 2/track 54).

Second, this alleviates the problem of fragmentation. As shown in Figure 5.5, only the last disk block of a file might contain a fragment. Consider what might happen if disk files are placed in consecutive locations. Imagine a file is stored in 3 consecutive blocks. A second file is stored in 2 consecutive blocks immediately afterward. A third file is stored in 4 consecutive blocks after the second file. The second file is edited and enlarged. It no longer fits in the 2 blocks, and so is saved after the third file. Now, we have file 1 (3 blocks), freed space (2 blocks), file 3 (4 blocks), and file 2 (3 blocks). Unfortunately, that free space of 2 blocks may not be usable because new files might need more storage space. Thus, we have a 2-block fragment. It is possible that at some point, we would need the 2-block space to reuse that freed-up space, but it could remain unused for quite a long time. This is similar to what happens if you record several items on video or cassette tape and decide you no longer want one of the earlier items. You may or may not be able to fill in the space of the item you no longer want. Since we often edit and resave files, using consecutive disk blocks would likely create dozens or hundreds of unusable fragments across the file system. Thus, by distributing blocks, we can use any free block any time we need a new block.

Third, it is easier to maintain a description of the available free space in the file system by using the same mechanism to track disk blocks of a file. The free space consists of those disk blocks that either have not yet been used, or are of deleted files. Therefore, monitoring and recording what is free and what is used will not take extra disk space.

This leads to a natural question: if a disk file is broken into blocks that are scattered around the file system, how does the OS find a given block (for instance, the third block of a file)? Recall that the OS maintains a listing of the first disk block of each file. Each disk block contains as part of its storage the location of the next disk block in the file. The storage is a pointer. This creates what is known as a *linked list*. See Figure 5.6, where the OS stores a file pointer to the first block and each block stores a pointer to the next block in the file.

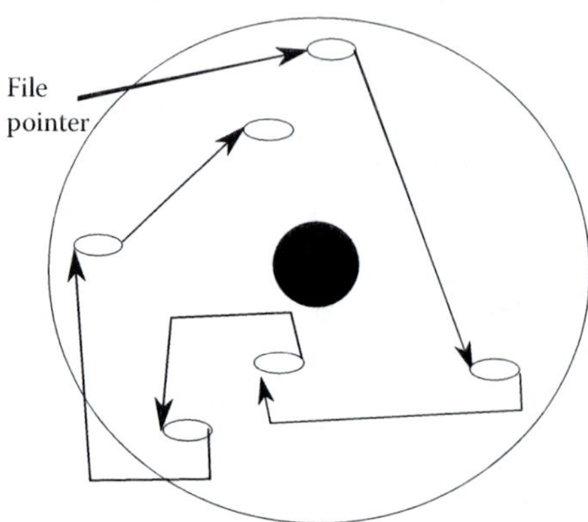

FIGURE 5.6 Pointers are used to determine next file block on disk.

To load a particular disk block, the OS first examines the directory listing to obtain the file's first pointer. The first block has a pointer to the second block, which has a pointer to the third block, and so forth. To reach block n, you must first go through $n - 1$ previous blocks, following those pointers. Following the linked list is inefficient because each block requires a separate disk access, which as noted previously, is one of the slower aspects of any computer. For a sequential file—that is, one that must be accessed in order from start to finish—accessing the file in this manner is not a drawback because the file is loaded, block by block, in order.

DISK UTILITIES

You might recall from Chapter 4 that part of the OS is made up of utility software. This software includes antiviral programs, for instance. Another class of utility programs is the disk utilities. Disk utilities can help a user or system administrator manage and maintain the file system by performing tasks that are otherwise not available. Below is a list of common disk utilities.

Disk defragmentation: although the use of disk blocks, scattered around the file system, prevents the creation of fragments, fragments can still arise. Disk defragmentation moves disk blocks across the disk surfaces in order to move them together and remove free blocks from between used blocks. This can make disk performance more efficient because it moves blocks of a file closer together so that, when the file is accessed wholly, the access time is reduced because all blocks are within the same general area on a surface. Whereas a Windows file system may occasionally benefit from defragmentation, supposedly the Linux file system will not become inefficient over time.

File recovery: if you delete a file and then empty the recycle bin, that file is gone, right? Not so. File recovery programs can try to piece together a file and restore it—as long as the disk blocks that made up the file have not been written over with a new file.

Data backup: many users do not back up their file system, and so when a catastrophic error arises, their data (possibly going back years) get lost. Backups are extremely important, but they are sometimes difficult to manage—do you back up everything or just what has changed since the last backup? A backup utility can help you manage the backup process.

Many files can be accessed randomly—that is, any block might be accessed at any time. To accommodate this, some OSs, including Windows, use a file allocation table (FAT). The FAT accumulates all of the pointers of every disk block into a single file, stored at the beginning of the disk. Then, this file is loaded into memory so that the OS need only search memory to find the location of a given block, rather than searching the disk itself. Searching memory for the proper block's disk location is far more efficient than searching disk. Figure 5.7 illustrates a portion of the FAT. Notice that each disk block is stored in the FAT, with its successor block. So, in this case, we see that the file that contains block 150 has a next block at location 381, whereas the file that contains block 151 has a next location at 153 and after 153, the next location is 156, which is the last location in the file. Block 152 is a bad sector and block 154, 155, and 732 are part of another file. EOF stands for "end of file".

Free space can also be stored using a linked list where the first free block is pointed to by a special pointer in the file system's partition, and then each consecutive free block is pointed at by the previous free block. To delete a file, one need only change a couple of pointers. The last free space pointer will point to the first block in the deleted file, and the entry in the directory is changed from a pointer to the first block, to null (to indicate that the file no longer exists). Notice that you can easily reclaim a deleted file (undelete a file) if the deleted files' blocks have not been reused. This is because file deletion does not physically delete a file from the disk but instead adds the file's disk blocks to the list of free blocks. See the discussion of file recovery in the side bar on the previous page.

The pointers from the file system (directory) to the physical location in the disk drive are sometimes referred to as *hard links*. That is, the linking of the file name to location is a hard link. Linux lets you have multiple hard link pointers to a file. Some OSs also offer *soft links*. In a soft link, the pointer from the file system points not to the physical location of the file, but to another entry in the directory structure. The soft link, also known as a symbolic link in Linux and a shortcut in Windows, points to the original entry. For instance, if you create file f1.txt in directory /home/foxr, and later set up a soft link called f2.txt in /home/foxr/cit130/stuff, then the directory /home/foxr/cit130/stuff has an entry f2.txt whose pointer points to /home/foxr/f1.txt, and not to the first block of the file itself. See Figure 5.8 for an example where two hard links, called File1 and File2, point to the first block of the file, whereas a soft link, called File3, points at the hard link File2.

With either kind of link, *aliases* are created. An alias means that a particular entity (file in this case) can be referenced through multiple names. Aliases can be dangerous because someone might change an item without realizing that they are changing another item. For instance, if I were to alter f1.txt, then f2.txt changes as well since f2.txt is merely a pointer to f1.txt. The situation is worse when the original file is deleted. The OS does not necessarily

File allocation table (portion)

Block	150	151	152	153	154	155	156
Next location	381	153	Bad	156	155	732	EOF

FIGURE 5.7 Portion of a file allocation table.

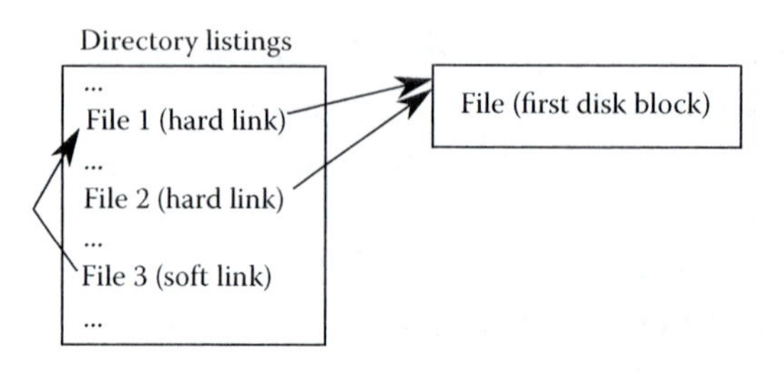

FIGURE 5.8 Hard links and soft links.

check to see if there are soft links pointing to a file that is being deleted. The result is that the file no longer exists and yet a soft link points to it. In this case, following the soft link yields an error. One can delete the soft link without deleting the physical file itself or causing a problem, but deleting the original file without deleting the soft link can lead to errors.

Disk access is a great deal slower than CPU speed or access to main memory. Disk access time is a combination of several different factors. The address must be mapped from a logical position to a disk block through the FAT followed by mapping to the physical location in terms of surface, sector, and track. This is relatively quick as this mapping is done by the CPU accessing memory instead of disk. Now, the read/write must be positioned over the proper location on the disk. This requires moving the read/write head to the proper track (known as *seek time*) and waiting for the proper sector to be spun underneath the read/write head (known as *rotational delay* or rotational latency). Now, the read/write head can begin reading or writing the magnetic charges on the disk. This action is more time consuming that the access performed in DRAM (dynamic RAM). As the bits are being read or written, information must be transferred between disk and memory. This is known as *transfer time*. The combination of mapping (usually negligible), seek time, rotational delay, and transfer time is on the order of several milliseconds (perhaps 5 to 30 ms). A millisecond is a thousandth of a second. Recall that our processors and static RAM (SRAM) operate at the nanosecond rate (a billionth of a second) and DRAM is perhaps 25 to 100 times slower, but still in the dozens of nanosecond range. Therefore, the disk access time is a factor of millions of times slower than CPU and memory response rates!

LINUX FILE SYSTEM

Linux File Space

The Linux file space is set up with default root-level directories, which are populated with basic files when you install Linux. This allows you to know where to find certain types of files. These default directories are:

- /bin: common operating system programs such as ls, grep, cp, mv.
- /sbin: similar to /bin, but the programs stored here are of Linux commands that are intended for the system administrator, although many may be available to normal

users as well. Note: /sbin is often not part of a user's path, so the user may need to type the full path name in for the command, e.g., /sbin/ifconfig.

- /etc: configuration files specific to the machine (details are provided later in this section).
- /root: the home directory for user root. This is usually not accessible to other users on the system.
- /lib: shared libraries used during dynamic linking; these files are similar to dll files in Windows.
- /dev: device files. These are special files that help the user interface with the various devices on the system. Some of the important devices are listed below.
  - /dev/fd0—floppy disk (if available)
  - /dev/hda0—master IDE drive on the primary IDE controller
  - /dev/ht0—first IDE tape drive
  - /dev/lp0—first parallel printer device
  - /dev/null—not a device but instead a destination for program output that you do not want to appear on the monitor—in essence, sending anything to /dev/null makes it disappear
  - /dev/pcd0—first parallel port CD ROM drive
  - /dev/pt0—first parallel port tape
  - /dev/random and /dev/urandom—random number generators (urandom has potential problems if used to generate random numbers for a cryptography algorithm)
  - /dev/sda0—the first SCSI drive on the first SCSI bus
  - /dev/zero—this is a simple way of getting many 0s

  Note: for devices ending in 0, if you have other devices of the same type, you would just increase the number, for instance, lp1, lp2 for two additional printers, or pcd1 for a second CD ROM drive.
- /tmp: temporary file storage for running programs that need to create and use temporary files.
- /boot: files used by a bootstrap loader, e.g., LILO (Linux Loader) or GRUB (Grand Unified Boot loader).
- /mnt: mount point used for temporarily partitions as mounted by the system administrator (not regularly mounted partitions).

- /usr, /var, /home: mount points for the other file systems.
  - /home—the users' file space
  - /var—run-time data stored by various programs including log files, e-mail files, printer spool files, and locked files
  - /usr—various system and user applications software, with subdirectories:
    - /usr/bin: many Linux user commands are stored here although some are also stored in /bin or /usr/local/bin.
    - /usr/sbin: system administration commands are stored here.
    - /usr/share/man, /usr/share/info, /usr/share/doc: various manual and documentation pages.
    - /usr/include: header files for the C programming language.
    - /usr/lib: unchanging data files for programs and systems.
    - /usr/local: applications software and other files.
- /proc: this is a peculiar entry as it is not actually a physical directory stored in the file system but instead is kept in memory by the OS, storing useful process information such as the list of device drivers configured for the system, what interrupts are currently in use, information about the processor, and the active processes.

There are numerous important system administration files in /etc. Some of the more significant files are listed here:

- Startup scripts
  - /etc/inittab: the initial startup script that establishes the run-level and invokes other scripts (in newer versions of Linux, this script has been replaced with the program /etc/init)
  - /etc/rc.d/rc0.d, /etc/rc.d/rc1.d, /etc/rc.d/rc2.d, etc: directories of symbolic links that point to startup scripts for services, the listings in each directory dictate which services start up and which do not start up at system initialization time based on the run-level (See the section Forms of Process Management in Chapter 4 and Chapter 11 for additional details)
  - /etc/init.d: the directory that stores many of the startup scripts
    - These startup scripts are discussed in Chapter 11.
- User account files
  - /etc/passwd: the user account database, with fields storing the username, real name, home directory, log in shell, and other information. Although it is called the password file, passwords are no longer stored there because this file is readable

by anyone, and placing passwords there constituted a security risk; so the passwords are now stored in the shadow file.

- /etc/shadow: stores user passwords in an encrypted form.
- /etc/group: similar to /etc/passwd, but describes groups instead of users.
- /etc/sudoers: list of users and access rights who are granted some privileges beyond normal user access rights.
  - These files are discussed in Chapter 6.

- Network configuration files
  - /etc/resolv.conf: the listing of the local machine's DNS server (domain name system servers).
  - /etc/hosts: stores lists of common used machines' host names and their IP addresses so that a DNS search is not required.
  - /etc/hosts.allow, /etc/hosts.deny: stores lists of IP addresses of machines that are either allowed or disallowed log in access.
  - /etc/sysconfig/iptables-config: the Linux firewall configuration file, set rules here for what types of messages, ports, and IP addresses are permissible and impermissible.
  - /etc/xinetd: the Internet service configuration file, maps services to servers, for instance, mapping telnet to 23/tcp where telnet is a service and 23/tcp is the port number and server that handles telnet; this is a replacement for the less secure inetd configuration file.
    - Some of these files are discussed in more detail in Chapters 11 and 12.
- File system files
  - /etc/fstab: defines the file systems mounted at system initialization time (also invoked by the command mount –a).
  - /etc/mtab: list of currently mounted file systems, updated automatically by the mount and umount commands, and used by commands such as df.
  - /etc/mime.types: defines file types; it is the configuration file for the file and more commands so that these commands know how to treat the given file type.
- Message files
  - /etc/issue: contains a short description or welcoming message to the system, the contents are up to the system administrator.
  - /etc/motd: the message of the day, automatically output after a successful login, contents are up to the system administrator and is often used for getting information to every user, such as warnings about planned downtimes.

- User startup scripts
    - /etc/profile, /etc/bash.rc, /etc/csh.cshrc: files executed at login or startup time by the Bourne, BASH, or C shells. These allow the system administrator to set global defaults for all users. Users can also create individual copies of these in their home directory to personalize their environment.
        - Some of these files are discussed in more detail in Chapter 11.
- /etc/syslog.conf: the configuration file that dictates what events are logged and where they are logged to (we cover this in detail in Chapter 11).
- /etc/gdm: directory containing configuration and initialization files for the Gnome Display Manager (one of the Linux GUI systems).
- /etc/securetty: identifies secure terminals, i.e., the terminals from which root is allowed to log in, typically only virtual consoles are listed so that it is not possible to gain superuser privileges by breaking into a system over a modem or a network.
- /etc/shells: lists trusted shells, used by the chsh command, which allows users to change their login shell from the command line.

Although different Linux implementations will vary in specific ways such as startup scripts, types of GUIs, and the types and names of services, most Linux dialects have the above directories and files, or similar directories and files. See Figure 5.9, which shows

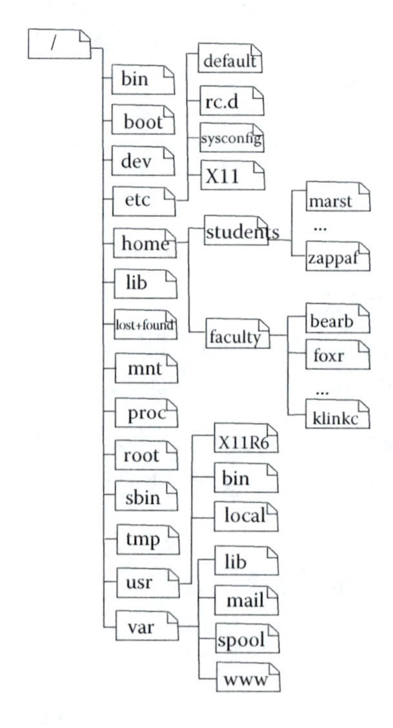

FIGURE 5.9 Typical Linux directory structure.

the typical arrangement of directories along with some of the initial subdirectories. The /home directory, of course, will differ based on the population of users. The commonality behind the Linux file system structure helps system administrators move from one version of Linux to another. However, with each new release of a version of Linux, changes are made and so the system administrator must keep up with these changes.

The Linux file system consists not only of the hard disk but also mountable (removable) devices. In order to access a device, it must be physically attached, but it must also be logically mounted through the mount command. Some mounting is done for you at the time you boot your system by having a mount file called /etc/fstab (file system table). To unmount something from the file system, use umount. To mount everything in the fstab file, use mountall. Note: you will not be able to edit fstab or execute mount, umount, or mountall unless you are root. Another file, /etc/mtab, stores a table of those partitions that are *currently* mounted.

Linux Partitions

In order to ensure that the various directories have sufficient room, system administrators will typically partition the hard disk into three or more areas, each storing a different section of the file system. This allows specific areas of the file space to be dedicated for different uses such as the kernel, the swap space, and the user directory space. Partitions provide a degree of data security in that damage to one partition may not impact other partitions (depending on the source of the damage). It is also possible in Linux to establish different access controls on different partitions. For instance, one partition might be set as read-only to ensure that no one could write to or delete files from that partition. One can establish disk quotas by partition as well so that some partitions will enforce quotas, whereas others do not. Partitions also allow for easy mounting and unmounting of additional file spaces as needed or desired.

The root partition (/) will store the Linux /boot directory that contains the boot programs, and enough space for root data (such as the root's e-mail). The largest partition will most likely be used for the /home directory, which contains all of the users' directories and subdirectories. Other directories such as /var and /usr may be given separate partitions, or they may be part of / or /home. The swap partition is used for virtual memory. It serves as an extension to main (DRAM) memory so that only currently and recently used portions of processes and data need to reside in DRAM. This partition should be big enough to store the executable code of all currently executing processes (including their data). It is common to establish a swap space that is twice the size of DRAM. Swap spaces can also be distributed across two (or more) disks. If the swap space was itself divided among two hard disk drives, it could provide additional efficiency as it would be possible to handle two page swappings at a time (refer back to page swapping in Chapter 4).

Partitioning the file space is accomplished at OS installation time, and therefore, once completed, the partitions and their sizes are fixed. This is unfortunate as you perform the installation well before you have users to fill up the user's partition, and therefore you must make an educated guess in terms of how large this partition should be. At a later point, if you want to alter the partition sizes you may have to reinstall the OS, which would mean that all user accounts and directories would be deleted. You could, if necessary, back up all OS files and user accounts/directories, reinstall the OS, change the partition size, and then restore the OS files and user accounts/directories. This would be time consuming (probably take several hours at a minimum) and should only be done if your original partitioning was inappropriate. Fortunately, Linux and Unix do offer another solution, the use of a dynamic partition resizer. However, using one is not something to do lightly. For instance, if a power outage or disk crash occurs during resizing, data may be permanently destroyed. On the other hand, it is much simpler than saving the file system, reinstalling the OS, and restoring the file system.

There are a number of Linux commands that deal with the file system. The df command (display file system) will show you all of the partitions and how full they are in both bytes and percentage of usage. As mentioned above, the mount and umount commands allow you to mount and unmount partitions from the file system.

The quotaon and quotaoff commands allow you to control whether disk quotas are actively monitored or not. The system administrator can establish disk quotas for users and groups. If disk quotas are on for a given file system, then the OS will ensure that those users/groups using that file system are limited to the established quotas.

The file command will attempt to guess at what type of file the given file is. For instance, it might return that a file is a text file, a directory, an image file, or a Open Office document. The find instruction performs a search of the file system for a file of the given name. Although find is a very useful Linux instruction, it is not discussed further because it is a rather complicated instruction.

The utility fsck (file system check) examines the file system for inconsistent files. This utility can find bad sectors, files that have been corrupted because they were still open when the system was last shut down, or files whose error correction information indicates that the file has a problem (such as a virus). The fsck utility not only finds bad files and blocks, but attempts to repair them as well. It will search for both logical and physical errors in the file system. It can run in a mode to find errors only, or to find and try to fix errors. The fsck program is usually run at boot time to ensure a safe and correct file system. If errors are found, the boot process is suspended, allowing fsck to find and attempt to fix any corruption found.

Linux Inodes

Inodes are the Unix/Linux term for a file system component. An inode is not a file, but is a data structure that stores information about a file. Upon installing Linux and partitioning the file system, the OS generates a number of inodes (approximately 1% of the total file system space is reserved for inodes). Notice that there is a preset number of inodes for a system, but it is unlikely that the system will ever run out of available inodes. An inode will store information about a single file; therefore, there is one inode for each file in the system. An inode for a created file will store that file's user and group ownership, permissions, type of file, and a pointer to where the file is physically stored, but the inode does not store the file's name, and is not the file itself. Instead, in the directory listing, a file's name has a pointer to the inode on disk and the inode itself points to the file on disk. If a file has a hard link, then both the file name and the hard link point to the inode. If a file has a soft link, the file name points to the inode whereas the soft link points to the file name.

In order to obtain an inode's number, the ls (list) command provides an option, –i. So, ls –i filename returns that file's inode number. You can obtain the file's information, via the inode, by using ls –l.

Files are not stored in one contiguous block on a disk, as explained earlier in this chapter, but instead are broken into blocks that are scattered around the file system. Every file is given an initial number of blocks to start. The inode, aside from storing information about the file, contains a number of pointers to point to these initial blocks. The first pointer points at the first disk block, the second pointer points at the second disk block where the first disk block stores the first part of the file, the second disk block stores the second part of the file, etc.

If a file needs additional blocks, later inode pointers can be created so that the inode pointer points to a block of pointers, each of which point to the additional disk blocks. Such a pointer block is called an *indirect block*. There can be any number of levels added, for instance, an inode pointer might point to an indirect block, which itself points to an indirect block that has pointers to actual disk blocks. Figure 5.10 illustrates this concept, showing the hierarchical structure of disk blocks, inodes, and pointers in Linux. The inode frees the Linux OS from having to maintain a FAT (as used in Windows).

Aside from storing information about files, inodes are used to store information on directories and symbolic links. A symbolic link is merely an alternate path to reach a file. You often set up symbolic links if you wish to be able to specify access to a given file from multiple starting points. To create a symbolic link, use ln –s filename1 filename2. Here, filename1 is the preexisting file and filename2 is the symbolic link. Typically, one of these two file names contains a path so that the file and symbolic link exist in two separate directories (it makes little sense to have a symbolic link in the same directory

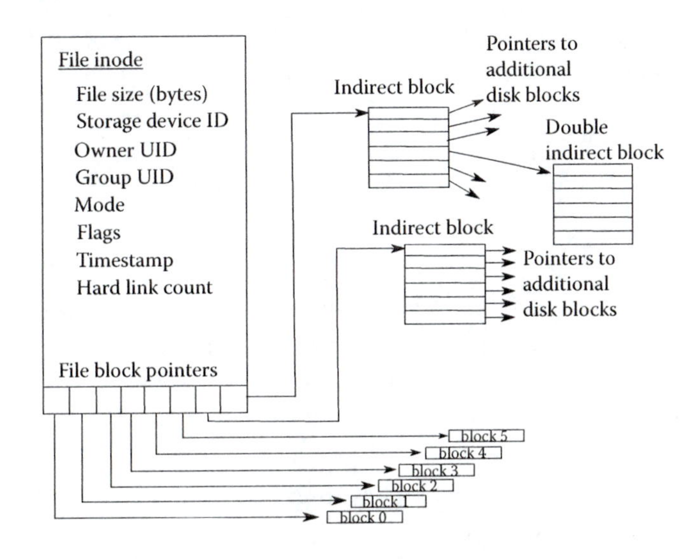

FIGURE 5.10 Hierarchical inode structure in Linux.

as the file you are linking to). For instance, if there is an important executable program in the directory /usr/share/progs called foo and you want to easily access this from your home directory, you might change to your home directory and then issue the command ln –s /usr/shar/progs/foo foo. This creates the symbolic link, foo, in your home directory that references the file /usr/shar/progs/foo.

One must be careful when dealing with symbolic links because if someone deletes or moves the original file, the link is no longer valid. If the file foo were deleted from /usr/shar/progs, then typing ~/foo would result in an error.

COMPUTER VIRUSES

Any discussion of the file system would be incomplete without discussing computer viruses and their dangers. A computer virus is a program that can replicate itself. Most commonly, a computer virus is also a malicious program in that its true purpose is to damage the computer on which it is stored, or spy on the users of that computer.

How do you get infected? A virus hides inside another executable program. When that program is executed, the virus code executes. The virus code typically comes in two parts, the first is to replicate itself and store its own executable code inside other programs or data files. The second part is the malicious part.

Often a virus waits for a certain condition to arise (older viruses might wait for a certain day such as Friday the 13th, or a time, or an action such as the user attempting to delete a given file) before the malicious portion executes. The malicious part of the virus might delete data or copy data and send it to a third party (this data might include information stored in cookies such as your browser history and/or credit card numbers).

Today, the word "virus" applies to many different types of malicious software: spyware, adware, Trojan horses, worms, rootkits. However, the term virus really should only apply to self-replicating software, which would not include spyware, adware, or Trojan horses. The group of software combined is now called malware.

How do you protect yourself? The best way is to avoid the Internet (including e-mail)! But this is impractical. Instead, having an up-to-date antiviral/antimalware program is essential. Disabling cookies can also help. Also, avoid opening attachments from people you do not know. It is also helpful to have some disaster recovery plan. One approach is to create backups of your file system—in the case of a bad infection, you can reinstall the OS and restore your files. If you are a Linux or Mac user, you might feel safe from malware. Most malware targets Windows machines because Windows is the most common OS platform, but in fact malware can affect any platform.

A hard link, created by the ln command without the –s parameter, is a pointer to the file's inode, just as the original name in the directory points to the file's inode, unlike the symbolic link that merely points to the file's name in the directory. Hard links can only be used to link to files within the same file system. A link across mounted file systems (partitions) must be a symbolic link. This limits hard link usefulness. Hard links are uncommon, but symbolic links are very common. If you do an ls –l, you will see symbolic links listed with a file type (the first letter in the listing) of 'l' to indicate soft link, and with the name filename -> true location. The filename is the name you provided to the link, whereas true location is the path and file name of the item being pointed to by the link.

WINDOWS FILE SYSTEM

The Windows file system layout is simpler than that of Linux, which makes it more complicated when you have to find something because you will find far more files in any one directory than you tend to find in Linux. The Windows layout has been fairly stable since Windows 95. The C: partition is roughly equivalent to Linux' / root directory. A: and B: are names reserved for floppy disks (which typically are not available in computers purchased in recent years) and D: is commonly assigned to the computer's optical drive. Other partitions can be added (or the main hard disk can be partitioned into several "drives", each with its own letter).

Underneath the root of the file system (C:), the file system is divided into at least three directories. The OS is stored under the Windows folder. This directory's subfolder System32 contains system libraries and shared files, similar to Linux' /usr/lib. Most of the application software is located under Program Files. In Windows 7, there is a separate directory C:\Program Files (x86) to separate 64-bit software (Program Files) from older 32-bit software (Program Files (x86)). In Linux, most of the application software is under /usr. User directories are commonly stored under C:\Users (whereas in Linux, these directories are underneath /home). There may be a C:\Temp directory that is similar to Linux' /tmp.

A large departure between Windows and Linux takes place with the user directories. Under C:\Users, default folders are set up for various types of files, Desktop, Downloads, Favorites, Links, My Documents, My Music, My Pictures, My Videos, Searches. In Linux, for the most part, the user decides where to place these items. In Windows, the specific folders are established by the software, unless overridden by the user. There are also soft links so that "My Documents" actually refers to C:\Users\youraccount\My Documents. By providing these default directories and soft links, it helps shelter the user from having to remember where files have been stored, or from having to understand the file space.

Many of the Windows operations dealing with the file system are available through GUI programs. Some of these are found through the Control Panel, others are available through OS utilities such as Norton Utilities and McAfee tools. The Control Panel includes tools for system restoration (which primarily restores system settings but can also restore file system components), backup and restore for the file system, and folder options to control what is displayed when you view folder contents. And, of course, the default file manager program is Windows Explorer. There are also DOS commands available. The most significant DOS command is chkdsk, a utility that serves a similar purpose as fsck does in Linux. Other commands include defrag (disk defragmentation utility), diskpart that allows you to repartition a disk or perform related administrative services on the disk (such as assigning it a drive letter or attributes), and find, similar to the Linux find program.

When considering the Windows file system, one might feel that it is intuitively easy to use. On the other hand, the Windows approach might feel overly restrictive. For instance, if a user wishes to create a CIT 130 folder to collect the various data files related to that course, a Linux user might create /home/foxr/cit130. Underneath this directory, the user might place all of the files in a flat space, or might create subdirectories, for instance, Pictures, Documents, Searches, and Videos. In Windows, the user would have to create cit130 folders underneath each of the established folders in order to organize all material that relates to CIT 130, but the material is distributed across numerous folders. Only if the user is wise enough to override the default would the user be able to establish a cit130 folder under which all items could be placed.

MOVING AROUND THE FILE SYSTEM

Here, we look at specific Linux and DOS commands for moving around their respective file systems. In the following examples, we see the OS commands, the results of those commands and some comments describing what the operations are doing. We will focus first on Linux, which is slightly more complicated. Mastering Linux will make it easier to master DOS. Assume that we have the structure of directories, subdirectories, and files in Figure 5.11. Items in boxes are files, and all other items are directories. For DOS, replace home with Users.

Linux

What follows is a Linux shell session (commands entered in a Linux shell and the results), along with explanations of the commands. Comments are given beneath some of the commands. Linux commands appear after $ symbols, which we will assume is the Linux prompt.

Command/Result	Explanation
`$ pwd`	print working directory
`/home/foxr`	
`$ cd CIT130`	change to the CIT130 subdirectory
`$ ls`	list the contents
`HW LABS`	

There are two items in this directory; both are printed

`$ ls -l`	perform a long listing of the contents
`drwxr-xr-x 2 foxr foxr 1024 Jan 20 03:41 HW`	
`drwxr-xr-x 4 foxr foxr 4096 Jan 21 17:22 LABS`	

The long listing provides more details including the creation date and time (in military time, thus 17:22 instead of 5:22 P.M.). For now, we concentrate only on the first item, which looks like a random sequence of characters. The initial ‘d’ describes that the item is a directory. The remaining letters and hyphens provide the access rights (permissions). We will cover that topic in Chapter 6.

`$ cd HW`	change to the HW directory
`$ cp h1.txt /home/foxr/CIT130/LABS`	copy the h1.txt file to the LABS directory

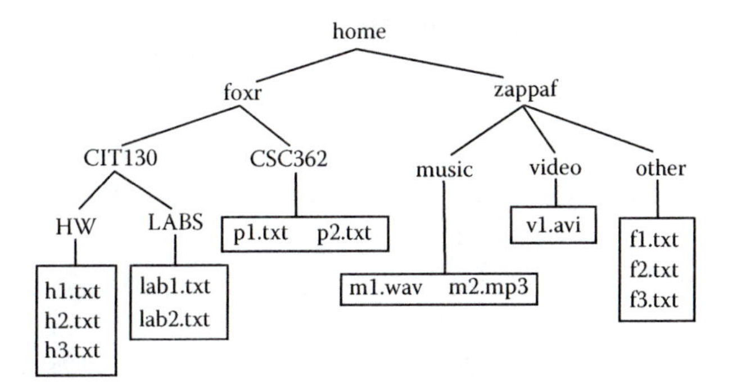

FIGURE 5.11 Layout for examples in this section.

This instruction uses an absolute path, that is, to specify the directory LABS, we start at the root level of Linux (the first /).

`$ cd ..`	move up one level (/home/foxr/CIT130)
`$ mv LABS/lab1.txt .`	move lab1.txt to here

The . means "this directory". Notice that the specification of the LABS subdirectory is made as a relative path, that is starting at this point.

`$ ls ../../CSC362`	list the contents of the CSC362 directory
`p1.txt p2.txt`	

The ../../CSC362 is a relative path starting at this point, but moving up two levels (../..) and down one level (CSC362).

`$ cd ~/CSC362`	change to your CSC362 directory
`$ rm *.*`	delete everything here

The ~ symbol represents the current user's home directory. So in this case, ~ means /home/foxr. The * as used in rm is a wildcard symbol. The wildcard * means "anything". Using *.* means "anything that contains a period". This command then will delete all files that have a period in their name (files with extensions, as in p1.txt, a file such as foo would not be selected because it lacks a period).

The rm command is set up to execute rm –i, which means "interactive mode". This will list each item and ask you if you are sure you want to delete it, one at a time, for example:

rm: remove regular file 'p1.txt'? y

rm: remove regular file 'p2.txt'? y

rm: remove regular file 'lab1.txt'? n

The responses provided by the user, in this case, will cause rm to delete p1.txt and p2.txt but not lab1.txt (which was moved here from the CIT130/LABS directory).

`$ cd ..`	move up to foxr
`$ rmdir CSC362`	delete the entire CSC362 directory
`rmdir: CSC362: Directory not empty`	

Because the directory is not empty, it cannot be deleted. We could delete the directory if we first deleted the remaining item, lab1.txt.

Alternatively, we could issue the command rm –r CSC362. The –r option means "recursive delete". A recursive delete deletes all of the contents of the directory before deleting the directory, recursively (so that if the directory had a subdirectory, the contents of the subdirectory are deleted first, followed by the contents of the directory including the subdirectory itself). So, for instance, if we issued rm –r ~foxr/CIT130, the pattern of deletions would operate as follows: h1.txt, h2.txt, h3.txt, HW (the directory), h1.txt, lab2.txt (recall that we previously copied h1.txt from the HW directory to here and that we moved the lab1.txt file to another directory), LABS (the directory), and finally the CIT130 directory.

The –r option can be very powerful but also very dangerous. It is best to never use rm –r unless you do rm –ir. The –i portion of the option forces rm to seek confirmation before deleting anything. It is especially important never to use rm –r as system administrator unless you are absolutely sure of what you are doing. Imagine that you did cd / (move to the top of the file system) followed by rm –r *.*. You would delete everything (including the rm program itself)!

Note: cp also has a recursive mode, cp –r, to recursively copy any subdirectories of a directory but mv does not have a recursive mode.

`$ cd /`	move to the top of the file system
`$ cd home`	move down to the home directory
`$ ls`	list the contents
`foxr zappaf`	
`$ cd zappaf`	move to zappaf's directory
`$ cp music/*.* ~`	copy files in music to your directory

Again, the *.* means "everything with a period". So, this command copies all files in zappaf's music directory that have a period in their name to ~, which denotes your home directory.

`$ mv videos/*.* ~`	move everything
`mv: cannot move '/home/zappaf/videos/v1.avi' to './v1.avi': Permission denied`	

Although you might have permission to access zappaf's directory and the contents, you do not have access to move because move in essence deletes files from that user's directory. We will explore permissions in Chapter 6.

`$ cd ~`	return to your home directory
`$ mkdir STUFF`	create a directory called STUFF
`$ ln -s videos/v1.avi STUFF/video`	create a symbolic link

The symbolic link is stored in the subdirectory STUFF (/home/foxr/STUFF), and the link is called video. The link points to or links to /home/zappaf/videos/v1.avi.

`$ ls STUFF`	list contents of the STUFF directory
`video`	

The ls command merely lists items by name, without providing any detail. If we want greater detail, we must use the –l option of ls (for "long listing").

```
$ ls –l STUFF
lrwxrwxrwx 1 foxr foxr  16 Jan 21 17:13   video -> /home/zappaf/videos/v1.avi
```

With the long listing, we see that video is a link to the location /home/zappaf/videos/v1.avi. The "l" character at the beginning of the line is the file type (link) and the -> on the right indicates where the link is pointing to.

DOS

We limit our look at the DOS commands because they are similar and because there are fewer options. Here we step through only a portion of the previous example.

Command/Result	Explanation
`C:`	switch to C: partition/drive
`C:\> cd Users\foxr`	move to foxr directory
`C:\Users\foxr> dir`	list the contents

Notice that when you switch directories, your prompt changes. There is no equivalent of the pwd Linux command to list your current directory. The dir command provides the contents of the directory. This is excerpted below. Some details are omitted.

```
Wed 11/23/2011  10:53 AM <DIR>     .
Wed 11/23/2011  10:53 AM <DIR>     ..
Wed 11/23/2011  10:55 AM <DIR>     Desktop
```

Wed 11/23/2011 10:55 AM <DIR> Documents

Wed 11/23/2011 10:55 AM <DIR> Downloads

...

0 File(s)

15 Dir(s)

`C:\Users\foxr> cd Documents`	change to Documents directory
`C:\Users\foxr\Documents> copy *.* \Users\zappaf`	copy everything to zappaf's directory
`C:\Users\foxr\Documents> cd ..`	move up 1 level
`C:\Users\foxr> cd Downloads`	move down to Downloads
`C:\Users\foxr\Downloads> move *.* \Users\zappaf`	move instead of copy
`C:\Users\foxr> cd \Users\zappaf`	switch to zappaf's directory

Notice that there is no equivalent of ~ in DOS, so you have to specify either a relative or absolute path to get to zappaf. From \Users\foxr\Downloads to zappaf, you could use this relative path: cd ..\..\zappaf.

`C:\Users\zappaf> dir`	list zappaf's directory contents

This should now include all of the files found under foxr's Documents and Downloads. The listing is omitted here.

C:\Users\zappaf> del *.*

C:\Users\zappaf>

Notice that the deletion took place. You were not prevented from deleting zappaf's files. This is because proper permissions to protect zappaf's directory and files was not set up!

We conclude this example with these comments. As in Linux, to create and delete directories, you have the commands mkdir and rmdir. You can also create hard and soft links using the command mklink. As you may have noticed, the * wildcard symbol also works in DOS. DOS commands also permit options. The del command can be forced to prompt the user by using del /P and to delete without prompting using /Q (quiet mode). Unlike Linux, del does not have a recursive mode.

FILE SYSTEM AND SYSTEM ADMINISTRATION TASKS

Although it is up to individual users to maintain their own file system space, there are numerous tasks that a system administrator might manage. The first and most important task of course is to make sure that the file system is available. This will require mounting the system as needed. The Linux commands mount and umount are used to mount a new partition to the file system and unmount it. The commands require that you specify both the file system and its mount point. A mount point is the name by which users will reference it. For instance, if you do a df –k, you will see the various mount points including /var and /home. The command mount –a will mount all of the file system components as listed in the special file /etc/fstab (file system table). This command is performed automatically at system initialization time, but if you have to unmount some partitions, you can issue this command (as root) to remount everything. The /etc/fstab file includes additional information that might be useful, such as disk quotas, and whether a given file space is read-only or readable and writable. All currently mounted partitions are also indicated in the file /etc/mtab.

Setting up disk quotas is a complex operation but might be useful in organizations where there are a lot of users and limited disk space. You might establish a disk usage policy (perhaps with management) and implement it by setting quotas on the users.

Another usage policy might include whether files should be encrypted or not. Linux has a number of open source encryption tools, and there are a number of public domain tools available for Windows as well.

Remote storage means that a portion of the file system is not local to this computer, but accessible over a network instead. In order to support remote storage, you must expand your file system to be a network file system. NFS is perhaps the most common form of this, and it is available in Unix/Linux systems. NFS is very complex. You can find details on NFS in some of the texts listed in Further Reading.

These last topics, although useful, will not be covered in this text but should be covered in a Linux system administration text.

FURTHER READING

Many of the same texts referenced in Further Reading in Chapter 4 cover material specific to Linux, Unix, Windows, and Mac OS file systems. Additionally, the OS concepts books from the same section all have chapters on file systems. More detailed material on Windows, Unix, and Linux file systems can be found in these texts.

- Bar, M. *Linux File Systems.* New York: McGraw Hill, 2001.
- Callaghan, B. *NFS Illustrated.* Reading, MA: Addison Wesley, 2000.
- Custer, H. *Inside the Windows NT File System.* Redmond, WA: Microsoft Press, 1994.
- Kouti, S. *Inside Active Directory: A System Administrator's Guide.* Boston: Addison Wesley, 2004.

- Leach, R. *Advanced Topics in UNIX: Processes, Files and Systems.* Somerset, NJ: Wiley and Sons, 1994.
- Moshe, B. *Linux File Systems.* New York: McGraw Hill, 2001.
- Nagar, R. *Windows NT File System Internals.* Amherst, NH: OSR Press, 2006.
- Pate, S. *UNIX Filesystems: Evolution, Design and Implementation.* New Jersey: Wiley and Sons, 1988.
- Von Hagen, W. *Linux Filesystems.* Indianapolis, IN: Sams, 2002.

On the other hand, these texts describe file systems from a design and troubleshooting perspective, which might be more suitable for programmers, particularly OS engineers.

- Carrier, B. *File System Forensic Analysis.* Reading, MA: Addison Wesley, 2005.
- Giampaola, D. *Practical File System Design.* San Francisco: Morgan Kaufmann, 1998.
- Harbron, T. *File Systems: Structure and Algorithms.* Upper Saddle River, NJ: Prentice Hall, 1988.
- Kerrisk, M. *The Linux Programming Interface: A Linux and UNIX System Programming Handbook.* San Francisco: No Starch Press, 2010.
- Pate, S. and Pate, S. *UNIX Filesystems: Evolution, Design and Implementation.* Hoboken, NJ: Wiley and Sons, 2003.
- Tharp, A. *File Organization and Processing.* New York: Wiley and Sons, 1998.

REVIEW TERMS

Terminology introduced in this chapter:

Absolute path
Alias
Archive
Block
Cylinder
Defragmentation
Directory
FAT
File
File system
Folder
Fragment
Hard link
Indirect block
inode
Lossless compression
Lossy compression
Mounting

Mount point
NFS
Partition
Platter
Pointer
Read/write head
Recursive delete
Relative path
Rotational delay
Sector
Seek time
Soft link
Spindle
Surface
Track
Transfer time

REVIEW QUESTIONS

1. How does a file differ from a directory?
2. What makes the file space hierarchical?
3. Why are disk files broken up into blocks and scattered across the disk surface(s)?
4. Imagine that a file block is located at disk block 581132. What does this information tell you?
5. What is a pointer and what is a linked list?
6. What is a FAT? Where is the FAT stored?
7. Why is the FAT loaded into memory?
8. How much slower is disk access than the CPU?
9. The Linux operating system does not maintain a FAT, so how are disk blocks accessed?
10. What is an indirect block inode?
11. What would happen if your Linux file system were to run out of inodes?
12. How does the Linux file system differ from the Windows file system?
13. In the Linux file system, where do you find most of the system configuration files?
14. In the Linux file system, where do you find the system administration commands (programs)? The general Linux commands? The user (applications) programs?
15. Which of the Linux items listed under /dev are not physical devices?
16. If a disk partition is not mounted, what are the consequences to the user?
17. What do each of these Linux commands do? cd, cp, ln, ls, mkdir, pwd, rm, rmdir

18. What do each of these DOS commands do? cd, copy, del, dir, mklink, move
19. What is the difference between ls and ls –l in Linux?
20. What is the difference between rm –i, rm –f, and rm –r in Linux?
21. Why does DOS not need a pwd command?
22. What must be true about a directory in Linux before rmdir can be applied?
23. If you set up a symbolic link to a file in another user's directory, what happens if the user deletes that file without you knowing about it? What happens if you delete your symbolic link?
24. What does ~ mean in Linux? Is there a similar symbol in DOS?
25. What does .. mean in both Linux and DOS?
26. What is the difference between an absolute path and a relative path?
27. Refer back to Figure 5.11. Write an absolute path to the file p1.txt.
28. Refer back to Figure 5.11. Write an absolute path to the file m2.mp3.
29. Refer back to Figure 5.11. Assuming that you are in the subdirectory LABS, write a relative path to the file v1.avi. Write an absolute path to the file v1.avi.
30. Refer back to Figure 5.11. Assuming that you are in the subdirectory CSC362, write a Linux copy to copy the file p1.txt to the directory HW.
31. Refer back to Figure 5.11. Assuming that you are in the subdirectory CSC362, write a Linux move command to move the file p1.txt to the directory other (under zappaf).

DISCUSSION QUESTIONS

1. In the Windows operating system, the structure of the file system is often hidden from the user by using default storage locations such as My Documents and Desktop. Is this a good or bad approach for an end user who is not very knowledgeable about computers? What about a computer user who is very familiar with the idea of a file system?
2. Do you find the finer breakdown of the Linux file system's directory structure to be easier or harder to work with than Windows? Why?
3. What are some advantages to separating the file system into partitions? Do you find the Windows approach or the Linux approach to partitions to be more understandable?
4. Now that you have interacted with a file system through both command line and GUI, which do you find more appealing? Are there situations where you find one approach to be easier than the other? To be more expressive?

5. The discussion of the FAT (in Windows) and inodes (in Linux) no doubt can be confusing. Is it important for an IT person to have such a detailed understanding of a file system? Explain.

6. A hard disk failure can be catastrophic because it is the primary storage media that contains all of our files and work. Yet the hard disk is probably the one item most apt to fail in the computer because of its moving parts. Research different ways to backup or otherwise safeguard your hard disk storage. Describe some of the approaches that you found and rank them in terms of cheapest to most expensive cost.

7. Within your lifetime, we have seen a shift in technology from using magnetic tape to floppy disk as storage to hard disk, and in between, the rise and fall of optical disks. Although we still use optical disks, they are not nearly as common today for data storage as they were 10 years ago. Research the shift that has taken place and attempt to put into perspective the enormous storage capacity now available in our computers and how that has changed the way we use computers and store files.

CHAPTER 6

Users, Groups, and Permissions

In Chapter 4, the idea of protection and security was introduced. In order to support these operating system requirements, computer accounts are created. Each account comes with certain access rights. These rights can be altered in operating systems such as Windows and Linux such that the owner of a resource can dictate who can use that resource. This chapter examines user accounts, account creation, groups and group creation, passwords, and permissions.

The learning objectives of this chapter are to

- Describe the role of the user account and the group.
- Introduce the mechanisms to create and maintain user and group accounts in Windows and Linux.
- Explain how to change file and directory permissions in both Windows and Linux.
- Discuss password management including the concept of the strong password.
- Introduce the Linux sudo command.

USERS

Recall that a computer system is more than a computer; it is a collection of devices, software, and users. If the computer system in question is to be used by more than one user, then the computer system has to offer some degree of protection. Protection ensures that the resources granted to one user do not interfere with other users or their resources. Resources include memory storing user processes, running applications, shared data, and files. Different operating systems offer different approaches to protection. Here, we will look at the most common protection mechanism—the user account. Along with establishing user accounts, we examine the use of permissions so that users can control access to their resources.

Aside from protection, user accounts are useful for a number of other purposes. They provide security in that only authorized users are allowed to use the computer and its available resources. They provide a means by which we can gather statistics on computing usage. For instance, as a system administrator, you can inspect how the number of users is impacting CPU performance, whether there is adequate main memory, and to what extent the users are using the file space. This can, in turn, provide support for asking management to spend money on further resources. Tracking computer usage supports both performance monitoring and accounting.

A *user* is an agent who uses a computer. Most commonly, we will consider humans to be users. However, in some situations, a user might be a software agent. In most operating systems, a user requires an account to use the computer. A user account is a set of data, directories, and privileges granted to a user, as set up by a system administrator.

Typically, a user account will be coupled with a user name and a user password. These are used to *authenticate* the user so that the user can be granted access by the operating system. Authentication is most often accomplished using a login whereby the user enters both the user name and password, which are then compared in a database of stored username/passwords. Some more advanced systems may couple this with biometric or keycard access whereby the user must swipe a card through a card reader or use some biometric such as voice print or fingerprint. These are far less common. In some cases, the biometric is used in place of a password.

One key concern with proper protection of a computer system is to ensure that users use *strong passwords*. It is a fairly simple matter for a hacker to devise a program that will test all dictionary words as passwords in an attempt to break into a user account. Therefore, strong passwords are those that do not appear in a dictionary. Requirements for a strong password are typically that they contain at least one non-alphabetic character or a combination of upper- and lower-case letters and are at least eight characters in length. Additionally, strong passwords might require that the user change the password frequently (every 1 to 3 months) without repeating previous passwords. Other restrictions might include that a password not repeat previous passwords' characters. For instance, if your password was abcdef12, then changing the password to abcdef13 might not be permissible because it is too similar to the previous password.

Multiuser systems require user accounts so that the operating system knows who the current user is and can match up the user with proper permissions. A single-user system does not require any user accounts because it is assumed that the same user will always use that system. Today, although most people have their own computers, it is still common to create user accounts for a few reasons. First, most operating systems permit remote access and therefore to protect against other agents from attempting to utilize the single user's resources, establishing a login is essential. Second, all systems need to be administered by a system administrator. Even if there is a single user who uses the system, the need to differentiate between the user acting as an end user and the user acting as system administrator is critical. If the user wants to, for instance, delete a bunch of files and accidentally includes system files in the list, the operating system would deny the request if the current user is the normal user. But if there was no user account, only a system administrator account, then the operating system would not question the deletion leading to perhaps a catastrophic mistake by the user.

There are at least two types of user accounts. The typical user is an end-user, whose account gives them access to their own file space and permission to use shared resources, including application software. Their accounts usually do not give them access to system resources, particularly system administration programs. The other type of user is then the system administrator, known as root in Linux and Unix. Root has access to everything—every program, every file, every resource, no matter who the owner of that resource is. Some operating systems provide intermediate level accounts where a user is given more resources than the minimum while not being at the level of a system administrator. In Linux and Unix, root can permit other users to have some additional access rights. This is done through the program sudo, covered in Miscellaneous User Account Topics. As will be discussed in Setting Up User Accounts, software is also given user accounts. In the case of software accounts, access is actually more restricted than typical end-user accounts, for instance, by not having a home directory and/or not having a login shell.

SETTING UP USER ACCOUNTS

Here, we look at how to create user accounts in both Linux and Windows. Note that only a system administrator will be able to set up user accounts.

Linux

An initial user account is created when you install the Linux OS. This account is a necessity so that the first user of the system will not have to log into the graphical user interface (GUI) as root. Doing so can be dangerous as any window that the user were to open would have root privileges rather than ordinary user privileges. Therefore, by the time you are ready to use Linux, there will be one user account already created. From there, however, you would have to create others.

In Linux, there are two ways to set up user accounts: through the GUI and through the command line. Both approaches require that the person setting up the accounts be the system administrator. The GUI is shown in Figure 6.1.

The User Manager is fairly simple to use. You can add or delete users, and alter the properties of users. You can also add, delete, and alter groups (groups are discussed in Role of a Group). Adding a user requires that you specify, at a minimum, the user's user account name, the user's name (a string which will typically be the first and last names), and an account number. The account number is used to index the user in the user file, /etc/passwd. This account number will typically be greater than 99 as the first 99 numbers are reserved for software accounts. The default account number is 1 greater than the last user added to the system. However, a system administrator may want to have a numbering scheme whereby numbers do not follow sequentially. For instance, in a university, faculty accounts may be assigned numbers between 100 and 200, graduate students numbers between 201

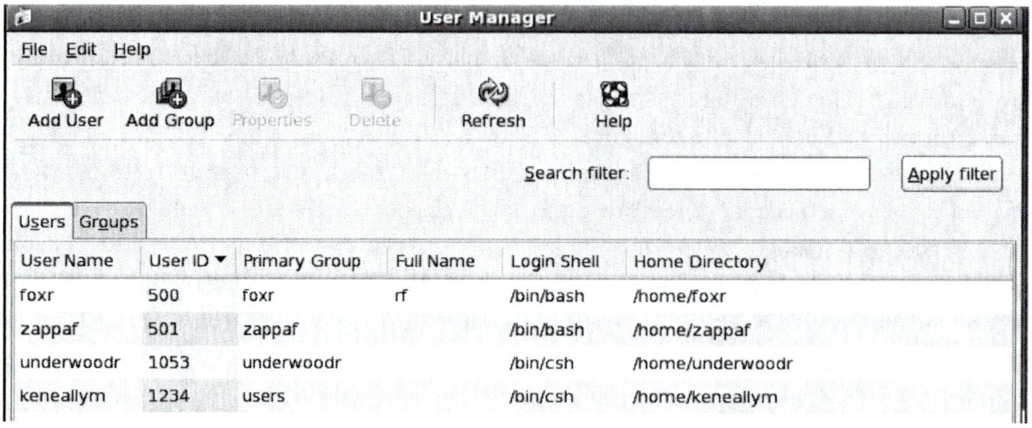

FIGURE 6.1 Linux user manager.

and 500, and undergraduate students numbers greater than 500. In such a case, the next user you add may not be in the same category as the previously added user and therefore you would not want to just add 1 to the last user's number to obtain a new number.

In Figure 6.1, you can see that four users have already been created. For instance, the first user has an account name of foxr, a user number (UID) of 500, a login shell of /bin/bash, and a home directory of /home/foxr. As a system administrator, you would create account policies such as the form of account names (a common policy is last name—first initial or last name—first initial and a number if that particular combination has been used, so that for instance Robert Fox might be foxr1). Other policies would include the default locations of directories, default shells, and whether and to what extent to specify the person's full name. Again, in a university setting, faculty directories may be placed in one location and student directories in another.

Figure 6.2 shows the Add User window. Most of the entries are obvious. Here, the new user is Tommy Mars and is being given the username marst. Notice that you must enter an initial user password and confirm it. This enforces proper security so that accounts are not created without passwords. However, this could be a hassle for a system administrator who has to create dozens or hundreds of accounts. As we will see later, the command-line approach to creating accounts is preferable in such cases. The login shell and home directory default to /bin/bash and /home/*username*, respectively. A system administrator can alter what appears as defaults. Additionally, upon entering a new user, you can change these as needed. Login shells available also include csh, ksh, sh, and tsch (these are covered in Chapter 9), and you can also select nologin, which means that the user does not have a login shell upon logging in. This is often specified when the user is a piece of software. You can also override the default UID and specify a different number as desired. For instance, if you use one set of numbers for faculty and another for students, you might override the default value for a new student account.

One item in Figure 6.2 requires explanation. Notice the second checkbox, "Create a private group for the user." When a user account is created, a group that shares the user's

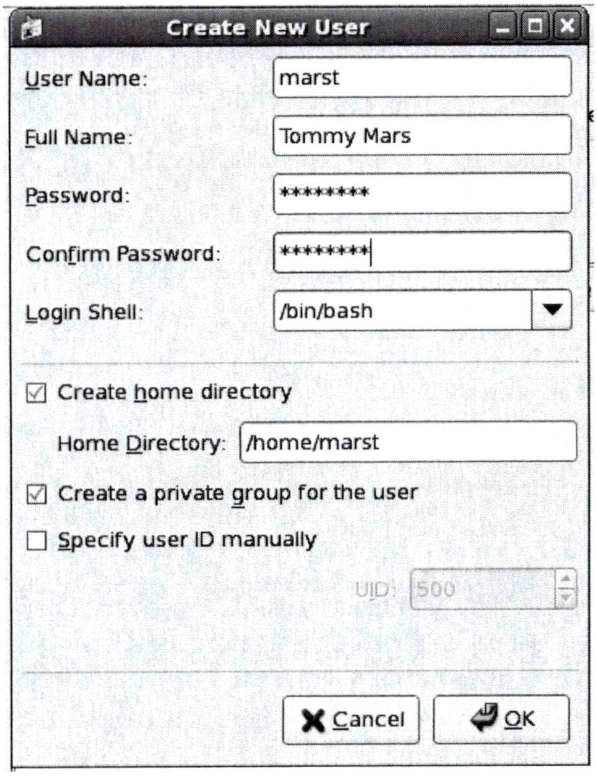

FIGURE 6.2 Create new user window.

account name is also created. For instance, user foxr will also have a group foxr. The private group will be a group that contains exactly one user. If you turn that checkbox off, the user account is created but a private group is not created. We discuss groups in Role of a Group.

Deleting or changing a user requires selecting the user from the list of users in the User Manager window. Once selected, you can either select Properties to change user properties or Delete User to delete the user. You may change the user's user name, full name, password, directory, or login shell. You can also specify an expiration date for the account, lock the password (so that the user is not allowed to change it), specify account expiration information (i.e., establish a date by which the user must change their password), and adjust the groups that the user is in. We will talk about password expirations at the end of this section. If you select Delete, to delete a user, you are also asked whether the user's home directory, mail file, and temporary files should also be deleted or not. One might want to save these directories and files for archival and/or security purposes.

Although the GUI User Manager is easy to use, many system administrators will choose to use the command line useradd instruction instead. The reason is that a shell script can be written to create a number of new user accounts automatically by using a file that lists the new users' names. In addition, the useradd instruction does not require the specification of an initial password, thus shortening the process. Of course, without an initial password, the user account can easily be broken into and therefore some initial password should be created. However, a separate shell script could take care of that. The password instruction is discussed later.

The useradd instruction is easy to use if you want to specify just a few properties, as with the GUI. The basic form of useradd is:

```
useradd [-u uid [-o]] [-g group] [-G group, …] [-d home]
[-s shell] [-m] username
```

The notation here requires a little explanation. The items listed inside of [ ] are optional. Therefore, the only required parameter for useradd is the username. The simplest form of this command would appear like this:

```
useradd marst
```

If you do not specify any other parameter aside from username, the UID defaults to being 1 greater than the last user, the directory defaults to /home/*username*, and the shell defaults to the default shell (typically bash). The –m parameter forces Linux to create a home directory. The reason why this is optional is that some user accounts should not have home directories, particularly those created for software. Software accounts will be explained in more detail later.

To specify a non-default ID, use –u followed by the ID number. The –o parameter is used to permit non-unique ID numbers. There is little reason to permit duplicate IDs, and this could become a problem. However, the option is still available and might be used, for instance, if you want users to share an account without sharing the same login/password, or if you want to give a user multiple account names. Such a command might look like this:

```
useradd -u 500 -o foxr2
```

which would create a second account that would use the same UID as foxr (from Figure 6.1).

To override the default private group, –g allows you to specify a different private group whereas –G allows you to insert this user into a number of other groups. You might notice in Figure 6.1 that keneallym has been inserted into the default group users as he was not provided his own private group. The parameters –d and –s are respectively used to specify home directories and shells other than the default. Two other parameters of note are –p, which allows you to specify an initial password, and –c, which allows you to specify the name field. The –c parameter is used to insert a "comment" into the user account file, /etc/passwd, but the comment is primarily used to specify the user's name.

Figure 6.3 demonstrates the useradd command, using several of the parameters. In this case, we are adding the user Ruth Underwood. Her login shell is csh (c-shell) rather than bash. Her home directory is the default directory. We are giving her a user ID

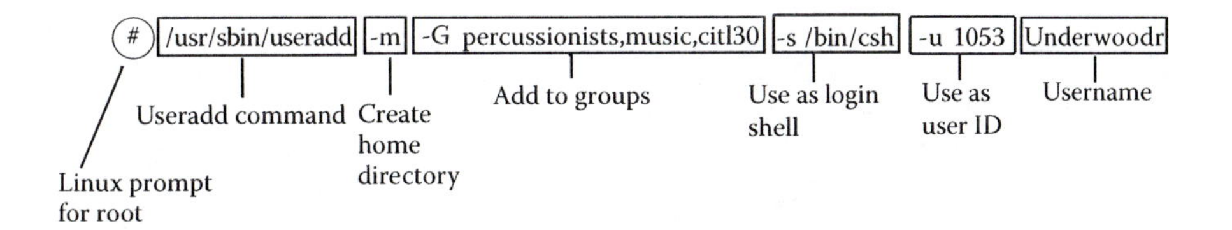

FIGURE 6.3 Linux useradd command.

number outside of the ordinary sequence, and assigning her to several groups. The # symbol is not part of the command but instead the prompt for Linux' system administrator (root). Note that we specify the full path to the instruction because /usr/sbin is probably not in root's path. Also notice that there is no space placed between the groups (percussionists, music, cit130).

As you add a user, the user's account information is added to the file /etc/passwd. The information in this file includes for each account, the account name, ID number, group ID number, name (comment field), home directory, and login shell. The group ID number is the ID number of the user's private account. For actual users (humans), ID and private group numbers are usually the same.

Interestingly, missing from the /etc/passwd file is the user's password. In the past, this file included all user passwords, stored in an encrypted format. But the /etc/passwd file can be read by anyone. This created a security risk. Even though passwords were encrypted, one could see the length of a password. For instance, if my password were eight characters long and I noticed that someone else's encrypted password was the same length as my encrypted password, I might infer that their password was also eight characters in length, and that could help me crack their password. Now, the /etc/passwd file stores the character 'x' for passwords, and all passwords are stored in the file /etc/shadow. The shadow file is accessible only by root.

Figure 6.4 shows portions of the /etc/passwd file. This listing shows three segments of the password file. The accounts listed in the figure are of operating system accounts, other software accounts, and user accounts. Operating system accounts include root (naturally, the most important account), bin, daemon, adm, and other Linux software accounts such as mail, news, uccp, ftp. Application software accounts include apache, for instance. Each account in the passwd file is on a separate line, and within a line, the individual pieces of information for each user are separated by colons.

Most of the software accounts specify /sbin/nologin to prevent the software from having a login shell (this is done for security purposes), and have directories other than under /home. For instance, root's directory is /root; lp, mail, and apache have directories under /var; and news has a directory under /etc. The /var directory is largely used for software storage and log files. Also notice that all of the software accounts have UIDs less than 100 (below 42 in this case), whereas the human users account numbers start at 500. In every case, the password is merely x, indicating that the password now resides in the /etc/shadow

```
root:x:0:0:root:/root:/bin/bash
bin:x:1:1:bin:/bin:/sbin/nologin
daemon:x:2:2:daemon:/sbin:/sbin/nologin
adm:x:3:4:adm:/var/adm:/sbin/nologin
lp:x:4:7:lp:/var/spool/lpd:/sbin/nologin
sync:x:5:0:sync:/sbin:/bin/sync
shutdown:x:6:0:shutdown:/sbin:/sbin/shutdown
halt:x:7:0:halt:/sbin:/sbin/halt
mail:x:8:12:mail:/var/spool/mail:/sbin/nologin
news:x:9:13:news:/etc/news:
uucp:x:10:14:uucp:/var/spool/uucp:/sbin/nologin
operator:x:11:0:operator:/root:/sbin/nologin
games:x:12:100:games:/usr/games:/sbin/nologin
gopher:x:13:30:gopher:/var/gopher:/sbin/nologin
ftp:x:14:50:FTP User:/var/ftp:/sbin/nologin
nobody:x:99:99:Nobody:/:/sbin/nologin
nscd:x:28:28:NSCD Daemon:/:/sbin/nologin
apache:x:48:48:Apache:/var/www:/sbin/nologin
mailnull:x:47:47::/var/spool/mqueue:/sbin/nologin
smmsp:x:51:51::/var/spool/mqueue:/sbin/nologin
hsqldb:x:96:96::/var/lib/hsqldb:/sbin/nologin
sshd:x:74:74:Privilege-separated_SSH:/var/empty/sshd:/sbin/nologin
foxr:x:500:500:rf:/home/foxr:/bin/bash
zappaf:x:501:501::/home/zappaf:/bin/bash
underwoodr:x:1053:1053::/home/underwoodr:/bin/csh
keneallym:x:1234:100::/home/keneallym:/bin/csh
marst:x:1235:1235:Tommy_Mars:/home/marst:/bin/bash
```

FIGURE 6.4 Excerpts of the /etc/passwd file.

file. All entries also have a comment field. In some cases, the comment field merely lists the user name (root, bin, daemon) and in other cases a true name (FTP User for FTP, NSCD Daemon for nscd, X Font Server for xfs, and rf for foxr). Users zappaf, underwoodr, and keneallym have an empty comment field, which appears as ::, that is, nothing occurs between the two colon symbols.

Entries in the /etc/shadow file are somewhat cryptic because they are encrypted. For instance, two of the users in Figure 6.4 (foxr and zappaf) might have entries as follows:

```
foxr:$1$KJlIWAtJ$aE1YjCp9i5j924vlUXx.V.:15342:0:38:31:::
zappaf:$1$G2nb6Kml$2pWqFx8EHC5LNypVC8KFf0:15366:0:14:7:::
```

The listings are of the user name followed by the encrypted password. This is followed by a sequence of information about password expiration. In the above example, there are four numbers, but there could be as many as six. These numbers are in order:

- The number of days since January 1, 1970 that the password was last changed
- The minimum number of days required between password changes (this might be 7 if, for instance, we require that passwords not be changed more often than once per week)

- The maximum number of days that a password is valid (99999 is the default, here foxr's password is valid for 38 more days whereas zappaf's is only valid for 14 more days)
- The number of days until a warning is issued to alert the user that the password needs to be changed
- The number of days after password expiration that an account will become disabled
- The number of days since January 1, 1970 that the account has been disabled

These last two numbers do not appear in the example above because they had not been established for these two users.

What we see here is that foxr changed his password at some point in the recent past and zappaf did not, so zappaf's password expires sooner than foxr's. In 7 days, zappaf will be warned to reset the password. You can see through simple subtraction (14 – 7 for zappaf and 38 – 31 for foxr) that the system warns a user 7 days in advance of password expiration. If the user does not reset the password, the account becomes disabled.

In Linux, the passwd instruction is used to change a passwd. For a typical user, the command is just passwd. The user is then prompted for their current password to ensure that the user is allowed to change the password (imagine if you walked away from your computer and someone came up and tried to change your password without your knowledge!), followed by the new password, entered twice to confirm the new password. The passwd program will inform you of whether a password is a bad password because it is too short or because the characters of the password match or come close to matching a dictionary word.

As a system administrator, you can change any user's passwd with the command passwd username, for instance, passwd foxr. As root, you are not required to enter the user's current password before changing it. This can be useful for creating initial passwords or for changing mass passwords because of a policy change. But it can also be very dangerous as you may change someone's password inadvertently, and if you enter the passwd command without the username, you will be changing the root password.

The passwd command has many useful parameters for a system administrator. With these, the administrator can enforce deadlines for changing passwords. The –f parameter forces the user to change the password at the next login. With –w days (where days is a positive integer), the warn field is set so that the user is warned in that many days to change their password. With –x max, the maximum field is set that establishes the number of days that the password remains valid before it must be changed. When combined, you are able to establish that a user must change a password in some number of days and be warned about it before that. For instance, passwd –w 7 –x 30 foxr would require that foxr change his password within 30 days, and if the password is not changed in the next 23 days, foxr receives a warning. The –l parameter locks the password entry so that it cannot be changed.

HACKING

The term hacking is meant to convey a programmer, someone who "hacks" code. In fact, hacking has several definitions. Another is applied to the hobbyist who likes to experiment with something new. But many people view a hacker as a person who attempts to break into computers by guessing user passwords (more appropriately, this is called cracking). There are many ways to break into a user's account. We look at some of them here.

Password cracking: trying a variety of passwords such as no password at all (just the enter key), dictionary words, the user's name, or the name of the user's spouse, children, pets, parents, etc.

Social engineering: trying to get the user to give you their password, for instance, by calling on the phone and saying "This is Bob from IT, our server crashed and in order to restore your account, I need your password." This is also called phishing.

Dumpster diving: many people write their passwords on post-it notes. A quick look around the person's workplace or their trashcan may yield a password!

IP spoofing: by changing your IP address, you may be able to intercept Internet messages intended for someone else. It is possible (although not very likely) that passwords might be found this way.

The usermod command is used to modify a user's account. You can use this command to change the user's comment, home directory, initial group, other groups (See the section Role of a Group), login shell, user ID, password, account expiration date, and the number of days after a password expires before the account is permanently disabled. Two other options allow you to lock or unlock the user's password.

As mentioned earlier, human users and software are given accounts. You might ask why software would be given an account. The reason is because of the use of permissions to control access to files and directories. This is covered in Permissions. Consider, however, that a piece of software will need to read and write files to a particular directory. We want to make sure that other users cannot write to that directory or to those files for security purposes (we may also want to restrict other users from reading the directory and files). When a user starts a process, the process is owned by that user. And therefore, if a file was not owned by that user, then the file may not be readable or writable to that program. If, instead, the program has its own account that owns the files, then the program can access those files even though the user who ran the program cannot. Therefore, in Linux, there are two types of user accounts, those for human users and those for software. It is common, as described earlier, that software accounts will not have a login shell (because the program will never log in) and may not have a home directory.

Windows

Early versions of Windows were intended to be single user systems, so there was no need to create user accounts. With Windows NT, a networked version of Windows, this changed. Now, user accounts are incorporated into each version of Windows. As with Linux, an initial user account is created during OS installation so that the user of the computer can log in as a non-administrator. Although the initial user account does not need a password, it is strongly recommended that the account have a password (and similarly for all future accounts created).

To create additional accounts, you must act as Administrator. You would do this through the User Accounts window, available through the control panel. The User Account function is made available through a GUI as shown in Figure 6.5.

Similar to the Linux User Manager, you have the ability to add, remove, or alter the properties of users. Properties here, however, are limited to group placement. By default, there is only one group, Administrators. However, you can create additional groups as needed. In Figure 6.5, there are three accounts: the system administrator account (Administrator); foxr, who is also an administrator; and heggeju, who is in the group HelpLibraryUpdaters. This account is used by a member of IT to automatically update the computer's system as needed. The advanced tab at the top of the User Account window permits one to change passwords, add groups, and alter group memberships. It is likely that if you are running Windows 7 on a home computer that you will have two accounts: Administrator and the account that you created when you installed the operating system. If there are multiple users who share the computer, you may or may not desire to give each user a separate account. By having separate accounts, you create not only file space for each individual user (under the C:\Users directory), but you also can set file permissions so that different users are able to access different files and run different programs. If you are dealing with a Windows 7 machine that is shared among users in a work environment, then it is likely that each user has an account within the organization's domain (domains are not covered

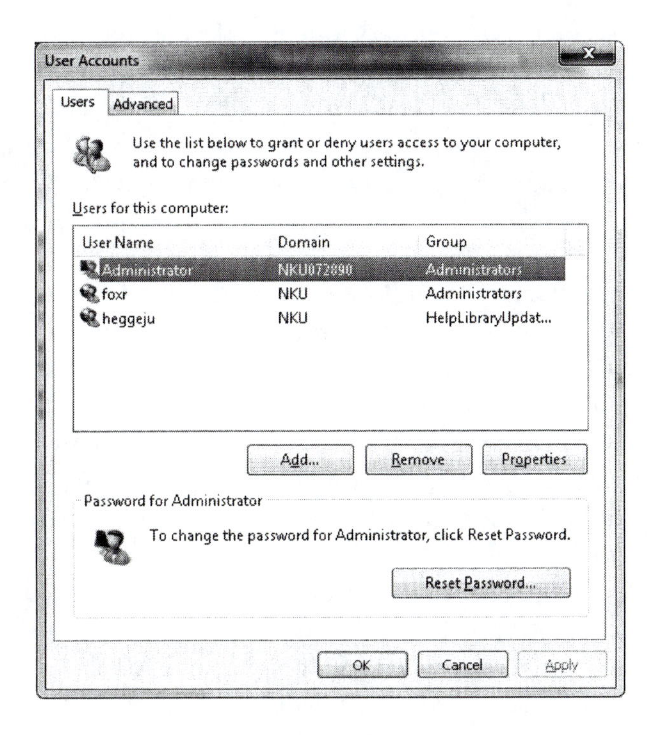

FIGURE 6.5 Windows user account GUI.

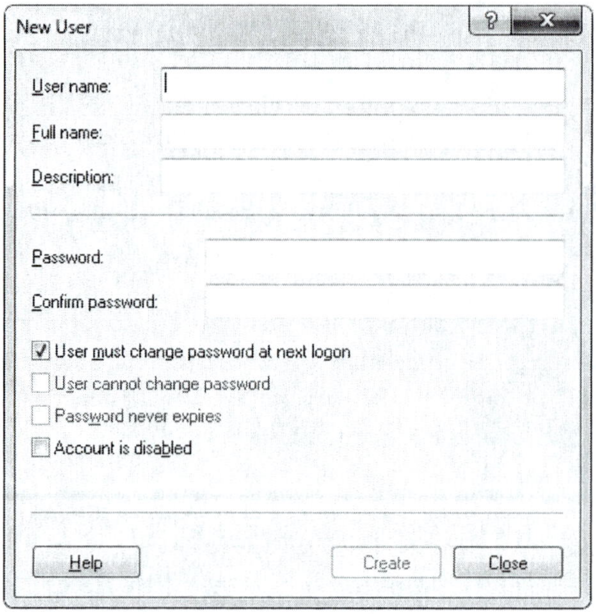

FIGURE 6.6 New user window.

in this text) and you are able to add an account for them to any of the computers in the domain.

Users can also be added through the Computer Management tool. Right click on the Computer shortcut icon and select Manage. From the Computer Management tool, select System Tools and then Local Users and Groups. Right click on the Users icon and select New User. The New User window appears, as shown in Figure 6.6. Notice here that a password is required, unlike adding a new user through the window in Figure 6.5.

ROLE OF A GROUP

An individual user creates and manages files. It is often desirable to share the files with others. However, there are some files where the access should be limited to specific users. The ability to control who can access a file and what types of access rights those users are granted is accomplished through permissions. In Linux, for instance, permissions can be established for three different classes of users: the file's owner, those users who share the same group as the file's owner, and the rest of the world. This requires that we have an entity called a group. Both Linux and Windows use groups so that access does not have to be restricted solely to the owner of the file and the rest of the world.

A group is merely a name (and an ID number) along with the list of users who are a part of that group. A file can be owned by both a user and a group. Permissions can be established so that the owner has certain access rights, and group members have different access rights. In this way, for instance, an owner might be able to read and write a file, whereas group members can read the file only and all other users have no access to the file.

Both Linux and Windows permit the creation of groups. In Linux, group creation can be accomplished through the User Manager GUI (see Figure 6.1, the Group tab is used) or through the groupadd instruction from the command line. As with users, groups are given ID numbers. When creating a group, like adding a user, you may specify the group ID number, or it can default to being 1 greater than the last created group. In Windows, group creation is handled through the Computer Manager window. Once a group is created, individual users can be added to the group.

In Linux, users can be added or removed from groups through the User Management tool or from the command line. To add a user to an existing group, you can use either useradd with the –G parameter, or usermod (modify user). An example is usermod –G cit130 zappaf. This will add zappaf to the group cit130. Group information is stored in the file /etc/group. This file stores, for each group, the group name, the group ID number, a password for the group (as with /etc/passwd, this merely appears as an 'x'), and the list of usernames who belong to that group. One could edit this file to change group memberships, although it is safer to use the usermod instruction. In Windows, users can be added or removed from groups through the Computer Manager Users and Groups selection, as shown in Figure 6.7.

In Linux, when creating a new user account, you are able to specify whether a private account should be generated for that user. This is typically the case. Creating a private group causes a new group whose name is equal to that of the user name to be created and added to the /etc/group file. The only member of that group will be the user. For instance, there will be a group called foxr with one member, foxr, and a group called zappaf with one member, zappaf. Any file created by a user, by default, is owned by that user's private group. It is up to the file owner to change the group ownership. If foxr were to create a file f1.txt,

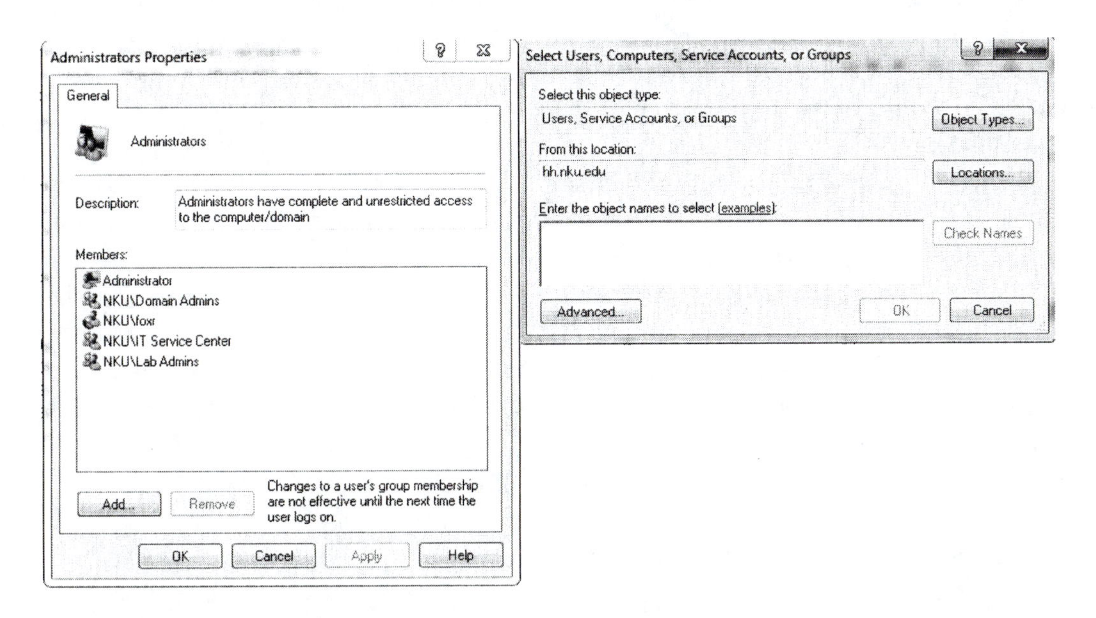

FIGURE 6.7 Adding (or removing) users to groups in Windows.

it would initially be owned by user foxr and the group foxr. If foxr wanted to share the file with members of the cit130 group, foxr would have to change group ownership. The Linux command chown allows the owner of a file to change the owner to another user, and chgrp allows the owner of the file to change the group owner to another group.

These two Linux commands work as follows:

```
chown newuser filename
chgrp newgroup filename
```

For instance, if you have a file foo.txt owned by root, and as root you want to change it to be owned by foxr, the command would be chown foxr foo.txt. You might change the group ownership as well, for instance, as chgrp cit130 foo.txt. Notice that this command is changing group ownership not to foxr's private group, but to cit130, presumably a group of more than one user.

If you want to change both owner and group, this can be done using chown alone:

```
chown newuser:newgroup filename
```

If you do not own a file, you would not be able to change the user or group ownership. However, if you are root, you can change the ownership or group of any file.

PERMISSIONS

Access control lists (ACLs) are used in some operating systems to implement protection of the system resources. For every resource, there is a list that describes for every user in the system, what their access rights are for that resource. Users with no access will not be listed. For instance, some file f1.txt might have an ACL of foxr:read,write,execute; zappaf:read,execute; underwoodr:read. The advantage of an ACL is that you are able to specify different access rights for every user. The disadvantages are that the ACLs require storage space that grows with each user that you add to the list, and the rather time-consuming nature of changing access rights. For instance, imagine that a new user, dukeg, is added to the system. We decide to give dukeg read access to every file in the system. Adding dukeg:read to every ACL, if not automated through some script, could be an unpleasant chore for an administrator.

Linux simplifies the use of an ACL by having three types of permissions—read, write, and execute—applied to three classes of users: the file's owner, the file's group, and the rest of the world. These permissions are placed on every system file and directory. Users have the option to alter any or all permissions from their default settings when the file /directory is created.

If a file is readable, it can be viewed or copied. If a file is writeable, it can be written to, overwritten (modified), or deleted. If a file is executable, it can be executed. We usually would not worry about the executability of nonprograms. Shell scripts are an exception. A shell script is not executed directly; it is instead executed by an interpreter. This should imply that the shell script need only be readable by the interpreter but, in fact, it needs to be both readable and executable. The interpreter would not be the owner of the file nor in the file's group, so it would require read and execute access for the rest of the world.

Permissions on directories have a slightly different meaning than the permissions of files. If a directory is readable, then you can view its contents (i.e., do an ls). If a directory is not readable, not only can you not view its contents, but you would not be able to view the contents of individual files in that directory. If a directory is writable, then you are able to save files into that directory as well as copy or move files into that directory. If a directory is executable, then you can cd into it. If a directory were readable but not executable by you, then although you could use ls on the directory, you would not be able to cd into it.

The three classes of users in Linux permissions are the resource's owner, the resource's group, and everyone else in the world who can log into the Linux machine. We will refer to these as user (u), group (g), and other (o). This is unfortunately a little confusing because owner is represented by 'u' rather than 'o'. For each of these three levels, there are three possible access rights that can be supplied: read (r), write (w), and execute (x). These letters are important as we will see below. Files and directories will default to certain permissions. In order to change permissions, you must own the file (or be root) and use the chmod (change mode) command. This is covered in the Linux subsection below.

In Windows, file access is slightly different. Access rights are full control, modify, read & execute, read, write, and special permissions. Full control is given to the owner who then has the ability to change the access rights of others. Modify means that the file can be written over, that is, the save command will work. Read & execute would be applied to an executable file, whereas read would be applied to a nonexecutable. Write allows one to save anew through save as, or resave. Any user or group can be given any combination of the above access rights. Access rights can be set to "Allow" or "Deny". We look at modifying Windows permissions in the Windows subsection.

Linux

As stated, Linux has three levels of permission: owner (u for user), members of the group (g), and the rest of the world (o for others). Each level can be set to any combination of readable (r), writable (w), and executable (x). You are able to view the permissions of a file (or directory) when you use the command ls –l. The –l option means "long form". The permissions are shown as a nine-character set showing for the owner if readable is set, writable is set, executable is set, and the same for group and world. For instance, a file that is listed as rwxr-x--x means that it is readable, writable, and executable by the owner, readable and executable by members of the group, and executable only by the world. The - indicates that the particular permission is not set. When doing an ls –l, this list of nine characters is preceded by a tenth character that represents the file type. Typically, this is either a hyphen (file) or 'd' (directory), and as we saw in Chapter 5, the letter 'l' means a link.

Figure 6.8 shows a typical long listing. The listing includes, aside from the permissions, the owner and group of the file, the file's size, last modification date, and file's name. Note that the ~ at the end of some file names indicates that it is a backup file. When you modify a file using certain editors (such as Emacs), the old file is still retained but labeled as an old file with the ~ character. Therefore, foo2.txt is a newer version and foo2.txt~ is an older version. Saving foo2.txt again would copy foo2.txt into foo2.txt~ to make room for the new version.

Let us take a look at a couple of items in Figure 6.8 to gain a better understanding of the information supplied by the long listing. First, Desktop has permissions of "drwxr-xr-x". The 'd' indicates that this is a directory. The directory is owned by foxr and in foxr's private group. The user foxr has rwx permissions (read, write, execute). Therefore, foxr can cd into the directory, ls the directory, read files from the directory (presuming those files are readable by the owner), write to the directory, and resave files to the directory (again, presuming those files are writable by the owner). Other users, whether in foxr's private group (of which there would probably be no one) or the rest of the world have read and execute access to this directory but not write access. All other items shown in Figure 6.8 are files because their permissions start with the character '-'. The file foo2.txt is readable and writable by foxr only. No other user (aside from root) can read or write this file. The file foo3.txt is readable and writable by foxr and users in the group foxr (again, there are probably none other than foxr). It is read-only by the rest of the world.

To change permissions, you must either be the owner of a file or the system administrator. The command to change permissions is chmod (change mode). The command receives two parameters: the changes in permission and the name of the file. Changing the permissions can be done in one of three ways.

First, you can specify what changes you want by using a notation like this: u+w,g-x. The first letter in each group represents the type of user: user (u), group (g), other (o), recalling that user means owner and other means the rest of the world. The plus sign indicates that you are adding a permission, and the minus sign indicates that you are removing a permission. The second letter in the group is the permission being added or removed. In the above example, the owner is adding writability to himself and removing executability from the group. You can list multiple permission changes for each type of user such as through u-wx,g-rwx. This would remove both write and execute from the owner's permissions and remove all permissions (rwx) from the group.

Second, you can specify the entire set of permissions by using the letters (u, g, o, r, w, x) but assigning the permissions through an equal sign instead of the plus and/or minus. For instance, we might use u=rwx,g=rx,o=x to give a file all of the permissions for owner, read and execute for the group, and execute only for the world. To specify no permission for a level, you immediately follow the equal sign with a comma, or leave it blank for o

```
drwxr-xr-x 2 foxr foxr 1024 Aug 23  2011 Desktop
-rw------- 1 foxr foxr  138 Aug 24  2011 foo2.txt
-rw------- 1 foxr foxr  138 Jun  4 09:09 foo2.txt~
-rw-rw-r-- 1 foxr foxr   20 Jun  4 09:11 foo3.txt
-rw-rw-r-- 1 foxr foxr  276 Jun  4 09:11 foo.txt
-rw-rw-r-- 1 foxr foxr  138 Jun  4 09:10 foo.txt~
-rw-r--r-- 1 foxr foxr   17 Oct  3  2011 script1
```

FIGURE 6.8 Long listing in Linux.

(other). For instance, u=rw,g=,o= would give the owner read and write access, and group and world would have no permissions.

You can combine the first and second methods as long as they are consistent. You would tend to use the first method if you want to change permissions slightly and the second method if you want to reset permissions to something entirely different (or if you did not know what the permissions were previously). Note that for both of these approaches, the letter 'a' can be used to mean all users (u, g, and o). So, for instance, a+r means everyone gains read access, or a=x means everyone is reset to only have execute access.

The third approach is similar to the second approach in that you are replacing all access rights with a new set; however, in this case you specify the rights as a three-digit number. The three digits will correspond to the rights given to user, group, and other, respectively. The digit will be between 0 and 7 depending on which combination of read, write, and execute you wish to assign. To derive the digit, think of how the file permissions appear using the notation rwxrwxrwx. Change each group of 3 (rwx) into three binary bits where a bit is 1 if you want that access right granted and a 0 if you do not want that access right granted. For instance, rwxr-x--x would be rewritten as 111 101 001. Now, convert each of these 3-bit numbers to decimal (or octal): 111 is 7, 101 is 5, 001 is 1, giving us 751. Therefore, the three-digit number matching the access rights rwxr-x--x is 751.

Another way to view the 3-valued number is to simply assign the values of 4 for readability, 2 for writability, and 1 for executability. Now, add up the values that you want for each of owner, group, and world to create your 3-valued number. If you want to combine readable, writable, and executable, this will be 4 + 2 + 1 = 7. If you want to permit only readable and writable, this will be 4 + 2 = 6. If you want to permit only readable and executable, this will be 4 + 1 = 5. If you want readability only, this will be 4. And so forth. You compute the sum for each of owner, group, and world. So the value 640 would be readable and writable for owner, readable for group, and nothing for world. The value 754 means that the owner would have all three types of access (r, w, x), the group would have read and execute permission and the world would have read access only.

The following are some other common values you might find for files and directories:

660—read and write for owner and group, nothing for the world

755—read, write, execute for owner, read and execute for group and world

711—read, write, and execute for owner, execute for group and world

444—a read-only file for everyone

400—read access for owner, nothing for group and world

000—no access for anything (only system administrator could access such a file)

The following are some values you would probably not see for files and directories:

557—there is no reason why the world would have greater access than the owner

220—a "write-only" file makes little sense without being able to open or read the file

420—this file is read only for the owner but write only for the group

777—although this is an allowable permission, it is dangerous to give the world write access

Furthermore, it is rare not to make a directory executable, whereas non-executable files are rarely given execute access. So we would tend to see 755 or 775 for directories and 644 or 664 for files.

Each of chmod, chown, and chgrp can alter multiple files at a time by listing each file. For instance, chown zappaf foo1.txt foo2.txt foo3.txt, will change the owner of all three files to the user zappaf.

Linux can also be enhanced with a feature called Security-Enhanced Linux (SELinux), which allows permissions to be allocated in a much finer grained manner. SELinux goes above and beyond the permissions discussed thus far and incorporates the use of ACLs to allow much greater levels of control and security. SELinux maintains a completely separate set of users and roles than the ones we have discussed so far. SELinux is much more complex than standard Linux file permissions, and the details of its use are beyond the scope of this text.

Windows

The Windows operating system uses a similar idea for permissions as in Linux. Changing permissions is not as easy (at least when you have a large number of files to change) and does not provide quite the amount of flexibility. To view the level of permissions on a file, from the file manager, right click a file's icon and select properties. From the properties window, select the Security tab. You will see something like what is shown in Figure 6.9.

From the security tab on a file's properties, you will see the various groups that have access to the file. They will include System, Administrators, and Users. In this case, we also see Authenticated Users. These are roughly equivalent to the system administrator and the world. Authenticated Users might constitute a group. Permissions are limited to full control, modify, read and execute, read, write, and special. If you give a user full control access, then you are allowing that user to change permissions of others. Changing permissions requires first that you select the Edit… button. From a new pop-up window, you can add or remove permissions by clicking in check boxes. You can also add or remove specific users and groups from the list. Thus, Windows gives you a way to add permissions for specific users without having to deal with group creation. See Figure 6.10.

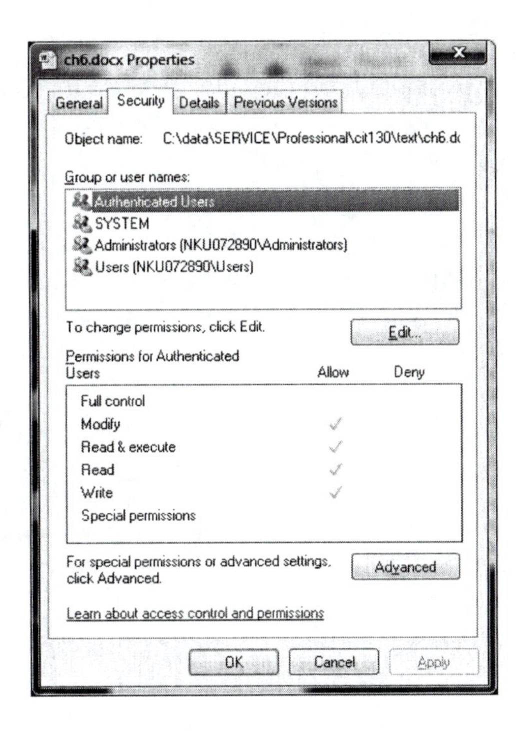

FIGURE 6.9 Security tab to change permissions.

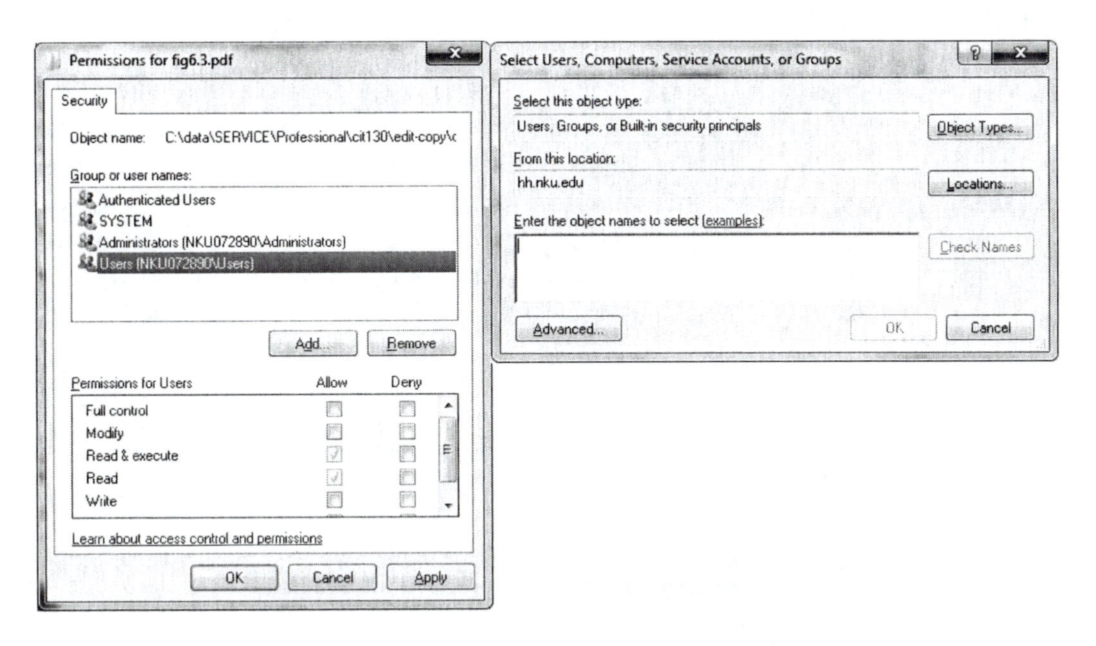

FIGURE 6.10 Changing permissions (left) and adding or removing users and groups (right).

MISCELLANEOUS USER ACCOUNT TOPICS

Consider the following situation. You are currently logged into your Windows 7 machine. You are editing a few files. One of your family members needs to quickly access a file of their own and e-mail it to a friend. Since that file is owned by that family member and they have secured the permissions to be readable only by their own account, you cannot simply let them use the computer for a minute. Instead, you would have to save your files, log out, let them log in, edit their file, e-mail it, log out, and then you would have to log back in and reopen your files. This task could take several minutes. Instead, Windows 7 has a *switch user* facility (this feature was first included in Windows XP and available in Vista). To switch user, go to the Start button, select the Shut down menu and select Switch user (Figure 6.11). This allows you to interrupt your own session, let your family member log in and do their work, and then return to your account. When you return to your account, you will find it exactly as you left it, with all of your windows still open and your programs running. It is probably best to save your work before switching users, however, in case your family member accidentally or mistakenly shuts the computer down rather than logging off. The switch user facility does not save any of your files.

In Linux, switching users is done in a much simpler fashion. If you are working in a shell window, you use the command su username. You will be asked to log in as the new user. Although you now are the new user in terms of access rights, interestingly, you will be still be in whatever working directory you were in before issuing the su command. Therefore, you might follow the su command with a cd ~ command to move to the new user's home directory. To exit from the new shell session, type exit and you will be back to your previous log in (and the previous directory). If you issue the command su without a user name, you will be attempting to switch to root, the system administrator account. Note that if you are currently root, you can switch to *any* user without having to enter their password.

As has been noted before, every operating system has a special account, the system administrator (Administrators in Windows, root in Linux and Unix, and sometimes referred to as superuser in other operating systems). This account has access to every system resource. In Windows, there are a set of administrator programs and tools that are not available to ordinary users. In Linux, many of the administrator programs are only available to root, and these are often found either in /sbin or /usr/sbin. Among these programs are useradd, userdel, usermod, groupadd, groupdel, and groupmod (all of these are

FIGURE 6.11 Switching users.

in/usr/sbin). If you do an ls –l on these files, you will see that they all have permissions of rwxr-x---, which means that only the owner and group members can read or execute these programs. The owner is root and the group is root, so only root can run them. There are other system programs that are only executable by root in both of these directories.

Consider in Linux that you might want to provide some form of permission to a subset of users by using groupadd. So you wish to create a group, add the subset of users to that group, and then change permissions on your file giving those users access. But if you look at the groupadd program (located in/usr/sbin), you will see that it is owned by root and in root's private group. Furthermore, its permissions are "-rxwr-x---" meaning that it is not accessible to the average Linux user (only root or someone in root's private group). Yet, as a system administrator, you would not want to encourage people to make their files open to the world in such cases, and there does not seem to be another option unless the user can convince the system administrator to use groupadd to create the new group. Linux does offer another approach, although this might be deemed a compromise between forcing users to set dangerous permissions and having the system administrator create groups on request. This approach is to permit one or a select group of users the ability to run specific system programs. This is done through the program sudo and a file called /etc/sudoers.

In order to use sudo, there are two steps. First, as root, you edit the sudoers file. While the file already has a number of entries, you would add your own, probably at the bottom of the file. An entry is of the form:

```
username(s) host = command
```

There can be more than one user listed if you wish to give the same access to multiple users. Usernames will be separated by commas. You can also specify members of a group by using

```
%groupname or %gid (the ID number of the group).
```

The value for host will either be localhost, all, or a list of host (computer) names. The word localhost means the machine that the sudoers file is stored on. This restricts users to having to be logged in to that machine and restricts the command to only be executed on that machine. The word all is used to indicate any machine on the network. The command can be a list of commands that are separated by spaces. The command must include the full path name, for example,

```
/sbin/mount or /usr/sbin/useradd.
```

The second step is to issue the sudo command. The sudo command is the word sudo followed by the actual command that the user wishes to issue. For instance, if you have been given sudoer privilege to run useradd, then you would perform the operation as

```
sudo /usr/sbin/useradd -d/home/cit130 -G cit130,students -m newguy
```

Here, in addition to the useradd command (which defines a different home directory location and adds the new user to two existing groups), the command is preceded by sudo. The sudo program, before attempting to execute the command, requires that the user enter their password. If the password matches, then the sudo program compares the username and command to the entries in the /etc/sudoers file. If a match is found, the command is executed as if the user was in fact root; otherwise, the user is given an error message such as:

```
foxr is not in the sudoers file. This incident will be reported.
```

Sudo access should not be granted on a whim. In fact, it should not be granted at all unless the user has been trained and the organization provides a policy that clearly states the roles and requirements of granting some system administration access. One general exception to this rule is the following sudoer entry:

```
%users localhost =/sbin/shutdown -h now
```

The shutdown command shuts down the machine. This command is issued by the GUI when the user selects shutdown. Therefore, all users will need to be able to shut down the computer whether by GUI or command line. The users group is a group of all users, automatically adjusted whenever users are added to the system.

User accounts permit tailored environments, known as shells in Linux, often referred to as the desktop in Windows. Information describing the tailored environment is stored in the user's home directory. Upon logging in, the file(s) will be loaded into the shell or GUI environment. In this way, the user is able to create an environment for how things look, where things will be located, and shortcuts for executing commands. In Windows, this file, or profile, is saved automatically as the user adjusts the desktop. In Linux, there are several different files that can store profile information. For instance, if the user is using the Bash shell as their login shell, the user would most likely edit the .bashrc file stored in their home directory. We will explore Bash in more detail in Chapter 9.

FURTHER READING

There are no specific books that discuss user accounts or groups, or file permissions. Instead, those are topics in texts of various operating systems. For such texts, see Further Reading section of Chapter 4.

REVIEW TERMS

Terminology introduced in this chapter:

- Access control list
- Account name
- Account
- Execute access

Group

Group access

Owner

Owner access

Password

Permission

Read access

Root account

SELinux

Shell

Strong password

Superuser

Sudoers

Sudo

Switching user

User

World

World access

Write access

REVIEW QUESTIONS

1. Why are user accounts needed in computer systems? Are user accounts needed if only one person is going to be use the computer?
2. How does a normal user account differ from a software account?
3. What is the difference between a user account and an administrator account?
4. Examine a Linux/etc/shadow file. Can you figure out the length of any of the passwords given their encrypted form?
5. Why do we need groups?
6. How do you create a new user account in Linux from the command line?
7. How do you change ownership of a file in Linux from the command line?
8. How do you change group ownership of a file in Linux from the command line?
9. What does the usermod command allow you to do in Linux?
10. What is a file permission?
11. What is the difference between read access and execute access in a directory?
12. What does execute access mean for a file that is not an executable program?
13. What does the sudoers file store? What does the sudo command do?
14. Why would we allow all users in a Linux computer system the ability to issue the shutdown command?

REVIEW PROBLEMS

1. Provide Linux chmod instructions using the ugo= approach to match the following permissions:

```
foo.txt            -rwxr-xr--
bar                drwx--x--x
stuff.txt          -rw-r--r--
notstuff.txt       -r-x--x---
another.txt        -rwxrw-r--
temp               drwxrwxr-x
```

2. Repeat #1 using the 3-valued approach in Linux.
3. Assume all items in a directory are currently set at 644. Provide Linux chmod instructions using the ugo+/– approach to match the following permissions.

```
foo.txt            -rwxr-xr--
bar                drwx--x--x
stuff.txt          -rw-r--r--
notstuff.txt       -r-x--x---
another.txt        -rwxrw-r--
temp               drwxrwxr-x
```

4. Repeat #3 using the 3-value approach in Linux.
5. What is the difference between the Linux instruction useradd and usermod?
6. What is the difference between the Linux instruction chown and chgrp? How can you use chown to change both file owner and file group?
7. Write a Linux passwd command to force user marst to change his password within 10 days, issuing a warning in 7 days if the password has not been changed by then.
8. How would you lock an account using the Linux passwd command? How would you force a user to change their password in the next login using the Linux passwd command?

DISCUSSION QUESTIONS

1. Organizations have different policies for computer usage by their users. Some questions that an organization may want to generate polices for include: How long should an account exist after the user has left the organization? Should users have disk quotas? If so, should all users have the same quota size? Can users store personal files in

their storage space? Develop your own policies assuming that the organization is a small private university or college. How would your policies differ if it was a large public university?

2. Repeat #1 for a small (10–20 employees) company. How would your policies differ if it was a large (1000+ employees) company?

3. Do you find yourself with numerous computer accounts and different passwords for each? If so, do you have a system for naming and remembering your passwords? For instance, do you use the same password in each account? Do you write your passwords down? Do you use people's (or pets') names as passwords? Why are these bad ideas? Do you have any recommendations to help people who might have a dozen different computer accounts or more? As a system administrator, should you advise your colleagues on how to name and remember their passwords?

4. In Linux, the /etc/passwd file is automatically modified whenever you use the useradd, userdel, or usermod instruction. Why should you always use these instructions rather than directly modifying the passwd file yourself? What might happen if you modified the file directly that would not happen if you used the proper Linux command(s)?

5. Although the passwords are stored in an encrypted form, why was leaving the passwords in the /etc/passwd file a threat to system security? Attempt to explain how a user could use the encrypted password information to hack into other users' accounts.

6. Explain why groups are an important unit to help better support system permissions in an operating system. What would be the consequences on a user if groups were not available?

7. Find a Linux system administrator reference and examine the commands available to the system administrator that are not available to ordinary users. From the list, can you identify any commands that you might want to make available to ordinary users, or even a few select users, via the sudo command? If so, explain which commands and why or in what circumstances they should be available via sudo.

CHAPTER 7

History of Computers

Most people take computers for granted today without even noticing their impact on society. To gain a better understanding of our digital world, this chapter examines the history of computing: hardware, software, and users. This chapter begins with the earliest computing devices before moving into the development of the electronic, digital, programmable computer. Since the 1940s, computers fall into four "generations," each of which is explored. The evolution of computer software provides three overlapping threads: changes in programming, changes to operating systems, changes to application software. The chapter also briefly considers how computer users have changed as the hardware and software have evolved.

The learning objectives of this chapter are to

- Compare computer hardware between the four computer generations.
- Describe the impact that integrated circuits and miniaturization have played on the evolution of computer hardware.
- Describe shifts in programming from low level languages to high level languages including concepts of structured programming and object-oriented programming.
- Explain how the role of the operating system arose and examine advances in software.
- Discuss the changes in society since the 1950s with respect to computer usage and computer users.

In all of human history, few inventions have had the impact on society that computers have had. Perhaps language itself, the ability to generate and use electricity, and the automobile have had similar impacts as computers. In the case of the automobile, the impact has been more significant in North America (particularly in the United States) than in other countries that rely on mass transit such as trains and subways. But without a doubt, the computer has had a tremendous impact on most of humanity, and the impact has occurred in a shorter time span than the other inventions because large-scale computer usage only dates back perhaps 25–30 years. In addition to the rapid impact that computers have had, it is certainly the case that no other human innovation has improved as dramatically as the computer. Consider the following simple comparisons of computers from the 1950s to their present-day counterparts:

- A computer of the 1950s cost hundreds of thousands of 1950s dollars, whereas computers today can be found as cheaply as $300 and a "typical" computer will cost no more than $1200.
- A computer of the 1950s could perform thousands of instructions per second, whereas today the number of instructions is around 1 billion (an increase of at least a million).
- A computer of the 1950s had little main memory, perhaps a few thousand bytes; today, main memory capacity is at least 4 GB (an increase of perhaps a million).
- A computer of the 1950s would take up substantial space, a full large room of a building, and weigh several tons; today, a computer can fit in your pocket (a smart phone) although a more powerful general-purpose computer can fit in a briefcase and will weigh no more than a few pounds.
- A computer of the 1950s would be used by no more than a few people (perhaps a dozen), and there were only a few hundred computers, so the total number of people who used computers was a few thousand; today, the number of users is in the billions.

If cars had progressed like computers, we would have cars that could accelerate from 0 to a million miles per hour in less than a second, they would get millions of miles to a gallon of gasoline, they would be small enough to pack up and take with you when you reached your destination, and rather than servicing your car you would just replace it with a new one.

In this chapter, we will examine how computers have changed and how those changes have impacted our society. We will look at four areas of change: computer hardware, computer software, computer users, and the overall changes in our society. There is a separate chapter that covers the history of operating systems. A brief history of the Internet is covered in the chapter on networks, although we will briefly consider here how the Internet has changed our society in this chapter.

EVOLUTION OF COMPUTER HARDWARE

We reference the various types of evolution of computer hardware in terms of generations. The first generation occurred between approximately the mid 1940s and the late 1950s. The second generation took place from around 1959 until 1965. The third generation then lasted until the early 1970s. We have been in the fourth generation ever since. Before we look at these generations, let us briefly look at the history of computing before the computer.

Before the Generations

The earliest form of computing was no doubt people's fingers and toes. We use decimal most likely because we have 10 fingers. To count, people might hold up some number of fingers. Of course, most people back in 2000 B.C. or even 1000 A.D. had little to count. Perhaps the ancient shepherds counted their sheep, and mothers counted their children. Very few had much property worth counting, and mathematics was very rudimentary. However, there were those who wanted to count beyond 10 or 20, so someone invented a counting device called the abacus (see Figure 7.1). Beads represent the number of items counted. In this case, the abacus uses base 5 (five beads to slide per column); however, with two beads at the top of the column, one can either count 0–4 or 5–9, so in fact, each column represents a power of 10. An abacus might have three separate regions to store different numbers, for instance: the first and third numbers can range from 0 to 9999, and the middle number can be from 0 to 99999. We are not sure who invented the abacus or how long ago it was invented, but it certainly has been around for thousands of years.

Mathematics itself was very challenging around the turn of the millennium between BC and AD because of the use of Roman numerals. Consider doing the following arithmetic problem: 42 + 18. In Roman numerals, it would be written as XLII + XVIII. Not a very easy problem to solve in this format because you cannot simply line up the numbers and add the digits in columns, as we are taught in grade school.

It was not until the Renaissance period in Europe that mathematics began to advance, and with it, a desire for automated computing. One advance that permitted the improvement in mathematics was the innovation of the Arabic numbering system (the use of digits 0–9) rather than Roman numerals. The Renaissance was also a period of educational improvement with the availability of places of learning (universities) and books. In the 1600s, mathematics saw such new concepts as algebra, decimal notation, trigonometry, geometry, and calculus.

In 1642, French mathematician Blaise Pascal invented the first calculator, a device called the Pascaline. The device operated in a similar manner as a clock. In a clock, a gear rotates, being moved by a pendulum. The gear connects to another gear of a different size. A full revolution of one gear causes the next gear to move one position. In this way, the first gear would control the "second hand" of the clock, the next gear would control

FIGURE 7.1 An abacus. (Courtesy of Jrpvaldi, http://commons.wikimedia.org/wiki/File:Science_museum_030.jpg.)

the "minute hand" of the clock, and another gear would control the "hour hand" of the clock. See Figure 7.2, which demonstrates how different sized gears can connect together. For a mechanical calculator, Pascal made two changes. First, a gear would turn when the previous gear had rotated one full revolution, which would be 10 positions instead of 60 (for seconds or minutes). Second, gears would be moved by human hand rather than the swinging of a pendulum. Rotating gears in one direction would perform additions, and rotating gears in the opposite direction would perform subtractions. For instance, if a gear is already set at position 8, then rotating it 5 positions in a clockwise manner would cause it to end at position 3, but it will have passed 0 so that the next gear would shift one position (from 0 to 1), leaving the calculator with the values 1 and 3 (8 + 5 = 13). Subtraction could be performed by rotating in a counterclockwise manner.

In 1672, German mathematician and logician Gottfried Leibniz expanded the capability of the automated calculator to perform multiplication and division. Leibniz's calculator, like Pascal's, would use rotating gears to represent numbers whereby one gear rotating past 10 would cause the next gear to rotate. However, Leibniz added extra storage locations (gears) to represent how many additions or subtractions to perform. In this way, the same number could be added together multiple times to create a multiplication (e.g., 5 * 4 is just 5 added together 4 times). Figure 7.3 shows both Pascal's (a) and Leibniz's (b) calculators. Notice the hand crank in Leibniz's version to simplify the amount of effort of the user, rather than turning individual gears, as with the Pascaline.

In 1801, master weaver Joseph Marie Jacquard invented a *programmable* loom. The loom is a mechanical device that allows threads to be interwoven easily. Some of the threads are raised, and a cross-thread is passed under them (but above other threads). The raising

FIGURE 7.2 Gears of a clock. (Courtesy of Shutterstock/mmaxer.)

(a)

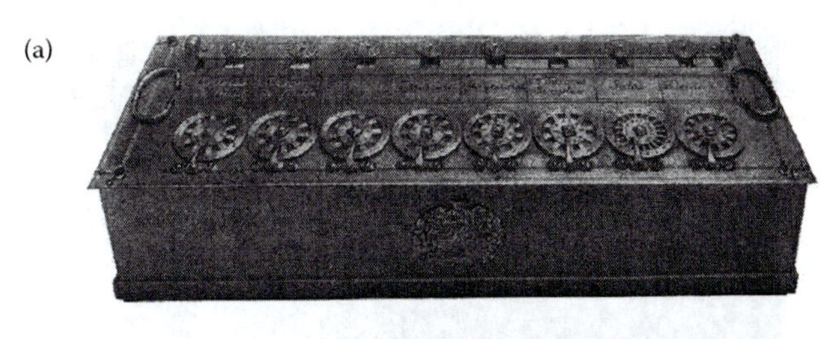

(b)

FIGURE 7.3 (a) Pascal's calculator (Scan by Ezrdr taken from J.A.V. Turck, 1921, *Origin of Modern Calculating Machines*, Western Society of Engineers, p. 10. With permission.) and (b) Leibniz's calculator. (Scan by Chetvorno, taken from J.A.V. Turck, 1921, *Origin of Modern Calculating Machines*, Western Society of Engineers, p. 133. With permission.)

of threads can be a time-consuming process. Hooks are used to raise some selection of threads. Jacquard automated the process of raising threads by punch cards. A punch card would denote which hooks are used by having holes punched into the cards. Therefore, a weaver could feed in a series of cards, one per pass of a thread across the length of the object being woven. The significance of this loom is that it was the first programmable device. The "program" being carried out is merely data that dictates which hooks are active, but the idea of automating the changes and using punch cards to carry the "program" instructions would lead to the development of more sophisticated programmable mechanical devices. Figure 7.4 illustrates a Jacquard loom circa end of the 1800s. Notice the collection of punch cards that make up the program, or the design pattern.

In addition to the new theories of mathematics and the innovative technology, the idea behind the binary numbering system and the binary operations was introduced in 1854 when mathematician George Boole invented two-valued logic, now known as Boolean logic.* Although this would not have a direct impact on mechanical-based computing, it would eventually have a large impact on computing.

In the early 1800s, mathematician Charles Babbage was examining a table of logarithms. Mathematical tables were hand-produced by groups of mathematicians. Babbage knew that this table would have errors because the logarithms were computed by hand through a tedious series of mathematical equations. These tables were being computed by a new idea,

* As with most discoveries in mathematics, Boole's work was a continuation of other mathematicians' work on logic including William Stanley Jevons, Augustus De Morgan, and Charles Sanders Peirce.

FIGURE 7.4 Jacquard's programmable loom. (Mahlum, photograph from the Norwegian Technology Museum, Oslo, 2008, http://commons.wikimedia.org/wiki/File:Jacquard_loom.jpg.)

difference equations. Difference equations consist solely of additions and subtractions, but each computation could be very involved. Babbage hit on the idea that with automated calculators, one could perhaps program a calculator to perform the computations necessary and even print out the final table by using a printing press form of output. Thus, Babbage designed what he called the *Difference Engine*. It would, like any computer, perform input (to accept the data), processing (the difference equations), storage (the orientation of the various gears would represent numbers used in the computations), and output (the final set of gears would have digits on them that could be printed by adding ink). In 1822, he began developing the Difference Engine to compute polynomial functions. The machine would be steam powered. He received funding from the English government on the order of £17,000 over a 10-year period.

By the 1830s, Babbage scrapped his attempts at building the Difference Engine when he hit upon a superior idea, a general-purpose programmable device. Whereas the Difference Engine could be programmed to perform one type of computation, his new device would be applicable to a larger variety of mathematical problems. He named the new device the

Analytical Engine. Like the Difference Engine, the Analytical Engine would use punch cards for input. The input would comprise both data and the program itself. The program would consist of not only mathematical equations, but branching operations so that the program's performance would differ based on conditions. Thus, the new device could make decisions by performing looping (iteration) operations and selection statements. He asked a fellow mathematician, Lady Ada Lovelace, to write programs for the new device. Among her first was a program to compute a sequence of Bernoulli numbers. Lovelace finished her first program in 1842, and she is now considered to be the world's first computer programmer. Sadly, Babbage never completed either engine having run out of money. However, both Difference Engines and Analytical Engines have been constructed since then using Babbage's designs. In fact, in 1991, students in the UK constructed an Analytical Engine using components available in the 1830s. The cost was some $10 million. Figure 7.5 is a drawing of the Difference Engine on display at London's Science Museum.

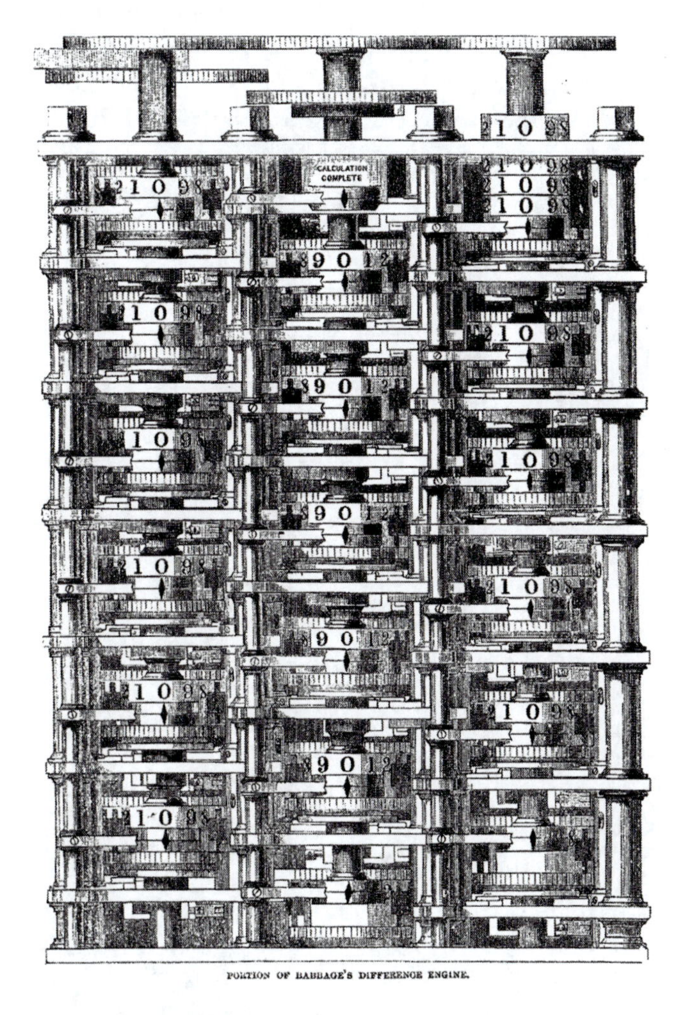

FIGURE 7.5 A difference engine. (From *Harper's New Monthly Magazine*, 30, 175, p. 34. http://digital.library.cornell.edu/cgi/t/text/pageviewer-idx?c=harp;cc=harp;rgn=full%20text;idno=harp0030-1;didno=harp0030-1;view=image;seq=00044;node=harp0030-1%3A1. With permission.)

Babbage's "computer" operated in decimal, much like the previous calculator devices and was mechanical in nature: physical moving components were used for computation. Gears rotated to perform additions and subtractions. By the late 1800s and into the 1900s, electricity replaced steam power to drive the rotation of the mechanical elements. Other mechanical elements and analog elements (including in one case, quantities of water) were used rather than bulky gears. But by the 1930s and 1940s, relay switches, which were used in the telephone network, were to replace the bulkier mechanical components. A drawing of a relay switch is shown in Figure 7.6. A typical relay switch is about 3 cm^2 in size (less than an inch and a half). The relay switch could switch states more rapidly than a gear could rotate so that the performance of the computing device would improve as well. A relay switch would be in one of two positions and thus computers moved from decimal to binary, and were now referred to as *digital* computers.

First Generation

Most of the analog (decimal) and digital computers up until the mid 1940s were special-purpose machines—designed to perform only one type of computation (although they were programmable in that the specific computation could vary). These included devices to

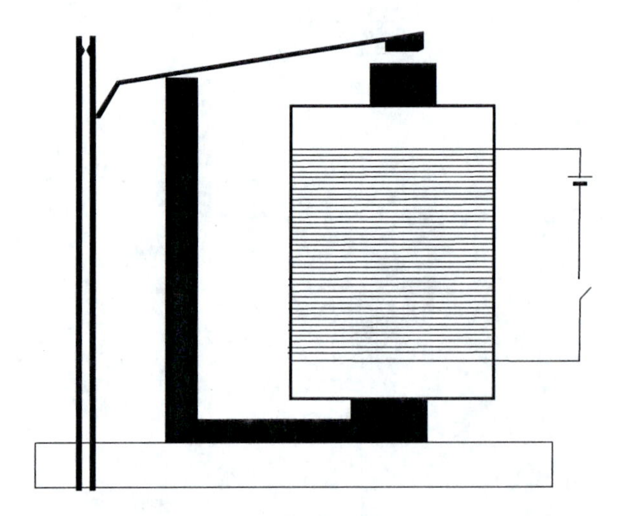

FIGURE 7.6 Electromagnetic relay switch. (Adapted from http://commons.wikimedia.org/wiki/File:Schema_rele2.PNG.)

TABLE 7.1 Early Computers

Name	Year	Nationality	Comments
Zuse Z3	1941	German	Binary floating point, electromechanical, programmable
Atanasoff-Berry	1942	US	Binary, electronic, nonprogrammable
Colossus Mark 1	1944	UK	Binary, electronic, programmable
Harvard (Mark 1)	1944	US	Decimal, electromechanical, programmable
Colossus Mark 2	1944	UK	Binary, electronic, programmable
Zuse Z4	1945	German	Binary floating point, electromechanical, programmable

compute integral equations and differential equations (note that this is different from the previously mentioned difference equations).

By the time World War II started, there was a race to build better, faster, and more usable computers. These computers were needed to assist in computing rocket trajectories. An interesting historical note is that the first computers were not machines—they were women hired by the British government to perform rocket trajectory calculations by hand! Table 7.1 provides a description of some of the early machines from the early 1940s.

The Allies also were hoping to build computers that could crack German codes. Although completed after World War II, the ENIAC (Electronic Numerical Integrator and Computer) was the first digital, general-purpose, programmable computer, and it ended all interest in analog computers. What distinguishes the ENIAC from the computers in Table 7.1 is that the ENIAC was *general-purpose*, whereas those in Table 7.1 were either special purpose (could only run programs of a certain type) or were not electronic but electromechanical. A general-purpose computer can conceivably execute any program that can be written for that computer.

Built by the University of Pennsylvania, the ENIAC was first made known to the public in February of 1946. The computer cost nearly $500,000 (of 1940s money) and consisted of 17,468 vacuum tubes, 7200 crystal diodes, 1500 relays, 70,000 resistors, 10,000 capacitors, and millions of hand-soldered joints. It weighed more than 30 tons and took up 1800 ft^2. Data input was performed by punch cards, and programming was carried out by connecting together various electronic components through cables so that the output of one component would be used as the input to another component (Figure 7.7). Output was produced using an offline accounting machine. ENIAC's storage was limited to about 200 digits. Interestingly, although the computer was a digital computer (which typically means a binary representation), the ENIAC performed decimal computations. Although the ENIAC underwent some upgrades in 1947, it was in continuous use from mid 1947 until October 1955. The ENIAC, with its use of vacuum tubes for storage, transistors, and other electronics for computation, was able to compute at the rate of 5000 operations per second. However, the reliance on vacuum tubes, and the difficulty in programming by connecting

FIGURE 7.7 Programming the Electronic Numerical Integrator and Computer (ENIAC). (Courtesy of http://commons.wikimedia.org/wiki/File:Eniac.jpg, author unknown.)

components together by cable, led to a very unreliable performance. In fact, the longest time the ENIAC went without a failure was approximately 116 hours.

The ENIAC, and other laboratory computers like it, constitute the *first generation* of computer hardware, all of which were one-of-a-kind machines. They are classified not only by the time period but also the reliance on vacuum tubes, relay switches, and the need to program in machine language. By the 1940s, transistors were being used in various electronic appliances. Around 1959, the first computers were developed that used transistor components rather than vacuum tubes. Transistors were favored over vacuum tubes for a number of reasons. They could be mass produced and therefore were far cheaper. Vacuum tubes gave off a good deal of heat and had a short shelf life of perhaps a few thousand hours, whereas transistors could last for up to 50 years. Transistors used less power and were far more robust.

(a)

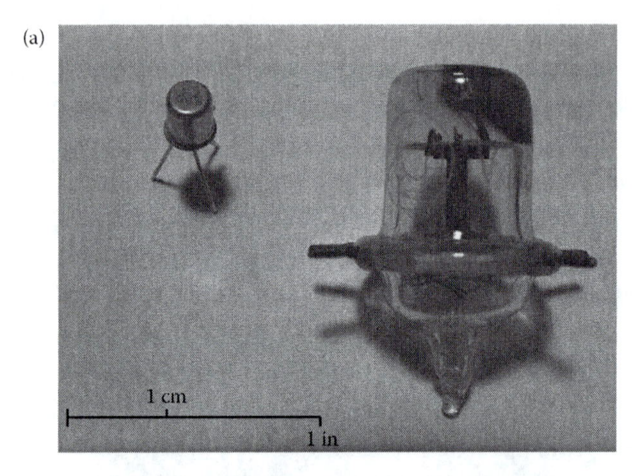

(b)

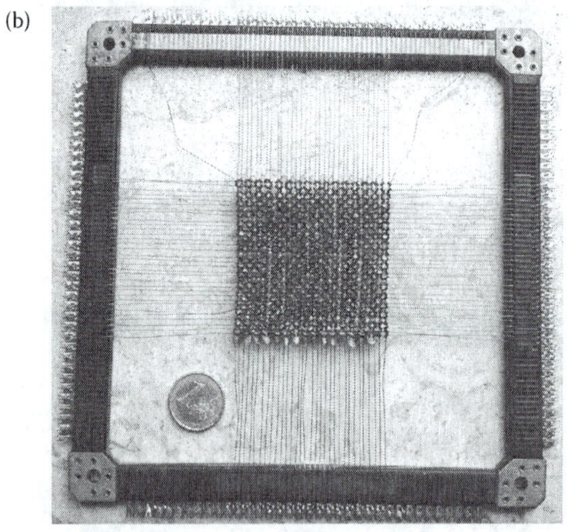

FIGURE 7.8 The vacuum tube and transistor (a) (Courtesy of Andrew Kevin Pullen, http://commons.wikimedia.org/wiki/File:955ACORN.jpg.) and magnetic core memory (b) (Courtesy of HandigeHarry, http://commons.wikimedia.org/wiki/File:Core_memory.JPG.)

Second and Third Generations

Around the same time, magnetic core memory was being introduced. Magnetic core memory consists of small iron rings of metal, placed in a wire-mesh framework. Each ring stores one bit by having magnetic current rotate in either clockwise or counterclockwise fashion.

It was these innovations that ushered in a new generation of computers, now referred to as the *second generation*. Figure 7.8 illustrates these two new technologies. Figure 7.8a provides a comparison between the vacuum tube and transistor (note the difference in size), and Figure 7.8b shows a collection of magnetic core memory. The collection of magnetic cores and wires constitute memory where each ring (at the intersection of a horizontal and a vertical wire) stores a single bit. The wires are used to specify which core is being accessed. Current flows along one set of wires so that the cores can retain their charges. The other set of wires is used to send new bits to select cores or to obtain the values from select cores.

The logic of the computer (controlling the fetch–execute cycle, and performing the arithmetic and logic operations) could be accomplished through collections of transistors. For instance, a NOT operation could be done with two transistors, an AND or OR operation with six transistors, and a 1-bit addition circuit with 28 transistors. Therefore, a few hundred transistors would be required to construct a simple processor.

By eliminating vacuum tubes, computers became more reliable. The magnetic core memory, although very expensive, permitted computers to have larger main memory sizes (from hundreds or thousands of bytes to upward of a million bytes). Additionally, the size of a computer was reduced because of the reduction in size of the hardware. With smaller computers, the physical distance that electrical current had to travel between components was lessened, and thus computers got faster (less distance means less time taken to travel that distance). In addition, computers became easier to program with the innovation of new programming languages (see the section Evolution of Computer Software).

FIGURE 7.9 IBM 7094 mainframe. (From National Archives and Records Administration, record # 278195, author unknown, http://arcweb.archives.gov/arc/action/ExternalIdSearch?id=278195&jScript=true.)

More computers were being manufactured and purchased such that computers were no longer limited to government laboratories or university research laboratories. External storage was moving from slow and bulky magnetic tape to disk drives and disk drums. The computers of this era were largely being called *mainframe* computers—computers built around a solid metal framework. All in all, the second generation found cheaper, faster, easier to program, and more reliable computers. However, this generation was short-lived. Figure 7.9 shows the components of the IBM 7094 mainframe computer (circa 1962) including numerous reel-to-reel tape drives for storage.

During the 1950s, the silicon chip was introduced. By 1964, the first silicon chips were used in computers, ushering in the *third generation*. The chips, known as printed circuits or integrated circuits (ICs), could incorporate dozens of transistors. The IC would be a pattern of transistors etched onto the surface of a piece of silicon, which would conduct electricity, thus the term semiconductor. Pins would allow the IC, or chip, to be attached to a socket, so that electrical current could flow from one location in the computer through the circuit and out to another location. Figure 7.10 shows both the etchings that make up an IC and the chip itself with pins to insert the chip into a motherboard. The chip shown in Figure 7.10b is a typical chip from the late 1960s.

ICs would replace both the bulkier transistors and magnetic core memories, so that chips would be used for both computation and storage. ICs took up less space, so again the distance that current had to flow was reduced even more. Faster computers were the result. Additionally, ICs could be mass produced, so the cost of manufacturing a computer

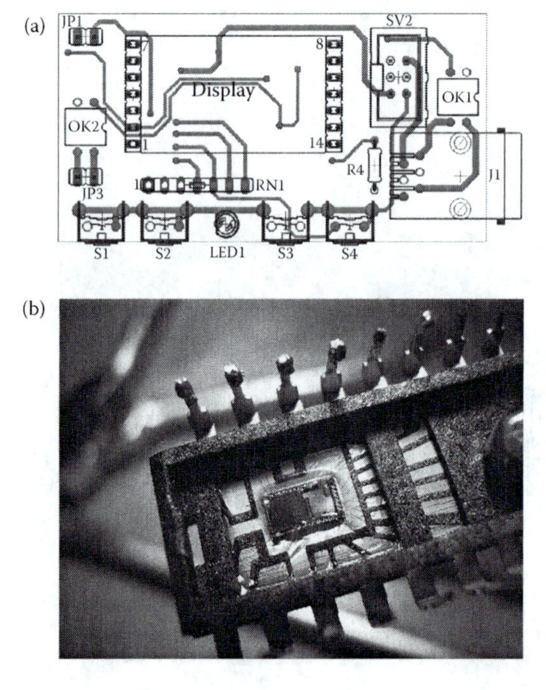

FIGURE 7.10 An integrated circuit (a) (Courtesy of Martin Broz, http://commons.wikimedia.org/wiki/File:Navrh_plosny_spoj_soucastky.png.) and a silicon chip (b) (Courtesy of Xoneca, http://commons.wikimedia.org/wiki/File:Integrated_circuit_optical_sensor.jpg.)

was reduced. Now, even small-sized organizations could consider purchasing a computer. Mainframe computers were still being produced at costs of perhaps $100,000 or more. Now, though, computer companies were also producing *minicomputers* at a reduced cost, perhaps as low as $16,000. The minicomputers were essentially scaled-down mainframes, they used the same type of processor, but had reduced number of registers and processing elements, reduced memory, reduced storage, and so forth, so that they would support fewer users. A mainframe might be used by a large organization of hundreds or thousands of people, whereas a minicomputer might be used by a small organization with tens or hundreds of users.

During the third generation, computer companies started producing *families* of computers. The idea was that any computer in a given family should be able to run the same programs without having to alter the program code. This was largely attributable to computers of the same family using the same processor, or at least processors that had the same instruction set (machine language). The computer family gave birth to the software development field as someone could write code for an entire family and sell that program to potentially dozens or hundreds of customers.

Organizations were now purchasing computers with the expectation that many employees (dozens, hundreds, even thousands) would use it. This created a need for some mechanisms whereby the employees could access the computer remotely without having to go to the computer room itself. Computer networks were introduced that would allow individual users to connect to the computer via *dumb* terminals (see Figure 7.11). The dumb terminal was merely an input and output device, it had no memory or processor. All computation and storage took place on the computer itself. Operating systems were improved to handle multiple users at a time. Operating system development is also described in Evolution of Computer Software.

FIGURE 7.11 A dumb terminal, circa 1980. (Adapted from Wtshymanski, http://en.wikipedia.org/wiki/File:Televideo925Terminal.jpg.)

Fourth Generation

The next major innovation took place in 1974 when IBM produced a single-chip processor. Up until this point, all processors were distributed over several, perhaps dozens, of chips (or in earlier days, vacuum tubes and relay switches or transistors). By creating a single-chip processor, known as a *microprocessor*, one could build a small computer around the single chip. These computers were called *microcomputers*. Such a computer would be small enough to sit on a person's desk. This ushered in the most recent generation, the *fourth generation*.

It was the innovation of the microprocessor that led to our first personal computers in the mid 1970s. These computers were little more than hobbyist devices with little to no business or personal capabilities. These early microcomputers were sold as component parts and the hobbyist would put the computer together, perhaps placing the components in a wooden box, and attaching the computer to a television set or printer. The earliest such computer was the Mark 8. Apple Computers was established in 1976, and their first computer, the Apple I, was also sold as a kit that people would build in a box. Unlike the Mark 8, the Apple I became an overnight sensation (Figure 7.12).

Although the early microcomputers were of little computing use being hobbyist toys, over time they became more and more popular, and thus there was a vested interest in improving them. By the end of the 1970s, both microprocessors and computer memory capacities improved, allowing for more capable microcomputers—including those that could perform rudimentary computer graphics. Modest word processing and accounting software, introduced at the end of the 1970s, made these computers useful for small and

FIGURE 7.12 Apple I built in a wooden box. (Courtesy of Alpha1, http://commons.wikimedia.org/wiki/File:Apple_1_computer.jpg.)

TABLE 7.2 Linear versus Exponential Increase

1000	1000
1200	2000
1400	4000
1600	8000
1800	16,000
2000	32,000
2200	64,000
2400	128,000

mid-sized businesses. Later, software was introduced that could fill niche markets such as desktop publishing, music and arts, and education. Coupled with graphical user interfaces (GUI), personal computers became an attractive option not just for businesses but for home use. With the introduction of a commercialized Internet, the computer became more than a business or educational tool, and today it is of course as common in a household as a car.

The most significant change that has occurred since 1974 can be summarized in one word: *miniaturization*. The third-generation computers comprised multiple circuit boards, interconnected in a chassis. Each board contained numerous ICs and each IC would contain a few dozen transistors (up to a few hundred by the end of the 1960s). By the 1970s, it was possible to miniaturize thousands of transistors to be placed onto a single chip. As time went on, the trend of miniaturizing transistors continued at an *exponential* rate.

Most improvements in our society occur at a slow rate, at best offering a linear increase in performance. Table 7.2 demonstrates the difference between an exponential and a linear increase. The linear improvement in the figure increases by a factor of approximately 20% per time period. This results in a doubling of performance over the initial performance in about five time periods. The exponential improvement doubles each time period resulting in an increase that is *orders-of-magnitude* greater over the same period. In the table, an increase from 1000 to 128,000 occurs in just seven time periods.

What does this mean with respect to our computers? It means that over the years, the number of transistors that can be placed on a chip has increased by orders of magnitude rather than linearly. The improvements in our computers have been dramatic in fairly short periods. It was Gordon Moore, one of the founders of Intel, who first noticed this rapidly

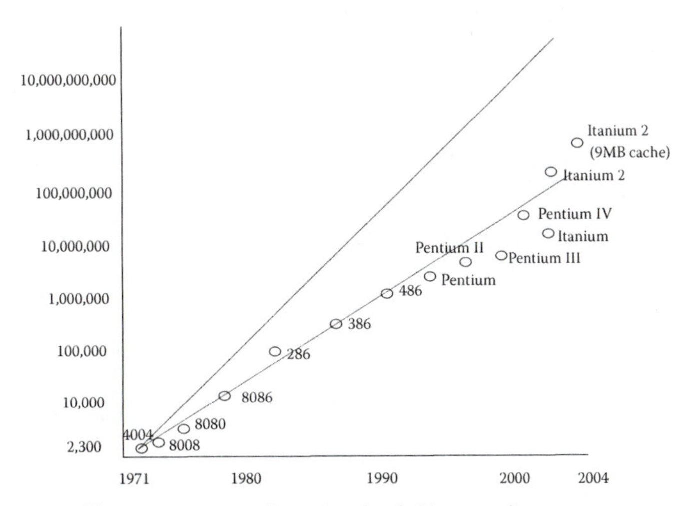

FIGURE 7.13 Charting miniaturization—Moore's law. The scale on the left increases exponentially, not linearly. (Adapted from Wgsimon, http://commons.wikimedia.org/wiki/File:Moore_Law_diagram_%282004%29.jpg.)

increasing "transistor count" phenomenon. In a 1965 paper, Moore observed that the trend from 1958 to 1965 was that the number of transistors on a chip was doubling every year. This phenomenon has been dubbed "Moore's law." Moore predicted that this trend would continue for at least 10 more years.

We find, in fact, that the degree of miniaturization is a doubling of transistor count roughly every 18 to 24 months, and that this trend has continued from 1965 to the present. The graph in Figure 7.13 illustrates the progress made by noting the transistor count on a number of different processors released between 1971 and 2011. It should be reiterated that Moore's law is an observation, not a physical law. We have come to rely on the trend in miniaturization, but there is certainly no guarantee that the trend will continue forever. In fact, there were many engineers who have felt that the rate of increase would have to slow down by 2000. Once we reached 2000 and Moore's law continued to be realized, engineers felt that 2010 would see the end of this trend. Few engineers today feel that Moore's law will continue for more than a few more years or a decade, and so there is a great deal of active research investigating new forms of semiconductor material other than silicon.

What might it mean if Moore's law was to fail us? Consider how often you purchase a car. Most likely, you buy a car because your current car is not road-worthy—perhaps due to accidents, wear and tear from excessive mileage, or failure to keep the car up to standards (although in some cases, people buy new cars to celebrate promotions and so forth). Now consider the computer. There is little wear and tear from usage and component parts seldom fail (the hard disk is the one device in the computer with moving parts and is likely to fail much sooner than any other). The desire to purchase a new computer is almost entirely made because of obsolescence.

What makes a computer obsolete? Because newer computers are better. How are they better? They are faster and have more memory. Why? Because miniaturization has led to a greater transistor count and therefore more capable components that are faster and have larger storage. If we are unable to continue to increase transistor count, the newer computers will be little better or no better than current computers. Therefore, people will have less need to buy a new computer every few years. We will get back to this thought later in the chapter.

Moore's law alone has not led us to the tremendously powerful processors of today's computers. Certainly, the reduction in size of the components on a chip means that the time it takes for the electrical current to travel continues to lessen, and so we gain a speedup in our processors. However, of greater importance are the architectural innovations to the processor, introduced by computer engineers. These have become available largely because there is more space on the chip itself to accommodate a greater number of circuits. Between the excessive amount of miniaturization and the architectural innovations, our processors are literally *millions* of times more powerful than those of the ENIAC. We also have main memory capacities of 8 GB, a number that would have been thought impossible as recently as the 1980s.

Many architectural innovations have been introduced over the years since the mid 1960s. But it is the period of the past 20 or so years that has seen the most significant advancements. One very important innovation is the *pipelined* CPU. In a pipeline, the fetch–execute cycle is performed in an overlapped fashion on several instructions. For instance, while instruction 1 is being executed, instruction 2 is being decoded and instruction 3 is being fetched. This would result in three instructions all being in some state of execution at the same time. The CPU does not execute all three simultaneously, but each instruction is undergoing a part of the fetch–execute cycle. Pipelines can vary in length from three stages as described here to well over a dozen (some of the modern processors have 20 or more stages). The pipelined CPU is much like an automotive assembly line. In the assembly line, multiple workers are working on different cars simultaneously.

Other innovations include parallel processing, on-chip cache memories, register windows, hardware that speculates over whether a branch should be taken, and most recently, multiple cores (multiple CPUs on a single chip). These are all concepts studied in computer science and computer engineering.

The impact of the microprocessor cannot be overstated. Without it, we would not have personal computers and therefore we would not have the even smaller computing devices (e.g., smart phones). However, the fourth generation has not been limited to just the innovations brought forth by miniaturization. We have also seen immense improvements in secondary storage devices. Hard disk capacity has reached and exceeded 1 TB (i.e., 1 trillion bytes into which you could store 1 billion books of text or about 10,000 high-quality CDs or more than 1000 low-resolution movies). Flash drives are also commonplace today and provide us with portability for transferring files. We have also seen the introduction of long-lasting batteries and LCD technologies to provide us with powerful laptop and notebook computers. Additionally, broadband wireless technology permits us to communicate practically anywhere.

MACHINES THAT CHANGED THE WORLD

In 1992, PBS aired a five-part series on the computer called *The Machine That Changed the World*. It should be obvious in reading this chapter that many machines have changed our world as our technology has progressed and evolved. Here is a list of other machines worth mentioning.

Z1—built by Konrad Zuse in Germany, it predated the ENIAC although was partially mechanical in nature, using telephone relays rather than vacuum tubes.

UNIVAC I—built by the creators of the ENIAC for Remington Rand, it was the first commercial computer, starting in 1951.

CDC 6600—the world's first supercomputer was actually released in 1964, not the 1980s! This mainframe outperformed the next fastest computer by a factor of 3.

Altair 8800—although the Mark 8 was the first personal computer, it was the Altair 8800 that computer hobbyists initially took to. This computer was also sold as a computer kit to be assembled. Its popularity though was nothing compared to the Apple I, which followed 5 years later.

IBM 5100—the first commercially available laptop computer (known as a portable computer in those days). The machine weighed 53 lb and cost between $9000 and $19,000!

IBM Simon—In 1979, NTT (Nippon Telegraph and Telephone) launched the first cellular phones. But the first smart phone was the IBM Simon, released in 1992. Its capabilities were limited to mobile phone, pager, fax, and a few applications such as a calendar, address book, clock, calculator, notepad, and e-mail.

TABLE 7.3 Milestone Comparisons

Year	Automotive Milestone	Computing Milestone	Year
6500 B.C.	The wheel	Abacus	4000? B.C.
1769	Steam-powered vehicles	Mechanical calculator	1642
1885	Automobile invented	Programmable device	1801
1896	First automotive death (pedestrian hit by a car going 4 mph)	First mechanical computer designed (analytical engine)	1832
1904	First automatic transmission	First digital, electronic, general purpose computer (ENIAC)	1946
1908	Assembly line permits mass production	Second-generation computers (cheaper, faster, more reliable)	1959
1911	First electric ignition	Third-generation computers (ICs, cheaper, faster, computer families)	1963
1925	About 250 highways available in the United States	ARPANET (initial incarnation of the Internet)	1969
1940	One quarter of all Americans own a car	Fourth-generation computers (microprocessors, PCs)	1974
1951	Cars reach 100 mph at reasonable costs	First hard disk for PC	1980
1956	Interstate highway system authorized (took 35 years to complete)	IBM PC released	1981
1966	First antilock brakes	Macintosh (first GUI)	1984
1973	National speed limits set in the United States, energy crisis begins	Stallman introduces GNUs	1985
1977	Handicapped parking introduced	Internet access in households	1990s
1997	First hybrid engine developed	Cheap laptops, smart phones	2000s

Today, we have handheld devices that are more powerful than computers from 10 to 15 years ago. Our desktop and laptop computers are millions of times more powerful than the earliest computers in the 1940s and 1950s. And yet, our computers cost us as little as a few hundred to a thousand dollars, little enough money that people will often discard their computers to buy new ones within just a few years' time. This chapter began with a comparison of the car and the computer. We end this section with a look at some of the milestones in the automotive industry versus milestones in the computer industry (Table 7.3). Notice that milestones in improved automobiles has taken greater amounts of time than milestones in computing, whereas the milestones in the automobile, for the most part, have not delivered cars that are orders-of-magnitude greater as the improvements have in the computing industry.

EVOLUTION OF COMPUTER SOFTWARE

The earliest computers were programmed in machine language (recall from Chapter 2 that a machine language program is a lengthy list of 1s and 0s). It was the engineers, those building and maintaining the computers of the day, who were programming the computers. They were also the users, the only people who would run the programs. By the second generation, better programming languages were produced to make the programming task easier for the programmers. Programmers were often still engineers, although not necessarily the same engineers. In fact, programmers could be business people who wanted software to perform operations that their current software could not perform. In the mid 1960s, IBM personnel made a decision to stop producing their own software for their hardware families. The result was that the organizations that purchased IBM computers would have to go elsewhere to purchase software. Software houses were introduced and the software industry was born.

The history of software is not nearly as exhilarating as the history of hardware. However, over the decades, there have been a number of very important innovations. These are briefly discussed in this section. As IT majors, you will have to program, but you will most likely not have to worry about many of the concerns that arose during the evolution of software because your programs will mostly be short scripts. Nonetheless, this history gives an illustration of how we have arrived where we are with respect to programming.

Early computer programs were written in machine language, written for a specific machine, often by the engineers who built and ran the computers themselves. Entering the program was not a simple matter of typing it in but of connecting memory locations to computational circuits by means of cable, much like a telephone operator used to connect calls. Once the program was executed, one would have to "rewire" the entire computer to run the next program. Recall that early computers were unreliable in part because of the short lifetime and unreliability of vacuum tubes. But add to that the difficult nature of machine language programming and the method of entering the program (through cables), and you have a very challenging situation. As a historical note, an early programmer could not get his program working even though the vacuum tubes were working, his logic was correct, and the program was correctly entered. It turned out that a moth had somehow gotten into a relay switch so that it would not pass current. This, supposedly, has led to the term *bug* being used to mean an error. We now use the term debugging not only to refer to removal of errors in a program, but just about any form of troubleshooting! See Figure 7.14.

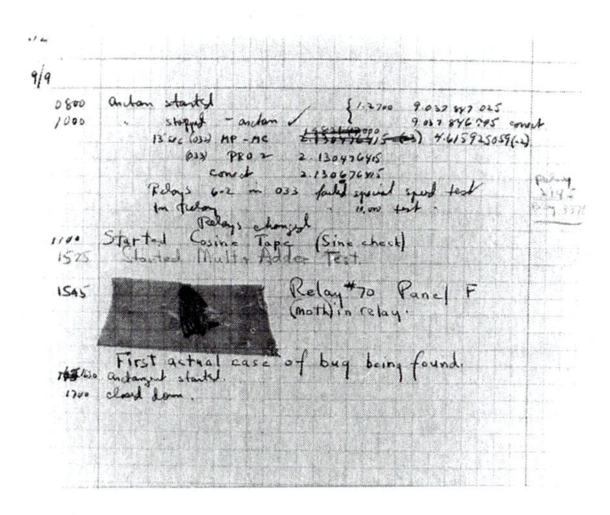

FIGURE 7.14 The first computer bug. (Courtesy of the U.S. Naval Historical Center, Online library photograph NH 96566-KN.)

Recall at this time, that computer input/output (I/O) was limited mostly to reading from punch cards or magnetic tape and writing to magnetic tape. Any output produced would be stored onto tape, unmounted from the tape drive, mounted to a printer, and then printed out. Since a computer only ran one program at a time and all input and output was restricted in such a manner, there was no need for an operating system for the computer. The programmer would include any I/O-specific code in the program itself.

The programming chore was made far easier with several innovations. The first was the idea of a language translator program. This program would take another program as input and output a machine language version, which could then be run on the computer. The original program, often known as the *source code*, could not be executed in its original form. The earliest language translators were known as *assemblers*, which would translate an assembly program into machine language. Assembly language, although easier than machine language, still required extremely detailed, precise, and low-level instructions (recall the example from Chapter 2). By 1959, language translation improved to the point that the language converted could be written in a more English-like way with far more powerful programming constructs. The improved class of language translator was known as a *compiler*, and the languages were called *high-level* languages. The first of these language translators was made for a language called FORTRAN. The idea behind FORTRAN was that the programmer would largely specify mathematical formulas using algebraic notation, along with input, output, and control statements. The control statements (loops, selections) would be fairly similar to assembly language, but the rest of the language would read more like English and mathematics. The name of the language comes from FORmula TRANslator. FORTRAN was primarily intended for mathematic/scientific computing. A business-oriented language was also produced at roughly the same time called COBOL (Common Business Oriented Language). Other languages were developed to support artificial intelligence research (LISP), simulation (Simula), and string manipulations (SNOBOL).

```
C code
     for(i=1;i<=x;i++)
          sum=sum+i;

Assembly Language code                                 Machine Language code
01231407    mov    dword ptr [i], 1                    C7 45 EC 01 00 00 00
0123140E    jmp    main+69h (1231419h)                 EB 09
01231410    mov    eax, dword ptr [i]                  8B 45 EC
01231413    add    eax, 1                              83 C0 01
01231416    mov    dword ptr [i], eax                  89 45 EC
01231419    mov    eax, dword ptr [i]                  8B 45 EC
0123141C    cmp    eax, dword ptr [x]                  3B 45 F8
0123141F    jg     main+7Ch (123142Ch)                 7F 0B
01231421    mov    eax, dword ptr [sum]                8B 45 E0
01231424    add    eax, dword ptr [i]                  03 45 EC
01231427    mov    dword ptr [sum], eax                89 45 E0
0123142A    jmp    main+60h (1231410h)                 EB E4
```

FIGURE 7.15 A comparison of high level, assembly and machine code.

Let us compare high level code to that of assembly and machine language. Figure 7.15 provides a simple C statement that computes the summation of all integer values from 1 to an input value, x. The C code is a single statement: a for loop. The for loop's body is itself a single assignment statement.

When this single C instruction is assembled into assembly language code (for the Intel x86 processor), the code is 12 instructions long. This is shown in the figure as three columns of information. The first column contains the memory address storing the instruction. The second column is the actual operation, represented as a mnemonic (an abbreviation). The third column contains the operands (data) that the instruction operates on.

For instance, the first instruction moves the value 1 into the location pointed to by a variable referenced as dword ptr. The fourth instruction is perhaps the easiest to understand, add eax, 1 adds the value 1 to the data register named the eax. Each assembly instruction is converted into one machine language instruction. In this case, the machine language (on the right-hand side of the figure) instructions are shown in hexadecimal. So, for instance, the first mov instruction (data movement) consists of 14 hexadecimal values, shown in pairs. One only need examine this simple C code to realize how cryptic assembly language can be. The assembly language mnemonics may give a hint as to what each operation does, but the machine language code is almost entirely opaque to understanding. The C code instead communicates to us with English words like for, along with variables and mathematical notation.

PROGRAMMERS WANTED!

Computer scientist and software engineer are fairly recent terms in our society. The first computer science department at a university was at Purdue University in 1962. The first Ph.D. in Computer Science was granted from the University of Pennsylvania in 1965. It was not until the 1970s that computer science was found in many universities. Software engineering was not even a term in our language until 1968.

Today, most programmers receive computer science degrees although there are also some software engineering degree programs. But who were the programmers before there were computer scientists? Ironically, like the story of Joe from Chapter 1, the computer scientist turned IT specialist, early programmers were those who learned to program on their own. They were engineers and mathematicians, or they were business administrators and accountants. If you knew how to program, you could switch careers and be a computer programmer.

Today, computer programmer is a dying breed. Few companies are interested in hiring someone who knows how to program but does not have a formal background in computing. So, today, when you see programming in a job listing, it will most likely require a computer science, information systems, computer engineering, or IT degree.

Although assembly language is not impossible to decipher for a programmer, it is still a challenge to make sense of. High level language code, no matter which language, consists of English words, mathematical notation, and familiar syntax such as a semicolon used to end statements. The C programming language was not written until 1968; however, the other high level languages of that era (FORTRAN, COBOL, etc.) were all much easier to understand than either machine or assembly language. The move to high level programming languages represents one of the more significant advances in computing because without it, developing software would not only be a challenge, the crude programming languages would restrict the size of the software being developed. It is unlikely that a team of 20 programmers could produce a million-line program if they were forced to write in either machine or assembly language.

Into the 1960s, computers were becoming more readily available. In addition, more I/O resources were being utilized and computer networks were allowing users to connect to the computer from dumb terminals or remote locations. And now, it was not just the engineers who were using and programming computers. In an organization that had a computer, any employee could wind up being a computer user. These users might not have understood the hardware of the computer, nor how to program the computer. With all of these added complications, a program was required that could allow the user to enter simple commands to run programs and move data files in such a way that the user would not have to actually write full programs. This led to the first operating systems.

In the early 1960s, the operating system was called a *resident monitor*. It would always be resident in memory, available to be called upon by any user. It was known as a monitor because it would monitor user requests. The requests were largely limited to running a program, specifying the location of the input (which tape drive or disk drive, which file(s)), and the destination of the output (printer, disk file, tape file, etc.). However, as the 1960s progressed and more users were able to access computers, the resident monitor had to become more sophisticated. By the mid 1960s, the resident monitor was being called an operating system—a program that allowed a user to operate the computer. The operating system would be responsible for program scheduling (since multiple programs could be requested by several users), program execution (starting and monitoring the program during execution, terminating the program when done), and user interface. The operating system would have to handle the requests of multiple users at a time. Program execution was performed by multiprogramming at first, and later on, time sharing (now called multitasking).

Operating systems also handled user protection (ensuring that one user does not violate resources owned by another user) and network security. Throughout the 1960s and 1970s, operating systems were text-based. Thus, even though the user did not have to understand the hardware or be able to program a computer, the user was required to understand how to use the operating system commands. In systems such as VMS (Virtual Memory System) run on DEC (Digital Equipment Corporation) VAX computers and JCL (Job Control Language) run on IBM mainframes, commands could be as elaborate and complex as with a programming language.

With the development of the personal computer, a simpler operating system could be applied. One of the most popular was that of MS-DOS, the disk operating system. Commands were largely limited to disk (or storage) operations—starting a program, saving a file, moving a file, deleting a file, creating directories, and so forth. Ironically, although the name of the operating system is DOS, it could be used for either disk or tape storage! The next innovation in operating systems did not arise until the 1980s. However, in the meantime....

Lessons learned by programmers in the 1960s led to a new revolution in the 1970s known as *structured programming.* Statements known as GOTOs were used in a large number of early languages. The GOTO statement allowed the programmer to transfer control from any location in a program to anywhere else in the program. For a programmer, this freedom could be a wonderful thing—until you had to understand the program to modify it or fix errors. The reason is that the GOTO statement creates what is now called *spaghetti code.* If you were to trace through a program, you would follow the instructions in sequential order. However, with the use of GOTO statements, suddenly after any instruction, you might have to move to another location in the program. Tracing through the program begins to look like a pile of spaghetti. In structured programming, the programmer is limited to high level control constructs such as while loops, for loops, and if–else statements, and is not allowed to use the more primitive GOTO statement. This ushered in a new era of high level languages, C and Pascal being among the most notable.

In the 1980s, another innovation looked to rock the programming world. Up until the mid 1980s, a programmer who wanted to model some entity in the world, whether a physical object such as a car, or an abstract object such as a word process document, would use individual variables. The variables would describe attributes of the object. For the car, for instance, variables might include age, gas mileage, type of car, number of miles, and current Blue Book value. A better modeling approach was to define classes of entities called *objects.* Objects could then be spawned by a program, each object being unique and modeling a different physical object (for instance, given a car class, we could generate four different cars). Objects would then interact with each other and with other types of objects. The difference between the object-oriented approach and the older, variable-based approach, is that an object is a stand-alone entity that would be programmed to handle its own internal methods as well as messages received from other objects. And with classes defined, a programmer could then expand the language by defining child classes. Through a technique called inheritance, a programmer is able to take a previous class and generate a more specific class out of it. This provides a degree of code reuse in that programmers could use other programmers' classes without having to reinvent the code themselves. The notion of

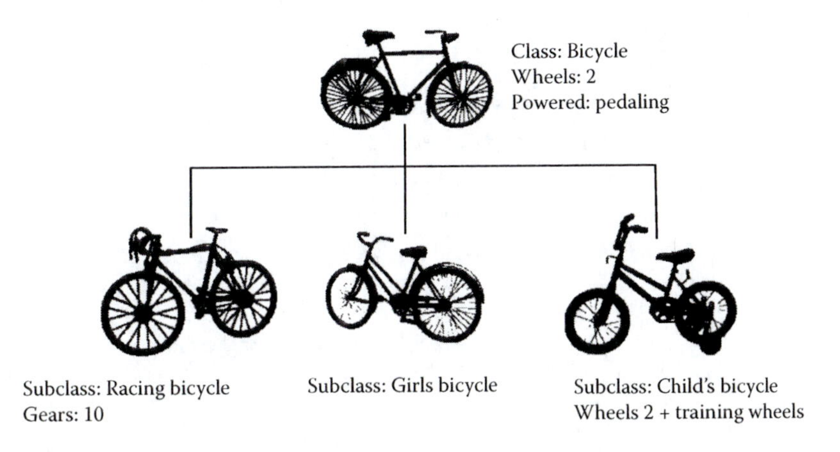

FIGURE 7.16 Object-oriented inheritance. (Adapted from Shutterstock/photo-master.)

inheritance is illustrated in Figure 7.16, where a bicycle class is the basis for several more specific types of bicycles. For instance, all bicycles in the hierarchy represent modes of transportation that contain two wheels powered by a human pedaling. However, there are subtypes of bicycles such as a racing bike, a girl's bike, or a bike with training wheels.

Object-oriented programming (OOP) was initially introduced in the language Smalltalk in the 1970s and early 1980s. In the mid 1980s, a variant of Smalltalk's object-oriented capabilities was incorporated into a new version of C, called C++. C++ became so popular that other object-oriented programming languages (OOPL) were introduced in the 1990s. Today, nearly every programming language has OOP capabilities.

Hand in hand with the development of OOPLs was the introduction of the first windowing operating system. To demonstrate the use of OOP, artificial intelligence researchers at Xerox Palo Alto California (Xerox Parc) constructed a windows-based environment. The idea was that a window would be modeled as an object. Every window would have certain features in common (size, location, background color) and operations that would work on any window (moving it, resizing it, collapsing it). The result was a windows operating system that they would use to help enhance their research. They did not think much of marketing their creation, but they were eager to demonstrate it to visitors. Two such visitors, Steven Jobs and Steven Wozniak (the inventors of the Apple personal computers), found this innovation to be extraordinary and predicted it could change computing. Jobs and Wozniak spent many years implementing their own version. Jobs was involved in the first personal computer to have the GUI, the Apple Lisa. But when it was released in 1983 for $10,000 per unit, very few were sold. In fact, the Lisa project became such a mess that Jobs was forced off of the project in 1982 and instead, he moved onto another Apple project, the Macintosh. Released in 1984 for $2500, the Macintosh was a tremendous success.*

A windows-based interface permits people to use a computer by directing a pointer at menu items or using clicking and dragging motions. With a windowing system, one does not need to learn the language of an operating system such as VMS, DOS, or Unix, but instead, one can now control a computer intuitively after little to no lessons. Thus, it was the GUI that opened up computer usage to just about everyone. Today, graphical programming and OOP are powerful tools for the programmer. Nearly all software today is graphical in nature, and a good deal of software produced comes from an object-oriented language.

Another innovation in programming is the use of an interpreter rather than a compiler. The compiler requires that the components of the software all be defined before compilation can begin. This can be challenging when a software project consists of dozens to hundreds of individual components, all of which need to be written before compilation. In a language such as Java, for instance, one must first write any classes that will be called upon by another class. However, if you simply want to test out an idea, you cannot do so with

* For those of you old enough, you might recall the first Apple Macintosh commercial. Airing during Super Bowl XVIII in January 1984 for $1.5 million, the commercial showed a number of similarly dressed men walking through drab hallways and seated in an auditorium. They were listening to a "Big Brother"-like figure. A young female athlete in a white tank top ran into the hall and threw a hammer at the screen. Upon impact, the screen exploded and the commercial ended with the caption that "...you'll see why 1984 won't be like '1984.'" The commercial was directed by Ridley Scott. You can find the commercial on YouTube.

incomplete code. The Lisp programming language, developed at the end of the 1950s for artificial intelligence research, used an interpreter. This allowed programmers to test out one instruction at a time, so that they could build their program piecemeal.

The main difference between a compiled language and an interpreted language is that the interpreted language runs inside of a special environment called the interpreter. The programmer enters a command, then the interpreter converts that command to machine language and executes it. The command might apply the result of previous commands by referring to variables set by earlier commands, although it does not necessarily have to. Thus, interpreted programming relies on the notion of a *session*. If the programmer succeeds in executing several instructions that go together, then the programmer can wrap them up into a program. In the compiled language, the entire program is written before it can be compiled, which is necessary before it can be executed. There are many reasons to enjoy the interpreted approach to programming; however, producing efficient code is not one of them. Therefore, most large software projects are compiled.

As an IT administrator, your job will most likely require that you write your own code from time to time. Fortunately, since efficiency will not necessarily be a concern (your code will be relatively small, perhaps as few as a couple of instructions per program), you can use an interpreted environment. In the Linux operating system, the shell contains its own interpreter. Therefore, writing a program is a matter of placing your Linux commands into a file. These small programs are referred to as *scripts*, and the process of programming is called shell scripting. In Windows, you can write DOS batch files for similar results.

Scripting goes beyond system administration, however. Small scripts are often written in network administration, web server administration, and database server administration. Scripts are also the tool of choice for web developers, who will write small scripts to run on the web server (server-side scripts), or in the web browser (client-side scripts). Server-side scripts, for instance, are used to process data entered in web-page forms or for generating dynamic web pages by pulling information out of a database. Client-side scripts are used to interact with the user, for instance, by ensuring that a form was filled out correctly or through a computer game. Many of the interpreted languages today can serve as scripting languages for these purposes. Scripting languages include perl, php, ruby, python, and asp. We will visit programming and some of these scripting languages in more detail in Chapter 14.

EVOLUTION OF THE COMPUTER USER

Just as computer hardware and computer software have evolved, so has the computer user. The earliest computer users were the engineers who both built and programmed the computers. These users had highly specialized knowledge of electronics, electrical engineering, and mathematics. There were only hundreds of such people working on a few dozen laboratory machines. By the 1950s, users had progressed to include programmers who were no longer (necessarily) the engineers building the computers.

In the 1960s and through the 1970s, users included the employees of organizations that owned or purchased time on computers. These users were typically highly educated, but perhaps not computer scientists or engineers. Some of the users were computer operators, whose job included such tasks as mounting tapes on tape drives and working with the computer hardware in clean rooms. But most users instead worked from their offices on dumb terminals. For this group of users, their skills did not have to include mathematics and electronics, nor programming, although in most cases, they were trained on the operating system commands of their computer systems so that they could enter operating system instructions to accomplish their tasks. The following are some examples from VMS, the operating system for DEC's VAX computers, which were very popular in the 1970s and 1980s (excerpted from a tutorial on VMS).

```
ASSIGN DEV$DISK:[BOB]POSFILE.DAT FOR015
COPY SYS$EXAMPLES:TUT.FOR TUT.FOR
DEL SYS$EXAMPLES:TUT.FOR
PRINT/FORM = 1/QUEUE = SYS$LASER TUT.FOR
SHOW QUEUE/ALL SYS$LASER
```

As the 1970s progressed, the personal computer let just about anyone use a computer regardless of their background. To control the early PCs, users had to know something of the operating system, so again, they wrote commands. You have already explored some MS-DOS commands earlier in the text. Although the commands may not be easy to remember, there are far fewer commands to master than in the more complex mainframe operating systems such as VMS.

With the release of the Apple Macintosh in 1984, however, the use of operating system commands became obsolete. Rather than entering cryptic commands at a prompt, the user instead controlled the computer using the GUI. All of the Macintosh software was GUI-based such that, for the first time, a user would not have to have any specialized knowledge to use the computer. It was intuitively easy. The Microsoft Windows operating system was introduced a year later, and by the 1990s, nearly all computers were accessible by windows-based GUI operating systems and applications software. In an interesting turn of events, it was often the younger generations who easily learned how to use the computer, whereas the older generations, those who grew up thinking that computers were enormous, expensive, and scary, were most hesitant to learn to use computers.

The most recent developments in computer usage is the look and feel of touch screen input devices such as smart phones and tablet PCs. Pioneered by various Apple products, the touch screen provides an even more intuitive control of the computer over the more traditional windows-style GUI. Scrolling, tapping, bringing up a keyboard, and using your fingers to move around perhaps is the next generation of operating systems. It has already been announced that Microsoft plans on adopting this look and feel to their next generation of desktop operating system, Windows 8.

Today, it is surprising to find someone who does not know how to use a computer. The required skill level remains low for using a computer. In fact, with so many smart phones on the planet, roughly half of the population of the planet can be considered computer users.

To understand computer fundamentals, one must know some basic computer literacy. To understand more advanced concepts such as software installation, hardware installation, performance monitoring, and so forth, even greater knowledge is needed. Much of this knowledge can be learned by reading manuals or magazines. It can also be learned by watching a friend or family member do something similar. Some users learn by trial and error. Today, because of the enormous impact that computers have had on our society, much of this knowledge is easily accessible over the Internet as there are web sites that both computer companies and individuals create that help users learn. Only the most specialized knowledge of hardware repair, system/network administration, software engineering, and computer engineering require training and/or school.

IMPACT ON SOCIETY

You are in your car, driving down the street. You reach a stop light and wait for it to change. Your GPS indicates that you should go straight. You are listening to satellite radio. While at the stop light, you are texting a friend on your smart phone (now illegal in some states). You are on your way to the movies. Earlier, you watched the weather forecast on television, telling you that rain was imminent. While you have not touched your computer today, you are immersed in computer technology.

- Your car contains several processors: the fuel-injection carburetor, the antilock brakes, the dashboard—these are all programmable devices with ICs.
- Your GPS is a programmable device that receives input from a touch screen and from orbiting satellites.
- Your satellite radio works because of satellites in orbit, launched by rocket and computer, and programmed to deliver signals to specific areas of the country.
- Your smart phone is a scaled-down computer.
- The movie you are going to see was written almost certainly by people using word processors. The film was probably edited digitally. Special effects were likely added through computer graphics. The list of how computers were used in the production of the film would most likely be lengthy, and that does not take into account marketing or advertising of the movie.
- The weather was predicted thanks to satellites (again, placed in orbit by rocket and computer), and the weather models were run on computers.
- Even the stop light is probably controlled by a computer. See the sensors in the road?

In our world today, it is nearly impossible to escape interaction with a computer. You would have to move to a remote area and purposefully rid yourself of the technology in order to remove computers from your life. And yet you may still feel their impact through bills, the post office, going to country stores, and from your neighbors.

Computer usage is found in any and every career. From A to Z, whether it be accounting, advertising, air travel, the armed forces, art, or zoology, computers are used to assist us and even do a lot of the work for us.

But the impact is not limited to our use of computers in the workforce. They are the very basis for economic transactions. Nearly all of our shopping takes place electronically:

inventory, charging credit cards, accounting. Computers dominate our forms of entertainment whether it is the creation of film, television programs, music, or art, or the medium by which we view/listen/see the art. Even more significantly, computers are now the center of our communications. Aside from the obvious use of the cell phone for phone calls, we use our mobile devices to see each other, to keep track of where people are (yes, even spy on each other), to read the latest news, even play games.

And then there is the Internet. Through the Internet, we shop, we invest, we learn, we read (and view) the news, we seek entertainment, we maintain contact with family, friends, and strangers. The global nature of the Internet combined with the accessibility that people have in today's society has made it possible for anyone and everyone to have a voice. Blogs and posting boards find everyone wanting to share their opinions. Social networking has allowed us to maintain friendships remotely and even make new friends and lovers.

The Internet also provides cross-cultural contact. Now we can see what it's like to live in other countries. We are able to view those countries histories and historic sites. We can watch gatherings or entertainment produced from those countries via YouTube. And with social networking, we can find out, nearly instantaneously what news is taking place in those countries. As a prime example, throughout 2011, the Arab Spring was taking place. And while it unfolded, the whole world watched.

To see the impact of the computer, consider the following questions:

1. When was the last time you wrote someone a postal letter? For what reason other than sending a holiday card or payment?

2. When was the last time you visited a friend without first phoning them up on your cell phone or e-mailing them?

3. When was the last time you read news solely by newspaper?

You will study the history of operating systems, with particular emphasis on DOS, Windows, and Linux, and the history of the Internet in Chapters 8 and 12, respectively. We revisit programming languages and some of their evolution in Chapter 14.

FURTHER READING

There are a number of excellent sources that cover aspects of computing from computer history to the changing social impact of computers. To provide a complete list could quite possibly be as lengthy as this text. Here, we spotlight a few of the more interesting or seminal works on the topic.

- Campbell-Kelly, M. and Aspray, W. *Computer: A History of the Information Machine.* Boulder, CO: Westview Press, 2004.

- Campbell-Kelly, M. *From Airline Reservations to Sonic the Hedgehog: A History of the Software Industry.* Cambridge, MA: MIT Press, 2004.

- Ceruzzi, P. *A History of Modern Computing*. Cambridge, MA: MIT Press, 1998.
- Ceruzzi, P. *Computing: A Concise History*. Cambridge, MA: MIT Press, 2012.
- Daylight, E., Wirth, N. Hoare, T., Liskov, B., Naur, P. (authors) and De Grave, K. (editor). *The Dawn of Software Engineering: from Turing to Dijkstra*. Heverlee, Belgium: Lonely Scholar, 2012.
- Ifrah, G. *The Universal History of Numbers: From Prehistory to the Invention of the Computer*. New Jersey: Wiley and Sons, 2000.
- Mens, T. and Demeyer, S. (editors). *Software Evolution*. New York: Springer, 2010.
- Rojas, R. and Hashagen, U. (editors). *The First Computers: History and Architectures*. Cambridge: MA: MIT Press, 2000.
- Stern, N. *From ENIAC to UNIVAC: An Appraisal of the Eckert-Mauchly Computers*. Florida: Digital Press, 1981.
- Swedin, E. and Ferro, D. *Computers: The Life Story of a Technology*. Baltimore, MD: Johns Hopkins University Press, 2007.
- Williams, M. *History of Computing Technology*. Los Alamitos, CA: IEEE Computer Society, 1997.

There are a number of websites dedicated to aspects of computer history. A few are mentioned here.

- http://americanhistory.si.edu/collections/comphist/
- http://en.wikipedia.org/wiki/Index_of_history_of_computing_articles
- http://www.computerhistory.org/
- http://www.computerhope.com/history/
- http://www.computersciencelab.com/ComputerHistory/History.htm
- http://www.trailing-edge.com/~bobbemer/HISTORY.HTM

REVIEW TERMS

Terminology introduced in this chapter:

Abacus	Compiler
Analytical Engine	Difference Engine
Assembler	Dumb terminal
Bug	ENIAC

GUI	Minicomputer
Integrated circuit	OOPL
Interpreter	Relay switch
Magnetic core memory	Resident monitor
Mainframe	Structured programming
Mechanical calculator	Vacuum tube
Microcomputers	Windows operating system
Microprocessor	

REVIEW QUESTIONS

1. How did Pascal's calculator actually compute? What was used to store information?
2. What device is considered to be the world's first programmable device?
3. In what way(s) should we consider Babbage's Analytical Engine to be a computer?
4. What notable achievement did Lady Ada Augusta Lovelace have?
5. Which of the four computer generations lasted the longest?
6. In each generation, the hardware of the processor was reduced in size. Why did this result in a speed up?
7. What was the drawback with using vacuum tubes?
8. What technology replaced vacuum tubes for second generation computers?
9. At what point in time did we see a shift in users from those who were building and programming computers to ordinary end users?
10. Since both the third- and fourth-generation computers used integrated circuits on silicon chips, how did the hardware of the fourth generation differ from that of the third?
11. In the evolution of operating systems, what was the most significant change that occurred in the fourth generation?
12. In the evolution of operating systems, at what point did they progress from tackling one program at a time to switching off between multiple programs?
13. What is structured programming and what type of programming instruction did structured programming attempt to make obsolete?
14. How did object-oriented programming improve on programming?
15. How do computer users today differ from those who used computers in the 1960s? From those who used computers in the 1950s?

DISCUSSION QUESTIONS

1. In your own words, describe the improvements in computer hardware in terms of size, expense, speed, value (to individuals), and cost from the 1950s to today.
2. A comparison was made in this chapter that said "if cars had progressed like computers, ..." Provide a similar comparison in terms of if medicine had progressed like computers.
3. What other human achievements might rival that of the computer in terms of its progress and/or its impact on humanity?
4. We might look at language as having an equal or greater impact on humanity as computers. Explain in what ways language impacts us more significantly than computer usage. Are there ways that computer usage impacts us more significantly than language?
5. Provide a ranking of the following innovations, creations, or discoveries in terms of the impact that you see on our daily lives: fire, refrigeration, automobiles, flight, radio/television, computers (including the Internet).
6. Imagine that a 50-year-old person has been stranded on an island since 1980 (the person would have been 18 at that time). How would you explain first what a personal computer is and second the impact that the personal computer has had on society?
7. Describe your daily interactions with computers or devices that have computer components in them. What fraction of your day is spent using or interacting with these devices (including computers).
8. Attempt to describe how your life would be different without the Internet.
9. Attempt to describe how your life would be different without the cell phone.
10. Attempt to describe how your life would be different without any form of computer (these would include your answers to questions #8 and #9).
11. Imagine that computing is similar to how it was in the early 1980s. Personal computers were available, but most people either used them at home for simple bookkeeping tasks (e.g., accounting, taxes), word processing, and/or computer games, and businesses used them largely for specific business purposes such as inventory and maintaining client data. Furthermore, these computers were entirely or primarily text-based and not connected to the Internet. Given that state, would you be as interested in IT as a career or hobby? Explain.

CHAPTER 8

Operating Systems History

Chapter 7 covered the evolution of hardware and software. In this chapter, the focus is on the evolution of operating systems with particular emphasis on the development of Linux and Unix and separately, the development of PC-based GUI operating systems. The examination of Linux covers two separate threads: first, the variety of Linux distributions that exist today, and second, the impact that the open source community has had on the development of Linux.

The learning objectives of this chapter are to

- Describe how operating systems have evolved.
- Discuss the impact that Linux has had in the world of computing.
- Compare various Linux distributions.
- Describe the role that the open source community has played.
- Trace the developments of PC operating systems.

In this chapter, we look at the history and development of various operating systems (OSs), concentrating primarily on Windows and Linux. We spotlight these two OSs because they are two of the most popular in the world, and they present two separate philosophies of software. Windows, a product of the company Microsoft, is an OS developed as a commercial platform and therefore is proprietary, with versions released in an effort to entice or force users to upgrade and thus spend more money. Linux, which evolved from the OS Unix, embraces the Open Source movement. It is free, but more than this, the contents of the OS (the source code) are openly available so that developers can enhance or alter the code and produce their own software for the OS—as long as they follow the Open GNUs Licensing (GL) agreement. Before we look at either Windows or Linux, we look at the evolution of earlier OSs.

BEFORE LINUX AND WINDOWS

As discussed in Chapter 7, early computers had no OSs. The users of these computers were the programmers and the engineers who built the computers. It was expected that their programs would run with few or no resources (perhaps access to punch cards for input, with the output being saved to tape). The operations of reading from punch cards and writing to tape had to be inserted into the programs themselves, and thus were written by the users.

Several changes brought about the need for an OS. The first was the development of language translators. To run a program, a programmer would have to first compile (or assemble) the program from its original language into machine language. The programmer would have to program the computer for several distinct steps. First, a translation program (compiler or assembler) would have to be input from tape or punch card. Second, the source code would have to be input, again from tape or punch card. The translation program would execute, loading the source code, translating the source code, and saving the resulting executable code onto magnetic tape. Then the executable program would be input from tape, with data being input again from card or tape. The program would be executed and the output would be sent to magnetic tape to be printed later. Without an OS, every one of the input and output tasks would have to be written as part of the program. To simplify matters, the *resident monitor* was created to handle these operations based on a few commands rather than dozens or hundreds of program instructions. See Figure 8.1, which illustrates in Job Control Language (JCL; used on IBM mainframes) the instructions to copy a file from one location to another.

Another change that brought about the need for OSs was the availability of computers. Into the second generation, computers were less expensive, resulting in more computers being sold and more organizations having access to them. With the increase in computers came an increase in users. Users, starting in the second generation, were not necessarily programmers or engineers. Now, a typical employee might use a computer. The OS was a mechanism that allowed people to use a computer without necessarily having to understand how to program a computer.

Additionally, the improvement in computer hardware led to OSs in two ways. First, the improved speed led to programs requiring less time to execute. In the first generation, the computer user was also the engineer and the programmer. Switching from one program

```
//IS198CPY  JOB (IS198T30500),'COPY JOB', CLASS = L, MSGCLASS = X
//COPY01    EXEC  PGM = IEBGENER
//SYSPRINT  DD  SYSOUT = *
//SYSUT1    DD  DSN = OLDFILE, DISP = SHR
//SYSUT2    DD  DSN = NEWFILE,
//              DISP = (NEW, CATLG, DELETE),
//              SPACE = (CYL, (40,5), RLSE),
//              DCB = (LRECL = 115, BLKSIZE = 1150)
//SYSIN     DD  DUMMY
```

FIGURE 8.1 JCL instructions.

and thus one user to another was time consuming. With programs taking less time to execute, the desire to improve this transition led to OSs. Also, the improved reliability of computers (having abandoned short lifespan vacuum tubes) permitted longer programs to execute to completion. With longer programs came a need for handling additional computing resources and therefore a greater demand for an OS.

OSs grew more complex with the development of multiprogramming and time sharing (multitasking). This, in turn, permitted dozens or hundreds or thousands of users to use the computer at the same time. This increase in usage, in turn, required that OSs handle more and more tasks, and so the OSs became even more complex. And as discussed in Chapter 7, windowing OSs in the fourth generation changed how users interact with computers. What follows is a brief description of some of the early OSs and their contributions.

The earliest OS used for "real work" was GM-NAA I/O, written by General Motors for use on IBM 701 mainframe computers. Its main purpose was to automatically execute a new program once the previously executing program had terminated. It used batch processing whereby input was supplied with the program. It was a collection of an expanded resident monitor written in 1955 along with programs that could access the input and output devices connected to the mainframe.

Atlas Supervisor in 1957 for Manchester University permitted concurrent user access. It is considered to be the first true OS (rather than resident monitor). Concurrent processing would not generally be available until the early 1960s (see CTSS). Additionally, the Atlas was one of the first to offer virtual memory.

BESYS was Bell Operating System, developed by Bell Laboratories in 1957 for IBM 704 mainframes. It could handle input from both punch cards and magnetic tape, and output to either printer or magnetic tape. It was set up to compile and run FORTRAN programs.

IBSYS, from 1960, was released by IBM with IBM 7090 and 7094 mainframes. OS commands were embedded *in programs* by inserting a $ in front of any OS command to differentiate it from a FORTRAN instruction. This approach would later be used to implement JCL (Figure 8.1) instructions for IBM 360 and IBM 370 mainframe programs.

CTSS, or Compatible Time-Sharing System, released in 1961 by the Massachusetts Institute of Technology's (MIT) Computation Center, was the first true time-sharing (multitasking) OS. Unlike the Atlas Supervisor, one component of CTSS was in charge of cycling through user processes, offering each a share of CPU time (thus the name time sharing).

EXEC 8, in 1964, was produced for Remington Rand's UNIVAC mainframes. It is notable because it was the first successful commercial multiprocessing OS (i.e., an OS that runs on multiple processors). It supported multiple forms of process management: batch, time sharing, and real-time processing. The latter means that processes were expected to run immediately without delay and complete within a given time limit, offering real-time interaction with the user.

TOPS-10, also from 1964, was released by Digital Equipment Corporation (DEC) for their series of PDP-10 mainframes. TOPS-10 is another notable time sharing OS because it introduced shared memory. Shared memory would allow multiple programs to communicate with each other through memory. To demonstrate the use of shared memory, a multiplayer Star Trek–based computer game was developed called DECWAR.

MULTICS, from 1964, introduced dynamic linking of program code. In software engineering, it is common for programmers to call upon library routines (pieces of code written by other programmers, compiled and stored in a library). By using dynamic linking, those pieces of code are loaded into memory only when needed. MULTICS was the first to offer this. Today, we see a similar approach in Windows with the use of "dll" files. MULTICS was a modularized OS so that it could support many different hardware platforms and was scalable so that it could still be efficient when additional resources were added to the system. Additionally, MULTICS introduced several new concepts including access control lists for file access control, sharing process memory and the file system (i.e., treating running process memory as if it were file storage, as with Unix and Linux using the /proc directory), and a hierarchical file system. Although MULTICS was not a very commercially successful OS, it remained in operation for more than a decade, still in use by some organizations until 1980. It was one of the most influential OSs though because it formed a basis for the later Unix OS.

OS360 for IBM 360 (and later, OS370 for IBM 370) mainframe computers was released in 1966. It was originally a batch OS, and later added multiprogramming. Additionally, within any single task, the task could be executed through multitasking (in essence, there was multithreading in that a process could be multitasked, but there was no multitasking between processes). OS360 used JCL for input/output (I/O) instructions. OS360 shared a number of innovations introduced in other OSs such as virtual memory and a hierarchical file system, but also introduced its own virtual storage access method (which would later become the basis for database storage), and the ability to spawn child processes. Another element of OS360 was a data communications facility that allowed the OS to communicate with any type of terminal. Because of the popularity of the IBM 360 and 370 mainframes, OS360 became one of the most popular OSs of its time. When modified for the IBM 370, OS360 (renamed System/370) had few modifications itself. Today, OS360 remains a popular experimental OS and is available for free download.

Unics (later, UNIX) was developed by AT&T in 1969 (see the next section for details). Also in 1969, the IBM Airline Control Program (ACP) was separated from the remainder of IBM's airline automation system that processed airline reservations. Once separated, ACP, later known as TPF (Transaction Processing Facility) was a transaction processing OS for airline database transactions to handle such tasks as credit card processing, and hotel and rental car reservations. Although not innovative in itself, it provides an example of an OS tailored for a task rather than a hardware platform.

In the early 1970s, most OSs merely expanded upon the capabilities introduced during the 1960s. One notable OS, VM used by IBM, and released in 1972, allowed users to create virtual machines.

A HISTORY OF UNIX

The Unix OS dates back to the late 1960s. It is one of the most powerful and portable OSs. Its power comes from a variety of features: file system administration, strong network components, security, custom software installation, and the ability to define your own kernel programs, shells, and scripts to tailor the environment. Part of Unix's power comes from its flexibility of offering a command line to receive OS inputs. The command line allows the user or system administrator to enter commands with a large variety of options. It is portable because it is written in C, a language that can be compiled for nearly any platform, and has been ported to a number of very different types of computers including supercomputers, mainframes, workstations, PCs, and laptops. In fact, Unix has become so popular and successful that it is now the OS for the Macintosh (although the Mac windowing system sits on top of it).

Unix was created at AT&T Bell Labs. The original use of the OS was on the PDP-7 so that employees could play a game on that machine (called Space Travel). Unix was written in assembly language and not portable. Early on, it was not a successful OS; in fact, it was not much of a system at all. After the initial implementation, two Bell Laboratories employees, Ken Thompson and Dennis Ritchie, began enhancing Unix by adding facilities to handle files (copy, delete, edit, print) and the command-line interpreter so that a user could enter commands one at a time (rather than through a series of punch cards called a "job"). By 1970, the OS was formally named Unix.

By the early 1970s, Unix was redeveloped to run on the PDP-11, a much more popular machine than the PDP-7. Before it was rewritten, Ritchie first designed and implemented a new programming language, C. He specifically developed the language to be one that could implement an OS, and then he used C to rewrite Unix for the PDP-11. Other employees became involved in the rewriting of Unix and added such features as pipes (these are discussed in Chapter 9). The OS was then distributed to other companies (for a price, not for free).

By 1976, Unix had been ported to a number of computers, and there was an ever-increasing Unix interest group discussing and supporting the OS. In 1976, Thompson took a sabbatical from Bell Labs to teach at University of California–Berkeley, where he and UCB students developed the Berkeley Standard Distribution version of Unix, now known as BSD Unix. BSD version 4.2 would become an extremely popular release. During this period, Unix also adopted the TCP/IP protocol so that computers could communicate over network. TCP/IP was the protocol used by the ARPAnet (what would become the Internet), and Unix was there to facilitate this.

Into the 1980s, Unix's increasing popularity continued unabated, and it became the OS of choice for many companies purchasing mainframe and minicomputers. Although expensive and complex, it was perhaps the best choice available. In 1988, the Open Source Foundation (OSF) was founded with the express intent that software and OS be developed freely. The term "free" is not what you might expect. The founder, Richard Stallman, would say "free as in freedom not beer." He intended that the Unix user community would invest time into developing a new OS and support software so that anyone could obtain it, use it, modify it, and publish the modifications. The catch—anything developed and released could only be released under the same "free" concept. This required that all code be made available for free as source code. He developed the GPL (GNUs* Public License), which

* GNU stands for GNU Not Unix. This is a recursive definition that really does not mean anything. What Stallman was trying to convey was that the operating system was not Unix, but a Unix-like operating system that would be composed solely of free software.

stated that any software produced from GPL software had to also be published under the GPL license and thus be freely available. This caused a divide into the Unix community—those willing to work for free and those who worked for profit.

The OSF still exists and is still producing free software, whereas several companies are producing and selling their own versions of Unix. Control of AT&T's Unix was passed on to Novell and then later to Santa Cruz Operation, whereas other companies producing Unix include Sun (which produces a version called Solaris), Hewlett Packard (which sells HP-UX), IBM (which sells AIX), and Compaq (which sells Tru64 Unix). Unix is popular today, but perhaps not as popular as Linux.

A HISTORY OF LINUX

Linux looks a lot like Unix but was based on a free "toy" OS called Minix. Minix was available as a sample OS from textbook author Andrew Tanenbaum. University of Helsinki student Linus Torvalds wanted to explore an OS in depth, and although he enjoyed learning from Minix, he ultimately found it unsatisfying, so he set about to write his own. On August 26, 1991, he posted to the comp.os.minix Usenet newsgroup to let people know of his intentions and to see if others might be interested. His initial posting stated that he was building his own version of Minix for Intel 386/486 processors (IBM AT clones). He wanted to pursue this as a hobby, and his initial posting was an attempt to recruit other hobbyists who might be interested in playing with it. His initial implementation contained both a Bash interpreter and the current GNU's C compiler, gcc. Bash is the Bourne Again Shell, a variation of the original Bourne shell made available for Unix. Bash is a very popular shell, and you will learn about it in Chapter 9. He also stated that this initial implementation was completely free of any Minix code. However, this early version of an OS lacked device drivers for any disk drive other than AT hard disks.

Torvalds released version .01 in September of 1991 and version .02 in October having received a number of comments from users who had downloaded and tested the fledgling OS. On October 5, to announce the new release, Torvalds posted a follow-up message to the same newsgroup. Here, he mentions that his OS can run a variety of Unix programs including Bash, gcc, make (we cover make in Chapter 13), sed (mentioned in Chapter 10), and compress. He made the source code available for others to not only download and play with, but to work on the code in order to bring about more capabilities.

By December, Torvalds released version .10 even though the OS was still very much in a skeletal form. For instance, it had no log in capabilities booting directly to a Bash shell and only supported IBM AT hard disks. Version .11 had support for multiple devices and after version .12, Torvalds felt that the OS was both stable enough and useful enough to warrant the release number .95. Interestingly, at this point, Torvalds heard back from Tanenbaum criticizing Torvalds' concept of an OS with a "monolithic" kernel, remarking that "Linux is obsolete." Tanenbaum would be proved to be wrong with his comments. In releasing his first full-blown version, Linus Torvalds decided to adopt the GNU General Public License. This permitted anyone to obtain a free copy of the OS, make modifications, and publish those modifications for free. However, Torvalds also decided to permit others to market versions of Linux. This decision to have both free versions (with source code available for modification) and commercial versions turns out to be a fortuitous idea.

Within a few years, Linux group supporters numbered in the hundreds of thousands. Commercial versions were being developed while the user community continued to contribute to the software to make it more useful and desirable. Graphical user interfaces were added to make Linux more useable to the general computing populace.

Today, both Unix and Linux are extremely popular OS formats. Both are available for a wide range of machines from handheld devices to laptops and PCs to mainframes to supercomputers. In fact, Linux can run on all of the following machines/processors: Sun's Sparc, Compaq's Alpha, MIPS processors, ARM processors (found in many handheld devices), Intel's x86/Pentium, PowerPC processors, Motorola's 680x0 processors, IBM's RS6000, and others. One reason why the OSs are so popular is that most servers on the Internet run some version of Unix or Linux.

Here are a few interesting facts about Linux:

- The world's most powerful supercomputer, IBM's Sequoia, runs on Linux, and 446 of the top 500 supercomputers run Linux.
- Ninety-five percent of the servers used in Hollywood animation studios run Linux.
- Google runs its web servers in Linux.
- The OSs for Google's Android and Nokia's Maemo are built on top of the Linux kernel.
- Notably, 33.8% of all servers run Linux (as of 2009), whereas only 7.3% run a Microsoft server.
- Available application software for Linux include Mozilla Firefox and OpenOffice, which are free, and Acrobat Reader and Adobe Flash Player, which are proprietary.
- There are more than 300 distributions of Linux deployed today.
- The Debian version of Linux consists of more than 280 million lines of code, which, if developed commercially, would cost more than $7 billion; the original Linux kernel (1.0.0) had 176,250 lines of code.

On the other hand, as of January 2010, slightly more than 1% of all desktops run Linux.

Aside from the popularity of Linux, its cross-platform capabilities and the ability to obtain Linux versions and application software for free, there are several other appealing characteristics of Linux. First, Linux is a stable OS. If you are a Windows user, you have no doubt faced the frustration of seeing application software crash on you with no notice and little reason. There are also times when rebooting Windows is a necessity because the OS has been active for too long (reasons for the need to reboot Windows include corrupted OS pages and fragmented swap space). Linux almost never needs to be rebooted. In fact, Linux only tends to stop if there is a hardware failure or the user shuts the system down. Anything short of a system upgrade does not require a reboot. Furthermore, Linux is not susceptible to computer viruses. Although most viruses target the Windows OS, if someone were to write a virus to target Linux, it would most likely fail because of Linux' memory management, which ensures no memory violations (see Chapter 4).* Also, Linux is able to support software more efficiently. As an experiment, Oracle compared the performance of Oracle 9i when run on Windows 2000 versus Linux. The Linux execution was 25–30% more efficient.

In a survey of attendees at LinuxCon 2011, the results showed that the most popular distribution of Linux was Ubuntu (34%), followed by Fedora/Red Hat (28%). Of those surveyed, nearly half used Linux at home, work, and/or school, whereas 38% used Linux at work and 14% at home.

DIFFERENCES AND DISTRIBUTIONS

Figure 8.2 provides a timeline for the various releases between the original Unics OS through Unix, MacOS, and Linux releases (up to 2010). See Figure 8.3 for a look at the more significant Linux distribution releases. What are the differences between Unix and Linux? What are the differences between the various distributions of Linux (Debian, Red Hat, etc.)? Why are there so many releases? Why is this so confusing?

Let us consider how Linux and Unix differ. If you look at the two OSs at a shallow level, they will look very similar. They have nearly identical commands (e.g., ls, cd, rm, pwd), and most of those commands have overlapping or identical options. They have many of the same shells available, and the most common graphical user interfaces (GUIs) run on both platforms (KDE and Gnome). They have very similar top level directory structures. Additionally, both Linux and Unix are reliable (as compared to say Windows), portable to many different types of hardware, and on the small size (when compared to an OS like Windows). What are the differences then?

Although Linux is open source and many of the software products written for Linux and Unix are open source, commercial versions of Unix are not open source, and most versions of Unix are commercial products. There are standards established for the various Unix releases that help maintain consistency between versions. This does not guarantee that a software product written for one version of Unix will run on another, but it does help. For Linux, the Linux Standards Base project began in 2001 and is an ongoing attempt to help create standards for all Linux releases. The standards describe what should or should not be contained in a distribution. For instance, the standard dictates which version of the C++ gnus compiler and which C++ libraries should be included. In 2011, the Java compiler was removed from the standard. However, Linux programmers do not have to follow the Linux Standards Base. This can lead to applications software that may not run on a given distribution, patches that are required for one release but not another, and command line arguments which work in one distribution that may not work in another. This can be very frustrating for both programmers and users.

Other differences between Linux and Unix are subtle and perhaps not noticeable to end users. Some differences are seen by system administrators because configuration files

* Linux does not claim to be virus-free, but years of research have led to an operating system able to defeat many of the mechanisms used by viruses to propagate themselves.

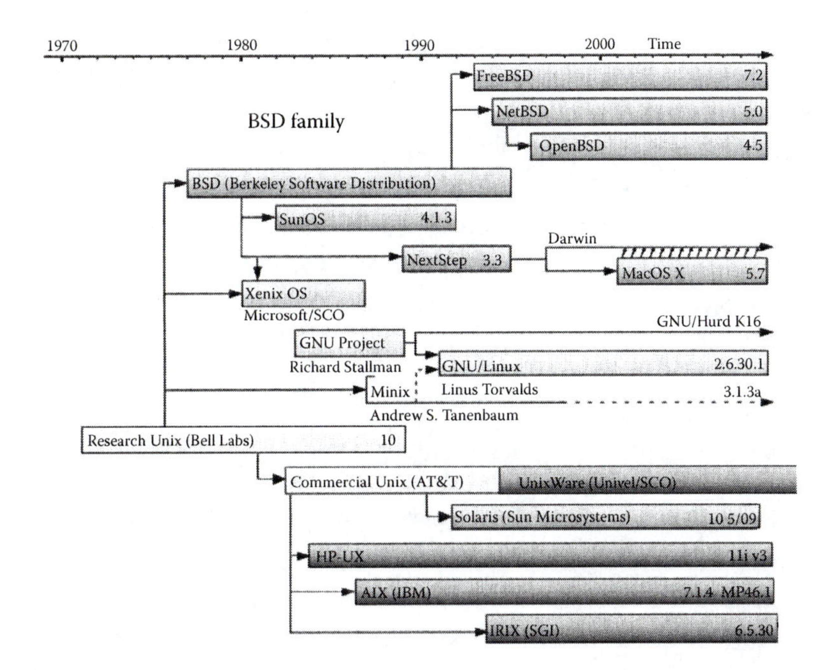

FIGURE 8.2 Unix/Linux timeline. (Courtesy of http://commons.wikimedia.org/wiki/File:Unix.png, author unknown.)

will differ (different locations, different names, different formats). A programmer might see deeper differences. For instance, the Linux kernel rarely executes in a threaded form whereas in Unix, kernel threads are common.

What about the Linux releases? As you can see in Figure 8.3, there are many different forms, or distributions, of Linux, and the various "spinoff" distributions have occurred numerous times over the past 20 years. For instance, as shown in the figure, SLS/Slackware Linux was used to develop S.u.S.E Linux, which itself was used to develop OpenSUSE.

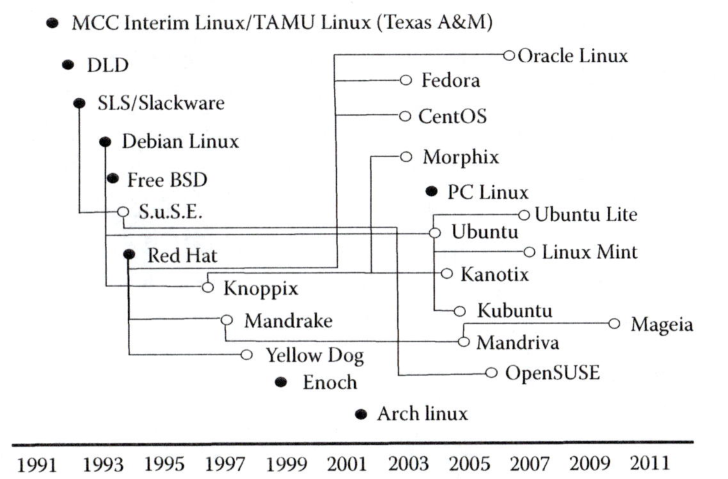

FIGURE 8.3 Key Linux distribution releases.

Debian Linux led to both Knoppix and Ubuntu, whereas Ubuntu led to Kubuntu, Linux Mint, and Ubuntu Lite. Red Hat was the basis for Mandrake, Yellow Dog, CentOS, Fedora, and Oracle Linux. Out of Mandrake came PC Linux and Mandiva.

The main two distributions of Linux are Debian and Red Hat, which along with Slackware (SLS), were the earliest distributions. Although Figure 8.3 shows the major distributions of Linux, there are dozens of other releases that are not shown.

There are far too many differences in the various distributions to list here. In fact, there are websites whose primary reason for existence is to list such differences. Here, we will limit ourselves to just four releases: Red Hat, Debian, Ubuntu, and SUSE.

- Red Hat—Released in 1994 by Bob Young and Marc Ewing, it is the leading version of Linux in terms of development and deployment. It can be found anywhere from embedded devices to Internet servers. Red Hat (the OS) is open source. However, Red Hat Enterprise Linux is only available for purchase, although it comes with support. Red Hat supports open source applications software, and in an interesting development, has acquired proprietary software and made it available as open source as well. Some of the innovations of Red Hat include their software management package, RPM (Red Hat Package Manager), SELinux (security enabled Linux), and the JBoss middleware software (middleware permits multiprocessing between applications software and the OS). Two Spinoffs of Red Hat are Fedora and CentOS, both of which were first released in 2004 and are free and open source. CentOS Linux is currently the most popular Linux distribution for web servers.

- Debian—Released in 1993 by Ian Murdock, Debian was envisioned to be entirely noncommercial, unlike Red Hat for which organizations can purchase and obtain support. Debian boasts more than 20,000 software packages. Debian has been cited as the most stable and least buggy release of any Linux. Debian also runs on more architectures (processors) than any other version of Linux. However, these two advantages may not necessarily be as good as they sound, as to support this the OS developers are more conservative in their releases. Each release is limited regarding new developments and features. In fact, Debian major releases come out at a much slower pace than other versions of Linux (every 1 to 3 years). Debian, like Red Hat, can run in embedded devices. Debian introduced the APT (Advanced Package Tool) for software management. Unlike Red Hat, the Debian OS can be run from removable media (known as a *live CD*). For instance, one could boot Debian from either CD or USB device rather than having Debian installed on the hard disk. Aside from Ubuntu (see below), there are several other Debian spinoffs, most notably MEPIS, Linux Mint, and Knoppix, all of which can be run from a boot disk or USB.

- Ubuntu—This was released in 2004 by Mark Shuttleworth and based on an unstable version of Debian, but has moved away from Debian so that it is not completely compatible. Unlike Debian, Ubuntu has regular releases (6 months) with occasional long-term support releases (every 3 to 5 years). Ubuntu, more than other Linux releases, features a desktop theme and assistance for users who wish to move from Windows to Linux. There are 3-D effects available, and Ubuntu provides excellent support for Wiki-style documentation. Shuttleworth is a multimillionaire and shared Ubuntu with everyone, including installation CDs for free (this practice was discontinued in 2011). As with other Debian releases, Ubuntu can be run from a boot disk or USB device.

- SUSE—Standing for Software und System-Entwicklung (German for software and systems development), it is also abbreviated as SuSE. This version of Linux is based on Slackware, a version of Linux based on a German version of Softlanding Linux Systems (SLS). Since SLS became defunct, SuSE is then an existing descendant of SLS. The first distribution of SuSE came in 1996, and the SuSE company became the largest distributor of Linux in Germany. SuSE entered the U.S. market in 1997 and other countries shortly thereafter. In 2003, Novell acquired SuSE for $210 million. Although SuSE started off independently of the Linux versions listed above, it has since incorporated aspects of Red Hat Linux including Red Hat's package manager and Red Hat's file structure. In 2006, Novell signed an agreement with Microsoft so that SuSE would be able to support Microsoft Windows software. This agreement created something of a controversy with the Free Software Foundation (FSF; discussed in the next section).

Still confused? Probably so.

OPEN SOURCE MOVEMENT

The development of Linux is a more recent event in the history of Unix. Earlier, those working in Unix followed two different paths. On one side, people were working in BSD Unix, a version that was developed from UC Berkeley from work funded by Defense Advanced Research Projects Agency (DARPA). The BSD users, who came from all over the world, but largely were a group of hackers from the West Coast, helped debug, maintain, and improve Unix over time. Around the same time in the late 1970s or early 1980s, Richard Stallman, an artificial intelligence researcher at MIT, developed his own project, known as GNU.* Stallman's GNU project was to develop software for the Unix OS in such a way that the software would be free, not only in terms of cost but also in terms of people's ability to develop the software further.

Stallman's vision turned into the free software movement. The idea is that any software developed under this movement would be freely available and people would be free to modify it. Therefore, the software would have to be made available in its source code format (something that was seldom if ever done). Anyone who altered the source code could then contribute the new code back to the project. The Unix users who followed this group set about developing their own Unix-like OS called GNU (or sometimes GNU/Linux). It should be noted that although Stallman's project started as early as 1983, no stable version of GNU has ever been released.

* GNU stands for GNU Not Unix, thus a recursive definition, which no doubt appeals to Stallman as recursion is a common tool for AI programming.

In 1985, Stallman founded the FSF. The FSF, and the entire notion of a free software movement, combines both a sociological and political view—users should have free access to software to run it, to study it, to change it, and to redistribute copies with or without changes.

It is the free software movement that has led to the Open Source Initiative (OSI), an organization created in 1998 to support open source software. However, the OSI is not the same as the FSF; in fact, there are strong disagreements between the two (mostly brought about by Stallman himself). Under the free software movement, all software should be free. Selling software is viewed as ethically and morally wrong. Software itself is the implementation of ideas, and ideas should be available to all; therefore, the software should be available to all. To some, particularly those who started or are involved in OSI, this view is overly restrictive. If one's career is to produce software, surely that person has a right to earn money in doing so. The free software movement would instead support hackers who would illegally copy and use copyrighted software.

The open source community balances between the extremes of purely commercial software (e.g., Microsoft) and purely free software (FSF) by saying that software contributed to the open source community should remain open. However, one who uses open source software to develop a commercial product should be free to copyright that software and sell it as long as the contributions made by others are still freely available. This has led to the development of the GNU General Public License (GPL). Although the GPL was written by Stallman for his GNU Project, it is in fact regularly used by the open source community. It states that the given software title is freely available for distribution and modification as long as anything modified retains the GPL license. Stallman called the GPL a copyleft instead of a copyright. It is the GPL that has allowed Linux to have the impact it has had. About 50% of Red Hat Linux 7.1 contains the GPL as do many of the software products written for the Linux environment. In fact, GPL accounts for nearly 65% of all free software projects.*

Today, the open source community is thriving. Thousands of people dedicate time to work on software and contribute new ideas, new components within the software, and new software products. Without the open source philosophy, it is doubtful that Linux would have attained the degree of success that is has because there would be no global community contributing new code. The open source community does not merely contribute to Linux, but Linux is the largest project and the most visible success to come from the community. It has also led to some of the companies rethinking their commercial approaches. For instance, Microsoft itself has several open source projects.

A HISTORY OF WINDOWS

The history of Windows in some ways mirrors the history of personal computers because Windows, and its predecessor, MS-DOS, have been the most popular OS for personal

* The GPL discussed here is actually the second and third versions, GPLv2 and GPLv3, released in 1991 and 2005, respectively. GPLv1 did not necessarily support the ideology behind FSF, because it allowed people to distribute software in binary form and thus restrict some from being able to modify it.

computers for decades. Yet, to fully understand the history of Windows, we also have to consider another OS, that of the Macintosh. First though, we look at MS-DOS.

MS-DOS was first released in 1981 to accompany the release of the IBM PC, the first personal computer released by IBM. IBM decided, when designing their first personal computer for the market, to use an open architecture; they used all off-the-shelf components to construct their PC, including the Intel 8088 microprocessor. This, in effect, invited other companies to copy the architecture and release similar computers, which were referred to as IBM PC Clones (or Compatibles). Since these computers all shared the same processor, they could all run the same programs. Therefore, with MS-DOS available for the IBM PC, it would also run on the PC Clones. The popularity of the IBM PC Clones was largely because (1) the computers were cheaper than many competitors since the companies (other than IBM) did not have to invest a great deal in development, and (2) there was more software being produced for this platform because there were more computers of this type in the marketplace. As their popularity increased, so did MS-DOS's share in the marketplace.

LEARNING MULTIPLE OPERATING SYSTEMS

As an IT professional, you should learn multiple OSs. However, you do not need to buy multiple computers to accomplish this. Here, we look at three approaches to having multiple systems on your computer.

The first is the easiest. If you have an Apple Macintosh running OS version X, you can get to Unix any time you want by opening up an xterm window. This is available under utilities. The xterm window runs the Mach kernel, which is based, at least in part, on FreeBSD and NetBSD versions of Unix. This is not strictly speaking Unix, but it looks very similar in nearly every way.

Another approach, no matter which platform of computer you are using, is to install a virtual machine. The VM can mimic most other types of machines. Therefore, you install a VM in your computer, you boot your computer normally, and then you run the VM software, which boots your VM. Through a virtual machine, you can actually run several different OSs. The drawbacks of this approach are that you will have to install the OSs, which may require that you purchase them from vendors (whereas Linux is generally free, Windows and the Mac OS are not). Additionally, the VM takes a lot of resources to run efficiently, and so you might find your system slows down when running a VM. A multicore processor alleviates this problem.

Finally, if you are using a Windows machine, and you do not want to use a virtual machine, then you are limited to dual booting. A dual boot computer is one that has two (or more) OSs available. Upon booting the computer, the Windows boot loader can be paused and you can transfer control to a Linux boot loader. The two most popular boot loaders are Lilo (Linux Loader) and GRUB (Grand Unified Boot loader). Thus, Windows and Linux share your hard disk space and although the default is to boot to Windows, you can override this to boot to Linux any time. Unlike the Macintosh or the virtual machine approach, however, you will have to shut down from one OS to bring up the other.

DOS was a single tasking, text-based (not GUI) OS. The commands were primarily those that operated on the file system: changing directories, creating directories, moving files, copying files, deleting files, and starting/running programs. As a single tasking OS, the OS would run, leaving a prompt on the screen for the user. The user would enter a command. The OS would carry out that command.

If the user command was to start a program, then the program would run until it completed and control would return as a prompt to the user. The only exceptions to this single tasking nature were if the program required some I/O, in which case the program would be interrupted in favor of the OS, or the user did something that caused an interrupt to arise, such as by typing cntrl+c or cntrl+alt+del. Most software would also be text-based although many software products used (text-based) menus that could be accessed via the arrow keys and some combination of control keys or function keys.

In the 1970s, researchers at Xerox Palo Alto (Xerox Parc) had developed a graphical interface for their OS. They did this mostly to simplify their own research in artificial intelligence, not considering the achievement to be one that would lead to a commercial product. Steve Jobs and Steve Wozniak, the men behind Apple computers, toured Xerox Parc and realized that a windows-based OS could greatly impact the fledgling personal computer market.

Apple began development of a new OS shortly thereafter. Steve Jobs went to work on the Apple Lisa, which was to be the first personal computer with a GUI. However, because of development cost overruns, the project lagged behind and Jobs was thrown off of the project. He then joined the Apple Macintosh group. Although the Lisa beat the Macintosh to appear first, it had very poor sales in part due to an extremely high cost ($10,000 per unit). The Macintosh would be sold 1 year later and caught on immediately. The Apple Macintosh's OS is considered to be the first commercial GUI-based OS because of the Mac's success. During the Macintosh development phase, Apple hired Microsoft to help develop applications software. Bill Gates, who had been developing MS-DOS, realized the significance of a windows OS and started having Microsoft develop a similar product.

Although MS-DOS was released in 1981, it had been under development during much of the 1970s. In the latter part of the 1970s, Microsoft began working on Windows, originally called Interface Manager. It would reside on top of MS-DOS, meaning that you would first boot your computer to MS-DOS, and then you would run the GUI on top of it. Because of its reliance on MS-DOS, it meant that the user could actually use the GUI, or fall back to the DOS prompt whenever desired. Windows 1.0 was released in 1985, 1 year after the release of the Apple Macintosh. The Macintosh had no prompt whatsoever; you had to accomplish everything from the GUI.

Microsoft followed Windows 1.0 with Windows 2.0 in 1987 and Windows 3.0 in 1990. At this point, Windows was developed to be run on the Intel 80386 (or just 386) processor. In 1992, Windows 3.1 was released and, at the time, became the most popular and widely used of all OSs, primarily targeting the Intel 486 processor. All of these OSs were built upon MS-DOS and they were all multiprogramming systems in that the user could force a switch from one process (window) to another. This is a form of cooperative multitasking. None (excluding a version of Windows 2.1 when run with special hardware and software) would perform true (competitive) multitasking.

Around the same time, Microsoft also released Windows NT, a 32-bit, networked OS. Aside from being faster because of its 32-bit nature, Windows NT was designed for client–server networks. It was also the first version of Windows to support competitive multitasking. Windows NT went on to make a significant impact for organizations.

The next development in Windows history came with the release of Windows 95. Windows 95 was the first with built-in Internet capabilities, combining some of the features of Windows 3.1 with Windows NT. As with Windows NT, Windows 95 also had competitive multitasking. Another innovation in Windows 95 was plug and play capabilities. Plug and play means that the user can easily install new hardware on the fly. This means that the computer can recognize devices as they are added. Windows 95 was developed for the Intel Pentium processor.

In quick succession, several new versions of Windows were released in 1998 (Windows 98), 2000 [Windows 2000 and Windows Millennium Edition (Windows ME)], 2001 Windows XP (a version based on NT), and 2003 (Windows Server). In addition, in 1996, Microsoft released Windows CE, a version of Windows for mobile phones and other "scaled down" devices (e.g., navigation systems). In 2007, Microsoft released their next-generation windowing system, Vista. Although Vista contained a number of new security features, it was plagued with problems and poor performance. Finally, in 2009, Windows 7 was released.

In an interesting twist of fate, because of Microsoft's large portion of the marketplace, Apple struggled to stay in business. In 1997, Microsoft made a $150 million investment in Apple. The result has had a major impact on Apple Computers. First, by 2005, Apple Macintosh had switched from the RISC-based Power PC process to Intel processors. Second, because of the change, it permitted Macintosh computers to run most of the software that could run in Windows. Macintosh retained their windows-based OS, known as Mac OS, now in version X. Unlike Windows, which was originally based on MS-DOS, Mac OS X is built on top of the Unix OS. If one desires, one can open up a Unix shell and work with a command line prompt. Windows still contains MS-DOS, so that one can open a DOS prompt.

FURTHER READING

Aside from texts describing the hardware evolution as covered in Chapter 7, here are additional books that spotlight either the rise of personal computer-based OSs (e.g., *Fire in the Valley*) or the open source movement in Linux/Unix. Of particular note are both *Fire in the Valley*, which provides a dramatic telling of the rise of Microsoft, and the *Cathedral and the Bazaar*, which differentiates the open source movement from the free software foundation.

- Freiberg, P. and Swaine, M. *Fire in the Valley: The Making of the Personal Computer.* New York: McGraw Hill, 2000.
- Gancarz, M. *Linux and the Unix Philosophy.* Florida: Digital Press, 2003.
- Gay, J., Stallman, R., and Lessig, L. *Free Software, Free Society: Selected Essays of Richard M. Stallman.* Washington: CreateSpace, 2009.
- Lewis, T. *Microsoft Rising: . . . and Other Tales of Silicon Valley.* New Jersey: Wiley and Sons, 2000.

- Moody, G. *Rebel Code: Linux and the Open Source Revolution*. New York: Basic Books, 2002.
- Raymond, E. *The Cathedral and the Bazaar: Musings on Linux and Open Source by an Accidental Revolutionary*. Cambridge, MA: O'Reilly, 2001.
- St. Laurent, A. *Understanding Open Source and Free Software Licensing*. Cambridge, MA: O'Reilly, 2004.
- Watson, J. *A History of Computer Operating Systems: Unix, Dos, Lisa, Macintosh, Windows, Linux*. Ann Arbor, MI: Nimble Books, 2008.
- Williams, S. *Free as in Freedom: Richard Stallman's Crusade for Free Software*. Washington: CreateSpace, 2009.

The website http://distrowatch.com/ provides a wealth of information on various Linux distributions. Aside from having information on just about every Linux release, you can read about how distributions differ from an implementation point of view. If you find Figure 8.3 difficult to read, see http://futurist.se/gldt/ for a complete and complex timeline of Linux releases.

REVIEW TERMS

Terminology introduced in this chapter:

Debian Linux

Free software movement

GNU GPL

GNU project

Linux

Live CD

Macintosh OS

Minix

MS-DOS

Open architecture

Open source

Open source initiative

Red Hat Linux

SuSE Linux

Windows

Ubuntu Linux

Unix

REVIEW QUESTIONS

1. What is the difference between a resident monitor and an operating system?
2. What is the relationship between Windows and MS-DOS?

3. In which order were these operating systems released: Macintosh OS, Windows, MS-DOS, Windows 7, Windows 95, Windows NT, Unix, Linux?
4. What is the difference between the free software movement and the open source initiative?
5. What does the GNU general purpose license state?

DISCUSSION QUESTIONS

1. How can the understanding of the rather dramatic history of operating systems impact your career in IT?
2. Provide a list of debate points that suggest that proprietary/commercial software is the best approach in our society.
3. Provide a list of debate points that suggest that open source software is the best approach in our society.
4. Provide a list of debate points that suggest that free software, using Stallman's movement, is the best approach in our society.
5. How important is it for an IT person to have experience in more than one operating system? Which operating system(s) would it make most sense to understand?

CHAPTER 9

Bash Shell and Editing

For most computer users, including introductory IT students, using Linux will be a new experience. In fact, using the command line is unfamiliar territory for most. This chapter focuses on the Bash shell, which is available in all Linux distributions. Specifically, the chapter covers a number of features that the Bash shell offers to make command line interaction easier. Topics include command line editing, the use of aliases, and the history list. Additionally, the Bash interpreter is explored. The chapter concludes with an examination of two text editors, vi and Emacs.

The learning objectives of this chapter are to

- Introduce shortcuts in Bash that simplify command line entry.
- Describe the role of the Bash interpreter.
- Introduce Linux command redirection.
- Introduce the mechanisms to personalize a Bash shell.
- Discuss the importance of text editors in Linux.
- Provide instructions on how to use both the vi and Emacs editors.

In this chapter, we focus on some Linux-specific topics. First, we examine the Bash shell in detail. We look at command-line entry and editing, variables, aliases, and how to personalize a Bash shell. The features that we look at all support the Linux user in using the command line rather than the graphical user interface (GUI). We conclude the chapter with a look at two common text editor programs used in the Unix/Linux world: vi and Emacs. Although the editor topics may look entirely unrelated to Bash, you will find that some of the keystrokes that you use in Emacs are the same as those you will use at the command-line prompt to edit commands.

SHELLS

A shell is an *interpreted* environment. This means that you enter commands and, upon hitting the enter key, those commands are interpreted. Interpreting a command requires that the command be translated from its input form into an executable form. An interpreted program is translated and executed line by line, whereas the more traditional compiled program is translated entirely first, and then executed. Because of the sophistication in the Linux command line interpreter, we are able to enter complex Linux commands that include multiple instructions, redirection of input and output, and shortcut operations that are unfolded into more complex instructions.

The Linux/Unix shell is *tailorable*. Users can add shortcut operations called aliases, use shortcut operations such as ~ (which indicates the user's home directory), define variable values to be used in commands, and even redefine the look of the user prompt itself. Tailoring the environment can be done from the command line or by editing a variety of files. There are a number of different things that can be tailored, as we will see in the next section.

As the shell is text-based, interacting in a shell can be challenging. Most users are not familiar with the underlying operating system (OS) commands because they only communicate with the OS through the GUI. On the other hand, interacting at the shell level not only provides a greater degree of flexibility and control over the OS, it also saves resources because the shell itself does not require, for example, complicated graphics commands or mouse tracking. You might think of the shell as a more primitive way to access a computer, but in reality you are communicating with the computer at a *lower* level because you are avoiding the GUI layer.

In Linux/Unix, a shell is automatically opened if you access a computer remotely without the GUI. This is, for instance, the case when you use telnet or ssh to log into the computer. You can also open a shell by starting a terminal window program. In CentOS, for instance, you can open a terminal window when you right click your mouse on the desktop or select Terminal Window from the Applications menu. You can also start a new terminal window from the command line prompt, for instance, by issuing the command gnome-terminal to start a Gnome terminal window, or typing xterm to start an X window. The X windowing system is older than Gnome. The Gnome terminal window permits GUI interaction since it has a number of menus, whereas the Xterm window does not (see Figure 9.1). The user can open any number of terminal windows—each is its own shell, its own interpreted environment. You can even run different shells in each window although the default shell will be the user's login shell. You can also start a new shell session from inside of a running shell. In such a case, the new shell (what we might refer to as the inner shell session) overrides anything previously defined. Upon exiting the inner shell, you resume with your earlier session.

One of the earliest Unix shells was called the Bourne Shell (sh), first released in 1977. Later shells included the C shell (csh), the Tenex C shell, also called the T shell (tcsh), the Korn shell (ksh), and the Almquist shell (ash). There are a number of other, lesser-known or lesser-used shells such as the Busybox, Hamilton C shell, fish, Perl shell, and zoidberg.

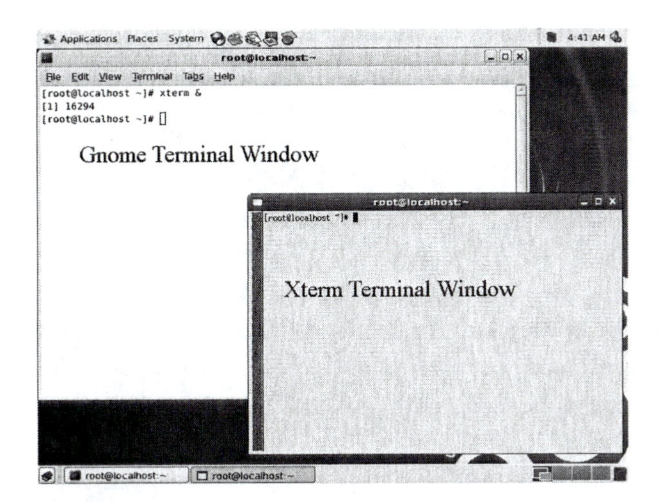

FIGURE 9.1 Terminal windows.

The names sh, csh, tcsh, ksh, and ash are the actual shell program names. You could, for instance, start a C shell by typing the command csh from the command line. This creates a new shell session as discussed in the previous paragraph. The Bash shell (bash) is actually known as the Bourne Again Shell, based on the Bourne shell. We will refer to it simply as bash. In the next three sections, we explore the bash shell in detail. It should be noted that many of the features found in Bash are also available in many of the other shells listed above.

BASH SHELL EDITING FEATURES

The Bash shell was written as a replacement for the Bourne shell. Bash was written to accompany the GNU OS, the open source community's attempt to build a Unix-like OS. It was later used as the default shell for Linux and Mac OS X. Bash has many of the same features from the original Bourne shell, but adds features that were created for such shells as csh and ksh. It also simplifies some of the syntax required in the Bourne shell so that, although it is a larger shell, it is easier to use.

In this section, we examine several features available in Bash that make entering and editing commands easier. These are

- History
- Command-line editing
- Tab completion
- Aliases
- Brace expansion
- Tilde expansion
- Wildcards

In the next section, we will examine other features of the Bash shell.

Every time you enter a command in your current Bash session, it is recorded in what is called the *history* list. The history list is available so that you can easily recall a previously entered command. To view the history list, type the command history. Figure 9.2 illustrates an example of a history list. History was first made available in csh but was found to be so useful and popular that it has been added to most shells since.

There are numerous ways to recall an instruction from the history list. From the command line, typing the up arrow key or control+p* will retrieve the previous instruction. Continuing to hit the up arrow key or cntrl+p will continue to step you backward through the history list. So, for instance, if you have entered the 10 commands shown in Figure 9.2, pressing cntrl+p three times will leave 'pwd' on the command line. Pressing enter then executes that command. To move forward through the history list once you have stepped backward, press the down arrow key or cntrl+n (p for previous, n for next).

Another way to recall an instruction from the history list is by typing !number, where number is the number of the entry in the history list (the exclamation mark is often referred to as bang). For instance, !2 will execute the "ls /home" instruction from Figure 9.2. To repeat the most recent command, enter !! (bang bang). You can also use the ! by specifying !chars, where chars are the first characters of the command you wish to repeat.

Let us assume that instruction 11 on the history list from Figure 9.2 is cp * /home/zappaf. If you then enter !c, it will perform the last instruction that starts with a 'c', the cp instruction. If you enter !cd, it will execute the last instruction that starts with 'cd', which would be instruction 5, cd ~. Similarly, !l (lower case 'L') repeats instruction 9 (ls –al) as does !ls and !ls – (i.e., bang followed by ls space hyphen). If we entered !ls /, it will repeat the second instruction in the list (ls /home).

The use of control+p and control+n are two of the many forms of command line editing available. The concept of command line editing was first made available in tcsh, where the commands were similar to the editing commands found in Emacs. Command line editing

```
 1. df -k
 2. ls /home
 3. su
 4. ls
 5. cd ~
 6. ls
 7. ls -l
 8. pwd
 9. ls -al
10. more .bashrc
```

FIGURE 9.2 A sample history list in Linux.

* This means to press the control key and while holding it, press the 'p' key. This notation will be used throughout this chapter along with escape+key, which means press the escape key (do not hold it down) and press the 'key' key. The difference between control and escape is that you hold down the control key, but you only press and release the escape key.

is performed on the text that is currently on the command line. Using a combination of control or escape and other keys, you can move the cursor along the command line, cut characters, and/or paste characters. The following is a list of the more useful editing operations. The notation here is c+char or m+char, where c+char means to hold the control key and press the character (e.g., c+n means "press the control key and while holding it down, press n"), and m+char means to press the escape key followed by the character (e.g., m+f means "press the escape key and then press f"). The reason we use 'm' for escape is that some people refer to the escape key as the "meta" key.

- c+n—display the next command from the history list on the command line
- c+p—display the previous command from the history list on the command line
- c+f—move cursor forward one character
- c+b—move cursor backward one character
- m+f—move cursor forward one word
- m+b—move cursor backward one word
- c+a—move cursor to the beginning of the line
- c+e—move cursor to the end of the line
- c+d—delete the character where the cursor is at
- c+k—kill (delete) the contents from the cursor to the end of the line
- c+w—delete all characters before the cursor to the beginning of the word
- c+y—take whatever was most recently deleted (using c+w or c+k) and yank it back, placed at the current cursor position (this is a paste operation)
- c+_—undo the last editing operation (delete or paste)

And, of course, you can use the backspace to delete the preceding character, del to delete the current character, the left and right arrow keys to move backward and forward, and the home and end keys to move to the beginning and ending of the line.

Once you learn these editing shortcuts, entering commands from the command line becomes at least a little easier. For instance, imagine that you wanted to copy five files from the current directory to the directory /home/zappaf, changing each file name. The five files are named file1.txt, file2.txt, etc., and you want to rename them to zappa1.txt, zappa2.txt, etc. The first command will be entered without the use of editing:

```
cp file1.txt /home/zappaf/zappa1.txt <enter>
```

The remaining four commands can be easily entered by editing this line. Press c+p to retrieve the instruction. Type c+a to move to the beginning of the line. Press m+f twice to move forward two words. The cursor is now at the period between file1 and txt. Press backspace to delete 1 and type 2 to insert a 2. Now, type c+e to get to the end of the line. Type m+b to move you to the beginning of 'txt' in zappa1.txt. Type c+b to move to the period, backspace to delete the 1 and 2 to insert 2. The command is now:

```
cp file2.txt /home/zappaf/zappa2.txt
```

Now, press the <enter> key to enter this revamped command. You would then repeat the steps above to repeat this instruction for the other files.

Another very useful Bash editing feature is called *tab completion*. As you enter a directory name or filename, if you reach a portion of that name that is unique, pressing the tab key will cause the interpreter to complete the name for you. Consider, for instance, that you are in a working directory that has the following files:

```
forgotten.txt   frank.txt   fresh.txt   functions.txt   funny.txt
other_stuff.txt
```

You want to view one of these files using less.* If you type *less fo<tab>*, the Bash interpreter will discover that only forgotten.txt starts with "fo" and therefore completes it for you. If instead you type *less fr<tab>*, Bash cannot complete this because it is unclear whether you want frank.txt or fresh.txt. In such a case, the interpreter emits a beep to let you know that the filename is not unique. If you were to type *less fra<tab>* then it can be completed to be frank.txt, or if you typed *less fre<tab>* it can be completed as fresh.txt. Pressing two tab keys will cause the Bash interpreter to list all matching filenames (or, if there are more than a few files, you will get a message that asks if all of the files should be listed). So, entering *less fr<tab><tab>* causes Bash to return with the list frank.txt and fresh.txt.

In Linux, an *alias* is a shortcut command. You define an alias so that when you type in your alias, it is replaced by the full command. Aliases can be entered any time at the command line, or they can be defined in other files (we will explore the use of these definition files in Personalizing Your Bash Shell). The form of an alias definition is

```
alias name=command
```

where name is the name of the shortcut and command is the Linux command that will replace the name. A simple example might be

```
alias add=/usr/sbin/useradd
```

* The less command is used to display the contents of the file to the terminal window, one screen at a time. At the end of each screen, you can move forward one screen, or forward or back by one line, or quit less and return to the command prompt. The command less, along with more and cat, are convenient ways to view the contents of a file, but less is more convenient than more or cat.

The add alias would be used to replace the lengthier instruction. If the Linux command includes blank spaces, the command must be placed inside of single quote marks (''). The name can be any string of characters (including digits and punctuation marks) as long as the characters selected are not reserved for other Linux purposes (for instance, %m can be used as an alias, but –m cannot).

In most cases, you define an alias to simplify typing. The example above allows you to shorten a command to something easier to type. Another example might be to reduce the length of an instruction that contains options. For instance, if you prefer that your ls commands always appear as a long listing, you could define the following alias:

```
alias ls='ls -l'
```

Notice the use of the quote marks because of the space between ls and –l. Also notice that the alias name can be the name of an already existing command.

An alias can also be useful if you cannot remember how to use the command. For instance, in the case of useradd, you may forget where it is stored. Or, in the case of a command that has a number of options, you may forget which ones to use. In the case of rm, you may forget to use the –i option, but to play safe, it is useful to always use rm with –i. Therefore, you might define the following alias.

```
alias rm='rm -i'
```

The ls command itself does not require parameters so that ls unfolds to ls –l. But both rm and useradd do require parameters. How do the parameters interact with the alias? For instance, if we wanted to delete foo.txt, we would type rm foo.txt. But now rm is an alias for 'rm –i'. In fact, the alias replacement works as you might expect. The alias is unfolded from rm to rm –i with foo.txt completing the instruction. So, rm foo.txt becomes rm –i foo.txt. Similarly, add –m –G cit130 bearb will unfold into /usr/sbin/useradd –m –G cit130 bearb. The alias is merely replaced before the command is executed. With the rm alias, you could specify even additional options. For instance, if you want to do a recursive deletion (rm –r), the command rm –r /foo unfolds to become rm –i –r /foo.

An alias can also be set up to handle common typos that a user may have. Imagine that, being a fast but not very careful typist, you often transpose two-letter commands. So, mv might come out as vm, rm as mr, and cp as pc. The two-letter combinations vm, mr, and pc are not Linux commands, and typing any of them will result in an error message at the command line (command not found). You could define an alias for each of these:

```
alias mr=rm
alias pc=cp
alias vm=mv
```

Notice in this case that no quote marks are required because the command does not include a space. Combining this idea with the previous example of rm, we could define these aliases:

```
alias rm='rm -i'
alias mr='rm -i'
```

What follows are a few other interesting ideas for aliases.

```
alias..='cd ..'                              A shortcut to move up a level
alias...='cd ../..'                          A shortcut to move up two levels
alias md=mkdir                               A shortcut to create a directory
alias egrep='grep -E'                        Enforcing a common option in grep
alias h='history 10'                         A shortcut to display a partial history
alias xt='xterm -bg black -fg white &'       A shortcut to generate a terminal window
```

If you define an alias at the command line, the alias exists only in that shell session. It would not exist in other terminal windows, nor if you started a new shell session (by typing bash for instance). For this reason, it is common to save aliases. We will discuss how to do this in Personalizing Your Bash Shell.

In addition to the shortcuts offered by aliases, two other shortcuts are available through tilde expansion and brace expansion. You have already seen tilde expansion as the tilde (~) is used to signify the user's home directory. The bash interpreter replaces ~ with the user's home directory, so for instance less ~/foo.txt becomes less /home/foxr/foo.txt. Tilde expansion can also be used to indicate other users' home directories with the notation ~username. For instance, you might use less ~zappaf/foo.txt to view a file in zappaf's home directory.

The brace expansion is used when you have a list of items that a command should be applied to. The braces, {}, will contain a list, where each item in the list is separated by a comma. As a simple example, imagine that you want to perform ls on several subdirectories in your home directory. In your home directory are foo1, foo2, and foo3, where foo2 contains two subdirectories, foo2a and foo2b. You could issue five separate ls commands (one each for foo1, foo2, foo2a, foo2b, foo3). Alternatively, you could list each of these in a single ls command: ls foo1 foo2 foo2/foo2a foo2/foo2b foo3. However, notice that for foo2a and foo2b, you have to list them as foo2/foo2a and foo2/foo2b. You could instead use brace expansion where you list each directory in a list inside of braces. The subdirectories could similarly be listed in an inner (or nested) set of braces. The command becomes: ls {foo1,foo2,foo2/{foo2a,foo2b},foo3}, which the Bash interpreter unfolds into five separate ls commands, ls foo1, ls foo2, ls foo2/foo2a, ls foo2/foo2b, and ls foo3.

Another form of expansion is called filename expansion. This takes place when you use wildcards among your filenames. A wildcard, much like in poker, can act as anything. The most common wildcard character is *, which in essence means "match anything". You would use the * when performing file system commands, such as ls. For instance, ls *.txt would list all files that end with the .txt extension, no matter what their name was. If the current directory contains the files fool.txt, foo2.txt, foo3.text, and bar.doc, the *.txt notation will match fool.txt and foo2.txt. It will not match foo3.text because txt is not the same as text, and it will not match bar.doc because doc does not match txt. However, the command ls *.* will match all four of these. On the other hand, if the directory also contained the file stuff, ls *.* will not match it because there is no period in stuff. The command ls * would match all five of the files (the four with a period and stuff).

The use of the wildcard is common for Linux users. There are other wildcard characters, such as ?, which means "match any single character". We will explore the wildcard characters in more detail in the next chapter when we look at regular expressions.

All of the various command line editing techniques provide the user with a powerful collection of shortcuts so that working with the command line is easier. However, with few exceptions, they all require practice if for no other reason than to remember that they exist. Introductory Linux users often forego using these techniques because they are not intuitive to the student just learning Linux. On the other hand, by learning these early, they allow users to be more efficient and therefore more effective in Linux.

EXPLORING THE BASH INTERPRETER

Now that we have learned some of the shortcuts and editing features available in the Bash shell, we turn to how the Bash interpreter operates. We first introduce three additional tools available to the user: variables, redirection, and obtaining help.

Variables are available in a Bash shell. A variable is a name given to a storage location. By using a variable, you can reference that variable from various commands rather than having to use the value directly. A simple example might be a variable that stores the number of licenses available for a software title. As users run the software, this variable is decremented, and as users terminate the software, the variable is incremented. If a user wishes to run the software and the value is 0, they are told that there are no free licenses currently available and that they must wait. In this case, the variable is shared among all users. In a Bash session, a variable is available to all of the software running in that shell. There are generally two types of Bash variables. User variables are those defined by the user and often only used via command line operations. There are also environment variables, established either by the OS, the Bash interpreter, or by some other running software. Such variables are usually available to all software. The Bash shell defines its own environment variables, many of which could be useful for the user. To see what environment variables have been set, enter the command env. Among the variables defined, you will probably see such things as HOSTNAME (the name of the computer), SHELL (the current shell, which for us will be /bin/bash), USER (your username), HOME (the user's home directory), and PWD (the current working directory). Notice that all of these environment variables are fully capitalized. This is a convention so that a user can easily determine if something is an environment variable or something else (such a user-defined variable or an alias).

To establish your own variable, just type an *assignment statement* on the command line. The format of an assignment statement is variable=value, where variable is the variable name and value is the value. The Bash shell defaults to storing strings, so value should be strings such as a person's name. If the string has a blank space, enclose the entire string in quote marks ("" or ''; we will differentiate between them later). The value can be an integer number; however, Bash will treat any number as a string unless you specify that it should be a number. This is done by enclosing the value inside of parentheses, as in age = (29).

To obtain the value stored in a variable, you must precede the variable name with a $. For instance, if first=Frank and last=Zappa, and you want to create a variable, fullname, you could issue the assignment statement fullname="$first $last". The variable fullname is then established as the value stored in the variable first, followed by a blank space, followed by the value stored in the variable last. If you were to do fullname="first last", then fullname is literally the name "first last" rather than the values stored in those variables. Notice that the quote marks are required because of the blank space. If you did not care about the blank space, you could also do fullname=$first$last as well as fullname="$first$last".

If the value in a variable is a number, and you want to perform some type of arithmetic operation on it, you have to enclose the entire operation in $((…)). Consider if, after storing 29 in age as shown above, you do newage=$age+1. Will you get 30 in newage? No; in fact, you get 29+1. Why? Quite literally, the command concatenates the items, the value stored in the variable age (29), the +, and the value 1. Concatenation combines strings together. But here, we want to add 1 to age. This would require the use of the ((...)) so the proper instruction is newage=$((age+1)). Notice that the $ precedes the ((…)) and does not appear immediately before the variable age. For those of you who have programmed before, you will probably be familiar with a reassignment statement—taking the value of a variable, changing it, and reassigning the variable to the new value. Thus, instead of newage=$((age+1)), you could also do age=$((age+1)).

If you had two numeric values, say X and Y, you could do Z=$((X+Y)). The arithmetic operations available are +, –, * (multiplication), / (division), and % (remainder, also known as mod or modulo).

Aside from assigning variables and using them in assignment statements, you can also view the contents of variables using the echo command. The echo command expects a list of items. Whatever is in that list is output. For instance,

```
echo Hi there, how are you?
```

will literally output "Hi there, how are you?" on the next line of the terminal window. Notice that you did not need to place "" around the list. You can also output values stored in variables as in echo $fullname or echo $age. You can combine literal text with variable values as in echo Hello $fullname. The echo statement permits but does not require that you place the list of items in quote marks. You can use "" or '', although there is one large difference. If you use '' and place a variable name inside of the single quote marks, you get the variable's name rather than the value. So for instance, echo "Hello $fullname" outputs Hello Frank Zappa (assuming fullname is storing Frank Zappa), whereas echo 'Hello $fullname' outputs Hello $fullname. And, of course, echo "Hello fullname" will just output Hello fullname because there was no $ before the variable.

One last comment about the echo statement. Imagine that you wanted to output a greeting to the user and the current date and time. Linux has the command date. If you were to use

```
echo "Hello $fullname, today is date"
```

you would literally get Hello Frank Zappa, today is date. By placing a Linux command inside of ` ` (backward quote marks), the command will execute before the echo statement. So,

```
echo "Hello $fullname, today is `date`"
```

will output something like Hello Frank Zappa, today is Thu Feb 9 17:54:03 EST 2012 (naturally, depending on the time and date when the command was entered). The command, placed in ` ` can also appear in this format: $(command). The previous echo command could instead be

```
echo "Hello $fullname, today is $(date)"
```

It is not very common that a user will need to define their own variables for such purposes as storing your name or age. Users can define variables that can be used in software that they write. More commonly though, variables are defined by software and used by software. There are several established environment variables in Linux. We saw a few of them earlier when we talked about the env command. Here are others of note.

- DISPLAY—the Linux name for standard output (defaults to :0.0).
- HISTSIZE—the number of commands that can be stored in the history list.
- PATH—a list of directories that are checked whenever the user enters a command; this allows the user to execute some programs or find some files without having to specify the entire path.
- PS1—defines the user's prompt, which might include special characters that represent the user's name, or the current directory, or the number of the command as it will be stored in the history list (see Table 9.1)
- PS2, PS3, and PS4 may be used to define other prompts that are used in various programs.

Of the variables listed above, you would most likely only alter PATH. The PATH variable stores a list of directories. Whenever you enter a Linux command or file name, if the item is not found in the current working directory, then the Linux interpreter checks for the item in every directory listed in the PATH variable. Only if it is not found in the current working directory or any of the PATH directories does it return an error message.

TABLE 9.1 Characters for the PS1 Variable to Specialize User Prompt

Character	Meaning
\d	Current date
\D{format}	Current date using the given format where format matches the format expected by the strftime command
\t	Current time in military format
\@	Current time in 12-hour format (AM/PM)
\H, \h	Full computer host name or first part of host name
\s	Shell name
\u	User name
\w	Current working directory
\!	Current command number (as it would appear in the history list)
$	Default user specifier (prompt)
@	Default separator, for instance \d@\u $
\$?	Status of the last command
\e[... \e[m	Change color of the prompt, the ... is the color code in the form #;##, where the first # is either 0 (light) or 1 (dark) and the second ## is the color (30 for black, 31 for red, 32 for green, 33 for brown, 34 for blue, 35 for purple, 36 for cyan)

In the Bash shell, the PATH variable is established initially by the system administrator, typical storing /usr/kerberos/bin:/usr/local/bin:/usr/bin:/bin:/usr/X11R6/bin. The ":" is used to separate each path. You should recognize many of these paths (e.g., /usr/local/bin, /usr/bin, /bin). Kerberos is used for network authentication and programs found in its bin directory are common network programs (e.g., ftp, telnet, rlogin). The X11R6 directory stores X11 (windowing) files.

If you wanted to add to your PATH variable, you could issue a reassignment statement as follows:

```
PATH=$PATH:newdirectory and
```

Here, you are setting PATH to be equal to what is currently in the PATH variable followed by a : and the new directory. So, PATH=$PATH:/home/foxr would add /home/foxr's home directory onto your path. You might do this with your own home directory so that, no matter where you are working in the Linux file system, your home directory will be tested along with the rest of the PATH directories for a file. You would, of course, substitute your own username for foxr's.

The PS1 variable stores how your prompt looks. By changing PS1, you change your prompt's appearance. Unless you know what you are doing, it is best to leave PS1 alone. Similarly, there is little need to change other environment variables.

The Bash interpreter will typically take input from the keyboard and send output to the terminal window. However, there are times when you want to redirect either input or output. For instance, you might want to change the input for a program to come from a file rather than the command line. Or, you might want to send the output to a file instead of the display. A third possibility is that you might want to take the output of one command and use it as the input to another command. These are all forms of *redirection*.

Redirection is performed using one of the following five sets of characters:

- < redirect the input to come from an input file
- > redirect the output to go to an output file, overwriting the file if it already exists
- >> redirect the output to be appended to an already existing file, or create the file if it does not exist
- << redirect the input to come from keyboard where the input will terminate with a special keyword that you specify after <<
- | redirect the output of one command to be the input of another—this is known as a *pipe*

The first four instances of file redirection merely override the standard input and/or output. For instance, the standard output for cat is the screen. Typing cat foo1.txt will output the contents of foo1.txt to your monitor. This is similar to the instructions more and less, but the output appears all at once, scrolling down and beyond the screen so that you only wind up seeing the last screen's worth. In more and less, the output pauses at the end of each screen. In using cat, you may decide to take the output and sent it to another file. You would do this using redirection as

```
cat foo1.txt > foo2.txt
```

The above instruction gives you the same result as if you did cp foo1.txt foo2.txt. However, imagine that you wanted to take the output of *several* files and copy them to another file. The cp command does not do this. Therefore, you could use cat and redirection as in

```
cat file1.txt file2.txt file3.txt > file4.txt
```

The file file4.txt would then be the concatenation of file1.txt, file2.txt, and file3.txt. If file4.txt already exists, this command *overwrites* it. If you were to use >> instead of >, then either the three files would be copied into the new file file4.txt if file4.txt did not previously exist, or they would be appended to the end of file4.txt if file4.txt did exist.

The input redirection < is easy to use. In essence, you want to replace input from the keyboard to come from a file instead. This is only useful if the expected input is coming from keyboard rather than file. So for instance, the wc (word count) program does not require < as the typical form of the instruction is wc filename. The most common use of the input redirection will be when you write your own shell scripts. This will be discussed in Chapter 14. For now, we will ignore using <.

The other form of input redirection, <<, in some ways is the opposite of <. Where < says “accept input from this file rather than standard input”, << says “accept input from standard input rather than a file". The << is of use when you want to force a program that accepts its input from file to accept your keyboard input instead. When using <<, the program will pause to let you enter items. How does the program know when your input is done? It is not merely pressing the <enter> key because you may want to have line breaks in your input. Instead, after the << symbols, you place a keyword. It is this keyword that will terminate input. The common example for using << is to create a file using cat. The notation for cat is generally cat filename(s). However, if you do cat << keyword, then input comes from the keyboard until you type in the keyword. Consider the following:

```
cat << done
```

This will result in a prompt appearing, >. At the prompt, you begin typing. You can type whatever you wish, including pressing the <enter> key. The cat program continues to accept your input until you type in done. At that point, cat has received all of its input and now it executes, which results in displaying the contents of the input to standard output, that is, what you have just typed in now appears on the screen. This is not very useful. However, you can also redirect the output. If you were to do:

```
cat << done > shoppinglist.txt
```

and you type in a shopping list, ending with the word "done", then whatever you have typed in now is stored in the file shoppinglist.txt. Thus, from cat, and using redirection, you can create your own textfiles. Try the following, where the > marks are displayed in your terminal window:

```
cat << done > shoppinglist.txt
            > bananas
            > apples
            > bread
            > milk
            > aspirin
            > done
```

You should find a file, shoppinglist.txt, that contains the text bananas, apples, bread, milk, aspirin, each on a separate line, but not the word done.

The most common use of redirection will be to send output to a file. For example, if you want to save file information, you could do one of the following:

```
ls -l > directory_listing.txt
ls -l >> directory_listing.txt
```

In the latter case, if the file already exists, rather than overwriting it, the command will append to the file.

The last form of redirection is called a *pipe*. The pipe redirects the output of one program to be the input to another program. As a simple example, you use ls to view the contents of the current directory. However, there are too many files, so many of them scroll past the screen too quickly for you to see. You want to force ls to pause in between screens. You can do this using the more or less command. So you use a pipe as in

```
ls | more
or    ls | less.
```

The instruction first executes ls, but instead of displaying the output, it redirects the output to the next program, more (or less).

You could also tackle this problem in a far more roundabout fashion by entering

```
ls > directory_listing.txt
```

and then entering

```
less directory_listing.txt
```

The pipe saves you from having to do two separate commands and saves you from creating a file that you really do not need.

REDIRECTION IN MS-DOS

Redirection is a very useful tool for users of an OS. MS-DOS has its own versions of redirection that are similar although not exactly like those of Linux.

To move the contents of a command to a file, use > as with Linux. The >> redirection performs append. In both cases, the notation is to place > or >> immediately before the file name with no blank space. So, for instance, sending the contents of a directory listing to a file dir.txt would be done by

```
dir >dir.txt
```

MS-DOS also permits redirection from file rather than keyboard by using <filename. There is no equivalent of <<.

MS-DOS also has a pipe, of sorts. Some people refer to it as a fake pipe. Like Linux, the pipe is noted with the | symbol, which appears between two instructions. However, in Linux, the two instructions surrounding a pipe execute simultaneously with the output of the first being channeled into the input for the second (the term simultaneous means that both programs begin at the same time although their execution is sequential). But in MS-DOS, the first instruction executes and the output is saved to a temporary file. Then, the second instruction is executed with input redirected to come from the temporary file rather than keyboard. So, the MS-DOS version of a pipe is, in essence, two separate instructions, connected through redirection. This is far less efficient.

In Linux, a pipe can be used as often as is needed. This could potentially lead to lengthy and very complicated instructions. Consider the following operation:

```
ls -l|sort -fr| less
```

In this case, you are obtaining a long listing of the current directory, sorting the list of items in reverse alphabetical order and ignoring case (it will not differentiate between upper and lower case letters) and then displaying the result using less. We will see other uses of the pipe in the next two chapters.

We wrap up this section by considering help features available in Bash. In fact, these features will be found in any Linux shell. The help described here relates to how to use various Linux commands. The most common form of help is the manual pages (man) for an instruction. To view a manual page, use the man command, which takes on the form man command, as in man sort or man find. There are manual pages for most but not every Linux command. The man pages contain a number of different types of information that help explain how to use the command. They include:

A synopsis of its usage (the syntax)

A description of the instruction

The options (parameters) available for the instruction

The authors of the program

Files related to the command and other commands related to or similar to this one

Let us take a look at some examples. First, we look at the rm instruction. When we do a man rm, we initially see the screen shown in Figure 9.3.

The first screen of the rm man page lists the instruction's name, a synopsis of its usage, a description, and the first two options. The bottom of the window shows ":" with the cursor following it. To control movement in the man page, we will use the space bar to move forward, screen by screen. But like the less command, we can also move backward and

```
RM(1)                        User Commands                        RM(1)

NAME
       rm - remove files or directories

SYNOPSIS
       rm [OPTION]... FILE...

DESCRIPTION
       This  manual  page  documents  the  GNU version of rm.  rm removes each
       specified file.  By default, it does not remove directories.

       If a file is unwritable, the standard input is a tty,  and  the  -f  or
       --force  option is not given, rm prompts the user for whether to remove
       the file.  If the response is not affirmative, the file is skipped.

OPTIONS
       Remove (unlink) the FILE(s).

       -f, --force
              ignore nonexistent files, never prompt

       -i, --interactive
:
```

FIGURE 9.3 First screen of the rm man page.

forward one line at a time using the up and down arrow keys. The letter 'q' quits (exits) the man page. The options (only a portion of which are shown in Figure 9.3) continue onto the next two screens. Figure 9.4 shows what the remainder looks like.

In examining the man page for rm, we can see that the options are –f, -i, -r or –R, and –v. We can also use --interactive in place of –i, or --force in place of –f. There are also options for --help and --version, and so forth.

The rm command is simple. A more complex man page occurs with the instruction find. In fact, the find man page is 50 screens long. One notable difference in the two commands' man pages is that find includes a list of examples to help illustrate how to use the command. Although examples are not common in man pages, you will find them when the command is challenging enough.

Another variation in man pages can be found with the man page for mount. Whereas rm and find only have one usage under synopsis, mount has four. They are as follows:

```
mount [-lhV]
mount -a [-fFnrsvw] [-t vfstype] [-o optlist]
mount [-fnrsvw] [-o options [,…]] device | dir
mount [-fnrsvw] [-t vfstype] [-o options] device dir
```

These four lines indicate that mount can be used with four different types of parameters. First, mount can be followed by one or more of l, h, or V as options. Second, mount –a can be followed by a list of options, a virtual file system type, and any additional options for this command. Third, mount can be followed by a list of options, additional options, and a device or directory (the | indicates that either a device or a directory should be provided). Finally, mount can be followed by a list of options, a virtual file system type, a list of options, a device, and a directory. Many instructions have multiple entries under synopsis. Another, but simpler, example can be found with the useradd command.

```
       prompt before any removal

--no-preserve-root do not treat '/' specially (the default)

--preserve-root
       fail to operate recursively on '/'

-r, -R, --recursive
       remove directories and their contents recursively

-v, --verbose
       explain what is being done

--help display this help and exit

--version
       output version information and exit

By default, rm does not remove directories.  Use the --recursive (-r or
-R) option to remove each listed directory, too, along with all of its
contents.

To remove a file whose name starts with a '-', for example '-foo',  use
:
```

FIGURE 9.4 Second screen of the rm man page.

The second form of help available is the help command. The help command's syntax is help commandname. However, very few of the Linux commands have help pages, so it is best to use man.

The final form of help is not necessarily help for how to *use* a command. Instead, it might help you find a command. One problem with Linux is remembering all of the commands by name. If you knew the command's name, then the man page can help you learn how to use it. But what if you know what you want to accomplish but do not know the name of the command? This is where *apropos* comes in. The command's syntax is apropos string where string is the string that you are searching for. The apropos command lists all of the commands that have the associated string found in the command's description.

For instance, there is a command that reports on the current usage of virtual memory. You cannot remember the name of the command. You type

```
apropos "virtual memory"
```

The apropos command returns the following:

```
mremap          (2)    - re-map a virtual memory address
vfork           (3p)   - create a new process; share virtual memory
vmstat          (8)    - Report virtual memory statistics
```

Bingo, the last item is what we were looking for! Although the quote marks around virtual memory are not necessary, without them, apropos will return anything that matches either virtual or memory. This would give us a lengthier list to examine. However, with the quote marks, apropos will only return exact matches. If we wanted to find commands whose descriptions had both words virtual and memory, but not necessarily in that order, we would not want to use the quote marks.

As another example, you want to find a command to delete a file. You type apropos delete file, which returns a lengthy list of commands. So you try apropos "delete file". This time, apropos returns:

```
delete file: nothing appropriate
```

indicating that apropos could not find a match to "delete file". Using apropos can be tricky but it is a valuable resource.

Now that we have seen the various features available in the Bash shell, we conclude this section with a brief look at how the interpreter works. You have entered a command on the command line. What happens now? The following steps are taken:

- The interpreter reads the input.
- The interpreter breaks the input into words and operators
    - operators being symbols such as <, >, |, ~, *

- If the instruction has any quotes, those quotes are handled.
- The instructions are parsed for any defined aliases, and if any aliases are found, alias expansion (replacement) takes place.
- The various words and operators are now broken up into individual commands (for instance, if there are redirection operators present then the commands are separated from these operators).
- Shell expansions take place if any are called for
  - Brace expansion
  - Tilde expansion
  - Variable values are assigned
- If any of the commands appears either in ` ` or $(), execute the command.
- Execute any called for arithmetic operations.
- Perform any redirections (including pipes).
- Perform file name expansion, that is, match wildcard characters.
- The command is executed and, upon completion, the exit status of the command (if necessary) is displayed to the terminal window (if no output redirection was called for).

That sounds like a lot but it happens very quickly. You can now see why using the command line in Linux can give you a great deal of flexibility and power over using a GUI.

PERSONALIZING YOUR BASH SHELL

As stated in both Bash Shell Editing Features and Exploring the Bash Interpreter, you are able to enter alias commands and define variables from the command line. Unfortunately, any such aliases or variables would only be known within that given shell session. What if you want to have a given alias or variable known in every shell session? There are a number of files available to the user that are loaded whenever a new shell session is started. If you place your alias or variable definitions in one of these files, then the alias or variable would be known in *every* shell session. So here, we look at the use of these files. Before we examine how to use these files, let us consider what happens when a shell session starts.

When a shell starts, it goes through an initialization process. The Bash interpreter executes several shell scripts. These are, in order, /etc/profile, /etc/bashrc, .bash_profile, .bash_login, .profile, and .bashrc. The first two scripts are set up by the system administrator. They provide defaults for all users, for instance, by establishing aliases that all users should have and PATH variables that all users should start with. The profile script is executed whenever any type of shell starts, whereas the bashrc script is executed only if a Bash shell is being started. The last four scripts are defined by the user in their home directory.

The .bash_profile script is commonly used to override anything established in the system scripts as well as personalize items (aliases, variables) for the user. The .bash_login file is only executed if the user is first logging in (remotely) as opposed to opening a new shell. The .profile script is executed whenever any type of shell is opened, not just Bash. The .bashrc script is executed when a new Bash shell is started in a new window. For simplicity, it is probably best to just modify the .bash_profile script as it is used whenever any Bash session starts.

If you examine these files, you will find a variety of entries. There are predefined aliases and variables. But there are also shell script operations. Since we will examine shell scripting in Chapter 14, we will ignore the contents here. The .bash_profile script contains the following two operations:

```
source ~./bashrc
source ~./bash_login
```

The source instruction is used to execute a script. The ~ indicates that the script is in your home directory. The notation ./ means "execute this script". Therefore, these two instructions will execute the two scripts, .bashrc and .bash_login. If these two instructions are found in your .bash_profile script, then in fact you are free to edit any of these three files, and they will all execute when a new Bash shell is started.

Let us consider what you might put into one of these files, say the .bashrc file. Already present will be the following two instructions:

```
export PS1 = "…" //whatever prompt you want goes inside the quote
                     marks
export PATH = "$PATH:~/:~/bin:~/scripts:/usr/sbin"
```

The word "export" signifies that the variable being defined should be known throughout the shell's environment and not just inside of this script. The PS1 variable, already defined, is being reset to a prompt of your own liking. For instance, you might use \d to specify the date or \w to specify the current working directory. The latter assignment statement takes the PATH variable, defined by the system administrator in the /etc/profile file, and appends additional directories to it. To see what your PATH variable is storing, you can type echo $PATH. If you wish to define further variables, you can do so, but you should make sure that any variables have the word export preceding them so that they are known in your Bash shell environment.

You may also find some aliases already defined. You may edit or add to these aliases as you desire. For instance, you might specify aliases that correct common typos, or you might specify aliases that will help you reduce the amount of typing, as described earlier in this chapter.

One last comment. If you ever edit one of or more of these files and you later discovered that you have made mistakes or do not like what you have done, you can copy new versions of each of these files. They can be found in the directory /etc/skel.

TEXT EDITORS

As a Linux user, you will often use a text editor instead of a word processor. This is certainly true of a system administrator who will often have to edit configuration files or write shell script files, all of which should be stored as normal text files. As shown in the previous section, you may also wish to edit one of your Bash initialization scripts such as .bashrc. Being a text file, you should use a text editor. Therefore, it is worth learning about the text editors available in Linux.

The problem with using word processors is that word processors, by default, store documents in a formatted way. For instance, using Word will result in a Word doc file (.doc or .docx) and using OpenOffice's word processor will result in an .odt file. You can override the file type by using "save as" and changing the file type to text, but that can be a hassle. In addition, to start up a word process such as Word takes a greater amount of time when compared to starting a simple text editor. When you have a small editing task that you want to do quickly, you will want to forego the use of a full-blown word processor and settle for a text editor instead.

In Linux, there are three common text editors. First, there is a GUI editor often called Text Editor, or gedit. Then there are two older, text-based editors: vi and Emacs. The vi editor is the older of the two editors and comes with all Linux installations. Emacs may or may not be part of your Linux although it is easy enough to install. Although gedit contains a GUI interface so that you can move the cursor with the mouse and use the mouse for highlighting text and selecting menu operations, you may find that gedit is not available or desirable. For instance, if you have logged into a Linux computer using a remote login program such as telnet or ssh, you will not have access to gedit. Also, both vi and Emacs load faster and therefore, if you want to do a quick edit, it may not be worth the time to run gedit. So, here, we examine both vi and Emacs. An interesting side note is that Linux (and Unix) users who learn vi first will love vi and hate Emacs, whereas those who learn Emacs first tend to love Emacs and hate vi. Nevertheless, all Linux users should learn both, and use their preferred editor whenever they need to perform quick text editing.

The vi Editor

The vi editor is the default Linux text editor. A newer version of vi is called vim (vi improved). The following description applies to both vi and vim.

Because it is text-based, the commands that you would find as menus in a GUI-based editor are performed as keystrokes. The editor uses three different modes—command mode, insert mode, and replace mode—so that keystrokes differ by mode, as will be discussed below. It is important to remember that the keystrokes that are used as commands are case-sensitive. For instance, 'o' and 'O' are both forms of insert, but insert at different locations.

Command mode is the default mode. Keystrokes are interpreted as commands. Many keystrokes can be preceded by a number such that the command is performed that many times. For instance, the dd command deletes the current line. 5dd deletes five consecutive lines. Insert mode is the mode that is most similar to a word processor; as you type, the characters are entered into your document at the point of the cursor. With each keypress, the cursor advances to the next position. The only keystroke command in this mode is the escape key, which exits this mode, returning you to command mode. Replace mode is similar to insert mode; however, as you enter keystrokes, the characters overwrite the characters already present. As with the insert mode, the escape key returns you to the command mode.

In order to know which mode you are in, look at the bottom of the editor where you will either see "--INSERT--", "--REPLACE--", or nothing (command mode). There are numerous commands to enter either insert or replace mode. These are listed in Table 9.2. The reason for the different commands is that each will position the cursor in a different location. For instance, 'i' positions the insert cursor at the current cursor location, 'o' inserts a blank line in the line following the cursor and inserts the cursor at the beginning of the blank line, and 'O' inserts a blank line in the line preceding the cursor. To toggle command mode to insert or replace, use one of the letters in Table 9.2, and to return to command mode, press the escape key.

Aside from positioning the cursor when switching to insert mode, there are numerous other ways to move the cursor. These are listed in Table 9.3. Notice that the keystrokes to move the cursor one position (down, up, left, right) should be done through the j, k, h, and l keystrokes rather than the arrow keys, which may or may not be mapped in vi/vim. Using G or #G (where # is a number) moves you to a new point in the file quickly; however, #G requires that you know the line number. The search command can be more useful.

TABLE 9.2 Mode Commands

Keystroke	Description
i	Enter insert mode at the immediate left of the cursor
a	Enter insert mode at the immediate right of the cursor
o	Insert a blank line after the current line and enter insert mode
O	Insert a blank line before the current line and enter insert mode
I	Enter insert mode at the beginning of the current line
A	Enter insert mode at the end of the current line
r	Replace one character with the next character entered
R	Enter replace mode and continue to replace (overwrite) characters (until escape is pressed)

TABLE 9.3 Cursor Movement and Search Commands

j, k, h, l	Move the cursor down/up/left/right one position (the arrow keys may not work)
w, b	Move the cursor forward/backward by one word
G	Move to the last line of the file
1G	Move to the first line of the file
*G	Move to line * (where * is a number, such as 5G or 50G)
H, L	Move to the top/bottom line of the screen
0, $	Move to the first/last character of the line
/text <enter>	Search forward for text (the text can include regular expressions)
/<enter>	Search for the next occurrence of the most recent search
?text <enter>	Search backward for text
? <enter>	Search for the previous occurrence of the most recent search
<pg up> or cntrl+f, <pg down> or ctrl+b	Move the file up or down one screen's worth in the window

You can mark text and then jump to marked text. This is useful for editing large documents. For instance, if a file contains a book, you might mark the beginning of each chapter. There are 36 marks available in vi. They are denoted using either a letter from a to z or digit from 0 to 9 (26 letters and 10 digits provides 36 distinct marks). Inserting a mark is done by typing m*char* in command mode, where char is one of a letter from a to z or digit from 0 to 9. Jumping to a marked position is done using '*char*. You can also jump to the previous mark by typing ''*char*. For instance, if you place the mark 'z' at the beginning of Chapter 2 and you are midway through Chapter 5, pressing 'z moves you to the beginning of Chapter 2 and ''z returns you to the point midway through Chapter 5 of where you were before typing 'z.

If the cursor is on a delimiter as used in the C/C++ or Java programming language, such as {,}, (,), [or], pressing % will move you to the other delimiter. For instance, if you are on a { that starts a function, pressing % moves you to the } that closes that function. This is useful if you are editing a program.

Table 9.4 displays various repeat and undo options. Many commands can be executed multiple times by preceding the keystroke with a number. 5j, for instance, will perform the 'j' keystroke five times, that is, it will move the cursor down five lines. The exception is when the number precedes the letter G, in which case the cursor is repositioned at the beginning of that line number; for instance, 5G moves the cursor to the beginning of line 5.

Obviously any text editor/word processor requires the ability to cut-and-paste or copy-and-paste. As with a GUI-based word processor, the idea is to select a sequence of characters and then cut or copy them into a buffer. From the buffer, they can be retrieved. In text-based word processors, the expression to paste is often referred to as yank (yanking text back from the buffer). Table 9.5 provides the commands for cutting, copying, and pasting.

When you start vi, you can either specify a file or open a new buffer. Once you are editing a vi session, you will need file commands. The file commands all begin with a ':', as listed in Table 9.6. Notice that some of these require a filename, which you would type after the command, as in :w *name* to save the contents of the buffer to the file named *name*. There will be a blank space between the command and the name.

Table 9.7 provides a few leftover, miscellaneous vi commands.

Of course, the only way to really familiarize yourself with vi is through practice. Given the somewhat bizarre keystrokes needed to control vi, you might question whether you would or should ever use it. However, as a system administrator or shell script writer, you would probably need to use vi (or Emacs) often. As stated earlier, those Linux users who learn vi first tend to use vi. You will see in the Emacs editor that many of the commands are perhaps a little easier to remember because many commands start with the letter of the command.

TABLE 9.4 Repeat and Undo Commands

u	Undo the previous command (may be used multiple times to undo several commands)
.	Repeat the last edit command
n.	Repeat the last edit command *n* times where *n* is a number, such as 5.
nk	Repeat command *k* *n* times, where *k* is a keystroke and *n* is a number, such as 5j or 5dd

TABLE 9.5 Delete, Copy, Paste, and Change Commands

x	Delete the next character
*n*x	Delete the next *n* characters where *n* is a number
dw	Delete the next word
*n*dw	Delete the next *n* words where *n* is a number
dd	Delete this line
*n*dd	Delete the next *n* lines starting with the current line where *n* is a number (also d*n*d)
D	Delete from the cursor to the end of this line
yy	Copy the current line into a buffer
*n*yy	Copy the next n lines starting with the current line where *n* is a number (also y*n*y)
p	Put (paste) any lines stored in the buffer below the current line
J	Join the next line to the end of the current line (remove the end-of-line)
cw	Change current word by replacing text as you type until user presses escape key
C	Change all words in current line until user pressers escape key

TABLE 9.6 File Commands

:w <enter>	Save the file
:w *name* <enter>	Save the file as *name*
:w! *name* <enter>	Overwrite the file *name* with the contents of the current session
:n,m *name* <enter>	Save lines n through m of the current session to the file *name*
:q <enter>	Exit vi
:q! <enter>	Exit vi without saving
:wq <enter> or :x <enter>	Save file and exit vi
:r *name* <enter>	Insert the file *name* into the current position of the current file (open a file if the vi session is empty)

TABLE 9.7 Miscellaneous Commands

:h	View the vi help file
Ctrl+T	Return to your file from the help file
:!*command* <enter>	Execute *command* (this is useful if you know a command by name but cannot remember the keystroke(s))
:. =	Display current line number bottom of screen
: =	Display total line numbers at bottom of screen
^g	Displays current line number and total line numbers at bottom of screen

The Emacs Editor

Emacs (which stands for Editor MACroS) is a text editor like vi. However, whereas vi has three modes (insert, replace, and command), Emacs only has one mode (insert). Commands are issued by using either control or escape in conjunction with other keys. For instance, to move one line forward, you would type control+n ('n' for next). Many of the commands use a letter that is the first letter of the command, and thus Emacs can be easier to learn than vi. Emacs is often a preferred choice of programmers because the Emacs environment can, if set up properly, execute program instructions. For instance, the Lisp language is often installed in Emacs and so a programmer can edit Lisp code and then execute it all from within Emacs.

Starting Emacs brings up an initial "splash" window (Figure 9.5a). As soon as you begin editing (typing characters or moving the cursor), the image in Emacs is replaced with a blank page (except for three lines of text; see Figure 9.5b). You will notice in Figure 9.5 that there are GUI commands. If you start Emacs inside of a GUI OS, you are able to control Emacs through the mouse (as well as through keystrokes). However, if you open Emacs from a text-based session (such as if you ssh into a Linux computer), then you can only use the keystrokes.

In Emacs, you edit a buffer. A buffer is an area of memory containing text—this is the session that you are currently editing. In Emacs, you can have many buffers open. This allows you to edit multiple documents at one time and cut/copy and paste between them. There are generally two buffers visible at any one time: the main buffer (the large window in either part of Figure 9.5) and the minibuffer (the bottom portion of the window in either part of Figure 9.5). The minibuffer is a location where you will enter commands, file names, and responses to questions (such as "Save file?"). You can actually open two buffers at a time (in which case, the large window is split in half), and in any single buffer, you can continue to split it into two buffers. We will discuss this later.

Emacs wraps characters onto the next line once you reach the right side of the buffer, like any word processor; however, you can see that the line is extended because the end of the line is denoted with a \ character. If you were on such a line and used the command to move to the end of the line, the cursor moves to the end of the extended line.

For instance, if you have:

```
Desperate nerds in high offices all over the world have been known to enact the m\
ost disgusting pieces of legislation in order to win votes (or, in places where they d\
on't get to vote, to control unwanted forms of mass behavior).
```

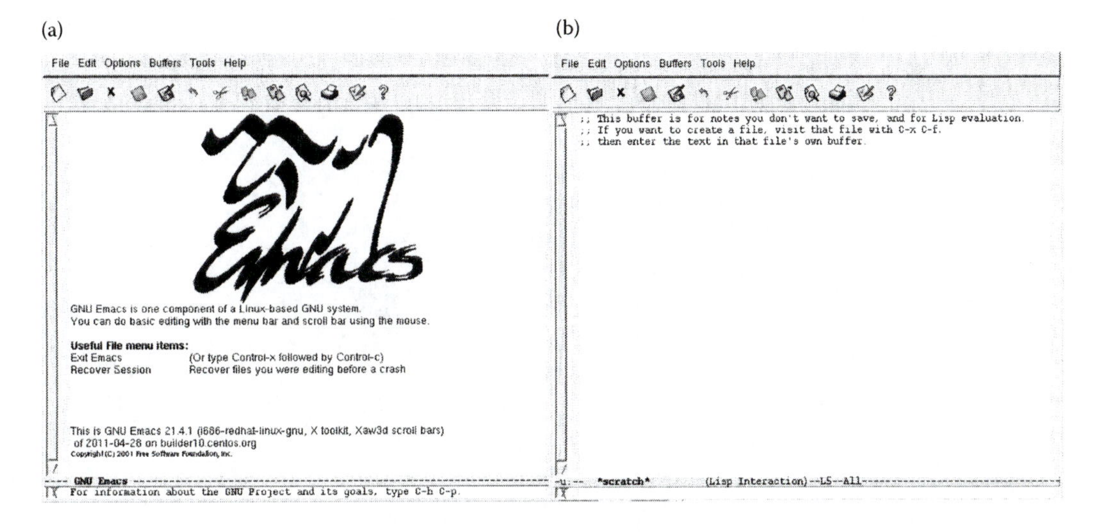

FIGURE 9.5 Emacs upon loading (a) and upon editing (b).

and the cursor is in the first line, then moving to the end of the line actually moves the cursor to the end of behavior). And not to the "m\".

When you enter any keystrokes, they are added to your current Emacs buffer at the position of the cursor, unless you are entering a command. To enter a command, you will either hold the control key down and type one or more characters (this will be denoted as c+*key*), or you will press and release the escape key and follow it with one or more characters (this will be denoted as m+*key*, m stands for meta, a term that some users use for the escape key). We begin with a look at the cursor movement commands, similar to what we saw in vi. However, in Emacs, most of these commands are denoted by their first letter, so, for instance, c+b moves the cursor backward one space. You may recognize some of these commands from the discussion on command line editing in bash covered in Bash Shell Editing Features. Table 9.8 provides the cursor movement commands.

Notice that you can mark a position, as with vi. However, you do not label marked locations (in vi you could label up to 36 of them). Here, Emacs will record up to 16 marks. But you can only return either to the most recent marked position (c+x c+x) or you can cycle through all marked positions by pressing c+u c+space, once per mark. If you continue to mark locations beyond the 16th, the earliest ones are forgotten.

Aside from using c+u to move through marks, the command c+u is used to mean "repeat the next movement 4 times". So, for instance, c+u c+n will move you ahead four times. Interestingly, c+u c+u repeats the next command 16 times. You can also use c+u n, where n is a number, which means "repeat the next command *n* times". You can also specify multidigit times (for instance, 15, 35, 200) by using m+digit m+digit. For instance, m+1 m+0

TABLE 9.8 Cursor Movement Commands

c+n	Move cursor one line down (next)
c+p	Move cursor one line up (previous)
c+f	Move cursor one position forward
c+b	Move cursor one position backward
c+a	Move cursor to the beginning of the line
c+e	Move cursor to the end of the line
c+v	Move cursor down one screen
m+v	Move cursor up one screen
m+f	Move cursor forward one word (to next blank)
m+b	Move cursor backward to beginning of word (or previous word)
m+a	Move cursor to beginning of previous sentence
m+e	Move cursor to beginning of next sentence
m+<	Move cursor to beginning of document
m+>	Move cursor to end of document
m+g *n*	Move to line number *n* where n is a number
c+space	Mark the current position
c+x c+x	Move to marked position (this also moves the mark to the position where you were just at)
c+s	Bring up minibuffer to enter search term, search forward (each c+s) afterward will search for next instance until you do another command
c+r	Search backward

c+f will move the cursor ahead 10 spaces, and m+3 m+2 c+n will move the cursor ahead 32 lines. If you reach the bottom of the file, the cursor will not advance any further, nor does it wrap around to the beginning.

As you type, the characters are inserted at the current cursor position. So Emacs is always in insert mode. The editing commands for deletion/cut and paste are provided in Table 9.9. Notice that the undo command will undo the most recent editing operation, but not a movement operation. So, for instance, if you delete a character, undo restores the character. If you delete a character, move the cursor, and select undo, then the deleted character is restored rather than being moved back to your previous location.

Note that text is appended to the yank buffer until you issue another command. For instance, if you were to delete five lines in a row, all five lines would be copied to the yank buffer. If you were to delete a line, type c+n to go to the next line, and delete that line, only the latest line is saved in the yank buffer because in between the two deletions, you performed another command. Successive deletions though append to the yank buffer.

For instance, consider the line:

```
The brown cow jumped over the silver spoon.
```

Assume the cursor is currently in the 'v' in the word "over". If you do a m+d, it deletes "ver". If you do another m+d, it deletes "the". If you do a c+y, "ver the" reappears. If the cursor is in the same place and you do m+<backspace> twice, it deletes "jumped ov" but not the "er". If you were to now do c+f followed by m+d, this deletes the "r" in "er". If you follow this with c+y, all that is yanked is "r" since the earlier "jumped ov" is now out of the buffer.

The file operations are shown in Table 9.10. The minibuffer is used to prompt the user for a name when using save as, open, and when you wish to exit Emacs with an unsaved

TABLE 9.9 Editing Commands

c+x u	Undo (you can also use c+_ and c+/)
c+d	Delete the next character (also the delete key)
m+d	Delete the next word (from the cursor on)
c+k	Delete the entire line (from the cursor on) (also m+k)
<backspace>	Delete previous character
m+<backspace>	Delete from the cursor backward to beginning of word
m+w	Save entire region from last marked spot to cursor (copy)
c+w	Delete entire region from last marked spot to cursor (cut)
c+y	Yank (paste) from the yank buffer (all of the above inserts characters into the yank buffer except for undo and c+d)
m+y	Same as c+y except that it replaces text with what is in the yank buffer rather than inserts text at the position of the cursor
c+t	Transposes current and previous character
m+u, m+l	Upper/lower cases all letters from cursor to end of word
m+c	Capitalizes current letter
m+%	Search and replace—you are prompted for both the search string and the replacement, and then for each match, you are asked whether to replace (y), skip (n), exit (q), replace all remaining instances (!) or replace once, and exit (.)

TABLE 9.10 File Commands

c+x c+s	Save file under current name
c+x c+w	Save as (you are prompted for a new name)
c+x c+f	Open a new file in a new buffer
c+x c+c	Exit Emacs (if the contents have not been saved but a filename exists, you are asked whether to save the contents before exiting)

document. Note that if you have never saved the buffer's contents, Emacs will exit without saving. If you have saved the buffer's contents to a file, but have made changes and not saved those changes, you are then prompted as whether to save the file under the existing name or a new name, or exit without saving.

The command m+x brings up a prompt in the minibuffer. From here, you can type a command and press the enter key. There are a large number of commands. If you are unsure of a command name, you can use Emacs' version of tab completion. Begin to type in the name of the command press the space bar. This will list in a buffer in the lower half of your window all commands that match. See Table 9.11, which discusses opening, closing, and movement between buffers. Note that the space bar for completion can also be used for file name completion. For instance, if you wish to open a file, you would tape c+x c+f. You are then prompted in the minibuffer to type the file name. You type the path /home/foxr/cit130/ but do not remember the file's name. Pressing the space bar would generate a list of all matching items, in this case, all files in that directory. Here are just some of the commands that you might enter into the minibuffer:

- append-to-file
- auto-save-mode
- calculator
- calendar
- capitalize-region, capitalize-word
- check-parens
- copy-file
- count-lines-page, count-lines-region

TABLE 9.11 Buffer Operations

c+x k	Close the current buffer (if unsaved material exists, you will be prompted to save it first)
c+x b	Switch to the most recent buffer other than the current buffer
c+x 2	Open a second window (under the current window) that takes up half the window's size to see a second buffer (note: doing this inside a halved window gives you two quarter sized windows, etc.)
c+x 1	Close the second window to a single window
c+x o	Switch from one buffer to another, including the minibuffer

- goto-line
- insert-file
- save-buffer
- shell (opens a Linux shell)
- undo

There are a number of programs that can be run from inside of Emacs. These include doctor (an artificial intelligence program), hanoi (Towers of Hanoi), phases-of-moon (to show you upcoming phases of the moon), and tetris. You would execute a program by using m+x name as in m+x doctor <enter> to run the doctor program. Inside of Emacs, you can execute Linux commands.

There are several built-in spell checking modes available using either the aspell or ispell programs. Flyspell (using aspell) highlights all misspelled words in the buffer. You can also use flyspell programming mode, which only highlights misspelled words in comments (for instance, finding all misspellings in the comments of a Java program). Ispell is used for the remainder of the modes. Ispell can check the highlighted word, or can complete the current word, or can spell check the current buffer or the current region. In ispell, the window splits into two buffers with lists of words to replace errors appearing in the lower buffer. You can also add words to the Ispell dictionary if desired. To run either spell checker, use M+x <command>, where <command> is one of the following:

- flyspell-mode
- flyspell-prog-mode
- ispell
- ispell-buffer
- ispell-region
- ispell-message
- ispell-change-dictionary (allows you to change your Ispell dictionary)

You can also use M+$ to check the spelling of the current word and M+x<tab> <esc> <tab> to complete the current word. When using ispell, if a word is selected as misspelled, you can do any of these actions:

- <space>—to skip the word (it is still considered misspelled, just ignored for now)
- r word <return>—will replace the highlighted word with the word you have entered
- digit—replace the word with the word selected by the digit entered from a list provided

- a—accept the word as is
- i—insert this word into your dictionary
- ?—show the entire list of options that can be used in Ispell

One last feature of Emacs is the ability to define and recall macros (this is also available in vi, although not discussed earlier). A macro is a program that you define as a series of keystrokes, to be executed more than once. You define the macro and then call upon it whenever you need it. To start a macro, type c+x (followed by the macro keystrokes followed by c+x). To execute the entered macro, type c+x e.

Let us consider an example. Imagine that you have a group of text and you want to place this text into an html table. The text currently is specified as # Ch. # p. #, where each # is a number. You want to place this in an html table where each row of the html table matches a row in the text table. The first # will be placed in one column, the Ch. # in a second column, and the p. # in a third column. An html table begins with <table> and ends with </table> and each row consists of: <tr><td>column 1</td><td>column 2</td><td>column 3</td></tr>. So we need to add to each row the <tr><td> before the first #, </td><td> before Ch. #, </td><td> before p. #, and </td></tr> to the end of the row. In order to accomplish our task above, we would do the following:

- Move the cursor to the first line of the table, type <table> <enter> and type c+x (
- Type c+a <tr><td>—this moves the cursor to the beginning of a line and adds the text <tr><td> to the beginning
- Type m+f—this moves the cursor passed the first word (in this case, the first number)
- Type </td><td>—to end the first column and start the second
- Type c+s . c+s . c+b c+b—this moves the cursor to immediately before p.
- Type </td><td>—to end the second column and start the third
- Type c+e </td></tr>—to move to the end of the row and end the column and row
- Type c+n—to move to the next line
- Type c+x)—to end the macro

If there were 15 rows for the table, we now do c+u 14 c+x e (execute the macro 14 more times) and finally type </table>. You can save macros and give them names so that you can define many macros and recall the macro you need. You can later edit the named macro and add to it or make changes.

Finally, Emacs can be confusing. If you are lost, you can bring up the help file in a new buffer. To bring up help, type c+h c+h, or if you want help on a specific command, type c+h command. If you are unsure of the command, type in the first letter or letters of the command. As with vi, the best way to learn Emacs is to practice!

FURTHER READING

Bash is only one of many shells available in Linux, although it is arguably the most popular today. There are numerous texts that provide details on the Bash shell and others. The following list spotlights command-line interaction with Linux (or Unix). You will find other texts describing shell programming listed in Chapter 14, and still other texts that contain material on various shells in the texts listed in the Further Reading section of Chapter 4.

- DuBois, P. *Using CSH & Tcsh*. Cambridge, MA: O'Reilly, 1995.
- Garrels, M. *Bash Guide for Beginners*. Palo Alto, CA: Fultus Corporation, 2004.
- Kiddle, O., Stephenson, P., and Peek, J. *From Bash to Z Shell: Conquering the Command Line*. New Jersey: Apress Media LLC, 2004.
- Myer, T. *MAC OS X UNIX Toolbox: 1000+ Commands for the Mac OS X UNIX*. Hoboken, NJ: Wiley and Sons, 2009.
- Newham, C. *Learning the Bash Shell*. Cambridge, MA: O'Reilly, 2005.
- Quigley, E. *UNIX Shells by Example with CDROM*. Upper Saddle River, NJ: Prentice Hall, 1999.
- Robbins, A. *Bash Pocket Reference*. Massachusetts: O'Reilly, 2010.
- Robbins, A. and Rosenblatt, B. *Learning the Korn Shell*. Massachusetts: O'Reilly, 2002.
- Shotts Jr., W. *The Linux Command Line: A Complete Introduction*. San Francisco, CA: No Starch Press, 2012.
- Sobell, M. *Practical Guide to Linux Commands, Editors and Shell Programming*. Upper Saddle River, NJ: Prentice Hall, 2009.

Although the best way to learn vi and Emacs is through practice and through help available from both the software and websites, there are also a few texts available.

- Artymiak, J. Vi(1) *Tips,: Essential vi/vim Editor Skills*, devGuide.net (self-published, see http://devguide.net/), 2008.
- Ayers, L. *GNU Emacs and XEmacs* (With CD-ROM) (Linux). Massachusetts: Muska & Lipman/Premier-Trade, 2001.
- Cameron, D., Elliot, J., Loy, M., Raymond, E., and Rosenblatt, B. *Learning GNU Emacs*. Massachusetts: O'Reilly, 2004.
- Oualline, S. *Vi iMproved (VIM)*. Indianapolis, IN: Sams, 2001.
- Robbins, A., Hannah, E., and Lamb, L. *Learning the vi and Vim Editors*. Massachusetts: O'Reilly, 2008.
- Stallman, R. *GNU Emacs Manual V 23.3*. Massachusetts: Free Software Foundation, 2011.

REVIEW TERMS

Terminology introduced in this chapter

Alias
Apropos
Bang (!)
Brace expansion
Command line
Command line editing
Command mode (vi)
Environment variable
Expansion
History
Insert mode (vi)
Interpreter
Macro (Emacs)
Man page
Minibuffer (Emacs)
PATH variable
PS1 variable
Pipe
Replace mode (vi)
Redirection
Shell
Tab completion
Tilde expansion
Wildcard

REVIEW QUESTIONS

1. Provide a definition for an interpreter.

2. What does the Bash interpreter do?

3. How do you reexecute the last instruction from the history list? How do you reexecute instruction 21 from the history list? How do you reexecute the last instruction that started with a 'c' from the history list?

4. How do you move successively backward through the history list so each instruction appears on the command line?

5. You have typed less abc<tab>. What does the <tab> key do? What would happen if you pressed <tab><tab> instead of a single <tab>?

6. Consider the following list of file names in the current directory:

```
forgotten.txt frank.txt fresh.txt functions.txt funny.txt other_
stuff.txt other_funny_stuff.txt
```

What would happen if you typed in "*less fun<tab>*"? What would happen if you typed in "*less fun<tab><tab>*"? What would happen if you typed in "*less funn<tab>*"?

7. From the command line, if you are in the middle of an instruction that you are editing, what does control+b do? What does control+f do? What does control+a do? What does control+e do? What does control+k do?

8. Define an alias so that the command del is equal to rm.

9. Define an alias so that rmall recursively deletes all files and subdirectories in the current directory.

10. Why might you define an alias rm for 'rm –i'?

11. How will the following instruction be interpreted? That is, what specific command(s) is(are) executed?

```
ls ~/{foo1,foo2/{foo3,foo4},foo5}
```

12. What does the * mean if you do ls *.txt? How about ls *.*? How about ls *? What is the difference between using * and ? in the ls command?

13. What does the $ do when placed in front of a variable?

14. What is the difference between NAME = Frank and NAME = $Frank?

15. Write an assignment statement to add the directories /usr/local/share/bin, /sbin, and /home/zappaf to your PATH variable.

16. Write an assignment statement to change your user prompt (PS1) to output your user name, an @ symbol, the current date, a space, the time in 12-hour format, a colon, the current command number, and then the $. For instance, it might look like this:

```
foxr@Mon Feb 13 2:03 EST 2012 pm:16$
```

17. The cat command will output the contents of a file to the screen. You can cat multiple files. Write a Linux command which will take the content of files foo1.txt, foo2.txt, and foo3.txt, and sort the lines in alphabetical order (using the sort command). This will require a pipe.

18. Repeat #17 but send the output to the text file sorted.txt.

19. Repeat #18 but append the output to the text file sorted.txt.

20. What does the command cat << foo do?

21. What does it mean if the man page for a command has multiple items listed under synopsis?

22. In what order are these files interpreted when you open a bash shell from inside of your Linux GUI? /etc/profile, /etc/bashrc, .profile, .bashrc, and .bash_profile?

23. If, as a user, you want to define an alias or a variable to be used in every bash session, where would you define it?

24. If, as a user, you want to define an alias or a variable for only the current bash session, where would you define it?

25. If, as a system administrator, you want to define an alias or a variable to be used in all users' bash sessions, where would you define it?

26. Why might you use vi or Emacs instead of a word processor or a GUI-based text editor?

27. What are the modes that vi uses? How do you switch from one mode to another?

DISCUSSION QUESTIONS

1. Explain the importance of being able to master the command line in Linux rather than relying on the GUI.

2. Following up on #1, list several duties of a system administrator that you would prefer to accomplish through the command line because it is either easier or because it provides you with greater flexibility and power.

3. It is common that students just learning to use Linux (or Unix) will avoid using command line editing, history, defining aliases, and other shortcuts available in the Bash shell. What are some motivating techniques that you might suggest so that students not only learn these shortcuts but choose to apply them consistently?

4. Similar to #3, students often forget how to, or refuse to, apply redirection commands instead writing several individual commands and even creating temporary files so that the output of one command can be used later. Describe some example problems that have helped motivate you to use and learn redirection.

5. Provide several reasons why you would use a text editor in Linux/Unix. Do similar situations exist in Windows?

6. What do you find to be the biggest challenge behind learning vi?

7. What do you find to be the biggest challenge behind learning Emacs?

8. The vi editor is a standard part of any Linux or Unix OS, but in many cases, Emacs must be installed separately. Why do you suppose Emacs is not a standard part of Linux/Unix? If you were a system administrator, would you install it if it were not specifically requested? Why or why not?

9. Run Emacs. Type m+x doctor <enter>. This starts the "Doctor" program (which originally was called Eliza). The program pretends to be a Rogerian psychotherapist. It is programmed to respond to your natural language (English) inputs with its own natural language comments or questions. It waits for you to enter a sentence and then press the enter key twice, and then it will respond to you. The program relies on a number of patterns that it searches for and if found, has built-in responses. For

instance, if it finds text such as "I like to listen to music", it identifies the word "like" and response by rearranging your statement into a question such as "why do you like to listen to music". The pattern found is "I like X", which is rearranged to become "why do you like X?" While running this program, first respond with reasonable statements that attempt to converse with Doctor. After you have played with it for a while, see if you can fool the program by entering nonsensical statements or ungrammatical statements. To exit the program, enter the text "bye" at the end.

CHAPTER 10

Regular Expressions

Chapter 9 focused on Linux. Chapter 10 also focuses on a Linux-related topic, that of regular expressions. Regular expressions provide a powerful tool for Linux users and administrators. With regular expressions, a user can search through text files not for specific strings but for strings that fit a particular pattern of interest. Linux offers the grep program, which performs such a search task given a regular expression. In this chapter, regular expressions are introduced along with numerous examples and an examination of grep. Because of the challenging nature in learning regular expressions, the reader should be aware that mastery of them only comes with an extensive experience and that this can be a difficult chapter to read and understand. It is recommended that the reader try out many of these examples. The chapter also examines the use of Bash wildcards.

The learning objectives of this chapter are to

- Describe regular expressions and why they are useful.
- Illustrate the use of each regular expression metacharacter.
- Provide numerous examples of regular expressions.
- Examine the grep program.
- Describe the use of wildcards in Bash and show how they differ from regular expressions.
- Combine ls and grep through redirection.

Consider a string of characters that contains only 1s followed by 0s, for instance, 111000, 100, and 10000. A regular expression can be used to specify such a pattern. Once written, a regular expression can be compared to a collection of strings and return those that match the pattern. A *regular expression* is a string that combines literal characters (such as 0 or 1) with *metacharacters*, symbols that represent options. With metacharacters, you can specify, for instance, that a given character or set of characters can match "any number of" or "at least one" time, or specify a list of characters so that "any one character" matches.

Regular expressions can be highly useful to either a user or system administrator when it comes to searching for files or items stored in files. In this chapter, we examine how to define regular expressions and how to use them in Linux. The regular expression is considered so useful that Linux has a built-in program called grep (global regular expression print), which is an essential tool for Linux users. Wildcards, a form of regular expressions, are available in Linux as well, although these are interpreted differently from their usage in regular expressions. Regular expressions have been built into some programming languages used extensively in Linux such as perl.

Let us consider two simple examples to motivate why we want to explore regular expressions. First, you, as a user, have access to a directory of images. Among the images are jpg, gif, png, and tiff formatted images. You want to list all of those under the tiff format. However, you are unsure whether other users will have named the files with a .tiff, .tif, .TIFF, .TIF, .Tiff, or .Tif extension. Rather than writing six different ls statements, or even one ls statement that lists each possible extension, you can use a regular expression. Second, as a system administrator, you need to search a directory (say /etc) for all files that contain IP addresses as you are looking to change some hardcoded IP addresses, but you do not remember which files to examine. A regular expression can be defined to match strings of the form #.#.#.#, where each # is a value between 0 and 255. In creating such a regular expression and using grep, you can see all of the matches using one command rather than having to examine dozens or hundreds of files.

In this chapter, we first examine the metacharacters for regular expressions. We look at dozens of examples of regular expressions and what they might match against. Then, we look at how some of the characters are used as wildcards by the Bash interpreter. This can lead to confusion because * has a different meaning when used as a regular expression in a program such as grep versus how the Bash interpreter uses it in an instruction such as ls. Finally, we look at the grep program and how to use it. Regular expressions can be a challenge to apply correctly. Although in many cases, their meaning may be apparent, they can often confound users who are not familiar with them. Have patience when using them and eventually you might even enjoy them.

METACHARACTERS

There is a set of characters that people use to describe options in a pattern. These are known as metacharacters. Any regular expression will comprise literal characters and metacharacters (although a regular expression does not require metacharacters). The metacharacter * means "match the preceding character 0 or more times"; so, for instance, a* means "zero or more a's". The regular expression 1010 matches only 1010 as it has no metacharacters. Since we will usually want our regular expressions to match more than one specific string, we will almost always use metacharacters. Table 10.1 provides the list of metacharacters. We will explore each of these in turn as we continue in this section.

We will start with the most basic of the symbols: * and +. To use either of these, first specify a character to match against. Then place the metacharacter * or + after the character to indicate that we expect to see that character 0 or more times (*) or 1 or more times (+).

TABLE 10.1 Regular Expression Metacharacters

<table>
<tr><th>Metacharacter</th><th>Explanation</th></tr>
<tr><td>*</td><td>Match the preceding character if it appears 0 or more times</td></tr>
<tr><td>+</td><td>Match the preceding character if it appears 1 or more times</td></tr>
<tr><td>?</td><td>Match the preceding character if it appears 0 or 1 time</td></tr>
<tr><td>.</td><td>Match any one character</td></tr>
<tr><td>^</td><td>Match if this expression begins a string</td></tr>
<tr><td>$</td><td>Match if this expression ends a string</td></tr>
<tr><td>[chars]</td><td>Match if the expression contains any of the chars in [ ]</td></tr>
<tr><td>[char$_i$-char$_j$]</td><td>Match if the expression contains any characters in the range from char$_i$ to char$_j$ (e.g., a–z, 0–9)</td></tr>
<tr><td>[[:class:]]</td><td>An alternative form of [ ] where the :class: can be one of several categories such as alpha (alphabetic), digit, alnum (alphabetic or numeric), punct, space, upper, lower</td></tr>
<tr><td>[^chars]</td><td>Match if the expression does not contain any of the chars in [ ]</td></tr>
<tr><td>\</td><td>The next character should be interpreted literally, used to escape the meaning of a metacharacter, for instance \$ means "match a $"</td></tr>
<tr><td>{n}</td><td>Match if the string contains n consecutive occurrences of the preceding character</td></tr>
<tr><td>{n,m}</td><td>Match if the string contains between n and m consecutive occurrences of the preceding character</td></tr>
<tr><td>{n,}</td><td>Match if the string contains at least n consecutive occurrences of the preceding character</td></tr>
<tr><td>{,m}</td><td>Match if the string contains no more than m consecutive occurrences of the preceding character</td></tr>
<tr><td>|</td><td>Match any of these strings (an "OR")</td></tr>
<tr><td>(…)</td><td>The items in … are treated as a group, match the entire sequence</td></tr>
</table>

For instance, 0*1* matches any string of zero or more 0s followed by zero or more 1s. This regular expression would match against these strings: 01, 000111111, 1, 00000, 0000000001, and the empty string. The empty string is a string of no characters. This expression matches the empty string because the * can be used for 0 matches, so 0*1* matches a string of no 0s and no 1s. This example regular expression would not match any of the following: 10, 00001110, 0001112, 00a000, or abc. In the first and second cases, a 0 follows a 1. In the other three cases, there are characters in the string other than 0 and 1.

The regular expression 0+1+ specifies that there must be at least one 0 and one 1. Thus, this regular expression would not match the empty string; neither would it match any string that does not contain one of the two digits (e.g., it would not match 0000 or 1). This expression would match 01, 00011111, and 000000001. Like 0*1*, it would not match a string that had characters other than 0 or 1 (e.g., 0001112), nor would it match a string in which a 0 followed a 1 (e.g., 00001110).

We can, of course, combine the use of * and + in a regular expression, as in 0*1+ or 0+1*. We can also specify literal characters without the * or +. For instance, 01* will match against a 0 followed by zero or more 1s—so, for instance, 0, 01, 01111, but not 1, 111, 1110, or 01a. Although * and + are the easiest to understand, their usefulness is limited when just specified after a character. We will find that * and + are more useful when we can combine them with [ ] to indicate a combination of repeated characters.

The ? is a variant, like * or +, but in the case of ?, it will only match the preceding character 0 or 1 time. This allows you to specify a situation where a character might or might not be expected. It does not, however, match repeatedly (for that, you would use * or +). Recall the .tiff/tif example. We could specify a regular expression to match either tiff or tif as follows: tiff?. In this case, the first three characters, tif, are expected literally. However, the last character, f, may appear 0 or 1 time. Although this regular expression does not satisfy strings such as TIFF (i.e., all upper-case letters), it is a start. Now, with ?, *, and +, we can control how often we expect to see a character, 0 or 1 time, 0 or more times, 1 or more times.

Unlike * and +, the ? places a limit on the number of times we expect to see a character. Therefore, with ?, we could actually enumerate all of the combinations that we expect to match against. For instance, 0?1? would match against only four possible strings: 0, 1, 01, and the empty string. In the case of 0*1*, there are an infinite number of strings that could match since "0 or more" has no upper limit.

Note that both the * and ? are used in Linux commands like ls as wildcards. In Bash and Wildcards, we will learn that in such commands, their meaning differs from the meanings presented here.

The . (period) can be used to match any single character. For instance, b.t could match any of these strings bat, bet, bit, but, bot, bbt, b2t. It would not match bt, boot, or b123t. The . metacharacter can be combined with *, +, and ?. For instance, b.*t will match any string that starts with a b, is followed by any number of other characters (including no characters) and ending with t. So, b.*t matches bat, bet, bit, but, bot, bbt, b2t, bt, boot, b123t, and so forth. The expression b.+t is the same except that there must be at least one character between the b and the t, so it would match all of the same strings except for bt. The regular expression b.?t would match bt or anything that b.t matches. The question mark applies to the . (period). Therefore, . is applied 0 or 1 time; so this gives us a regular expression to match either bt or b.t. It would not match any string that contains more than one character between the b and the t.

The next metacharacter is used to specify a collection or a list. It starts with [, contains a list, and ends with]. For example, [aeiou] or [123]. The idea is that such a pattern will match any string that contains any one of the items in the list. We could, for instance, specify b[aeiou]t, which would match any of bat, bet, bit, bot, and but. Or, we could use the [ ] to indicate upper versus lower case spelling. For instance, [tT] would match either a lower case t or an upper case T.

Now we have the tools needed to match any form of tif/tiff. The following regular expression will match any form of tif or tiff using any combination of lower- and upper-case letters: [tT][iI][fF][fF]?. The ? only applies to the fourth [ ] list. Thus, it will match either an upper- or lower-case t, followed by an upper- or lower-case i, followed by an upper- or lower-case f, followed by zero or one lower- or upper-case f.

The list specified in the brackets does not have to be a completely *enumerated* list. It could instead be a *range* such as the letters a through g. A range is represented by the first character in the range, a hyphen (-), and the last character in the range. Permissible characters for ranges are digits, lower-case letters, and upper-case letters. For instance, [0-9] would mean any digit, whereas [a-g] would mean any lower-case letter from a to g. That is, [a-g] is equivalent to [abcdefg]. You can combine an enumerated list of characters and a range, for instance [b-df-hj-np-tv-z] would be the list of lower-case consonants.

As an alternative to an enumerated list or range, you can also use the double brackets and a class. For instance, instead of [a-zA-Z], you could use [[:alpha:]], which represents the class of alphabetic characters. There are 12 standard classes available in the Linux regular expression set, as shown in Table 10.2. A nonstandard class is [[:word:]], which consists of all of the alphanumeric characters plus the underscore.

The list, as specified using [ ] or [[ ]], will match any single character if found in the string. If you wanted to match some combination of characters in a range, you could add *, +, ., or ? after the brackets. For instance, [a-z]+ means one or more lower-case letters.

Imagine that you wanted to match someone's name. We do not know if the first letter of the person's name will be capitalized but we expect all of the remaining letters to be in lower case. To match the first letter, we would use [A-Za-z]. That is, we expect a letter, whether upper or lower case. We then expect some number of lower-case letters, which would be [a-z]+. Our regular expression is then [A-Za-z][a-z]+. Should we use * instead of + for the lower-case letters? That depends on whether we expect someone's name to be a single letter. Since we expect a name and not an initial, we usually would think that a name would be multiple letters. However, we could also use [A-Za-z][a-z]* if we think a name might be say J.

What would [A-Za-z0-9]* match? This expression will match zero or more instances of any letter or digit. This includes the empty string (as * includes zero occurrences), any single letter (upper or lower case) or digit, or any combination of letters and digits. So each of these would match:

```
abc ABC aBc a12 1B2 12c 123456789 aaaaaa 1a2b3C4D5e
```

So what would not match? Any string that contained characters other than the letters and digits. For instance, a_b_c, 123!, a b c (i.e., letters and blank spaces), and a1#2B%3c*fg45.

TABLE 10.2 Regular Expression Classes

Class	Meaning
[[:alnum:]]	Alphanumeric—alphabetic character (letter) or digit
[[:alpha:]]	Alphabetic—letter (upper or lower case)
[[:blank:]]	Space or tab
[[:cntrl:]]	Any control character
[[:digit:]]	Digit
[[:graph:]]	Any visible character
[[:lower:]]	Lower-case letter
[[:print:]]	Any visible character plus the space
[[:punct:]]	Any punctuation character
[[:space:]]	Any whitespace (tab, return key, space, backspace)
[[:upper:]]	Upper-case letter
[[:xdigit:]]	Hexadecimal digit

Notice with the [ ] that we can control what characters can match, but not the order that they should appear. If we, for instance, require that an 'a' precede a 'b', we would have to write them in sequence using two sets of brackets, such as [aA][bB] to indicate an upper- or lower-case 'a' followed by an upper- or lower-case 'b'. We could also allow any number of them using [aA]+[bB]+ (or use * instead of +). This can become complicated if we want to enforce some combination followed by another combination. Consider that we want to create a regular expression to match any string of letters such that there is some consonant(s) followed by some vowel(s) followed by consonant(s). We could use the following regular expression:

```
[b-df-hj-np-tv-z]+[aeiou]+[b-df-hj-np-tv-z]+
```

Can we enforce a greater control on "1 or more"? The * and + are fine when we do not care about how many instances might occur, but we might want to have a restriction. For instance, there must be no more than five, or there must be at least two. Could we accomplish this using some combination of the ? metacharacter? For instance, to indicate "no more than five", could we use "?????"? Unfortunately, we cannot combine question marks in this way. The first question mark applies to a character, but the rest of the question marks apply to the preceding question mark.

We could, however, place a character (the period for "any character") followed by a question, and repeat this five times as in:

```
.?.?.?.?.?
```

This regular expression applies each question mark to a period. And since the question mark means "0 or 1", this is the same as saying any one, two, three, four, or five characters. But how do we force the characters to be the same character? Instead, what about [[:visib le:]]?[[:visible:]]?[[:visible:]]?[[:visible:]]?[[:visible:]]? Unfortunately, as with the period, the character in each [[:visible:]] can be any visible character, but not necessarily the same character.

Our solution instead is to use another metacharacter, in this case, {n,m}. Here, n and m are both positive integers with n less than m. This notation states that the preceding character will match between n and m occurrences. That is, it is saying "match at least n but no more than m of the preceding character." You can omit either bound to enforce "at least" and "no more than", or you can specify a single value to enforce "exactly".

For instance, 0{1,5}1* would mean "between one and five 0s followed by any number of 1s whereas [01]{1,5} means "between one and five combinations of 0 and 1". In this latter case, we would not care what order the 0s and 1s occur in. Therefore, this latter expression will match 0, 1, 00, 01, 10, 11, 000, 001, 010, 011, 100, 101, 110, 111, up to five total characters.

We would use {2,} to indicate "at least two" and {,5} to indicate "no more than five". With the use of {n,m}, we can now restrict the number of matches to a finite number. Consider 0{5}1{5}. This would match *only* 0000011111. However, [01]{5} would match any combination of five 0s and 1s.

It should be noted that the use of { } is only available when you are using the *extended regular expression set.* The program grep, by default, does not use the extended set of metacharacters. To use { } in grep, you would have to use extended grep. This is either egrep or grep –E. We will see this in more detail in The grep Program.

Let us combine all of the ideas we have seen to this point to write a regular expression that will match a social security number. The social security number is of the form ###-##-####, where each # is a digit. A regular expression to match such a number is given below:

```
[0-9][0-9][0-9][0-9][0-9][0-9][0-9][0-9][0-9]
```

This regular expression requires a digit, a digit, a digit, a hyphen, a digit, a digit, a hyphen, a digit, a digit, a digit, and a digit. Using {n}, we can shorten the expression to:

```
[0-9]{3}-[0-9]{2}-[0-9]{4}
```

What would a phone number look like? That depends on whether we want an expression that will match a phone number with an area code, without an area code, or one that could match a phone number whether there is an area code or not. We will hold off on answering this question for now and revisit it at the end of this section.

Let us try something else. How would we match an IP address? An IP address is of the form 0-255.0-255.0-255.0-255. The following regular expression is not correct. Can you figure out why?

```
[0-255].[0-255].[0-255].[0-255]
```

The IP address regular expression has two flaws; the first one might be obvious, the second is a bit more obscure. What does [0-255] mean? In a regular expression, you use the [ ] to indicate a list of choices. Choices can *either* be an enumerated list, such as [abc], or a range, such as [a-c]. The bracketed lists for this regular expression contain both a range and an enumerated list. First, there is the range 0–2, which will match 0, 1, or 2. Second, there is an enumerated list 5, 5. Thus, each of the bracketed items will match any of 0, 1, 2, 5, or 5. So the above regular expression would match 0.1.2.5 or 5.5.5.5 or 0.5.1.2. What it would not match are either 0.1.2.3 (no. 3 in the brackets) or 10.11.12.13 (all of the brackets indicate a digit, not a multicharacter value such as 13).

So how do we specify the proper enumerated list? Could we enumerate every number from 0 to 255? Not easily, and we would not want to, the list would contain 256 numbers! How about the following: [0-9]{1,3}. This expression can match any single digit from 0 to 9, any two digit numbers from 00 to 99, and any three-digit numbers from 000 to 999. Unfortunately, we do not have an easy way to limit the three-digit numbers to being 255 or less, so this would match a string such as 299.388.477.566. But for now, we will use this notation. So, let us rewrite our expression as

```
[0-9]{1,3}.[0-9]{1,3}.[0-9]{1,3}.[0-9]{1,3}
```

The second flaw in our original regular expression is the use of the . (period). Recall that . means "match any one character". If we use a period in our regular expression, it could match anything. So, our new regular expression could match 1.2.3.4 or 10.20.30.40 or 100.101.201.225, but it could also match 1a2b3c4 or 1122334 or 1-2-3-4, and many other sequences that are not IP addresses. Our problem is that we do not want the period to be considered a metacharacter; we want the period to be treated literally.

How do we specify a literal character? For most characters, to treat it literally, we just list it. For instance, abc is considered the literal string "abc". But if the character itself is a metacharacter, we have to do something special to it. There are two possibilities, the first is to place it in [ ]. This is fine, although it is not common to place a single character in [ ], so [0-9]{1,3}[.][0-9]{1,3}[.][0-9]{1,3}[.][0-9]{1,3} would work. Instead, when we want to specify a character that so happens to be one of the metacharacters, we have to "escape" its meaning. This is done by preceding the character with a \, as in \. or \+ or \{ or \$. So, our final answer (for now) is

```
[0-9]{1,3}\.[0-9]{1,3}\.[0-9]{1,3}\.[0-9]{1,3}
```

We will try to fix the flaw that permits three-digit numbers greater than 255 in a little while.

There are other uses of the escape character, these are called escape sequences. Table 10.3 provides a listing of common escape sequences. For instance, if you want to find four consecutive white spaces, you might use \s{4}. Or if you want to match any sequence of 1 to 10 non-digits, you could specify \D{1,10}.

The [ ] has another usage, although it can be a challenge to apply correctly. If you place a ^ before the enumerated list in the [ ], the list is now interpreted as meaning "match if none of these characters are present". You might use [^0-9]* to match against a string that contains no digits or [^A-Z]* to match a string that contains no upper-case letters. The expression [A-Z][a-z]+[^A-Z]* states that a string is expected to have an upper-case letter, some number of lower-case letters but no additional upper-case letters. This might be the case if we expect a person's name, as we would not expect, for instance, to see a name spelled as ZaPPa.

Why is [^...] challenging to use? To explain this, we must first reexamine what regular expressions match. Remember that a regular expression is a string used to match against

TABLE 10.3 Common Escape Sequences

\d	Match any digit
\D	Match any non-digit
\s	Match any white space
\S	Match any non-white space
\w	Match any letter (a-z, A-Z) or digit
\W	Match any non-letter/non-digit
\b	Match a word boundary
\B	Match any non-word boundary

another string. In fact, what the regular expression will match is a *substring* of a larger string. A substring is merely a portion of a string. For instance, if the string is "Frank Zappa", any of the following would be considered a substring: "Frank", "ank", "Zappa", "k Z", "ppa", "Frank Zappa", and even "" (the empty string).

Consider the expression 0{1,2}[a-zA-Z0-9]+. This regular expression will match any string that consists of one or two 0s followed by any combination of letters and digits. Now consider the following string:

```
0000abcd0000
```

As we have defined regular expressions earlier, this string should not match the expression because it does not have "one or two 0s", it literally has four 0s. However, the expression is only looking to match any substring of the string that has "one or two 0s followed by letters and digits". Since the string contains "0a", it matches. The regular expression *does not need to match* every character in the string; it only has to find some substring that does match.

Returning to the usage of [^…], let us look at an example. Consider [^A-Z]+. The meaning of this expression seems clear: match anything that does not contain capital letters. Now consider a string abCDefg. It would appear that the regular expression should not match this string. But the regular expression in fact says "do not match upper-case letters" but the string *also* contains lower-case letters. Therefore, the regular expression provided will match ab and efg from the string, so the string is found to match. What use is [^A-Z]+ then if it matches a string that contains upper-case letters? The one type of string this regular expression will not match is any string that *only* consists of upper-case letters. So, although it matches abCDefg, it would not match ABCDEFG. To make full use of [^…], we have to be very careful.

With this in mind, consider 0+1+. This will match 0001, 01111, 01, but it will also match 0101, and 000111aaa because these two strings do contain a sequence that matches 0+1+. So how can we enforce that the match should only precisely match the expression? For this, we need two additional metacharacters, ^ and $. The ^, as seen above, can be used inside of [ ] to mean "match if these are not found". But outside of the [ ], the ^ means to "match at the start of the string" and $ means to "match at the end of the string". If our regular expression is of the form ^expression$, it means to match only if the string is precisely of this format.

We might want to match any strings that start with numbers. This could be accomplished through the regular expression ^[0-9]+. We use + instead of *, because * could match "none of these", so it would match strings that do or do not start with digits. We might want to match any strings that end with a state abbreviation. All state abbreviations are two upper-case letters, so this would look like [A-Z][A-Z]$, or alternatively [A-Z]{2}$. If we wanted the state abbreviation to end with a period, we could use [A-Z][A-Z]\.$ and if we wanted to make the period optional, we could use [A-Z][A-Z]\.?$ Notice that using [A-Z][A-Z], we are also matching any two uppercase letters, so for instance AB and ZZ, which are not legal state abbreviations.

In general, we do not want to use both ^ and $ in an expression because it would overly restrict matching. To demonstrate the concepts covered so far, let us consider a file containing employee information. The information is, row by row, each employee's last name, first name, position, year of hire, office number, and home address.

Let us write a regular expression to find all employees hired in a specific year, say 2010. Our regular expression could just be 2010. However, just using 2010 could lead to erroneous matches because the string 2010 could appear as a person's office number or as part of an address (street address, zip code).

What if we want to find employees hired since 2000? A regular expression for this could be 20[01][0-9]. Again, this could match as part of an office number or address. If we use ^20[01][0-9]$ as our solution, we restrict matches to lines that consist solely of four-digit numbers between 2000 and 2019. No lines will match because no lines contain only a four-digit number. We must be more clever than this.

Notice that the year hired follows a last name, first name, and position. We could represent last name as [A-Z][a-z]+. We could represent first name as [A-Z][a-z]* (assuming that we permit a letter such as 'J' for a first name). We could represent a position as [A-Za-z]+, that is, any combination of letters. Finally, we would expect the year of hire, which in this case should be 20[01][0-9]. If each of the pieces of information about the employee is separated by commas, we can then construct our expression so that each part ends with a comma. See Figure 10.1. Notice that year of hire, as shown here, includes years in the future (up through 2019).

We could also solve this search problem from the other end of the row. The year hired will occur before an office number and an address. If we assume an office number will only be digits, and that an address will be a street address, city, state, and zip code, then to end the line, we would expect to see 20[01][0-9], [0-9]+, and an address. The address is tricky because it consists of several different components, a street address, a city, a state abbreviation (two letters), and a zip code. The street address itself might be digits followed by letters and spaces (e.g., 901 Pine Street) or it might be more involved. For instance, there may be periods appearing in the address (e.g., 4315 E. Magnolia Road), or it might include digits in the street name (e.g., 50 N. 10th Street). It could also have an apartment number that uses the # symbol (such as 242 Olive Blvd, #6). We could attempt to tackle all of these by placing every potential letter, digit, symbol, and space in one enumerated list, as in [A-Za-z0-9.#]+. The city name should just be letters (although it could also include spaces and periods) but we will assume just letters. The state is a two-letter abbreviation in upper-case letters, and the zip code should be a five-digit number (although it could also be in the form 12345-6789). Figure 10.2 represents the entire regular expression that could be used to end a string that starts with a year of hire of at least 2000, but only with a five-digit zip code (we see how to handle the nine-digit zip code below).

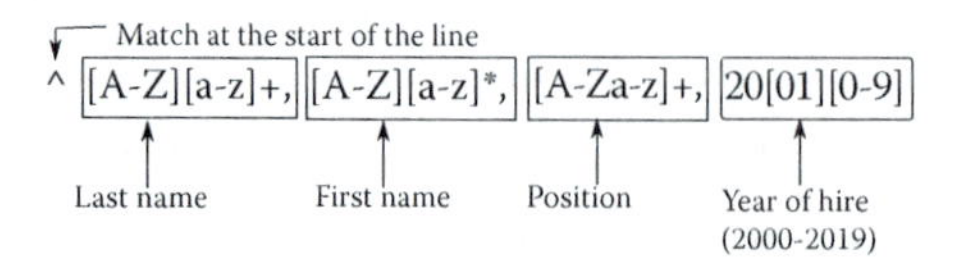

FIGURE 10.1 Representing "Year Hired Since 2000."

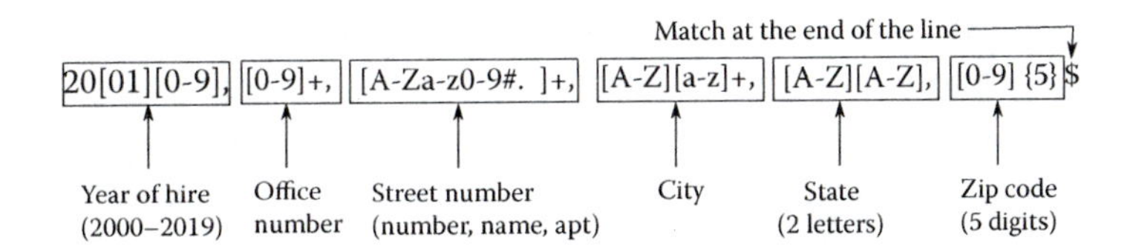

FIGURE 10.2 A possible regular expression to match a street address.

The last metacharacters are | and ( ). The use of these metacharacters is to provide sequences of characters where any single sequence can match. That is, we can enumerate a list of OR items. Unlike the enumerated list in [ ], here, we are enumerating *sequences*. Consider that we want to match any of these three 2-letter abbreviations: IN, KY, OH. If we used [IKO][NYH], it would match IN, KY, and OH, but it would also match IY, IH, KN, KH, ON, and OY. We enumerate sequences inside ( ) and places | between the options. So, (IN|KY|OH) literally will only match one of these three sequences.

With the use of ( ) and |, we can provide an expression to match both five-digit and nine-digit zip codes. The five-digit expression is [0-9]{5}. The nine-digit expression is [0-9]{5}-[0-9]{4}. We combine them using | and place them both in ( ). This gives us ([0-9]{5}|[0-9]{5}-[0-9]{4}).

Similarly, we can use ( ) and | to solve the IP address problem from earlier. Recall that our solution using [0-9]{1,3} would match three-digit numbers greater than 255, and thus would match things that were not IP addresses. We would not want to enumerate all 256 possible values, as in (0|1|2|3|...|254|255), but we could use ( ) and | in another way. Consider that [0-9] would match any one-digit number, and IP addresses can include any one-digit number. We could also use [0-9][0-9] because any sequence of 00-99 is a legitimate IP address, although we would not typically use a leading zero. So, we could simplify this as [0-9]{1,2}. However, this does not include the sequence 100–255. We could express 100–255 as several different possibilities:

```
1[0-9][0-9]—this covers 100-199
2[0-4][0-9]—this covers 200-249
25[0-5]—this covers 250-255
```

Figure 10.3 puts these options together into one lengthy regular expression (blank spaces are inserted around the "|" symbols to make it slightly more readable).

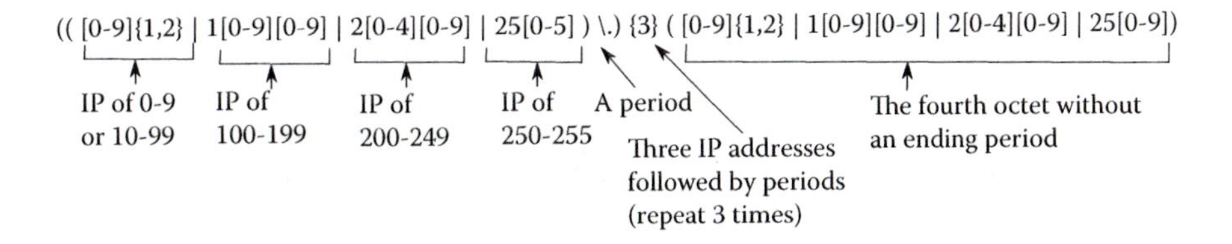

FIGURE 10.3 A solution to match legal IP addresses.

Let us wrap up this section with a number of examples. First, we will describe some strings that we want to match and come up with the regular expressions to match them. Second, we will have some regular expressions and try to explain what they match. Assume that we have a text file that lists student information of the following form:

```
Student ID (a 16 digit number), last name, first name, major,
minor, address.
```

Majors and minors will be three-letter codes (e.g., CSC, CIT, BIS, MIN) all in capital letters. A minor is required, but minors can include three blank spaces to indicate "no minor selected". The address will be street address, city, state, zip code.

We want to find all students who have majors of either computer science (CSC) or computer information technology (CIT). The obvious answer is to just search for (CSC|CIT). However, this does not differentiate between major and minor, we only want majors. Notice that the major follows the 16-digit number and the name. So, for instance, we might expect to see 0123456789012345, Zappa, Frank, CSC, … We could then write the regular expression starting at the beginning of the line:

```
^[0-9]{16}, [A-Z][a-z]+, [A-Z][a-z]+, (CSC|CIT)
```

We could shorten this. Since the minor is always preceded by a major, which is a capitalized three-letter block, what we expect to see before the major, but not the minor, is a first name, which is not fully capitalized. So we could reduce the expression as follows:

```
[A-Z][a-z]+, (CSC|CIT)
```

Here, we are requiring a sequence of an upper-case letter followed by lower-case letters (a name) followed by a comma followed by one of CSC or CIT. If the CSC or CIT matched the minor, the string preceding it would not include lower-case letters, and if the upper-case letter followed by lower-case letters matched a last name or street address, it would not be followed by either CSC or CIT.

We want to find all students who live in apartments. We make the assumption that apartments are listed in the address portion using either apt, apt., #, or apartment. We can use [.]? (or \.?) to indicate that the period after apt is optional. We can enumerate these as follows:

```
([Aa]pt[.]?|[Aa]partment|#)
```

We want to find all students who live in either Kentucky or Ohio. We will assume the state portion of the address is abbreviated. That would allow us to simply specify:

```
(KY|OH)
```

We would not expect KY or OH to appear anywhere else in an address. However, it is possible that a major or minor might use KY or OH as part of the three-letter abbreviation. If we wanted to play safe about this, we would assume that the state appears immediately before a zip code, which ends the string. For this, we could specify:

```
(KY|OH), [0-9]{5}$
```

or if we believe there will be a five-digit zip code and a four-digit extension, we could use:

```
(KY|OH), ([0-9]{5}|[0-9]{5}-[0-9]{4})$
```

Alternatively, since the state abbreviation will appear after a blank, and if it ends with a period, we could add the blank and period so that three-letter majors and minors will not match:

```
( KY\.| OH\.)
```

If we had a major, say SKY or OHM, it would not match these state abbreviations because of the space and period that surround the two letters.

SPAM FILTERS AND REGULAR EXPRESSIONS

So you want to build a spam filter to filter out unwanted e-mail. It is a simple task to write a program that will search through an e-mail (text file) for certain words: "Cash", "Act Now!", "Lose Weight", "Viagra", "Work at home", "You've been selected", and so forth. But spammers have fought back by attempting to disguise keywords.

Consider a spam e-mail advertising cheap and available Viagra tablets. The e-mail may attempt to disguise the word Viagra under a number of different forms: V1agra, V.i.a.g.r.a, Vi@gra, V!agra, ViaSEXYgra, and so forth. But regular expressions can come to our rescue here.

If we are only worried about letter replacements, we could try to enumerate all possible replacements in our regular expression, as in

```
[Vv] [iI!] [Aa@] [gG9] [Rr] [aA@]
```

What about a version where, rather than a common replacement (for instance, 3 for 'e' or 1 or ! for 'i'), the replacement character is unexpected? For instance, V##gra or Viag^^? Here, we have to be more careful with our regular expression. We could, for instance, try [Vv].{2}gra, [Vv]iag.{2}, and other variants, but now we have to be careful not to block legitimate words. For instance, [Vv]ia.{3} could match viable.

What about a version of the word in which additional characters are inserted, like V.i.a.g.r.a., ViaSEXYgra, or Via##gra? To tackle this problem, we can insert the notation .* in between letters. Recall that . means "any character" and * means "0 or more of them". So, V.*i.*a.*g.*r.*a.* would match a string that contains the letters 'V', 'i', 'a', 'g', 'r', 'a' no matter what might appear between them. Because of the *, we can also match strings where there is nothing between those letters.

Without regular expressions, spam filters would be far less successful. But, as you can see, defining the regular expressions for your filter can be challenging!

We want to find all students whose zip codes start with 41. This one is simple:

```
41[0-9]{3}
```

However, this could also match part of a student number (for instance, 0123412345 6789012), or it is possible that it could match someone's street address. Since we expect the zip code to end the string, we can remedy this with

```
41[0-9]{3}$
```

Or if we might expect the four-digit extension

```
(41[0-9]{3}|41[0-9]{3}-[0-9]{4})$
```

We want to find any student whose ID ends in an even number. We could not just use [02468] because that would match any string that contains any of those digits anywhere (in the ID number, in the street address, in the zip code). But student numbers are always 16 digits, so we want only the 16th digit to be one of those numbers.

```
[0-9]{15}[02468]
```

We could precede the expression with ^ to ensure that we are only going to match against student IDs (in the unlikely event that a student's street address has 16 or more digits in it!)

One final example. We want to find students whose last name is Zappa. This seems simple:

```
Zappa
```

However, what if Zappa appears in the person's first name or address? Unlikely, but possible. The last name always appears after the student ID. So we can ensure this matching the student ID as well: ^[0-9]{16}, Zappa

Let us look at this from the other side. Here are some regular expressions. What will they match against?

```
([^O][^H], [0-9]{5}$|[^O][^H], [0-9]{5}-[0-9]{4}$)
```

We match anything that does not have an OH, followed by a five- or nine-digit number to end the string. That is, we match any student who is not from OH.

```
[A-Z]{3},[ ]{4},
```

This string will find three upper-case letters followed by a comma followed by four blanks (the blank after the comma, and then no minor, so three additional blanks), followed by a comma. We only expect to see three upper-case letters for a major or a minor. So here, we are looking for students who have a major but no minor.

```
^0{15}[0-9]
```

Here, we are looking for a string that starts with 15 zeroes followed by any other digit. If student ID numbers were assigned in order, this would identify the first 10 students.

BASH AND WILDCARDS

Recall from Chapter 9 that the Bash interpreter performs multiple steps when executing an instruction. Among these steps are a variety of expansions. One of note is filename expansion. If a user specifies a wildcard in the filename (or pathname), Bash unfolds the wildcard into a list of matching files or directories. This list is then passed along to the program. For instance, with ls *.txt, the ls command does not perform the unfolding, but instead Bash unfolds *.txt into the list of all files that match and then provides the entire list to ls. The ls command, like many Linux commands, can operate on a single item or a list equally.

Unfortunately, the characters used to denote the wildcards in Bash are the same as some of the regular expression metacharacters. This can lead to confusion especially since most users learn the wildcards before they learn regular expressions. Since you learned the wildcards first, you would probably interpret the * in ls *.txt as "match anything". Therefore, ls *.txt would list all files that end with the .txt extension. The idea of collecting matches of a wildcard is called *globbing*. Recall that expansion takes place in a Bash command before the command's execution. Thus, ls *.txt first requires that the * be unfolded. The unfolding action causes the Bash interpreter to enumerate all matches. With *.txt, the matches are all files whose names end with the characters ".txt". Note that the period here is literally a period (because the period is not used in Bash as a wildcard). Therefore, the ls command literally receives a list of all filenames that match the pattern.

As discussed in Metacharacters, the * metacharacter means "match the preceding character(s) zero or more times." In the case of ls *.txt, the preceding character is a blank space. That would be the wrong interpretation. Therefore, as a Linux user or administrator, you must be able to distinguish the usage of wildcard characters as used in Bash to perform globbing from how they are used as regular expressions in a program such as grep.

The wildcard characters used in Bash are *, ?, +, @, !, \, [ ], [^...], and [[...]]. Some of these are included in the regular expression metacharacter set and some are not. We will examine the usage of the common wildcards in this section. Table 10.4 provides an explanation for each. Note that those marked with an [a] in the table are wildcards that are only available if you have set Bash up to use the extended set of pattern matching symbols. As we will assume this has not been set up in your Bash environment, we will not look at those symbols although their meanings should be clear.

We finish this section with some examples that use several of the wildcard symbols from Table 10.4. We will omit the wildcards that are from the extended set. For this example, assume that we have the following files and subdirectories in the current working directory. Subdirectories are indicated with a / before their name.

TABLE 10.4 Bash Wildcard Characters

*	Matches any string, including the null string
**	Matches all files and directories
**/	Matches directories
?	Matches any single character (note: does not match 0 characters)
+	Matches one or more occurrences (similar to how it is used in regular expressions)[a]
@	Matches any one of the listed patterns[a]
!	Matches anything except one of the list patterns[a]
\	Used to escape the meaning of the given character as with regular expressions, for instance * means to match against an *
[...]	Matches any of the enclosed characters, ranges are permitted when using a hyphen, if the first character after the [is either a – or ^, it matches any character that is not enclosed in the brackets
{...}	As with brace expansion in Bash, lists can be placed in { } to indicate "collect all", as in ls {c,h}* .txt, which would find all txt files starting with a c or h
[[:class:]]	As with regular expressions, matches any character in the specified class

[a] Only available if you have set Bash up to use the extended set of pattern matching symbols.

```
foo foo.txt foo1.txt foo2.dat foo11.txt /foo3  /fox  /foreign
/foxr  FOO  FOO.txt   FOO1.dat  FOO11.txt   foo5?.txt  /FOO4
```

See Table 10.5, which contains each example. The table shows the Linux ls command and the items from the directory that would be returned.

THE GREP PROGRAM

The grep program searches one or more text files for strings that match a given regular expression. It prints out the lines where such strings are found. In this way, a user or administrator can quickly obtain lines from text files that match a desired pattern. We hinted at

TABLE 10.5 Examples

ls Command	Items Returned
ls *.txt	foo.txt, foo1.txt, foo11.txt, FOO.txt, FOO11.txt, foo5?.txt
ls *.*	foo.txt, foo1.txt, foo2.dat, foo11.txt, FOO.txt, FOO1.dat, FOO11.txt, foo5?.txt
ls *	Will list all items in the directory
ls foo?.*	foo1.txt, foo2.dat
ls foo??.*	foo11.txt, foo5?.txt
ls *\?.*	foo5?.txt
ls *.{dat,txt}	Will list all items in the directory that end with either.txt or.dat
ls foo[0-2].*	foo1.txt, foo2.dat
ls *[[:upper:]]*.txt	FOO11.txt, FOO.txt
ls *[[:upper:]]*	FOO, FOO11.txt, FOO1.dat, FOO.txt,/FOO4
ls *[[:digit:]]*	Will list every item that contains a digit
ls foo[[:digit:]].*	foo1.txt, foo2.dat (it does not list foo11.txt because we are only seeking 1 digit, and it does not list foo5?.txt because we do not provide for the ? after the digit and before the period)

this back at the end of the section on Metacharacters when we looked at regular expressions as used to identify specific student records in a file. As grep can operate on multiple files at once, grep returns two things for each match, the file name and the line that contained the match. With the –n option, you can also obtain the line number for each match. When you use grep, depending on your regular expression and the strings in the file, the program could return the entire file if every line matches, a few lines of the file, or no lines at all.

The grep program uses the regular metacharacters as covered in Metacharacters. It does not include what are called the extended regular expression set, which include {, }, |. However, grep has an option, grep –E (or the program egrep), which does use the extended set. So, for the sake of this section, we will use egrep throughout (egrep and grep –E do the same thing).

The grep/egrep program works like this:

```
grep pattern filename(s)
```

If you want to use multiple files, you can either use * or ? as noted above in part 2, or you can list multiple file names separated by spaces. If your pattern includes a blank, you must enclose the pattern in '' or "" marks. It is a good habit to always use '' or "" in your regular expressions as a precaution.

In fact, the use of '' is most preferred. This is because the Bash interpreter already interprets several characters in ways that grep may not. Consider the statement grep !! filename. This statement seems straightforward, search filename for the characters !!. Unfortunately though, !! signals to the Bash interpreter that the last instruction should be recalled. Imagine that instruction was cd ~. Since the Bash interpreter unfolds such items as !! before executing the instruction, the instruction changes to grep cd ~ filename. Thus, grep will search filename for the sequence of characters cd ~.

Another example occurs with the $. You might recall that the $ precedes variable names in Bash. As with !!, the Bash interpreter will replace variables with their values before executing a command. So the command grep $HOME filename will be replaced with grep /home/username filename. To get around these problems, single quoting will cause the Bash interpreter to avoid any unfolding or interpretation. Double quoting will not prevent this problem because "$HOME" is still converted to the value stored in $HOME.

Grep attempts to match a regular expression to each line of the given file. This is perhaps not how we initially envisioned the use of regular expressions since we described a regular expression as matching to strings, not lines. In essence, grep treats each line of a file as a string, and looks to match the regular expression to any substring of the string. Either the pattern matches something on that line or it does not. The grep program will not search each individual string of a file (assuming strings are separated by white space). So, for instance, if a file had the line:

```
bat bait beat beet bet bit bite boot bout but
```

and we used the regular expression b.t, since b.t matches at least one item on the line, grep returns the entire line. Had we wanted to match each individual string on the line so that we only received as a response bat, bet, bit, but, we would have to resort to some other tactic.

One approach to using grep on each individual string of a file rather than each individual line would be to run a *string tokenizer* and pipe the results to grep. A string tokenizer is a program that separates every pair of items that have white space between them. Imagine that we have such a program called tokenizer. We could do

```
tokenizer filename | grep b.t
```

Notice in such a case, grep does not receive a filename(s), but instead the file information is piped directly to it.

Another thing to keep in mind is the need to use \ to escape the meaning of a character. We covered this earlier (Metacharacters) when we needed to literally look for a character in a string where the character was a metacharacter.

For instance, $ means "end of string", but if we were looking for a $, we would indicate this as \$. This is true in grep as it is in when specifying regular expressions in other settings. However, there are exceptions to the requirement for escaping the meaning (needing the \) in grep. For instance, if a hyphen is sought as one of a list of items, it might look like this: [!@#%&-=<>]. But recall that a hyphen inside of the [ ] is used to indicate a range; so instead we would have to specify this list as [!@#%&\-=<>]. The \- is used to indicate that the hyphen is sought literally, not to be used as a range. But there is an exception to this requirement. If the hyphen appears at the end of the list, there is no need to use \, so the list could be [!@#%&=<>-].

There are other instances where the escape character (\) is not needed. One is of a list of characters that include a $, but the $ is not at the end of the list of characters. If we intend to use $ to mean "end of string", grep does not expect to see any characters following it. Therefore, if we have $[0-9]+ (to indicate a dollar amount), grep treats the $ as a true dollar sign and not as the "end of string matching" metacharacter. The same is true of ^ if characters precede it. Finally, most characters lose their metacharacter meaning if found inside of [ ], so for instance we would not need to do [\$] or [\?] if we were searching for a $ or ?; instead, we could just use [$] or [?].

Let us examine grep (egrep) now. Figure 10.4 illustrates a portion of the results of applying the command egrep –n [0-9]{1,3}\.[0-9]{1,3}\.[0-9]{1,3}\.[0-9]{1,3} /etc/*. The regular expression here is the one we developed earlier to obtain four numbers separated by dots, that is, our four octets to obtain IP addresses. This regular expression could potentially return items that are not IP addresses (such as a match to the string 999.999.999.999), but no such strings are found in the /etc directory. Figure 10.4 is not the entire result because the grep instruction returns too many matches. You could pipe the result to less so that you can step through them all, or you could redirect the output to a file to print out or examine over time.

Notice that in the output we see the filename for each match, the line number of the match, and the line itself. You might notice that the /etc/Muttrc file does not actually contain IP addresses; instead, the matches are version numbers for the software (mutt-1.4.2.2). Even if we had used our "correct" regular expression to match IP addresses (from Figure

```
/etc/Muttrc:macro pager   <f1> "!less /usr/share/doc/mutt-1.4.2.2/manual.txt\n" "Show Mutt docume
ntation"
/etc/Muttrc:macro generic <f2> "!less /usr/share/doc/mutt-1.4.2.2/manual.txt\n" "Show Mutt docume
ntation"
/etc/Muttrc:macro index   <f2> "!less /usr/share/doc/mutt-1.4.2.2/manual.txt\n" "Show Mutt docume
ntation"
/etc/Muttrc:macro pager   <f2> "!less /usr/share/doc/mutt-1.4.2.2/manual.txt\n" "Show Mutt docume
ntation"
/etc/ntp.conf:restrict 127.0.0.1
/etc/ntp.conf:#restrict 192.168.1.0 mask 255.255.255.0 nomodify notrap
/etc/ntp.conf:#broadcast 192.168.1.255 key 42          # broadcast server
/etc/ntp.conf:#broadcast 224.0.1.1 key 42              # multicast server
/etc/ntp.conf:#multicastclient 224.0.1.1               # multicast client
/etc/ntp.conf:#manycastserver 239.255.254.254          # manycast server
/etc/ntp.conf:#manycastclient 239.255.254.254 key 42   # manycast client
/etc/ntp.conf:server     127.127.1.0     # local clock
/etc/ntp.conf:fudge      127.127.1.0 stratum 10
Binary file /etc/prelink.cache matches
/etc/resolv.conf:nameserver 172.28.102.11
/etc/resolv.conf:nameserver 172.28.102.13
/etc/resolv.conf:nameserver 10.11.0.51
/etc/resolv.conf:nameserver 10.14.1.10
/etc/resolv.conf.predhclient:nameserver 172.28.102.11
/etc/resolv.conf.predhclient:nameserver 172.28.102.13
/etc/resolv.conf.predhclient:nameserver 10.11.0.51
/etc/resolv.conf.predhclient:nameserver 10.14.1.10
/etc/termcap:# GNOME_Terminal 1.4.0.4 (Redhat 7.2)
```

FIGURE 10.4 A result from grep.

10.3), we would still have matched the entries in the Muttrc file. We could avoid this by adding a blank space before the IP address regular expression so that the – before 1.4.2.2 would cause those lines to not match.

The command whose results are shown in Figure 10.4 had to be submitted by root. This is because many of the /etc files are not readable by an end user. Had an end user submitted the grep command, many of the same items would be returned, but the following error messages would also be returned:

```
egrep: securetty: Permission denied
egrep: shadow: Permission denied
egrep: shadow-: Permission denied
egrep: sudoers: Permission denied
egrep: tcsd.conf: Permission denied
```

among others.

You might also notice a peculiar line before the /etc/resolv.conf lines. It reads “Binary file /etc/prelink.cache matches”. This is informing us that a binary file contained matches. However, because we did not want to see any binary output, we are not shown the actual matches from within that file. We can force grep to output information from binary files (see Table 10.6). The grep program has a number of useful options; some common ones are listed in Table 10.6. Of particular note are –c, -E, -i, -n, and –v. We will discuss –v later.

Let us work out some examples. We will use a file of faculty office information to search for different matches. This file, offices.txt, stores for each faculty, their office location (building abbreviation and office number), their last name, and then the platform of computer(s)

TABLE 10.6 grep Options

-a	Process a binary file as if it were a text file (this lets you search binary files for specific strings of binary numbers)
-c	Count the number of matches and output the total, do not output any matches found
-d read	Used to handle all files of a given directory, use recurse in place of read to read all files of a given directory, and recursively for all subdirectories
-E	Use egrep (allow the extended regular expression set)
-e *regex*	The regular expression is placed after –e rather than where it normally is positioned in the instruction; this is used to protect the regular expression if it starts with an unusual character, for instance, the hyphen
-h	Suppress the filename from the output
-i	Ignore case (e.g., [a-z] would match any letter whether upper or lower case)
-L	Output only filenames that have no matches, do not output matches
-m NUM	Stop reading a file after NUM matches
-n	Output line numbers
-o	Only output the portion of the line that matched the regular expression
-R, -r	Recursive search (this is the same as –d recurse)
-v	Invert the match, that is, print all lines that do not match the given regular expression

that they use. Each item is separated by a comma and the platform of computer might be multiple items. The options for platform are PC, Mac, Linux, Unix. Office locations are a two- or three-letter building designator, such as ST, MEP, or GH, followed by a space, followed by an office number, which is a three-digit number.

Write a grep command to find all entries on the third floor of their building (assuming their three-digit office number will be 3xx).

```
egrep '3[0-9][0-9]' offices.txt
```

Write a grep command that will find all entries of faculty who have PC computers.

```
egrep 'PC' offices.txt
```

What if there is a building with PC as part of its three-letter abbreviation? This would match. We could try this instead.

```
egrep 'PC$' offices.txt
```

This expression will only match lines where PC appears at the end of the line. However, if a faculty member has multiple computers and PC is not listed last, this will miss that person. Consider that a line will look like this: 123 ABC, Fox, PC, Mac. If we want to match PC, it should match after a comma and a space. This will avoid matching a building, for instance, 456 PCA or 789 APC. So our new grep command is

```
egrep ', PC' offices.txt
```

Notice that if the faculty member has multiple computers, each is listed as computer, computer, computer, and so forth. Therefore, if we have PC anywhere in the list, it will occur after a comma and a space. But if PC appears in a building, it will either appear after a digit and a space (as in 456 PCA) or after an upper-case letter (as in 789 APC).

Write a grep command that will find all entries of faculty who have more than one computer. Is there any way to include a "counter" in the egrep command? In a way, yes. Recall that we could control the number of matches expected by using {n, m}. But what do we want to actually count? Let us look at two example lines, a faculty with one computer and a faculty with more than one.

```
123 MEP, Newman, Mac
444 GH, Fox, PC, Linux
```

With one computer, the first entry only has two commas. With two computers, the second entry has three commas. This tells us that we should search for lines that have at least three commas. However, ',{3}' is not an adequate regular expression because that would only match a line in which there were at least three *consecutive* commas (such as 456 PCA, Zappa, PC,,,Mac). We should permit any characters to appear before each comma. In fact, in looking at our example lines, all we care about is whether there are spaces and letters before the comma (not digits since the only digits occur at the beginning of the line). Our regular expression then is

```
egrep '([A-Za-z]+,){3,}'
```

Notice the use of the ( ). This is required so that the {3,} applies to the entire regular expression (rather than just the preceding character, which is the comma). We could actually simplify this by using the . metacharacter. Recall that . can match any single character. We can add + to indicate one or more of any type of character. Since all we are interested in are finding at least three commas, we can use either '.+, .+, .+,' or we could use (.+,){3,}.

Write a grep command that will find all entries of faculty who have either a Linux or Unix machine. Here, we can use the OR option, as in (Linux|Unix). We could also try to spell this out using [ ] options. This would look like this: [L]?[iU]n[ui]x. The two grep commands are:

```
egrep '(Linux|Unix)' offices.txt
egrep '[L]?[iU]n[ui]x' offices.txt
```

Write a grep command that will find all entries of faculty who do not have a PC in their office. The obvious solution would be to use [^P][^C] in the expression. That is, find a string that does not have a P followed by C. Sadly, this expression, in egrep would return every line of the file. Why? Because the regular expression asks for any string that does not have PC. However, the way that grep works is that it literally compares a line for PC, and if the line is not exactly PC, then the line contains a match. If the line was 444 GH, Fox, PC,

Linux, this will match because there are characters on the line that are not PC, for instance, the '44' that starts the string. What we really want to have is a regular expression that reads "some stuff followed by no P followed by no C followed by some other stuff." Creating such a regular expression is challenging. Instead, we could simply use the regular expression PC and add the –v option to our grep command. That is,

```
egrep -v 'PC' offices.txt
```

This command looks for every line that matches 'PC' and then return the other lines of the file. Unfortunately, if we have an entry with an office of PCA or APC, that line would not be returned whether they have a PC or not. Therefore, we adjust the regular expression to be ', PC,' to avoid matching PC in the building name, so that the egrep command becomes

```
egrep -v ', PC' offices.txt
```

OTHER USES OF REGULAR EXPRESSIONS

With grep/egrep, we are allowed to use the full range of regular expression metacharacters. But in ls, we are limited to just using the wildcards. What if we wanted to search a directory (or a collection of directories) for certain files that fit a pattern that could be established by regular expression but not by wildcards? For instance, what if you wanted to list all files whose permissions were read-only? Recall from Chapter 6 that permissions are shown when you use the command ls –l. A read-only file would have permissions that contained r-- somewhere in the permissions list (we do not really care if the file is set as rwxrwxr-- or r-------- or some other variant just as long as r-- is somewhere in the permissions).

For ls, * has the meaning of "anything". So literally we want "anything" followed by r-- followed by "anything". Would ls –l *r--* accomplish this task for us? No. Let's see how this instruction would work. First, the Bash interpreter unfolds the notation *r--*. This means that the Bash interpreter obtains all names in the current directory that has anything followed by r followed by two hyphens followed by anything. For instance, foxr--text would match this pattern because of the r-- in the title. Once a list of files is obtained, they would be passed to the ls command, and a long listing would be displayed. Unfortunately, we have done this in the wrong order, we want the long listing first, and then we want to apply *r--*.

Our solution is quite easy. To obtain the long listing first, we do ls –l. We then pipe the result to egrep. Our instruction then becomes ls –l * | egrep 'r--' so that we obtain a long listing of all items in the current directory, and then we pass that listing (a series of lines) to egrep, which searches for any lines with r-- and returns only those. This command will return any line that contains r-- in the long listing. If there is a file called foxr--text, it is returned even if its permissions do not match r--. How can we avoid this? Well, notice that permissions are the first thing in the long listing and the filename is the last thing. We can write a more precise regular expression and include ^ to force it to match at the beginning of the line.

Permissions start with the file type. For this, we do not care what character we obtain, but it should only be a single character. We can obtain any single character using the period. We then expect nine more characters that will be r, w, x, or -. Of these nine characters, we will match on any r--. So we might use a regular expression (r--|[rwx-]{3}). This will match either r-- precisely or any combination of r, w, x, and – over three characters. Unfortunately, this will not work for us because it might match rwx or rw- or even ---. We could instead write this expression as (r--[rwx-]{6}|[rwx-]{3}r--[rwx-]{3}|[rwx-]{6}r--). Here, we require r-- to be seen somewhere in the expression. Now our command is rather more elaborate, but it prevents matches where r-- is found in the filename (or in the username or groupname). Figure 10.5 illustrates the solution; blank spaces are added around the "|" to help with readability.

You can combine ls –l and egrep to search for a variety of things such as files whose size is greater than 0, files that are owned by a specific user, or files created this year or this date. Can you think of a way to obtain the long listing of files whose size is greater than 0 using ls –l and egrep? This question is asked in this chapter's review problems (see questions 22 and 23).

There are a variety of other programs that use regular expressions beyond grep. The sed program is a stream editor. This program can be used to edit the contents of a file without having to directly manipulate the file through an editor. For instance, imagine that you want to capitalize the first word of every line of a textfile. You could open the file in vi or Emacs and do the editing by hand. Or, you could use the sed program. In sed, you specify a regular expression and a replacement string. The regular expression describes what string you are searching for. The replacement string is used in place of the string found. You can specify a replacement literally, or you can apply commands such as "upper case" or "lower case". You can remove an item as well.

The sed tool is very useful for making large substitutions quickly to a file. However, it requires a firm grasp of regular expressions. One simple example is to remove all html tags from an html file. One could define a regular expression as '<.*>' and replace it with nothing (or a blank space). Since anything in < > marks is an html tag, a single sed command could find and remove all of them. Another usage for sed is to reformat a file. Consider a file where information is stored not line by line, but simply with commas to delimit each item. You could replace commas with tab characters (\t) and/or new line characters (\n).

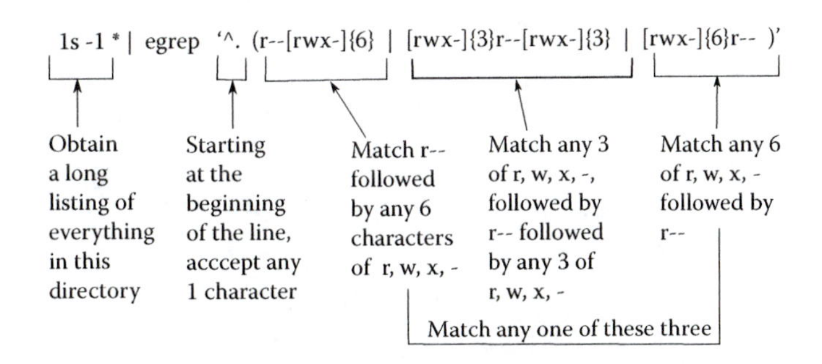

FIGURE 10.5 Solution to finding read only files in a directory using ls and egrep.

Another program of note is awk. The name of the program is the initials of the three programmers who wrote awk. Whereas sed searches for strings to replace, awk searches for strings to process. With awk, you specify pairs of regular expressions and actions. If a line matches a regular expression, that line is processed via the actions specified. One simple example of awk's use is to output specific elements of a line that matches a regular expression. In egrep, when a match is found, the entire line is output, but awk allows you to specify what you want to be output. In this way, awk is somewhat like a database program and yet it is far more powerful than a database because the matching condition is based on a regular expression.

Another example of using awk is to do mathematical operations on the matched items. For instance, imagine that a text file contains payroll information for employees. Among the information are the employees' names, hours worked this week, wages, and tax information. With awk, we can match all employees who worked overtime and compute the amount of overtime pay that we will have to pay. Or, we might match every entry and compute the average number of hours worked.

Regular expressions have been incorporated into both vi and Emacs. When searching for a string, you can specify a literal string, but you can also specify a regular expression. As with sed, this allows you to identify specific strings of interest so that you can edit or format them.

Finally, regular expressions have been incorporated into numerous programming languages. A form of regular expression was first introduced in the language SNOBOL (StriNg Oriented and SymBOlic Language) in the early 1960s. However, it was not until 1987 that regular expressions made a significant appearance in a programming language, and that language was Perl. Perl's power was primarily centered around defining regular expressions and storing them in variables. Perl was found to be so useful that it became a language that was used to support numerous Internet applications including web server scripting. Since then, regular expressions have been incorporated into newer languages including PHP, Java, JavaScript, the .Net platform, Python, Ruby, and Visual Basic.

We end this chapter with two very bad regular expression jokes.

> "If you have a problem, and you think the solution is using regular expressions, then you have two problems."

Q: What did one regular expression say to the other?
A: .*

FURTHER READING

Regular expressions are commonly applied in a number of settings whether you are a Linux user, a system administrator, a mathematician, a programmer, or even an end user using vi or Emacs. Books tackle the topic from different perspectives including a theoretical point of view, an applied point of view, and in support of programming. This list contains texts that offer practical uses of regular expressions rather than theoretical/mathematical uses.

- Friedl, J. *Mastering Regular Expressions.* Cambridge, MA: O'Reilly, 2006.
- Goyvaerts, J. and Levithan, S. *Regular Expressions Cookbook.* Cambridge MA: O'Reilly, 2009.
- Habibi, M. *Java Regular Expressions: Taming the java.util.regex Engine.* New York: Apress, 2003.
- Stubblebine, T. *Regular Expressions for Perl, Ruby, PHP, Python, C, Java and.NET.* Cambridge, MA: O'Reilly, 2007.
- Watt, A. *Beginning Regular Expressions (Programmer to Programmer).* Hoboken, NJ: Wrox, 2005.

Two additional texts are useful if you want to delve more deeply into grep, awk, and sed.

- Bambenek, J. and Klus, A. *Grep Pocket Reference.* Massachusetts: O'Reilly, 2009.
- Dougherty, D. and Robbins, A. *sed & awk.* Cambridge, MA: O'Reilly, 1997.

REVIEW TERMS

Terminology introduced in this chapter

Enumerated list	Regular expressions
Escape sequence	String tokenizer
Filename expansion	White space
Globbing	Wildcard
Metacharacters	

REVIEW QUESTIONS

1. What is a regular expression?
2. What does a regular expression convey?
3. What do you match regular expressions against?
4. What is the difference between a literal character and a metacharacter in a regular expression?
5. What is the difference between * and + in a regular expression?
6. What is the difference between * and ? in a regular expression?
7. What is the meaning behind a :class: when used in a regular expression? What does the class alnum represent? What does the class punct represent?

8. Why does the regular expression [0-255] not mean "any number from 0 to 255"?
9. How does the regular expression . differ from the regular expression [A-Za-z0-9]?
10. What does the notation {2,3} mean in a regular expression? What does the notation {2,} mean in a regular expression?
11. What does the escape sequence \d mean? What does the escape sequence \D mean?
12. What does the escape sequence \w mean? What does the escape sequence \W mean?
13. Does ^... have the same meaning as [^...] in a regular expression?
14. How does * differ between ls and grep?
15. Why does * differ when used in ls than in grep?
16. Which of the regular expression metacharacters are also used by the Bash interpreter as wildcard characters?
17. What is the difference between grep and egrep?
18. Provide some examples of how you might use regular expressions as a Linux user.
19. Provide some examples of how you might use regular expressions as a Linux system administrator.

REVIEW PROBLEMS

1. Write a regular expression that will match any number of the letters a, b, c in any order, and any number of them.
2. Repeat #1 except that we want the letters to be either upper or lower case.
3. Repeat #2 except that we want the a's to be first, the b's to be second, and the c's to be last.
4. Repeat #3 except that we only want to permit between 1 and 3 a's, 2 and 4 b's, and any number of c's.
5. Repeat #1 except that we want the letters to be either all upper case or all lower case.
6. Write a regular expression to match the letter A followed by either B or b followed by either c, d, e, or f in lower case, followed by any character, followed by one or more letter G/g (any combination of upper and lower case).
7. What does the following regular expression match against?

```
[a-z]+[0-9]*[a-z]*
```

8. What does the following regular expression match against? [A-Za-z0-9]+[^0-9]+

9. What does the following regular expression match against?

```
^[a-z]*$
```

10. Write a regular expression that will match a phone number with area code in ( ) as in (859) 572-5334 (note the blank space after the close paren).

11. Repeat #10 except that the regular expression will also match against a phone number without the area code (it will start with a digit, not a blank space).

12. Write a Linux ls command that will list all files in the current directory whose name includes two consecutive a's as in labaa.txt or aa1.c.

13. Write a Linux ls command that will list all files that contain a digit somewhere in their name.

14. Write a Linux ls command that will list all files whose name starts with the word file, is followed by a character and ends in .txt, for instance, file1.txt, file2.txt, filea.txt.

15. Repeat #14 so that file.txt will also be listed, that is, list all files that starts with the word file followed by 0 or 1 character, followed by .txt.

16. Write a grep command to find all words in the file dictionary.dat that have two consecutive vowels.

17. Repeat #16 to find all words that have a q (or Q) that is not followed by a u.

18. Repeat #16 to find all five-letter words that start with the letter c (or C).

19. Repeat #18 to find all five-letter words.

20. Using a pipe, combine a Linux ls and grep command to obtain a list of files in the current directory whose name includes two consecutive vowels (in any combination, such as ai, eo, uu).

21. Using a pipe, combine a Linux ls –l and grep command to obtain a list of files in the current directory whose permissions have at least three consecutive hyphens. For instance, it would find a file whose permissions are –rwxr----- or –rwxrwx--- but not –rwxr-xr--.

22. Using a pipe, combine a Linux ls –l and grep command to obtain a list of files in the current directory whose file size is 0.

23. Repeat #22 except that the file size should be greater than 0.

DISCUSSION QUESTIONS

1. As a Linux user, provide some examples of how you might apply regular expressions. HINT: consider the examples we saw when combining grep with ls commands. Also consider the use of regular expressions in vi and Emacs searches.
2. As a system administrator, provide some examples of how you might apply regular expressions.
3. Explore various programming languages and create a list of the more popular languages in use today, and indicate which of those have capabilities of using regular expressions and which do not.
4. One of the keys to being a successful system administrator is learning all of the tools that are available to you. In Linux, for instance, mastering vi and learning how to use history and command line editing are very useful. As you responded in #2, using regular expressions is another powerful tool. How do regular expressions compare to vi, history, command line editing with respect to being a successful system administrator? Explain.

CHAPTER 11

Processes and Services

The process and process management were introduced in Chapter 4. In this chapter, the user's control of the process is covered from the perspectives of a Windows 7 user and a Red Hat Linux user. Starting processes, monitoring processes, scheduling processes, and terminating processes are all covered. The chapter then examines the role of the service as a component of the operating system. Common services are examined followed by configuring services in Linux. Finally, automating service startup at system initialization is covered for both Windows 7 and Red Hat Linux.

The learning objectives of this chapter are to

- Describe the steps in starting, controlling/managing, and terminating processes in Windows 7 and Red Hat Linux.
- Introduce process scheduling through the Windows Task Scheduler and the Linux commands at, batch, and crontab.
- Discuss the role of the service in support of operating systems.
- Describe the process of configuring Linux services.
- Discuss how to control service startup in both Windows 7 and Linux.

The process was introduced in Chapter 4. In this chapter, we examine how users and system administrators control their processes. We also look at operating system (OS) services, which are background processes that can also be controlled by the system administrator.

STARTING A PROCESS

To run a program, the program must be an executable—that is, the program must be available in the computer's machine language. As most programs are written in a high level language (e.g., Java, C++), the programmer must translate the program from the high level language version into machine language. This is typically accomplished by using a program called a compiler. The compiler provides an executable version of the program as its output. You must compile the program for each platform that the program is intended to run on (e.g., a separate compilation is needed for a Macintosh, a Windows machine, a Linux machine, a mainframe).*

The first step in starting any process is to have the executable version of the program available. The executable program is stored somewhere in the file system (usually on hard disk, but it is also possible to run programs that are stored on optical disk, flash drive, etc.). When the user wishes to run a program, the OS must locate the executable program. In most OSs, the user indicates that a program should run through one of several possible approaches.

First, there may be shortcut icons. The shortcut icon is a graphical representation of a link to the program (recall soft links in Chapter 5). Double clicking on the shortcut causes the OS to start the program. Link's properties include the file's location in the file system. In Windows, you can view the properties of a shortcut by right clicking on the shortcut icon and selecting Properties.

See Figure 11.1 for an example of a shortcut to the program Gimp (an open source graphical editor). There are several tabs in the shortcut's properties window. The General

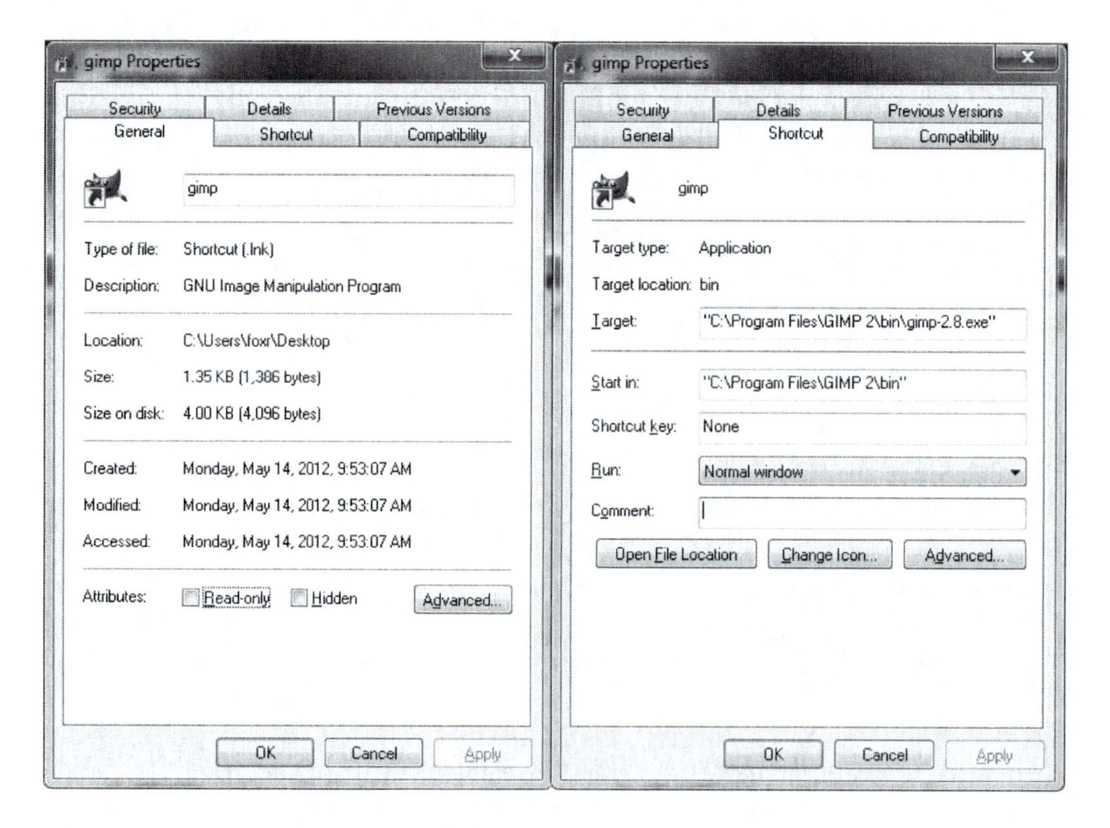

FIGURE 11.1 Two of the shortcut icon's property tabs.

* Some programming languages are interpreted. This means that programs of these languages run in an interpreted environment. The interpreter itself is the running program, and it must be an executable. The program that you are entering is entered in source code. The interpreter's job is to translate each entered instruction into machine language and execute it. Ruby, Python, Lisp, and the Bash shell scripting language are all examples of interpreted languages.

tab contains the location of the shortcut icon itself, whereas the Shortcut tab contains the location of the file's executable program in a box called Target. The shortcut itself is stored in the current user's Desktop, whereas the executable is located under the Program Files directory in a Gimp\bin subdirectory. Double clicking on the shortcut icon causes the OS to launch the program as stored in the Target location. Notice that the Target box is editable so that you can change the location of the executable if desired. The Security tab also contains useful information; it has the file permissions for the shortcut icon (these permissions should match the permissions of the executable itself).

To start a program, you can also double click on a data file created by that program. Usually, name extensions (e.g., .docx, .pdf, .mp3) are mapped to applications. Double clicking on the data file causes the OS to follow this mapping to the appropriate application software. That application software is started and the file is opened in the application software.

Most OSs also provide a program menu. Much like the shortcut icons, the program menu items store the target location of the executable program. In Windows, right clicking on the program name in the program menu and selecting properties brings up the same window as shown in Figure 11.1.

Linux has program menus and shortcut icons, just as in Windows. However, from Linux (as well as DOS) you can start a process from the command line. In reality, whether starting a program from a desktop icon, a program menu, or the command line, they all accomplish the same task in the same way. From the command line, however, you can provide optional and additional parameters. In most cases, the target location for the shortcut does not include parameters (although it could if desired). For instance, in Linux, you might specify vi –l someprogram.lsp. The –l option starts vi in "Lisp mode" to support editing Lisp programs. Additionally, by providing the filename in the command, vi opens with file already loaded.

Once you have issued your command to start a program, the OS takes over. First, it must locate the executable code stored in the file system. The directory listing will contain a pointer to the starting location of the program (disk block). Next, the OS must load the executable code into swap space, and copy a portion of the executable code (usually the first few pages) into memory. If memory has no free frames, the OS selects some frames to remove. The OS generates a page table for the process, and creates a process ID (PID) and a data structure that describes the process status and inserts it into a queue (ready or waiting, depending on processor load and process priority). The processor will eventually get to the process and begin executing it. This may be immediate or may take a few milliseconds depending on what other processes are being executed at the time.

In Linux (and Unix), processes are either parent processes or child processes (or both). A process that starts another process is known as the *parent* process. A process started by another process is known as a *child* process. As an example, if you are in a terminal window and you enter a command to start a process, say vi, then the terminal window process is the parent, and the process started, vi in this case, is the child. In Linux, only process 1, the process created when the system boots, is not a child process. All other process are child processes but may also be parent processes. The sidebar on the next page and the sidebar in Process Execution provide more detail on child and parent processes, and how they relate to each other.

In Windows, processes are not thought of as parents and children, although there may be a relationship if one process spawns another. In such a case, the spawned process is in the original process's process tree. We will use this to control how to kill off processes (covered later in this chapter).

PARENTS, CHILDREN, ORPHANS, AND ZOMBIES

As described in Chapter 4, there are processes and threads. However, we can also classify these entities under additional names. A process or thread may spawn another process or thread. The child is commonly a copy of the parent in that it is the same process, but it has its own data and its own goals. The parent and child may share resources, or the parent may limit the child's resources. The parent will spawn a child to have the child perform some subtask while the parent monitors the child. In fact, the parent may spawn numerous children and wait for them to finish their tasks, consolidating their efforts in the end.

As an example, the Apache webserver starts as a single parent process. It spawns children, each of which is tasked with handling incoming HTTP requests. If the parent finds that all of the children become busy with requests, it can spawn more children. Alternatively, if some children become idle, it may kill those children off.

In Unix and Linux, an orphan process is a child that still exists even after its parent has been killed. This may be a mistake, or it may be by design. Whichever is the case, an orphaned process is immediately adopted by a special process called init (a system process).

On the other hand, a child that has completed its task and has no reason to still exist may become a zombie process. In this case, the process is defunct—it no longer executes. However, because the parent has yet to obtain its results, the process remains in the process table until the parent can read the child's exit status.

PROCESS EXECUTION

During execution, the user is largely uninvolved with controlling the process' execution. The user can suspend a process, kill a process, move a process from foreground to background (or background to foreground), and change the process' priority. Otherwise, the process' execution is in the hands (or control) of the OS.

You might recall from Chapter 4 that a foreground process is one that can receive interaction with the user. The foreground actually represents two different notions. First, the foreground process is the process that is receiving interaction with the user. In both Windows and the graphical user interface (GUI) of a Linux OS, the foreground process is the process whose window is "on top" of other processes. This window is sometimes noted by having a different colored title bar. In Windows 7, you can tell the difference because the tab in the task bar for the process is somewhat illuminated, and the process' window itself has a red "X" in the upper right hand corner. Figure 11.2 compares the foreground versus background processes showing both the task bar and the windows (look in the upper right-hand corner of the Word document).

The other notion of foreground processes is more accurate. The foreground processes are those that are in the ready queue. That is, the foreground processes are those that the CPU is multitasking through. The background processes then are those that are of low enough

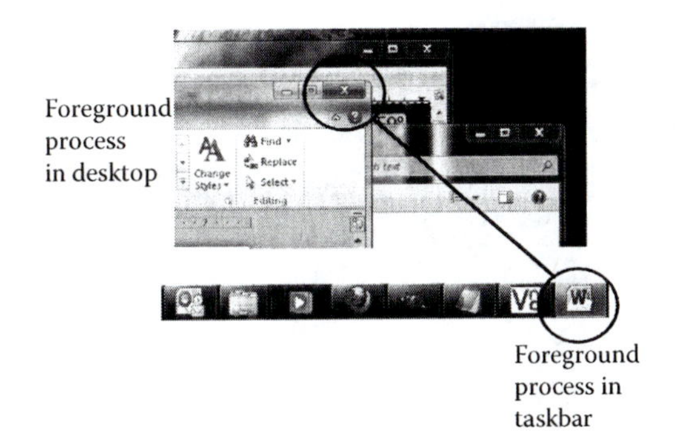

FIGURE 11.2 Foreground versus background in Windows 7.

priority that they wait until the CPU is freed of foreground processes before the CPU executes them. Here, we will primarily concentrate on the former notion of foreground.

To move a background process to the foreground in a GUI, select its tab, or if the window is visible on your desktop, click in the window. To move a foreground process to the background, click on the minimize button in its window, or select another process to move to the foreground.

Now, consider running multiple processes in a terminal window in Linux rather than in the GUI. Since you are using the terminal window, there are no tabs to select; there is no way to minimize a process. This would seem to indicate that in a terminal window, you can only run one process at a time. This, however, is fortunately not the case. The idea of foreground and background in a terminal window is controlled by a few simple operations.

First, you can launch a process in the foreground by just entering the command's name, as we have seen throughout the textbook. For instance, vi <enter> or grep … <enter> or top <enter> all launch the given program into the foreground for this terminal window. To start a process in the background, issue the command followed by an ampersand (&). For instance, if you feel that your grep command might take a lot of system resources and so you do not want it to interfere with the performance of the system, you might issue it as grep … & <enter>. The & forces the process to run in the background.

You can also force a process to switch from foreground to background or background to foreground. The commands are fg and bg, respectively. However, if a process is currently running a terminal window, you may not have access to the command line in order to issue the fg or bg command. To gain access to the command line prompt while a process is running in that window, type control+z. This suspends the running process and returns control of the shell to you.

For example, you issue the command top (top is explored in the next section). This is an interactive program. It fills the window with a display and updates itself so that you do not have access to the command line. Now, imagine that you press control+z. At this point, you would see:

```
[1]+ Stopped        top
```

This is telling you that job 1, top, is in a stopped state. At the command line, if you type jobs, you will see a list of all running and suspended jobs. If there is only one job, typing fg will resume that job in the foreground and typing bg will resume that job in the background.

Imagine that when you typed jobs, you instead saw the following:

```
> jobs
[1]- Stopped        top
[2]+ Stopped        vi
```

Here, there are two suspended jobs. If you type fg (or bg), the job selected is the one with the +, which indicates the most recently executing job. However, with the two numbers, you can also specify which job to move to the foreground or background by using fg n or bg n, where n is the job number. Both of these processes are interactive, so it makes little sense to move either to the background. However, you could easily resume one, suspend it, resume the other, suspend it, and resume the first, if for instance you were editing a document but wanted to, from time to time, see how the system was performing.

Consider, as another example, that you have two time-intensive jobs, and an editing job. You issue a find command that will search the entire file system, and a grep process that will have to search thousands of files. Both could take seconds or minutes. So you also want to edit a document in vi. You issue the following three commands:

```
> find …
  control+z
> bg 1
> egrep …
  control+z
> bg 2
> vi
```

Now you are editing in vi while both find and egrep run in the background. The steps here accomplish your task, but you could also more simply enter

```
> find … &
> egrep … &
> vi
```

Note that later in this chapter, we will examine another way to control a process—through its PID. The job number as used here is a different value. You obtain job numbers through the jobs command. You obtain PIDs through the ps or top command (covered in the next section).

In Linux, you can control a process' priority using a command called *nice*. The nice command refers to a process' *niceness*. Niceness, in Linux, is defined as how nice a process is toward other processes. The nicer a process is, the more it is willing to give up control of the CPU for other processes. In other words, a nicer process has a lower priority than a less nice process. The nice value ranges from –20 (least nice, or highest priority) to +19 (most nice, or lowest priority).

The nice command is issued when you launch a process from the command line, using the following format:

```
nice process-name -n value
```

This executes process-name establishing its initial nice value to *value*. If you do not issue a command using nice, it is given the default nice value of +10. Once a process is running, you can change its nice value using the exact same command.

For instance, if we launch some process foo issuing the command

```
nice foo -n 0
```

we can later adjust foo by issuing the command

```
nice foo -n 19
```

If you issue the nice command with no process or value, nice responds with the default nice value.

FORK, EXEC, PROCESS

In Linux, we use terms such as Fork and Exec. What do these mean? These are programs used to start child processes, that is, a process will call upon one of these programs when it must generate a child process. The difference is in how the commands react.

The fork command duplicates the parent and spawns the new process as a child. The child has a unique PID, has a parent PID of its parent, and its own resources including CPU time. It obtains its own memory, copying the parent's memory for its own initially. The child also has access to files opened by the parent. Once the fork command has generated the child, the parent can either continue executing, or enter a waiting state. In the latter case, the parent waits until the child (or children) terminate before resuming.

The exec command is similar to the fork except that the spawned child takes the place of the parent. In this case, the child takes on the parent's PID and all resources. The parent is terminated when the child takes over since the two processes cannot share the same PID.

PROCESS STATUS

A process' status is information about the process. It will include its state, the resources granted to it, and its PID. Both Windows and Linux allow you to investigate your processes. You can find, for instance, how much CPU time each process uses, how busy your processor is, and how much main memory and virtual memory are currently in use. You can also control your process' priority, adjusting the time the CPU will focus on the process during multitasking. Here, we look at the approaches used to probe the processor for process information.

In Windows, system performance and process status information is all available through the Windows Task Manager. The typical view is to look at the running *applications* (as opposed to the processes themselves). In Figure 11.3, we see that there are seven user applications currently running but 74 total processes (many of the processes are either system processes or were launched in support of the user applications). The CPU usage is 0% at the moment and physical memory is 28% (i.e., only 28% of memory is being used).

Under the Applications tab, there is little you can do other than start a new process, kill a process (End Task), or force the OS to switch from one process to another. In fact, by using "Switch To", you are merely moving a process from the background to the foreground while the current foreground process is moved to the background. If you select New Task…, a pop-up window appears. In this window, you enter the name of the command you wish to execute. The command must include the full path to the executable file, for instance "C:\Program Files\Internet Explorer\iexplore.exe". If you execute a process in this way, the process executes under administrator privileges.

The most important use of the Applications tab of the task manager is to kill a task that has stopped responding. Windows uses the expression "stopped responding" as a polite way to say that a process is deadlocked or otherwise has died but the OS cannot kill it off itself. To kill the process, highlight it in the list of processes and select End Task. If the application has unsaved data, you will be asked if you wish to save the data first.

The Processes tab provides far greater information than the applications tab. First, all of the processes are listed. Second, you see each process' current CPU and memory usage. You also see who owns the process (who started it). In Windows, you will commonly see one of four users: SYSTEM, LOCAL SERVICE, NETWORK SERVICE, or your own user

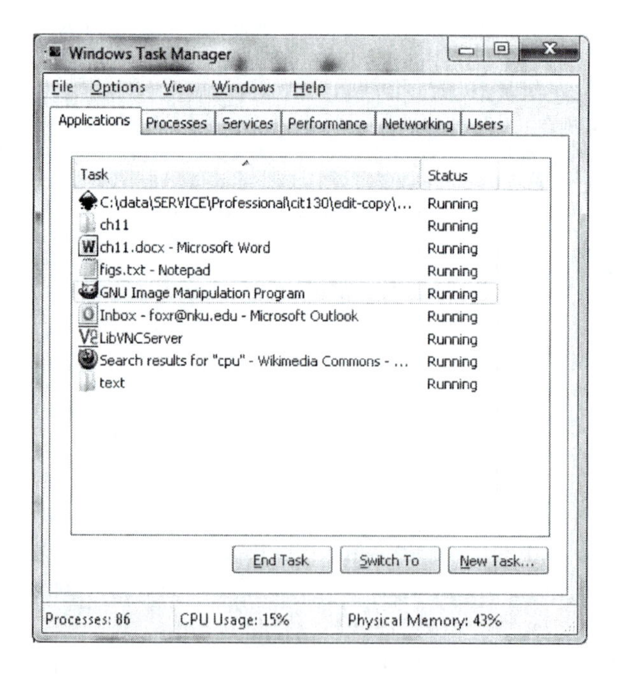

FIGURE 11.3 Windows Task Manager.

name. The first three indicate that either the OS started the process or that the process was spawned in support of other software (see Figure 11.4).

If you select any of the processes in the Processes tab, you can then right click on it and set the process' priority (raise or lower it) and affinity (which processors or cores are allowed to run the process). Priorities are limited to one of six values: low, below normal, normal, above normal, high, and real time. The default for a process is normal. Real time signifies that the process is so important that it must complete its task as it is presented, that is, that it should never be postponed. You can also end the process, end the process as well as all processes that it may have spawned (end process tree), debug the process, or bring up the process' properties. An example process properties window is shown in Figure 11.5. Here, you can see general information (type of file, description, location in the file system, size), change the process' security, obtain details about the process, or find out the process' compatibility.

The Services tab of the Task Manager is similar to the Processes tab except that it lists only services. The services are listed by PID, description, status, and group. If you click on any service, you can start or stop it. You can also obtain a list of all system services. We will explore services in detail in Services, Configuring Services, and Establishing Services at Boot Time.

The next two tabs of the Task Manager, Performance and Networking, provide system-specific information about resource utilization. Performance specifically covers CPU usage, main memory usage, virtual memory usage, and the current number of threads, processes, and handles running. A handle is a type of resource and might include open files, variables in memory, and pipe operations. Essentially, a handle is a pointer in a table

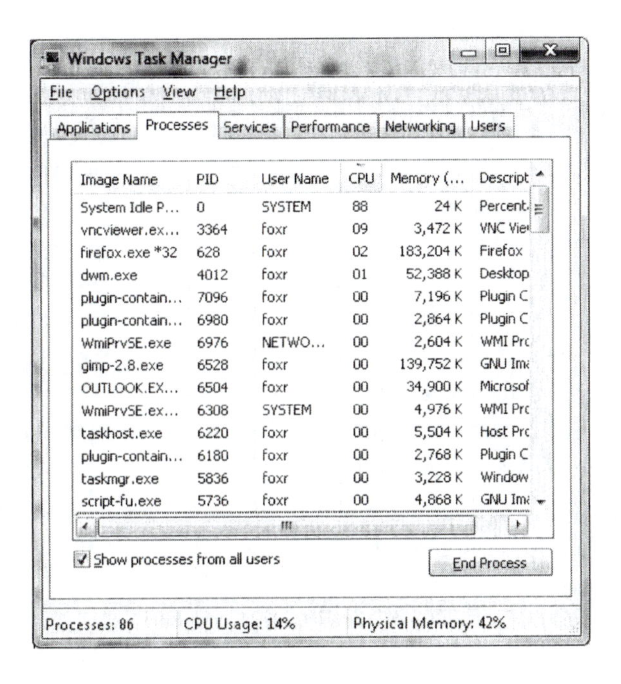

FIGURE 11.4 Processes in the Windows Task Manager.

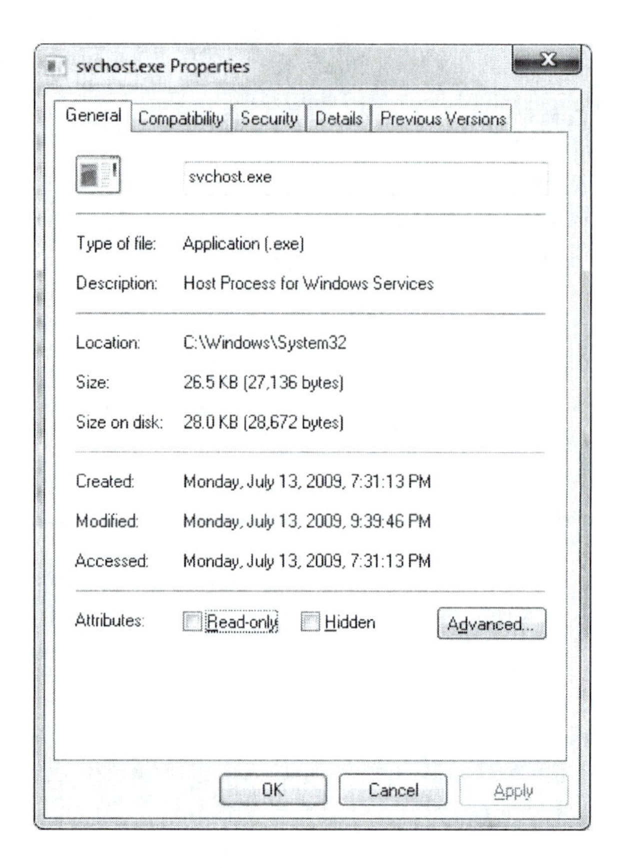

FIGURE 11.5 Process properties for svchost.exe.

to the actual entity that the handle represents. The Networking tab displays the current wireless and LAN connection usage (if any).

From the Performance tab, there is also a button to bring up Resource Monitor, which provides more details on CPU, Memory, Disk, and Network usage. Figure 11.6 displays both the Performance tab and the Resource Monitor window, displaying overall resource usage (notice the Resource Monitor has other tabs to isolate the usage of just CPU, memory, disk, or network). The final tab, Users, displays all users who are running processes. Typically, this will just be yourself, but it can show you if anyone else has remotely connected to your machine and is running any processes.

Using the resource monitor, you can select specific processes. This highlights their usage in the bottom-left portion of the window. In Figure 11.6, for instance, both firefox.exe (Mozilla Firefox web browser) and WINWORD.EXE (Microsoft Word) are highlighted and their memory usage is being displayed. Additional information is provided in this filter window, for instance, the amount of working memory, amount of memory in use that can be shared with other processes, and the amount of memory in use that is private to this process. Also being displayed in this filter window is the number of hard faults (page faults) that are occurring per second. You can similarly select the Disk or Network utilization for the selected processes.

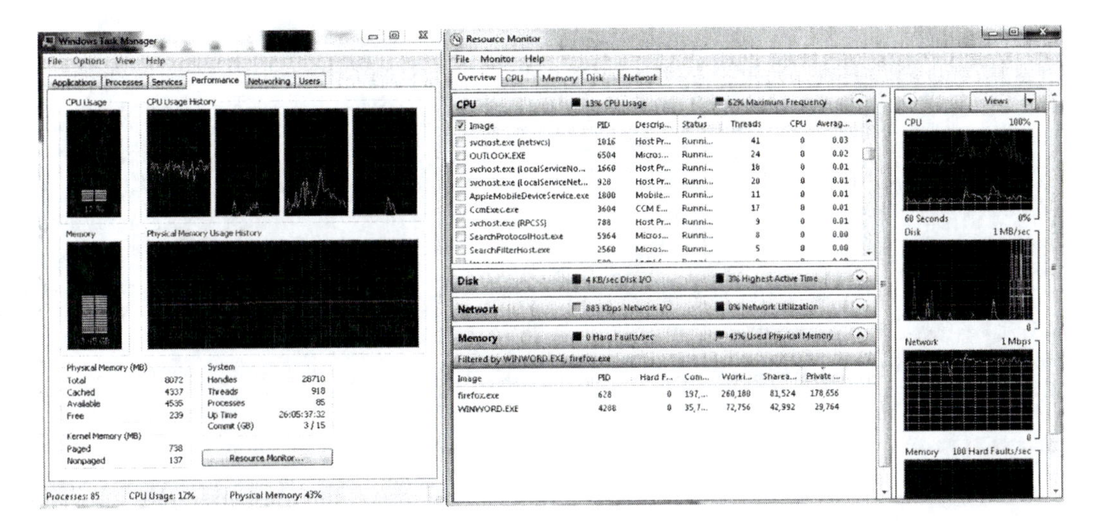

FIGURE 11.6 Performance tab and resource monitor window.

There are a number of different tools for obtaining performance information in Linux. The more traditional means of obtaining information about a process is through the ps (process status) program, executed at the command line. The ps command has a number of options that might be confusing, so we will look at the more common options here.

Using ps by itself shows you the active processes in the current window owned by you. This may not be of sufficient value to a user as there are many processes that would not be shown to you, those owned by other users (including root), and those started outside of the given window. Table 11.1 provides a list of the more common options. Notice that in many cases, you do not use the – symbol when providing the options in the ps command. This is a departure from Unix where the – symbol was required.

TABLE 11.1 Linux ps Command Options

a	Show all processes
c	Show true command name
e	Show environment after command
f	Show processes organized by parent/child relationship (this is displayed in ASCII-art as a hierarchy)
l	Display long format
m	Show all threads
o	Display in a user-defined format
p	Select by process ID
r	Output only currently running processes
S	Include dead child process data (summarized with parent)
t	Select by terminal window (tty)
T	Select processes on this terminal
U	Select processes of a specified user
u	Display user-oriented format
x	Select processes irrelevant of terminal (tty)

Perhaps the most common usages of ps are ps by itself and ps aux. The ps aux version gives you all processes of all users. This may be too much information, however. You can pipe the result to grep to obtain only those processes that match a given string. For instance, ps aux | grep foxr would display only those processes owned by foxr, but would show all processes, not just those limited to the given terminal window. A variation of the options aux is axf, which gives you the processes in the shape of a tree to show you which processes spawned other processes. For instance, from bash you might type xterm to open a terminal window, and from that terminal window, you might issue another command (say to compile the current file); thus, bash is the parent of xterm, which is the parent of the compile command.

The ps command, when supplied with the parameter u, gives a long listing of information that includes for each process, the process's owner, the PID, CPU usage, memory usage, terminal from which the command was issued, current process status, start time, and amount of CPU time that has elapsed. Note that obtaining the PID is particularly useful when you are faced with issuing other commands that impact a process. For instance, the kill command (covered in Terminating Processes) requires the PID for the process to be killed (terminated). Figure 11.7 demonstrates the ps command (top of figure) versus a

```
  PID TTY          TIME CMD
 5031 pts/1    00:00:00 su
 5034 pts/1    00:00:00 bash
 5261 pts/1    00:00:00 ps
USER       PID %CPU %MEM   VSZ   RSS TTY      STAT START   TIME COMMAND
root         1  0.0  0.1  2160   648 ?        Ss   May29   0:01 init [5]

root         2  0.0  0.0     0     0 ?        S<   May29   0:00 [migration/0]
root         3  0.0  0.0     0     0 ?        SN   May29   0:00 [ksoftirqd/0]
root         4  0.0  0.0     0     0 ?        S<   May29   0:00 [events/0]
root         5  0.0  0.0     0     0 ?        S<   May29   0:00 [khelper]
root         6  0.0  0.0     0     0 ?        S<   May29   0:00 [kthread]
root         9  0.0  0.0     0     0 ?        S<   May29   0:00 [kblockd/0]
root        10  0.0  0.0     0     0 ?        S<   May29   0:00 [kacpid]
root       168  0.0  0.0     0     0 ?        S<   May29   0:00 [cqueue/0]
root       171  0.0  0.0     0     0 ?        S<   May29   0:00 [khubd]
root       173  0.0  0.0     0     0 ?        S<   May29   0:00 [kseriod]
root       237  0.0  0.0     0     0 ?        S    May29   0:00 [khungtaskd]
root       238  0.0  0.0     0     0 ?        S    May29   0:00 [pdflush]
root       239  0.0  0.0     0     0 ?        S    May29   0:00 [pdflush]
root       240  0.0  0.0     0     0 ?        S<   May29   0:00 [kswapd0]
root       241  0.0  0.0     0     0 ?        S<   May29   0:00 [aio/0]
root       459  0.0  0.0     0     0 ?        S<   May29   0:00 [kpsmoused]
root       488  0.0  0.0     0     0 ?        S<   May29   0:00 [mpt_poll_0]
root       489  0.0  0.0     0     0 ?        S<   May29   0:00 [mpt/0]
root       490  0.0  0.0     0     0 ?        S<   May29   0:00 [scsi_eh_0]
root       493  0.0  0.0     0     0 ?        S<   May29   0:00 [ata/0]
root       494  0.0  0.0     0     0 ?        S<   May29   0:00 [ata_aux]
root       501  0.0  0.0     0     0 ?        S<   May29   0:00 [kstriped]
root       510  0.0  0.0     0     0 ?        S<   May29   0:13 [kjournald]
root       541  0.0  0.0     0     0 ?        S<   May29   0:00 [kauditd]
root       574  0.0  0.3  3228  1664 ?        S<s  May29   0:00 /sbin/udevd -d
```

FIGURE 11.7 ps versus ps aux.

portion of the output from the ps aux command (bottom of figure). As the ps aux command will display dozens of processes, it cannot be fully shown in one screen's worth.

Another command to obtain status information is top. While ps provides a snapshot of what is going on at the time you issue the command, top is interactive in that it updates itself and displays the most recent information, usually every 3 seconds. Because top is interactive, it remains in the window and you do not have access to the command line prompt while it is executing. The ps command is not interactive—it runs, displaying all active processes and then returns the user to the command line. Another difference between ps and top is that top only lists the processes that are most active at the moment. There are a number of parameters that can allow you to change what top displays. For the most part, unless you are trying to work with the efficiency of your system, you will find ps to be easier and more helpful to use. To exit from top, type ctrl+c. Figure 11.8 demonstrates the top command's output.

As with Windows, Linux also provides GUI tools for monitoring system performance. First, the System Monitor tool is akin to the Task Manager in Windows. The System Monitor has three tabs: to display processes, resource utilization, and the file system. Under the Processes tab, all processes are listed along with their status, CPU utilization, niceness, UID, and memory utilization. From here, you can stop or continue (wake up sleeping) processes, kill processes, or change their priority. The resources tab displays the CPU, memory, swap space, and network usage, both current and recent. The file systems tab displays the current mounted file system, similar to the df –k command. Figure 11.9 shows the processes and resources tabs.

```
top - 05:08:54 up 6 days, 21:01,  3 users,  load average: 0.52, 0.48, 0.45
Tasks: 130 total,   2 running, 128 sleeping,   0 stopped,   0 zombie
Cpu(s):  6.6%us,  3.6%sy,  0.0%ni, 83.4%id,  0.3%wa,  1.7%hi,  4.3%si,  0.0%st
Mem:    514908k total,   472176k used,    42732k free,   107128k buffers
Swap:  1020116k total,       80k used,  1020036k free,   205552k cached

  PID USER      PR  NI  VIRT  RES  SHR S %CPU %MEM    TIME+  COMMAND
 3473 foxr      15   0 47856 9928 8208 S 10.0  1.9  14:33.72 vino-server
 3086 root      15   0 34288  21m  10m S  0.7  4.4   0:21.85 Xorg
 3624 foxr      15   0 41252  12m 8560 R  0.3  2.5   0:01.84 gnome-terminal
    1 root      15   0  2160  648  560 S  0.0  0.1   0:01.35 init
    2 root      RT  -5     0    0    0 S  0.0  0.0   0:00.00 migration/0
    3 root      34  19     0    0    0 S  0.0  0.0   0:00.00 ksoftirqd/0
    4 root      10  -5     0    0    0 S  0.0  0.0   0:00.44 events/0
    5 root      10  -5     0    0    0 S  0.0  0.0   0:00.01 khelper
    6 root      10  -5     0    0    0 S  0.0  0.0   0:00.00 kthread
    9 root      10  -5     0    0    0 S  0.0  0.0   0:00.00 kblockd/0
   10 root      20  -5     0    0    0 S  0.0  0.0   0:00.00 kacpid
  168 root      15  -5     0    0    0 S  0.0  0.0   0:00.00 cqueue/0
  171 root      15  -5     0    0    0 S  0.0  0.0   0:00.00 khubd
  173 root      10  -5     0    0    0 S  0.0  0.0   0:00.00 kseriod
  237 root      20   0     0    0    0 S  0.0  0.0   0:00.00 khungtaskd
  238 root      20   0     0    0    0 S  0.0  0.0   0:00.00 pdflush
  239 root      15   0     0    0    0 S  0.0  0.0   0:00.17 pdflush
  240 root      10  -5     0    0    0 S  0.0  0.0   0:00.14 kswapd0
  241 root      15  -5     0    0    0 S  0.0  0.0   0:00.00 aio/0
  459 root      11  -5     0    0    0 S  0.0  0.0   0:00.00 kpsmoused
  488 root      10  -5     0    0    0 S  0.0  0.0   0:00.00 mpt_poll_0
  489 root      20  -5     0    0    0 S  0.0  0.0   0:00.00 mpt/0
```

FIGURE 11.8 The top command's output.

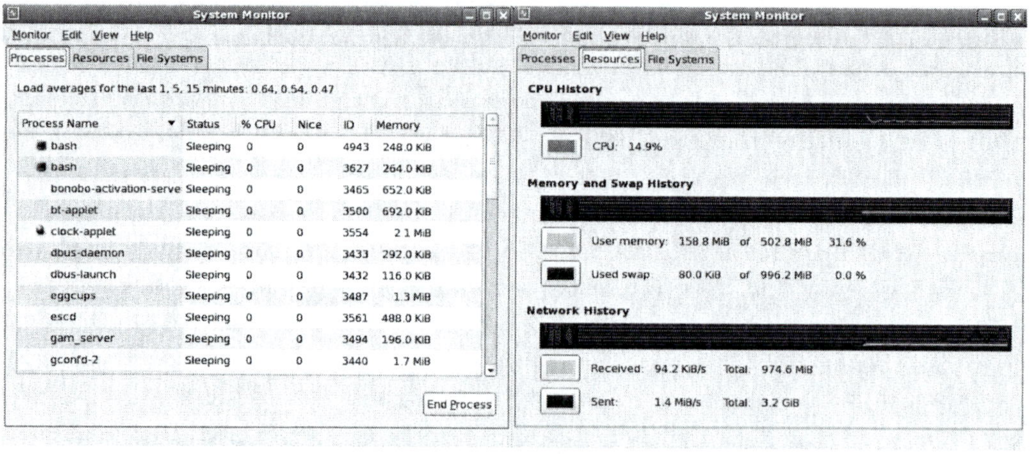

FIGURE 11.9 Two views of the System Monitor in Linux.

In addition to the Resource Monitor, you can also view the background and on-demand services through the Service Configuration GUI. There is also a System Log Viewer GUI available. Although this does not provide direct information about running processes, many processes provide run-time information in log files. Log files are discussed later in the chapter along with services. Both the Service Configuration Manager and the System Log Viewer tools require that you log in as root to use them. This is not true for the System Monitor.

SCHEDULING PROCESSES

The OS performs scheduling for us. There are a variety of forms of scheduling. In batch systems, scheduling was required to determine the order that the CPU would execute programs. As we tend not to use batch processing much today, this form of scheduling is one we can largely ignore. In multitasking, round-robin scheduling is the common approach to execute those processes in the ready queue. However, as discussed above, users can alter the behavior of round-robin scheduling by explicitly changing process priorities and/or moving processes to the background. Yet another form of scheduling that the OS performs is scheduling when processes are moved from a waiting queue to the ready queue. In personal computers, this is not typically relevant because all processes are moved to the ready queue. However, users may launch processes that place them into a waiting queue. Here, we focus on how to schedule processes in both Windows and Linux that indicate *when* a process should move to the ready queue.

Windows 7 provides the Task Scheduler program. See Figure 11.10. You are able to create scheduled actions to execute. Tasks can be scheduled for daily, weekly, or monthly activity, or a one-time occurrence. In any of these cases, you specify the starting time. You can also specify that a task should occur when the computer is next booted, when a specific user logs in, or when a type of event is logged. The action is one of starting a program or script, sending an e-mail, or displaying a message. As an example, you may set up a task to run your antiviral software each time your computer is booted, or every night at 3 AM. You

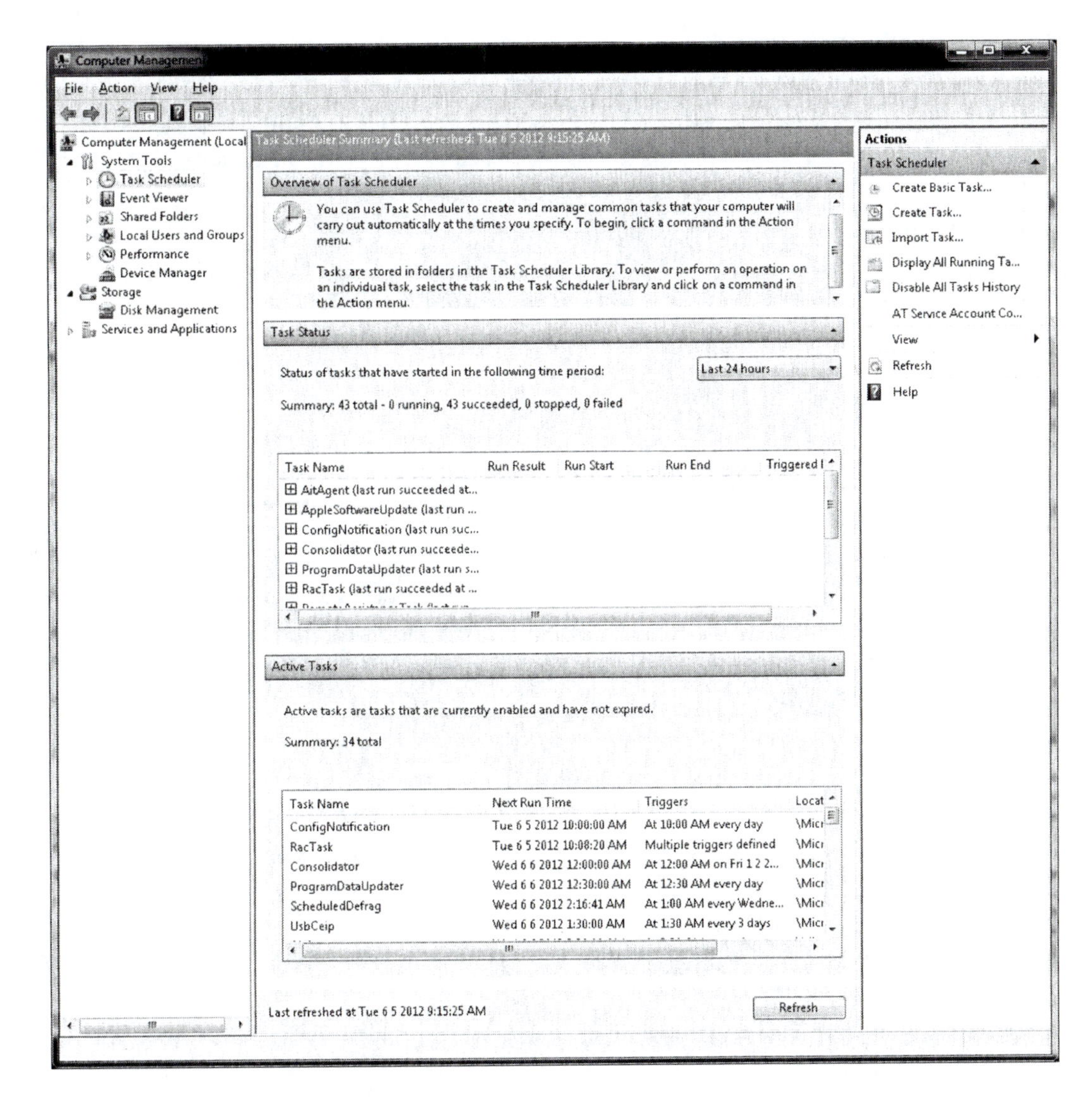

FIGURE 11.10 Windows Task Scheduler.

might similarly write a program to perform a complete backup and schedule that program to execute once per week.

In Linux, there are three scheduling programs of note. All three of these run processes in the background; therefore, the processes require input from files or sources other than the keyboard. This might require that the command(s) be issued with proper redirection. The three processes are crontab, at, and batch. The batch command allows you to schedule processes that will execute once the CPU load drops below 80%. The at command allows you to schedule processes to run at specific times. The at and batch commands perform one-time execution of the scheduled processes. The crontab command instead is used for recurring scheduling. As the at and batch commands are related, we discuss them together, but keep in mind that batch will cause the processes to run when CPU load permits execution, and at will run the processes only at the scheduled time.

Both at and batch can accept the command(s) to execute from the command line or from a file. The format of batch is merely batch or batch –f filename. The format for at is at TIME or at –f filename TIME. TIME formats are discussed later. If you use the –f option, the file must list, one line at a time, the process(es) to run. Each process should be a Linux command, including any necessary redirection—for instance, cat foo1.txt foo2.txt foo3.txt | sort >> foo4.txt. If you do not use the –f option, then the program (both at and batch) will drop you into a prompt that reads at >. From this prompt, you enter commands, one line at a time, pressing <enter> after each command. You end the input with control+d. When done, you are given a message such as

```
job 5 at 2012-03-01 12:38
```

which indicates the job number and the time it was entered (not the time when it will run or the PID it will run under).

The TIME indicator for at can take one of several formats. First, you can enter the time using a simple HH:MM format, such as 12:35 to run the process at 12:35. Without specifying AM or PM, the time is assumed to be military time*; thus 12:35 would be PM, whereas 00:35 would be AM, or alternatively 1:35 would be AM and 13:35 would be PM. You may include or omit the minutes, so you might specify 1:00 PM, 1 PM, 13:00 but not 13. There are also three special times reserved, midnight, noon, and teatime (4 PM).

The at command schedules the process(es) to run the next time the specified time is reached. For instance, if you specify noon and it is already 1:35 PM, the process(es) will run the next day at noon. You can also specify the date if it is not within the next 24 hours. Dates are provided using one of three notations, MMDDYY, MM/DD/YY, or DD.MM.YY, where MM is a two-digit month (as in 1 or 01 for January and 12 for December), DD is a two-digit date, and YY is a two-digit year. You can also specify either today or tomorrow in place of a date, as in 1:35 PM tomorrow.

An alternative to specifying a time and date is to use now + value. With now, you are able to specify how far into the future the process(es) should run. The value will consist of a count (an integer) and a time unit. You might specify now + 3 minutes, now + 2 hours, now + 7 days, now + 2 weeks, or now + 1 month. Notice that you are not allowed to specify seconds in the TIME value.

Once you have scheduled a task, you can inspect the scheduled task(s) by using atq. The atq instruction shows you all waiting jobs. The queue lists jobs scheduled by both at and batch. The atrm command allows you to remove jobs from the scheduling queue. If there is only a single job, atrm removes it. Otherwise, you must indicate the job number as in atrm 3 to remove the third scheduled job.

Both at and batch are programs to schedule jobs. The atd service (or daemon) monitors the scheduling queue to see if a process should be executed. It runs in the background, and

* Military time forgoes the AM or PM by using just a two-digit number. 00 represents midnight, 01–11 are 1 AM through 11 AM, 12 is 12 PM (noon). 13–23 represent 1 PM through 11 PM. For instance, 00:30 is 12:30 AM and 15:35 is 3:35 PM. To obtain the proper value, if it is after 12 noon, add 12 to the normal time (so that 3:35 PM becomes 15:35).

at every minute, compares the current time to the times of the scheduled tasks. Services are discussed starting in Services.

The other Linux scheduler is called crontab. This program would be used to schedule a *recurring* task, such as one that occurs every Monday at 1 AM or the first day of every month. In some ways it is like at, but it is more complicated. First, you must set up a cron file. The cron file specifies two things: the time/date/recurrence to be scheduled and the process to be scheduled. The time/date consists of five integer numbers or *. The five values, in order, represent the minute, hour, day, month, and day of week to be scheduled. Minute will be a number between 0 and 59, hour between 0 and 23 using military time, day of the month between 1 and 31, month between 1 and 12, and day of week between 0 and 6 (where 0 means Sunday, 6 means Saturday). The * is a wildcard, much as we saw in Bash, meaning "any time".

Here are some examples specifying the time portion of a crontab file:

```
15     3     1     *     *     –the first day of every month,
                                at 3:15 AM
0      14    *     *     0     –every Sunday at 2 PM
30     *     *     *     *     –every hour at half past
                                (e.g., 12:30, 1:30, 2:30)
0      0     12    31    *     –every December 31, at midnight
0      *     *     *     *     –every hour of every day
```

If you specify both a date (day and month) and day of the week, both are used. So, for instance:

```
0      0     15    *     0
```

will schedule for midnight every 15th AND every Sunday.

If you want to schedule multiple time periods other than specific dates and days of the week, you can list multiple entries for each time category, separating the entries with commas. For instance,

```
0,30   0,6,12,18    *     *     *
```

will schedule the task to occur daily at 12:00 AM, 12:30 AM, 6:00 AM, 6:30 AM, 12:00 PM, 12:30 PM, 6:00 PM, 6:30 PM. Another variant is if you want the recurrence to be within a specific time interval, for instance, every 15 minutes. This is specified using /15, as in

```
0/15   12    *     *     *
```

for 12:00 noon every day, recurring every 15 minutes until 1:00 PM.

You can also specify ranges such as

```
0      0     1-5   *     *
```

to execute every day from the 1st to the 5th of the month at midnight.

The second half of the crontab file is the list of commands to be executed. As the commands should fit on a single line in the file, you would either list a single command, or invoke a shell script. For instance, the entry:

```
0     0     1     *     *     ./somescript -a < foo.txt
```

will schedule the execution of .somescript –a < foo.txt to occur at midnight, the first of every month. The script somescript is located in the current directory, and it receives an option (–a) and input from foo.txt.

The reason that you are limited to one line for each scheduled task is that crontab allows files that contain multiple scheduled entries, one per line. Therefore, you could schedule numerous processes at different times using a single crontab file. Consider the following entries in a crontab file.

```
0     0     *     *     0     ./backup_script
0     0     *     *     2     ./cleanup_script
0/10  *     *     *     *     ./who_report
```

This requests that a backup program will run every Sunday at midnight, that a cleanup script will run every Tuesday at midnight, and that a script to report on who is logged in will run every 10 minutes of every hour of every day.

Writing the crontab file is the first step in scheduling tasks with crontab. The second is the issuing of the crontab process itself. The crontab process is invoked by the command crontab –f filename. Without this step, none of the tasks scheduled in the file will actually be scheduled. An additional option in crontab is –u username. This allows you to run the processes listed under a different user name; otherwise, the processes default to running under the user's account. This is particularly useful if you have switched to root before issuing crontab. You would not normally want a crontab job to run as root unless you knew specifically that the processes required root access. Therefore, as root, it is best to use crontab –u username –f filename, where username is your own account.

As with atq and atrm, you can examine the queue of crontab jobs and remove them. You would do this using crontab –l and crontab –r to list and remove the waiting jobs, respectively. You can also use crontab –e to edit the most recent crontab job. As the atd daemon is used to run at and batch jobs, the cron daemon is used to run crontab jobs.

One last note. In order to safeguard the system, you can specify which users are allowed to use crontab, at, and batch, and which users are not. There are several /etc files to control this. These are /etc/at.allow, /etc/at.deny, /etc/cron.allow, and /etc/cron.deny.

TERMINATING PROCESSES

There are multiple ways for a process to terminate. First, it may complete execution. Second, you may choose to stop it yourself. Third, the program may "die"—that is, abnormally abort its execution. Obviously, if a process terminates on its own once it completes execution, there is no need for the user to be concerned, nor are there any actions that the user must take. In the third case, there will usually be some kind of feedback to the user to indicate that a terminating error arose. In Windows, for instance, you may receive an error pop-up window. Or, the program's output window may just disappear from the screen. In Linux, programs that abnormally terminate usually leave behind a file called *core*. The core file is a "core dump", which is a snapshot of the process' working memory when it terminates. A programmer might examine the core file for clues as to the problem.

It is the second possibility that is troubling. The process remains "active" but is no longer making any progress toward completion. This might be caused by a deadlock situation (see Chapter 4), or child processes that have stopped responding, or a situation where the OS itself has lost pointers to the process. In any event, the user must discover that the process is not responding and decide what to do about it.

In Windows, to stop a process, you can just close the window. For processes that are not responding, this may not work and so you would invoke the Task Manager and kill the application or the process through this GUI. If the process is running in a terminal window (Linux) or DOS prompt window (windows), you can try ctrl+c to stop it. In Linux, you can also use the resource monitor to select and kill the process. You can also use the kill command from the command line. First, you must obtain the process' PID through the ps command. The kill command is kill *level* pid. The *level* determines how hard the OS should try to kill it. The highest level is –9 and should be used whenever possible. To kill a process, you must be the process owner or root. The command killall can kill all active processes. To use killall in this way, you must be root.

You may need to search through the running processes to see what is still running, and shut down processes that you no longer need. This is true in either Windows or Linux. Just because there is no icon pinned to the task bar does not mean that a process could not still be running. In Linux, processes are often difficult to keep track of when launched from a command line. An interesting historical note is that when MIT shut down their IBM 7094 mainframe in 1973, they found a low-priority process waiting that had been submitted in 1967! Whether this story is true is unknown.

Shutting down the entire system is an important step. You will need to shut down your computer after installing certain software (especially OS patches). Additionally, Windows 7 requires being shut down and restarted when it performs OS upgrades. Linux can perform upgrades to its OS without shutting itself down.

Shutting down your system also helps clean up your virtual memory—if you leave your system running for too long, it is possible that your virtual memory becomes fragmented causing poor performance. In Windows, it is important to reboot your OS occasionally (say at least once per week) because of corruption that can occur with the system as stored in virtual memory and main memory. Since Linux is more stable than Windows, reboots are less frequently needed.

How do you shut down your computer? Well, what you never want to do is merely shut it off. Both Windows and Linux have a shutdown sequence. It is important to follow it so that all processes and services can be shut down appropriately and so that all files will be closed. If you do not do this correctly, data files and system files can be corrupted resulting in problems in your OS over time.

In Windows, the shutdown process is taken care of through the Start button menu. The options are to shut down the system, reboot the system (restart), log off as current user, switch user while remaining logged in, or to put your system to sleep in either sleep mode or hibernate mode. Sleep mode is a power saver mode that allows you to start back up rapidly. Hibernate is a longer term mode—you would use hibernate when you do not expect to restart the computer again in the next hour or few hours. The shutdown routine is straightforward to use, although the process that shutdown will undertake is very complex in terms of the steps involved in actually shutting down both the OS and the hardware. Shutting down the machine can also be time consuming.

In Linux, you may use the GUI, or you can shut down the system through the command line by using the shutdown command. From the command line, shutdown accepts a parameter, -h, which causes the system to halt. Halting the system actually stops all of the machinery, whereas shutdown merely places you into a special mode that permits you to halt the system. In addition to –h, you must specify a time (in minutes) to denote when shutdown should occur. This gives you a "grace period" so that all users can log off in time. The command *shutdown –h 10* means to shutdown and halt the system in 10 minutes. You may also wish to send a message to users and processes to warn them of the imminent shutdown. The message might be something like "warning, system shutdown in 10 minutes, kill all processes NOW and log out!" The –r option not only shuts the system down, but reboots it afterward.

SERVICES

A *service* is a piece of OS software that provides a particular type of function. This makes it sound like any other OS component. There are several differences, however. First, a service may be running or stopped (or suspended) whereas the programs that make up the OS kernel remain running at all times. Second, a service is set up to work with any number of other agents—applications software, users, network communications, etc. Thus, a service might be thought of as software that can be called upon by unknown clients. The OS kernel instead handles requests from the user or running software, but not from unknown or remote clients. Third, services run in the background, without user intervention. Since services run in the background, services are not necessary for non-multitasking systems. For instance, consider a single tasking system, where either the OS is executing or the application software is executing. There is no background. But in a multitasking system, a service may be called upon from time to time.

Services may be started at boot time, they may be started based on some scheduling system, or they may be started (and stopped) by system administrators. In Linux, services are often referred to as *daemons* (pronounced demons), and their names often end with a d, for instance, the scheduling command, at, is controlled by a daemon known as atd. In Windows, they are known as Windows Services but unlike Linux, some Windows Services can run as a normal process, that is, execute in the foreground.

One advantage to having services in an OS is that a system administrator can tailor the environment by controlling which services run and which services do not run. For instance, there may be a need for a mail service if we expect that this computer will receive e-mail. However, if this is not the case, the e-mail service can be stopped so that it neither takes up memory space nor requires any CPU attention.

Additionally, the system administrator can control how the service runs. When a service is started, the service will likely read a *configuration file*. By altering the configuration file, you can control how that service works. We will explore some of the Linux configuration files in Configuring Services.

Another advantage of having services is that the services, if running, are placed in the background. Thus, a service only requires system resources when it is called upon to perform its service. So, for instance, a service does not wait in the ready queue and use CPU time. Only when the service must respond will it use the CPU.

In Windows, services have a couple of noteworthy features. First, they run in less privileged modes than the Administrator mode. For instance, they may run under Local Service or Network Service accounts. This provides the service with additional access rights over a typical user, but not the full access of the administrator. Also in Windows, services are compartmentalized so that they cannot influence other services. A service that has been compromised by virus or other form of intrusion will not impact other services. Finally, services are not allowed to perform operations on the file system, system registry, or computer network.

In Linux, most services are spawned by the init process, which is the first process started after Linux boots (see the section Establishing Services at Boot Time). Once the system has started, the system administrator is able to start (and stop) services on demand at any time, so those services started later are not spawned by init. Services are disassociated from terminals so that they run in the background not of a shell but of the entire OS. They are established as having the root directory as their working directory, and thus are not impacted by what file systems are mounted.

A service should not be confused with a *server*, although they are related concepts. We might say that all servers provide a service, but not all services are servers. One way to think of the distinction is that a server is a large-scale piece of software (probably many programs). The server may have a number of roles or accomplish many tasks within the service it provides. We also tend to think of a server (the software) running on a computer set aside to accomplish the service. We will discuss servers in more detail in Chapter 13.

There are a great number of services available in both Windows and Linux OSs. In Windows, one can view the active services, start and stop them, using the Task Manager. Figure 11.11 shows the Services tab of the Task Manager. Selecting a service allows you to then start or stop it.

In Figure 11.11, for instance, you can see such services as Netlogon, KeyIso, Power, and PlugPlay are running while services such as VaultSvc, EPS, and bthserv are stopped. The description column provides a brief description of the service, and for those running, the PID indicates which process started those services. In Figure 11.11, for instance, several processes were started by process 564. This is most likely the program lsass.exe. Process 708, which started the services Power and PlugPlay is svchost.exe. To obtain process PIDs, click on the process tab, and under view, select Select Columns. From here, you can select which columns should be viewable. PID is not viewable by default but you can make it viewable. Notice that services that are stopped do not have a PID listed.

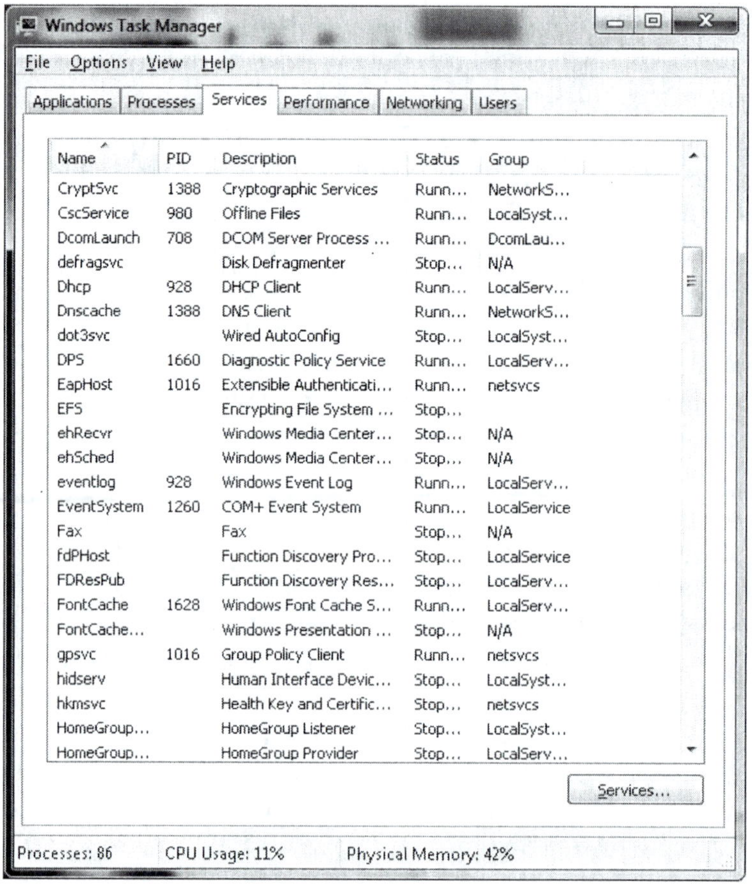

Windows Task Manager

File Options View Help

Applications | Processes | Services | Performance | Networking | Users

Name	PID	Description	Status	Group
CryptSvc	1388	Cryptographic Services	Runn...	NetworkS...
CscService	980	Offline Files	Runn...	LocalSyst...
DcomLaunch	708	DCOM Server Process ...	Runn...	DcomLau...
defragsvc		Disk Defragmenter	Stop...	N/A
Dhcp	928	DHCP Client	Runn...	LocalServ...
Dnscache	1388	DNS Client	Runn...	NetworkS...
dot3svc		Wired AutoConfig	Stop...	LocalSyst...
DPS	1660	Diagnostic Policy Service	Runn...	LocalServ...
EapHost	1016	Extensible Authenticati...	Runn...	netsvcs
EFS		Encrypting File System ...	Stop...	
ehRecvr		Windows Media Center...	Stop...	N/A
ehSched		Windows Media Center...	Stop...	N/A
eventlog	928	Windows Event Log	Runn...	LocalServ...
EventSystem	1260	COM+ Event System	Runn...	LocalService
Fax		Fax	Stop...	N/A
fdPHost		Function Discovery Pro...	Stop...	LocalService
FDResPub		Function Discovery Res...	Stop...	LocalServ...
FontCache	1628	Windows Font Cache S...	Runn...	LocalServ...
FontCache...		Windows Presentation ...	Stop...	N/A
gpsvc	1016	Group Policy Client	Runn...	netsvcs
hidserv		Human Interface Devic...	Stop...	LocalSyst...
hkmsvc		Health Key and Certific...	Stop...	netsvcs
HomeGroup...		HomeGroup Listener	Stop...	LocalSyst...
HomeGroup...		HomeGroup Provider	Stop...	LocalServ...

Services...

Processes: 86 | CPU Usage: 11% | Physical Memory: 42%

FIGURE 11.11 Services tab of the Task Manager.

At the bottom of the Services tab of the Task Manager Window is a Services... button. Clicking on this button starts the Services window (see Figure 11.12). In the Services window, you can view information about the service such as its description and whether it is started automatically or requires manual startup. You can also start the Services window by right clicking on the Computer desktop icon and selecting Manage, or by starting the Control Panel, selecting Administrative Tools, and from the list of tools, selecting Services.

From the Services window, right clicking on a service and selecting Properties (or selecting Properties from the Action menu or from the button bar) displays the service's properties. See Figure 11.13. There are four tabs in this window: General, Log On, Recovery, Dependencies.

The General tab provides the service's name, a description, its location in the file system, how it is started (in the case of DHCP Client, it is started automatically at boot time), its current status (if started, you can stop it and vice versa), and what parameters are supplied when it is started. DHCP Client does not start with any parameters.

Under Startup type, there are four options: Automatic (Delayed Start), Automatic, Manual, Disabled. These are self-explanatory except for the first one, which was introduced in Windows Vista. As not all automatic services are needed immediately upon

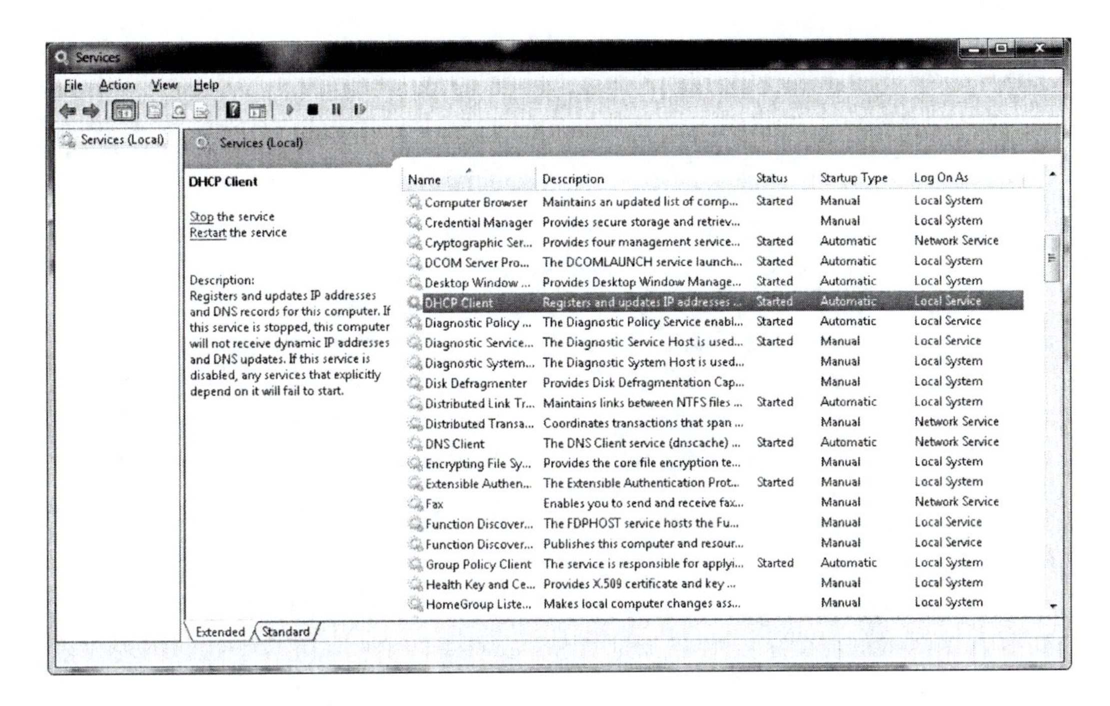

FIGURE 11.12 Windows Services.

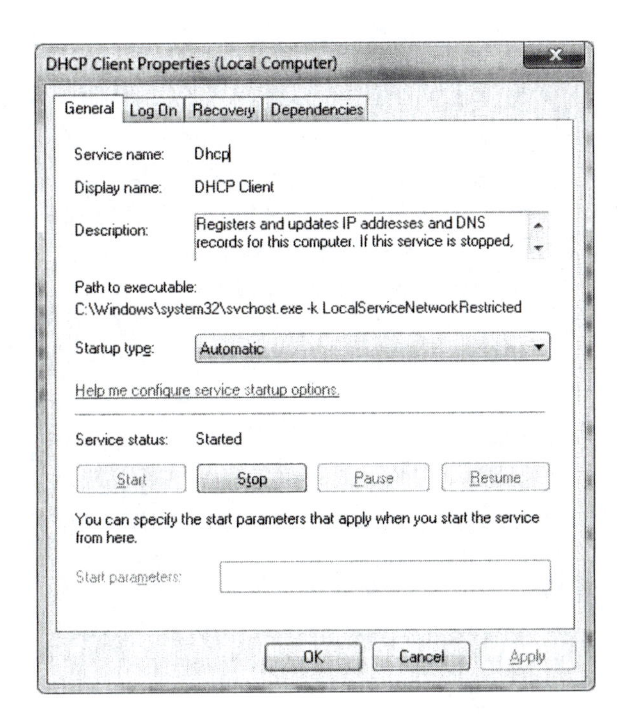

FIGURE 11.13 DHCP Properties window.

system boot, services can be categorized by their importance. Lower priority services can be started automatically upon boot, but postponed until other, more important services are started. Those denoted by "Delayed Start" do not get started until the initialization scripts, processes, and services run. The intention here is to provide a faster boot process for the user. Once booted, as the system has time, the remaining automatic services are started. As an example, the Windows Search and Windows Update services use this setting because their immediate service will not be needed.

Services may require access to system or network resources. If this is the case, the service requires a log in (account name and password). The log in type is indicated under the Log On tab (you can also see this in the Services window as the rightmost column, Log On As). For instance, DHCP will log on under the name Local Service. Other choices are Network Service and Local System. Local Service and Network Service will most likely require a log in account and password, whereas Local Service will not.

The Recovery tab indicates how the OS should attempt to recover if the service fails. In Windows, there are three levels of failure: first failure, second failure, and subsequent failures. The tab contains choices for each level, the choices being "take no action", "restart the service", "run a program" (in which case you indicate the program and any program parameters), and "restart the computer". Failures are counted since the last reset. By default, resets occur every day. The delay before a service restart may occur is also controllable. This value is set in minutes (from 0 upward).

The Dependencies tab lists the services on which this service depends, and the services that depend on this service. This indicates the services that need to be running for this service to run successfully, and if this service were to stop running, what other services would be impacted. For instance, the DHCP Client service relies on the Ancillary Function Driver for Winsock, NetIO Legacy TDI Support Driver, and Network Store Interface services, whereas WinHTTP Web Proxy Auto-Discovery Service relies on DHCP Client.

In Linux, the Service Configuration Tool (see Figure 11.14) lets you view, start, and stop services. This tool is a graphical means of interfacing with services that can also be controlled via the /etc/rc.d directories and command line instructions. Automatically starting services is explained in detail in Establishing Services at Boot Time.

When you select a service in this tool, you are given that service's description and any services that might rely on the service (i.e., dependencies). For instance, the syslog service requires that both syslogd and klogd be running, as shown in the figure. You are also given the "run level" for the service. This is described in Establishing Services at Boot Time as well. Notice that this tool offers two tabs, Background Services and On Demand Services. Most Linux services are intended to run in the background. However, you can also establish on demand services. These services will run when requested and will automatically end (exit the system) when there is nothing remaining for them to do.

In Linux, you can also view the status of any given service, start, stop, or restart it from the command line. There are two ways to control services from the command line. The common way is through the service command, which is under /sbin. The format of the command is /sbin/service *servicename command*, where *servicename* is the name of the service, for instance, syslog, and the *command* is one of start, stop, restart, and status. The

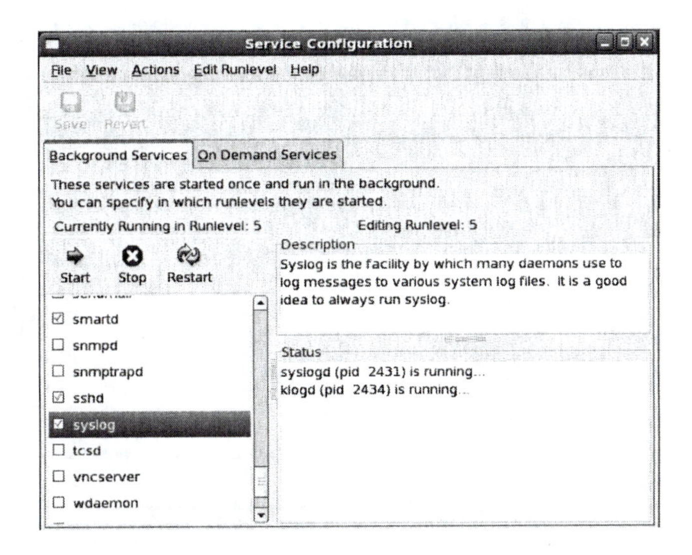

FIGURE 11.14 Linux Services in the Service Configuration window.

status command merely responds with the service's status: running, stopped, dead. The other command line approach to controlling a service is to issue a command directly to the service itself. The sshd service, for instance, is stored in /etc/init.d. Therefore, you can also issue the command /etc/init.d/sshd start (or restart, stop, or status).

There are a great number of services in both Windows and Linux, and their roles can differ widely. There is no easy categorization of all of the services. A general list of types of services along with some programs that handle these service categories is presented below.

Logging services. Log files play a critical role in determining the cause of unusual or erroneous events. Both Windows and Linux have services that regularly log events, whether generated by the OS or application software or from some other source. Windows Event Log stores and retrieves event information. Windows uses six levels of events: Audit Failure, Audit Success, Information, Warning, Error, and Critical. The Event Viewer program allows you to view events logged by Windows Event Log. Log entries can arise from applications, hardware events, Internet Explorer events, Media Center events, Security issues, System events, and others. Figure 11.15 demonstrates a view of the Event Logger, with the list of warnings enumerated, and Figure 11.16 shows some of the logged events from Applications. In Linux, syslogd and klogd share the logging chores, syslogd logs normal system and application events, whereas klogd logs specific kernel events. The Linux service auditd collects security-related events and records them in the audit.log files. The Linux service netconsole logs logging attempts. Other software might create their own log files, or they might call upon syslogd. The various log files in Linux are stored under /var/linux unless otherwise specified.

Scheduling services. Users, particularly system administrators, will commonly schedule processes to execute at certain times. For instance, an antiviral program might be scheduled to run every night at midnight, or a backup utility might be scheduled to run once

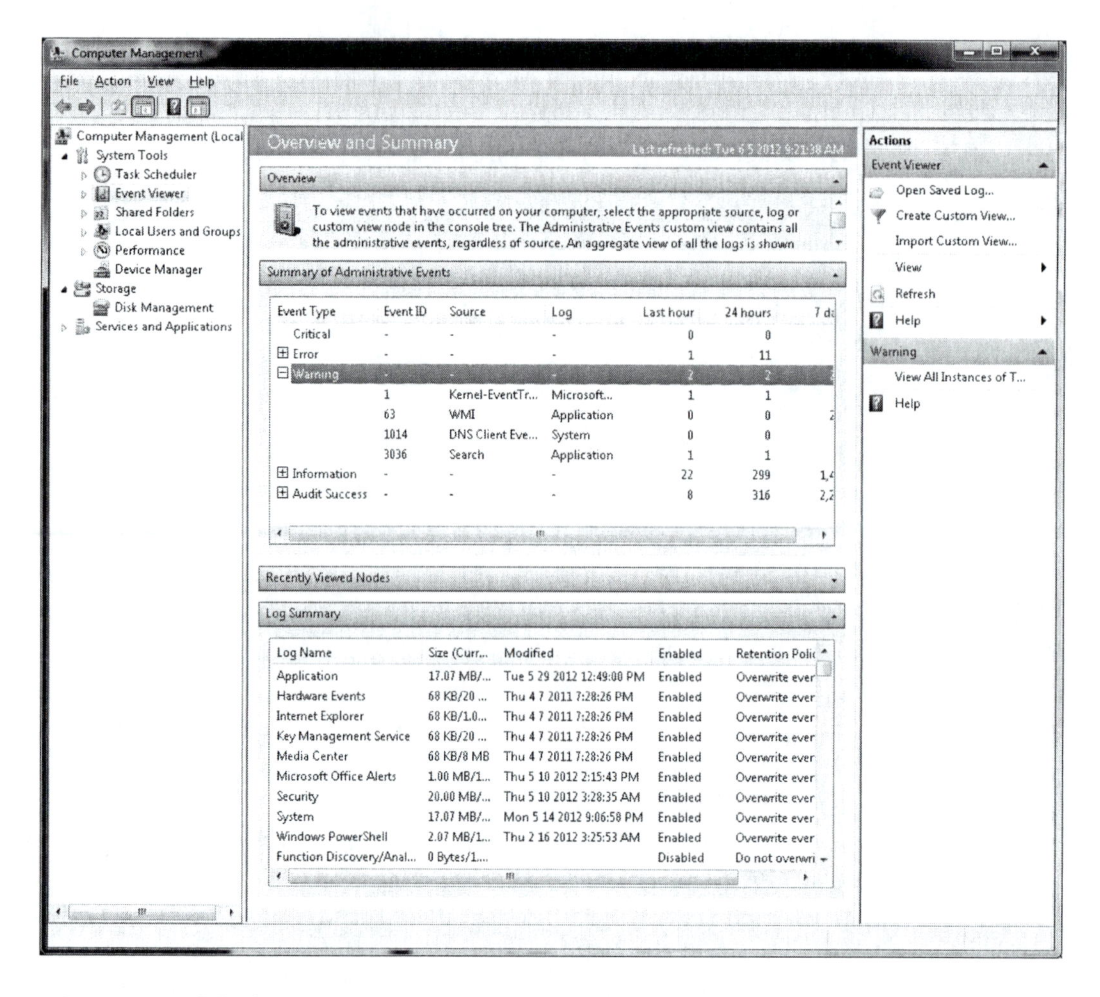

FIGURE 11.15 Windows Event Viewer.

Number of events: 5,150

Level	Date and Time	Source	Event ID	Task Category
Warning	Tue 6 5 2012 9:03:58 AM	Kernel-Even...	1	Logging
Warning	Tue 6 5 2012 8:44:35 AM	Search	3036	Gatherer
Warning	Mon 6 4 2012 8:44:35 AM	Search	3036	Gatherer
Warning	Sun 6 3 2012 8:44:35 AM	Search	3036	Gatherer
Warning	Sun 6 3 2012 4:24:46 AM	WMI	63	None
Warning	Sun 6 3 2012 4:24:46 AM	WMI	63	None
Warning	Sun 6 3 2012 4:24:46 AM	WMI	63	None
Warning	Sun 6 3 2012 4:24:46 AM	WMI	63	None
Warning	Sun 6 3 2012 4:24:46 AM	WMI	63	None
Warning	Sun 6 3 2012 4:24:46 AM	WMI	63	None
Warning	Sun 6 3 2012 4:24:45 AM	WMI	63	None
Warning	Sun 6 3 2012 4:24:45 AM	WMI	63	None
Warning	Sun 6 3 2012 4:24:45 AM	WMI	63	None
Warning	Sun 6 3 2012 4:24:45 AM	WMI	63	None
Warning	Sun 6 3 2012 4:24:45 AM	WMI	63	None
Warning	Sun 6 3 2012 4:24:45 AM	WMI	63	None
Warning	Sun 6 3 2012 4:24:45 AM	WMI	63	None

Event 1, Kernel-EventTracing
General Details

Event Properties - Event 3036, Search
General Details
The content source <SharePointWorkspaceSearch://{S-1-5-21-945550151-541155741-1648912389-35418}/> cannot be accessed.
Context: Application, SystemIndex Catalog
Details:
A server error occurred. Check that the server is available. (HRESULT : 0x80041206)

Log Name: Application
Source: Search Logged: Sun 6 3 2012 8:44:35 AM
Event ID: 3036 Task Category: Gatherer
Level: Warning Keywords: Classic
User: N/A Computer: NKU072890.hh.nku.edu
OpCode: Info
More Information: Event Log Online Help
Copy Close

FIGURE 11.16 Application warnings logged including a specific event.

per week at a specific time. The Windows scheduler is called Task Scheduler. It is both a scheduler and a service. Linux has several scheduling programs: at, batch, crontab. The Linux services that support scheduling are anacron, to execute processes scheduled for boot time, crond to execute processes scheduled by crontab, and atd to execute processes scheduled by at and batch.

SERVICE PACKS

Service packs are not related to services specifically, but because we are discussing services, it is useful to look at service packs here. The service pack is used by Microsoft to release updates of OSs. This permits Microsoft to fix errors, handle user complaints, and remove security holes between major releases of OSs.

Starting with NT, Microsoft has been releasing service packs for their Windows OSs. In most cases, service packs supersede previous service packs. For instance, to install service pack 2 you do not need to have installed service pack 1. However, this approach is not entirely consistent. XP service pack 3 relied on service pack 1 being previously installed.

Service packs used to be delivered on CD-ROM but today, service packs are available over the Internet for download and in fact your Windows 7 system can automatically download and install service packs. One day you may wake up and find that your OS runs a little differently—new features, tighter security, or a peculiar error no longer arises.

Service packs are not limited to Microsoft Windows. Other pieces of software now use this approach.

Network services. There are a number of different tasks involved in communicating over a computer network, so there are numerous network-related services. In Linux, there is the network service that configures all network interfaces at boot time. But then there are ISDN (Integrated Services Digital Network) for communication over the telephone system, nfsd (Network File System Daemon), nmbd (Network Message Block Daemon), ntpd (Network Time Protocol Daemon), snmptd (Simple Network Management Protocol Daemon), as well as services for a web server (httpd), ftp server (ftpd), ssh server (sshd), mail server (sendmail, fetchmail), and domain name server (named). In Windows, there are NTDS (Network Authentication Server), BITS (Background Intelligent Transfer Service), DNSCache and DNS (for DNS clients and servers), and NSIS (Network Store Interface Service) to collect routing information for active network interfaces, to name a few. In Windows, Workstation creates and maintains client network connections to remote servers. Other network services in Windows include Network Access Protection Agent, Network Connections, and Network Location Awareness.

Firewall and Internet services. In Linux, dnsmasq starts the DNS caching server and httpd is used to start and stop the default Apache web server. The services iptables and ip6tables control the default Linux firewall. The Linux nscd service handles password and group lookup attempts for running programs, caching results for future queries. This service is needed to support other network-based authentication services such as NIS and LDAP. In Windows, Base Filtering Engine manages the Windows Firewall and Internet Protocol security. DHCP Client registers and updates IP addresses and DNS records for the computer and DNS Client caches DNS names. If this service is stopped, name services continue through the DNS, but results are no longer cached for efficiency, and other services that explicitly depend on the DNS Client will fail. Similar to ncsd, the Windows service Netlogon maintains a secure channel between a computer and the domain controller for authenticating users and services.

File system. As with network services, there are many different tasks involved in file system services. In Linux, these include autofs and amd, which are used to automatically mount file system partitions; netfs, which mounts and unmounts any network file systems; and mdmonitor, which manages the software portion of a RAID storage device. In Windows, Block Level Backup Engine Service provides backup and recovery support on the file system.

Peripheral devices. Windows has a number of services to support peripheral devices. These include Audio Service to manage audio jack configurations, Bluetooth Support Service, and Plug and Play, which allows the computer to recognize newly attached devices with little or no user input. In Linux, CUPS (common Unix printing system) supports communication with printers. There are a number of services that all support BlueTooth management, including bluetooth, hidd, pand, and dund. The conman service performs console management. The gpm service adds mouse support for text-based Linux applications, for example, by permitting copy-and-paste actions via the mouse.

Miscellany. The Linux oddjobd service is interesting in that it supports applications that do not have their own privileges. For instance, if a program requires some operation to be performed, but does not have the proper privileges to execute the operation, the program can request it of the oddjobd service. In Windows, there are a number of diagnostic services available. There are also services to support such actions as parental controls. Both OSs provide power management services: ampd in Linux and Power in Windows. For instance, the ampd service monitors battery status and can be used to automatically shut down a computer when the battery is too low.

CONFIGURING SERVICES

When a service is started, it will usually read a configuration file. The configuration file specifies the properties by which the service will execute. For instance, the Linux firewall service will operate using a number of rules. These rules are specified in the configuration file. Typically, Windows services use .ini (initialization) files to store these specifications, although in some cases, they end in .cfg or have other or no extensions. In Linux, configuration files often end with the .conf extension although this is not always the case. Also, like Linux scripts, the use of # to begin a line in a configuration file indicates that the line is a comment and should be ignored by the software that reads the file.

To alter the behavior of a service, you must stop the service, edit the configuration file, and then restart the service. This should not be done lightly as a mistake in a configuration file might cause the service to malfunction or not function at all. Here, we concentrate on several Linux.conf files, examining their syntax and some of the changes that you might make. In Linux, most configuration files can be found in the /etc directory.

We will start with one of the Linux logging services, syslogd. This service logs events generated by application software and non-kernel portions of the OS. To be configured, syslogd needs to know three things:

1. What types of activity should be logged
2. What conditions require logging
3. Where the logging should occur

The file /etc/syslog.conf describes this information. As it is merely a text file, the system administrator is free to edit the file, which will then impact what gets logged and where. Each line of this file describes one form of logging event. A logging event will comprise one or more sources (software), the level of priority that would cause the event to be logged, and the action to take place should the source provide a message at the given priority level. Multiple sources can be listed on a single line if the sources would all share the same log file.

The format for any single entry is source.priority [;source.priority]* action. That is, each entry will be at least one source.priority, but can be any number of source.priority pairs as long as they are separated by semicolons. The use of the wildcard (*) in the syntax above means "0 or more copies" although we will see that we can use the * as the source, priority, and action. If used for either the source or the priority, the * indicates "any source" or "any priority level". For instance, one might use auth.* to indicate any authorization message, regardless of the priority level. Or, to indicate any message whose priority level is emergency, we could specify *.emerg. The entry *.* would log any event by any software.

The action portion of an entry is almost always a file name, indicating where the message should be logged. In most cases, the logs are saved under the /var/log directory. However, if the message log is in support of a particular piece of software, it is possible that a different location might be desired, such as a log subdirectory under the home directory of the software. Aside from specifying the location of the log file, an * can be used, which instead sends the message to all logged in users to an open terminal window. You can also control the terminal to output to by using /dev/tty1, for instance, to output to only terminal tty1s instead of any open window. Another alternate action is to pipe the log message to another piece of software. Such an entry might look like this: |exec /usr/local/bin/filter.

There are several default sources. These include auth (user authentication services), cron (the cronjob scheduler), daemon (all standard services), kern (the Linux kernel), lpr (printer server), mail (mail server), syslog (the syslogd service), and user (programs started by the user). The user programs are denoted as local0–local7, allowing for eight additional programs that can have customized logging. There are nine levels of priority that range from generating log messages only when the there is a total system failure to as simple as generating log messages for every activity of the given piece of software. The nine levels are shown in Table 11.2.

TABLE 11.2 Nine Priority Levels Used by syslog.conf

None	No priority
Debug	Debugging messages, used by programmers while testing their programs
Info	Informational messages about what the program is doing
Notice	Noteworthy events
Warning	Warnings about potential problems
Err	Errors that arise during execution
Crit	Messages that describe critical errors that will most likely result in the program terminating abnormally
Alert	Messages that will describe errors that not only result in a program terminating abnormally, but may also impact other running programs
Emerg	Messages about errors that may result in a system crash

Figure 11.17 shows the contents of the syslog.conf file. The first entry specifies that any kernel message should be sent to /dev/console, which is the console of the host computer (this message would not be sent to remote login consoles or other windows). The second entry specifies numerous messages: all informational messages and all mail, authpriv, and cron messages with no priority. These are all logged to /var/log/messages. The third entry requires that all other authpriv messages (above the "no priority" priority level) are logged to /var/log/secure. Can you make sense of the remaining entries? The use of the "–" in an action preceding a file name, for instance, -/var/log/maillog in the figure, indicates that messages should not be synchronized.

The Linux firewall requires a configuration file. This file is stored in /etc/sysconfig/iptables. Modifying this file allows you to add, change, and delete firewall rules. Each row of

```
# Log all kernel messages to the console.
# Logging much else clutters up the screen.
#kern.*                                                 /dev/console

# Log anything (except mail) of level info or higher.
# Don't log private authentication messages!
*.info;mail.none;authpriv.none;cron.none                /var/log/messages

# The authpriv file has restricted access.
authpriv.*                                              /var/log/secure

# Log all the mail messages in one place.
mail.*                                                  -/var/log/maillog


# Log cron stuff
cron.*                                                  /var/log/cron

# Everybody gets emergency messages
*.emerg                                                 *

# Save news errors of level crit and higher in a special file.
uucp,news.crit                                          /var/log/spooler

# Save boot messages also to boot.log
local7.*                                                /var/log/boot.log
```

FIGURE 11.17 Logging entries from /etc/syslog.conf.

this file is a rule that dictates a type of message that should be accepted or rejected by the firewall software. The rules have a peculiar syntax. They begin with the type of rule, an append rule (–A) or a flush rule (–F). The flush rule deletes all of the rules in the selected table, so most or all rules will begin with –A.

The next item in a rule is the name of the stream from which a message might arrive. Streams have generic names such as INPUT, OUTPUT, or FORWARD, or may be from a specific location such as fast-input-queue or icmp-queue-out. The examples below indicate the input stream RH-Firewall-1-INPUT.

Next are the specific conditions that the rule is to match. These may include ports, types of TCP/IP messages, source or destination IP addresses, or protocols. For instance, --dport 50 indicates a message over port 50, whereas –p udp --dport 5353 indicates a message over udp destination port 5353 and –d 1.2.3.4 indicates a destination IP address of 1.2.3.4.

The final entry in a rule is typically the action that the rule should apply when the rule matches a message. The most basic actions are ACCEPT, LOG, DROP, and REJECT. ACCEPT immediately stops processing the message and passes it on to the OS (i.e., past the firewall). The LOG action logs the message but continues processing rules in case other rules might match to specify further actions. DROP causes the message to be blocked, whereas REJECT blocks the message but also replies to the host sending the packet that the packet was blocked.

What follows are a few example rules:

```
-A RH-Firewall-1-INPUT --dport 50 -j ACCEPT
-A RH-Firewall-1-INPUT -p udp --dport 5353 -d 224.0.0.251 -j
ACCEPT
-A RH-Firewall-1-INPUT -p tcp -m tcp --dport 631 -j ACCEPT
```

Typically, the iptables file will end with a default rule. If previous rules specify messages to accept or log, then the following rule would be a "backstop" rule to reject all other messages.

```
-A RH-Firewall-1-INPUT -j REJECT --reject-with icmp-host-prohibited
```

The /etc/fstab file stores mount information. This is applied whenever the mount –a (mount all) command is issued. The configuration information lists each mount partition, mount point (directory), and specific information about that partition such as whether the partition has a disk quota on individual entries and whether the partition is read-only. For instance, one entry is

```
LABEL=/home /home ext3  defaults    1 2
```

This entry describes that the /home partition should be mounted at the directory /home, the partition's type is ext3, the partition uses the default options, and has values 1 and 2, respectively for the dump frequency (archiving schedule) and pass number (controls the order that the fsck program checks partitions). Options can include ro (read only), rw (read–write), auto or noauto (whether the partition should be automatically mounted at boot time or not), and usrquota (to establish quotas on each user directory). The ext3 type is common in Linux systems. However, if the partition is mounted over the network using the network file system (nfs), an additional option is to include the IP address using addr= #.#.#.#, where the # signs represent the octets of the IP address.

The /etc/resolv.conf file is one of the simplest configuration files. It stores the IP addresses of the domain name system servers (DNS) for the computer. Unlike the previous two example configuration files that comprised a number of rules in some unique syntax, this file consists of entries that look like:

```
nameserver 10.11.12.13
nameserver 10.14.15.16
```

The resolv.conf file is used by numerous Linux programs. We discuss the DNS in Chapter 12.

The mail server in Linux is sendmail. It has a number of configuration files. One, /etc/aliases, permits e-mail aliases to be established. For instance, addresses such as bin, daemon, adm, halt, and mail are aliased to root so that, if a message is sent to any of those locations, it is actually sent to root. Another sendmail configuration file is /etc/mail/sendmail.cf. However, it is advised that you should never directly edit this file. Instead, the file /etc/mail/sendmail.mc is a macro file. Editing it will allow you to generate a new sendmail.cf file. As sendmail is a very complex program, we will not examine this configuration file.

One last /etc configuration file worth noting is ldap.conf. LDAP is a server used to perform network-based authentication. That is, given a network of computers, an LDAP server can be used to authenticate a user on any of the network computers. Although LDAP is far beyond the scope of this text, its configuration file is worth mentioning. The format of the LDAP configuration file is unlike the previous files where each entry was a rule or instruction to the program. Instead, the LDAP configuration file contains *directives*. There are many forms of directives, each of which accomplishes a different task in initializing the LDAP server. For instance, the base directive specifies a "distinguished" name for searching. Another directive, uri, declares IP addresses of various types for the domain sockets. Other directives specify time limits, ports, filters, naming contexts, naming maps, and so forth. What follows is an example of an ldap.conf file:

```
uri ldap://127.0.0.1/
ssl no
tls_cacertdir/etc/openldap/cacerts
pam_password md5
```

Modifying a configuration file of a service does not automatically cause the service to accept those changes. In most cases, the service must be restarted. You can do this by either stopping the service and then starting it anew, or by using the restart command instead. In many cases, it is wise to stop the service before modifying the configuration file.

ESTABLISHING SERVICES AT BOOT TIME

We end this chapter by considering which services are started automatically at system initialization time and which are not. As described in Services, Windows services can be controlled through the Services tool. By selecting any service from this tool, you can start or stop the service. Through the service's properties window, you can specify the startup type. The choices are Automatic (Delayed Start), Automatic, Manual, Disabled. As an Administrator, you can change the startup type of any or all services as you desire. For instance, if it is felt that Parental Controls should always be running at boot time, you can change its startup type from Manual to Automatic. Similarly, Bluetooth Support Service is something that you might want running. However, as users will not require the Bluetooth service the instant that the system boots, you might set this to start by Automatic (Delayed Start). In this way, the service starts shortly after system initialization as time permits.

Linux similarly is set up to automatically start specific services at system initialization time, and the services that start can be adjusted by system administrators. However, unlike Windows, in Linux you can establish different startup services depending on the run level. There are seven run levels in Linux. These are shown in Table 11.3.

When Linux first boots, it runs the script /etc/inittab. One of the first things that inittab establishes is the run level. A run level of 5, for instance, starts the OS in full GUI mode, whereas run level of 3 is full text mode and run level of 1 is single user text mode. With the runlevel established, services are started based on the runlevel. The inittab script executes the script /etc/rc.d/rc. This script takes care of several startup activities, but among them, it iterates through the directory /etc/rc.d/rc#.d, where # is the runlevel (for instance, rc5.d for runlevel 5).

A look at /etc/rc.d/rc5.d shows a number of symbolic links. Each symbolic link has a name in the form of K##name or S##name. The K and S stand for "kill" and "start", respectively. This letter denotes whether the rc script will kill or start the service at startup. The number denotes the order in which the service is started (or stopped). See Figure 11.18, which lists the symbolic links for runlevel 5. A comparison between the /etc/rc.d/rc5.d and /etc/rc.d/rc3.d directories will show the services needed for the GUI versus those that are not, whereas a comparison between the /etc/rc.d/rc3.d and /etc/rc.d/rc2.d directories will show the services needed for network communication versus those that are not.

As each item in this directory is merely a symbolic link, the rc script follows the link to the actual service. All services are stored in the directory /etc/rc.d/init.d. Figure 11.19

TABLE 11.3 Linux Run Levels

Run Level	Description
0	Halt—shuts down all services when the system will not restart
1	Single-user mode—for system maintenance, operates as root without network capabilities
2	Multiuser mode without network—primarily used for maintenance and testing
3	Multiuser mode with network—text-based mode for normal operation
4	Not used
5	Multiuser mode with network and GUI—typical mode for normal operation
6	Reboot—shuts down all services for system reboot

```
K01dnsmasq          K50snmpd              S07iscsid         S19rpcgssd        S80sendmail
K02avahi-dnsconfd   K50snmptrapd          S08ip6tables      S22messagebus     S85gpm
K02NetworkManager   K69rpcsvcgssd         S08iptables       S25bluetooth      S90crond
K02oddjobd          K73ypbind             S08mcstrans       S25netfs          S90xfs
K05conman           K74nscd               S09isdn           S25pcscd          S95anacron
K05saslauthd        K74ntpd               S10network        S26acpid          S95atd
K05wdaemon          K85mdmpd              S11auditd         S26apmd           S96readahead_later
K10psacct           K87multipathd         S12restorecond    S26haldaemon      S97yum-updatesd
K10tcsd             K88wpa_supplicant     S12syslog         S26hidd           S98avahi-daemon
K15httpd            K89dund               S13cpuspeed       S26lm_sensors     S99firstboot
K20nfs              K89netplugd           S13irqbalance     S26lvm2-monitor   S99local
K24irda             K89pand               S13iscsi          S28autofs         S99smartd
K35vncserver        K89rdisc              S13portmap        S50hplip
K35winbind          K91capi               S14nfslock        S55sshd
K50ibmasm           S04readahead_early    S15mdmonitor      S56cups
K50netconsole       S05kudzu              S18rpcidmapd      S56rawdevices
```

FIGURE 11.18 The /etc/rc.d/rc5.d directory of startup services.

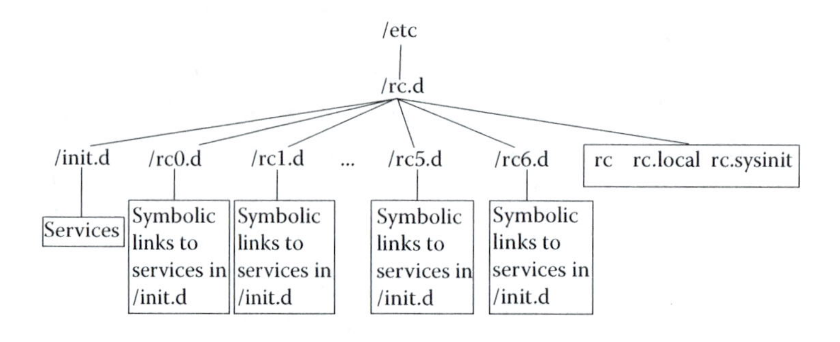

FIGURE 11.19 Structure of the /etc/rc.d directory.

illustrates the subdirectory structure for the startup scripts. The /etc/rc.d directory also contains two additional scripts: rc.sysinit and rc.local. The rc.sysinit script initializes a number of system settings, sets up the keyboard for use, established environment variables, and performs other hardware configuration steps. The rc.local script is available for the system administrator to further tailor the initial environment. It only executes once all other initialization scripts are executed. This file is initially nearly empty to start with.

As a system administrator, you can further tailor the run levels by merely changing symbolic link names. For instance, if it is deemed by the system administration that the Bluetooth service will not be needed in run level 5, the system administrator could rename the symbolic link S25bluetooth under /etc/rc.d/rc5.d to K25bluetooth. That is, rather than starting the bluetooth service, it is killed. Of course, if you are unsure about this, it is best to leave these symbolic links alone. You can always start or stop services at a later point by using /sbin/service bluetooth start/stop.

FURTHER READING

See Further Reading section in Chapter 4 for a list of texts that describe Windows 7, Linux, and Unix OSs. Process management is often covered as a chapter or two within these texts. Additionally, texts that discuss system administration for each OS will cover details on service configuration. You can find additional texts on configuring your Windows 7 and Linux OSs in certification study guides. The following texts specifically deal with configuring Linux.

- Crawley, D. *The Accidental Administrator: Linux Server Step-by-Step Configuration Guide*. Washington: CreateSpace, 2010.
- LeBlanc, D. and Yates, I. *Linux Install and Configuration Little Black Book: The Must-Have Troubleshooting Guide to Installing and Configuring Linux*. Scottsdale, AZ: Coriolis Open Press, 1999.

REVIEW TERMS

Terminology introduced in this chapter:

Child process
Configuration file
Core dump
Daemon
Delayed start (Windows)
Exec (Linux command)
Executing (process)
Executable
Firewall
Fork (Linux command)
Inittab (Linux)
Init.d (Linux)
Kill (a process)
Logging
Logon type (Windows)
Log file
Niceness
oddjobd (Linux)
Parent process
PID
Priority
Process properties
Process status
Process tree
Rc#.d (Linux)
Rc.sysinit (linux)
Recovery (Windows)
Resources
Run level (Linux)
Scheduling
Service
Service dependencies
Short cut icons
Shut down

Sleeping (process)

Spawn

Startup level (Windows)

Suspended (process)

Syslog (Linux)

Syslog.conf (Linux)

Task manager (Windows)

Terminating (process)

Waiting (process)

REVIEW QUESTIONS

1. In Windows, how can you move a process from the foreground to the background? In Linux from the command line, how can you move a process from the foreground to the background?
2. What is the difference between ps, ps a, and ps ax? Between ps aux and ps afx?
3. When starting a program in Linux, what happens if you end the command with an &?
4. What does the Linux jobs command do? How does it differ from ps?
5. In Linux, how does a parent generating a child through fork differ from a parent generating a child through exec?
6. If process A has a higher priority than process B, what does that mean with respect to both processes executing in a multitasking system?
7. If you were to increase a Linux process' niceness (make the nice value larger), does this raise or lower its priority?
8. You want to run a process at 3:45 PM tomorrow. How would you specify this using Linux' at command? What if it was 3:45 PM on March 1, 2013?
9. You want to run a process using at. How would you specify 3 hours from now? Three days from now?
10. You want a process to run recurring every Friday the 13th. How would do this? What if you wanted to specify every Friday at noon? What about the 15th of every month at 6:45 PM?
11. What does it mean to kill a process? Explain how you can kill a process in Windows and in Linux.
12. Why is it important to properly shut down a computer through the operating system's shut down routine rather than just turning it off?
13. What is a grace period when you use the Linux shutdown command?
14. How does a service differ from the operating system kernel?

15. How does a service differ from a server?

16. What does it mean that a service runs in the background?

17. In Windows, how can you determine which services are running? How can you stop a running service?

18. In Linux, how can you determine which services are running?

19. In Linux, what options can you specify in the services command?

20. In Linux, what is the difference between syslogd and klogd?

21. What is the difference between the anacron and crond services in Linux?

22. What is a configuration file? When you modify a configuration file, what happens to the running service? What do you have to do to have the changes made to the configuration file take effect?

23. What do the following entries in the syslog.conf file mean?

```
cron.*              /var/log/cron
authpriv.warn       |/usr/sbin/filter
*.emerg             *
```

24. What is the difference between the Crit, Alert, and Emerg priority levels for messages?

25. What does ACCEPT and REJECT mean when listed in rules in the iptables configuration file?

26. What do the K and S mean in the file names found in the /etc/rc5.d directories?

27. What is the difference between Linux run level 3 and 5? Why might you choose 3 instead of 5?

28. There are directories in/etc for each of rc0.d, rc1.d, …, rc.6.d. What do each of these directories represent? What is the number used for?

DISCUSSION QUESTIONS

1. You have started a process in Windows. It does not seem to be running correctly. What should you do in order to determine if the process is running, inspect its status, and either fix it or kill it?

2. Repeat question #1 but assume Linux instead of Windows.

3. Under what circumstances might you increase the priority of a process? Under what circumstances might you decrease the priority of a process?

4. As a system administrator, do you have the authority to change other users' process priorities? If your answer is yes, explain under what situations you should use this authority.

5. In Windows 7, bring up the resource monitor tool. Select the "Overview" tab. Explain the various types of information it is providing you.

6. Repeat #5 but select the "Memory" tab.

7. Repeat #5 but select the "Disk" tab.

8. Create a list of five tasks that you, as a system administrator, feel should be scheduled for off hours. For instance, you might want to perform an automated backup at 2 AM every Sunday night.

9. Explore the list of log files that are generated from the syslog daemon. As a system administrator, which of those log files do you feel you should inspect daily? Which should you inspect weekly? Which would you not inspect unless you had a particular reason to? Give an explanation for your answers.

10. As a user, how important is it for you to understand how a firewall works and to alter its configuration? As a system administrator?

11. As a system administrator, under what circumstances might you alter the services that start at boot time in Windows? In Linux?

12. As a follow-up to question #11, how would you alter the services that automatically start up at boot time in Windows? In Linux?

CHAPTER 12

Networks, Network Software, and the Internet

In this chapter, computer networking is covered. This chapter begins by describing the computer hardware that makes up the physical components of a computer network. Next, networks are discussed at an architectural level: classifications of networks and network protocols. Network software, including specific Linux software, is covered. Finally, the chapter examines the Internet: what makes it work, how it has grown, and what its future may be.

The learning objectives of this chapter are to

- Describe the role of network broadcast devices.
- Differentiate between types of network media.
- Compare network topologies and classifications.
- Discuss the role of each layer in the TCP/IP protocol stack and the OSI model.
- Explain IPv4 addressing and compare it to IPv6.
- Introduce popular forms of network software.
- Present a brief history of the Internet.
- Describe how communication takes place over the Internet.

A computer network is a collection of computers and computer resources (e.g., printers, file servers) connected in such a way that the computers can communicate with each other and their resources. Through computer networks, people can communicate, share data, share hardware, isolate and secure data, and provide a platform for easy data backup. Networks (particularly the Internet) also offer a means for commerce and sales. In fact, there are a number of different benefits that the Internet has provided.

Although networks are known to improve workplace efficiency through the sharing of data, resources, and communication, there are a number of costs associated with any computer network. There is the cost to purchase and set up the physical media and administer the network. However, that is not nearly the concern as the cost of securing the network properly, or the cost of having an insecure network. Additionally, through computer networks, most people have access to the Internet, and in a workplace, this could lead to undesirable behavior and inefficient use of time. Some of the threats to a computer network and activities required in securing a computer network are covered in Chapter 15.

Although a network can consist of many types of resources, it is the computer that is the primary tool used to communicate over a network. In a network, we define computers as being *local* or *remote*. A local computer is the computer that the user is using. That is, the user is physically present with that computer. A remote computer is a computer being accessed over the network. For instance, using a remote desktop connection or telnet [or ssh (secure shell)], you can log in to another computer. The computer you are logging into is referred to as the remote computer. A *host* is a type of computer that can be logged into from a remote location. This used to be an important distinction in that many personal computers could not be host computers. But today, that is not necessarily the case as most computers permit some form of remote login or remote desktop connection.

Computer networks can be viewed at a physical level (the connections or physical media over which communication is possible), a logical level (network topology), or a software level (the protocols and programs that allow the computers to communicate). This chapter examines some of the ideas behind computer networks at each of these levels.

NETWORKS AT A HARDWARE LEVEL

The physical level of a network defines how information is carried over the network. At this level, information is transmitted as bits over some type of media—a physical connection. The media transmits information as electrical current, electromagnetic waves, light pulses, or radio waves (sometimes at ultrahigh frequencies such as microwaves). The form of transmission is based on the type of media selected. The type of media is selected, at least in part, based on the distance between the resources on the network. The form of media also dictates to some extent the network's *bandwidth*. The bandwidth is the transfer rate permissible over the media, described as some number of bits per second (bps or b/s). Modern bandwidths are on the order of millions of bits per second (Mbits/second, Mbps). Older technologies such as computer MODEMs (described later) were limited to hundreds or thousands of bits per second, such as 56 Kbps.

The most common form of network connection used today is coaxial cable and fiber optic cable. In the past, the most common connection was through twisted wire pair, as used in much of the United States telephone network. We still find twisted wire pair used extensively because it is cheap and because so much of it is already in place. Both coaxial cable and twisted wire transmit information using electromagnetic waves, whereas fiber optic cable uses light pulses. For long distances, cable is too expensive, and so radio signals are sent via radio towers, cell phone towers, microwave towers, and bounced off of satellites

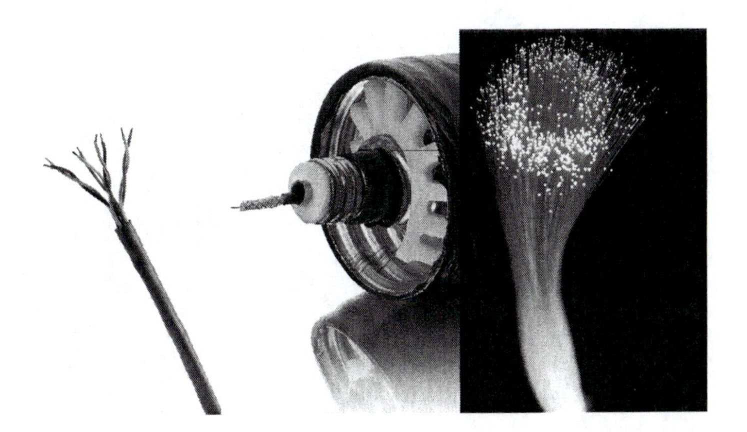

FIGURE 12.1 Forms of cable.

in orbit. See Figure 12.1 for a comparison of twisted wire pair (four in one cable in this figure), a coaxial cable, and dozens of strands of fiber optic cable.

At this physical level, the network is responsible for encoding or decoding the data into signals, modulating and demodulation signals, transmitting and receiving signals, and routing of signals. Transmission is the last step that the network performs when sending a message, and reception is the first step when receiving a message. Encoding/decoding requires translating the individual bits in the message from the form stored in the computer to the form that the network requires. The message, as stored in computer memory, consists of electrical charges (current) whereas when transmitted over fiber optic cable, the message will be a series of light pulses. If the physical media carries an analog signal rather than a digital signal, further translation is needed, known as *modulation*. When the signal is carried over the telephone line, a sequence of 1s and 0s is translated into a tone, to be broadcast. *Demodulation* translates from an analog signal to the original digital signal. Finally, routing steers the message from one network location to the next.

The form of routing depends on whether the network is *packet switched* or *circuit switched*. A circuit switched network requires that a full pathway, or circuit, be established before transmission can begin and maintained during the entire transmission. The telephone network is a circuit switched network. Because the path exists during the entire conversation, communication in either direction can occur simultaneously. In a packet switched network, a message's pathway is only established as it is sent. When a message is received at one location, if it is not the destination location, then the message is forwarded on to another location. The choice of pathway is based on network availability and amount of message traffic. Most computer networks are packet switched, with the Internet being the most well-known and commonly cited example. Figure 12.2 illustrates a network where the message is routed from location to location until it arrives at its destination. If the network was circuit switched, the route would be established in advance. If the network was packet switched, the route would be established one branch at a time. Thus, in packet switching, several messages between the same two resources could cross the network using different paths.

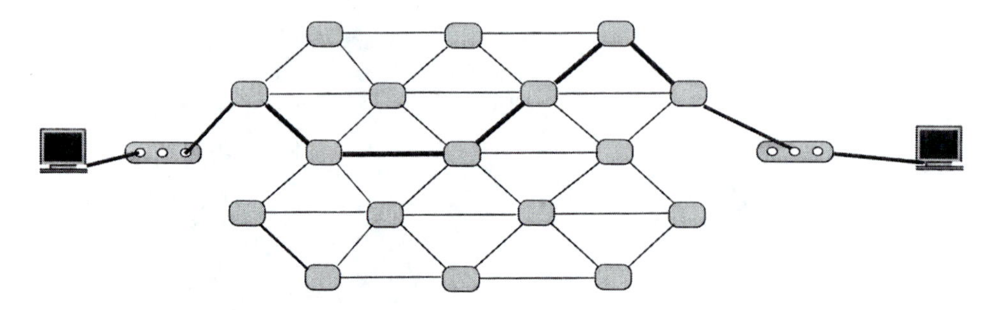

FIGURE 12.2 A path through a network.

Aside from the physical media that connects the network resources together, there are other devices used to broadcast messages from one location to another. The broadcast devices consist of *hubs, switches, routers,* and *gateways.* Collectively, these devices "glue" the network together by providing points where a network can connect to devices and to other networks.

A network hub is merely a device that connects multiple computers together. When there are multiple computers connected to a hub, a message received by the hub is made available to all computers in the hub. A destination address attached to the message indicates which computer the message is intended for, but it does not necessarily prevent other computers from picking up the message. A network switch is a more capable connection than a hub. For one, it records the local network addresses [Media Access Control (MAC) addresses] of all computers connected to the switch. A message is then only passed along to the computer that matches the destination address. Thus, the switch is able to utilize network bandwidth more efficiently.

The router is a device that connects multiple networks together. Therefore, the router is like a switch for switches. Although you can use a router to directly connect computers in a network, routers are typically used instead in specific locations in a network so that messages can be routed to other networks. Figure 12.3 is a network hub and Figure 12.4 demonstrates several network switches with devices attached. Externally, there is little to differentiate a hub from a switch from a router (other than perhaps the size). Internally, the switch has more hardware, including storage space for MAC addresses, than a hub. A router contains programmable routing tables and includes at least one input that comes from another network.

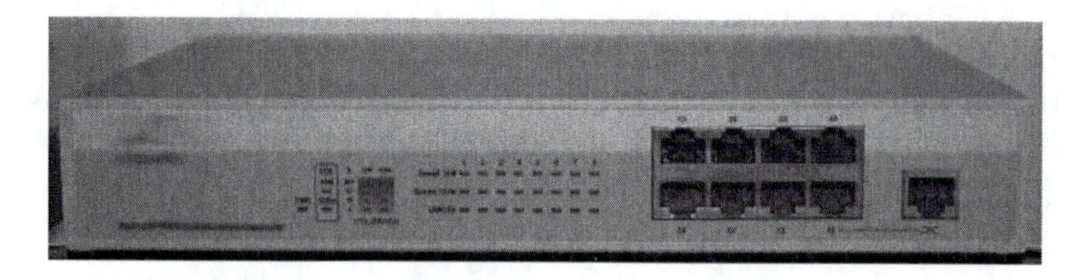

FIGURE 12.3 Network hub.

FIGURE 12.4 Network switch.

The switch and router use the message's destination address to select the line to route the incoming message to. A router has additional decision-making capabilities. For instance, if message traffic exceeds its capacity, the router may have to purposefully drop messages. Additionally, when there are multiple incoming messages, the router must select which to forward first. A routing decision is the destination network that a message is placed onto. This decision is generally left up to a routing table stored in the router's memory.

The network gateway is a router that connects networks of different types. That is, if there are two networks that use different protocols, the gateway is not only capable of routing messages from one network to the other but also of handling the differences in the messages themselves because of the different protocols (we discuss protocols in Networks at a Logical Level). Gateways appear at the edge of a network because they connect different types of networks. Gateways, unlike routers, switches, and hubs, are not core components within a network. For this reason, gateways may also serve as firewalls.

Figure 12.5 illustrates how these four broadcast devices differ. The hub merely passes any message to all devices, thus it is a shared communication (whether that was intended or not). This is shown in the top-left portion of Figure 12.5, where the first (leftmost) computer sends a message to the hub, which is then distributed to the remaining computers. On the other hand, the switch, shown in the bottom-left portion of Figure 12.5, passes a message on to only one destination device, so the communication is dedicated. The router, shown in the right side of Figure 12.5 connects local area networks together so that one hub or switch can be connected to another. In the figure, the second from the right computer in the top network is sending a message to the leftmost computer in the bottom network. In general, a router uses the message's destination address to determine which network to route the message onto. The gateway, which would also look like that shown in the right-hand side of Figure 12.5, serves the same purpose as the router except that it can convert from one protocol to another while passing a message from one network to another. The gateway then is used to connect different types of networks together, whereas the router connects networks of the same type (protocol) together.

Aside from communication over computer networks, users can communicate via the telephone system. In fact, this was the most common means of telecommunications for

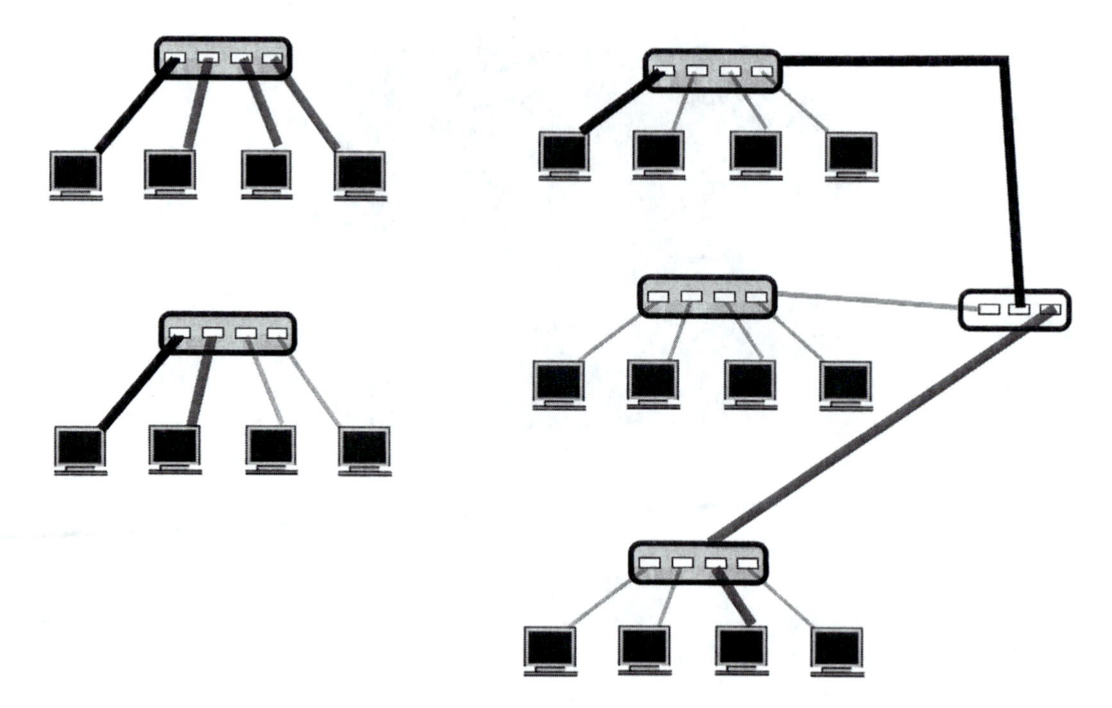

FIGURE 12.5 Hubs (upper left), switches (lower left), and routers (right).

decades. The telephone network is set up to pass signals in an analog form over twisted wire pair. Computer data are stored in a binary form, either as electrical current internally or magnetic charges stored on disk or tape. The binary data must first be converted into an analog form. This requires modulation (a digital-to-analog conversion). See Figure 12.6, where the sequence 1001 (or high, low, low, high current) must be translated into analog form, or a sound, in which the wave forms are closer together to represent 1s and further apart for 0s, which is heard by the human ear as different tones. In this form, the data can be transmitted over the telephone lines. The receiving device must convert the analog signal back into a digital form. This is demodulation (analog-to-digital conversion). A MODEM is a device that performs MOdulation and DEModulation. A user would connect the MODEM to the telephone line (for instance, a telephone wall jack), then place a phone call to a destination that also has a MODEM. Next, the telephone handset would be inserted into a MODEM cradle, as shown in Figure 12.7. The computers at both ends can now communicate with each other. After communication, the handset would be placed back on the telephone to end the phone call. Today, MODEMs are built into the computer

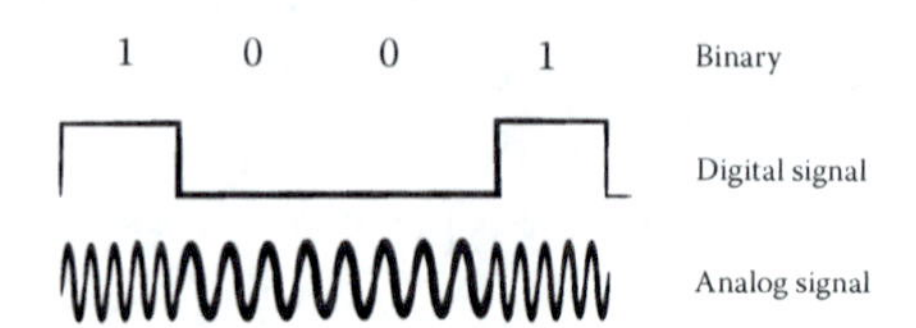

FIGURE 12.6 Signal modulation.

FIGURE 12.7 MODEM cradle.

so that you can plug your computer directly into the telephone wall jack although more commonly, a wireless card lets your computer communicate to your wireless MODEM, which can be placed at nearly any point in the house.

NETWORKS AT A LOGICAL LEVEL

There are many ways that the computer resources can be connected together in a computer network. The various layouts are collectively called network *topologies*. The topology chosen dictates the cost of the network as well as the amount of time it might take for a message to be routed to the destination. In addition, the topology can impact the reliability of the network. Figure 12.8 demonstrates a variety of different topologies. In the figure, computers and computer resources (file servers, printers, hubs, switches, routers, CD ROM towers, tape drives, etc.) are all denoted as circles. The lines that link the nodes together are the connections in the network, typically made up of some sort of cable (as

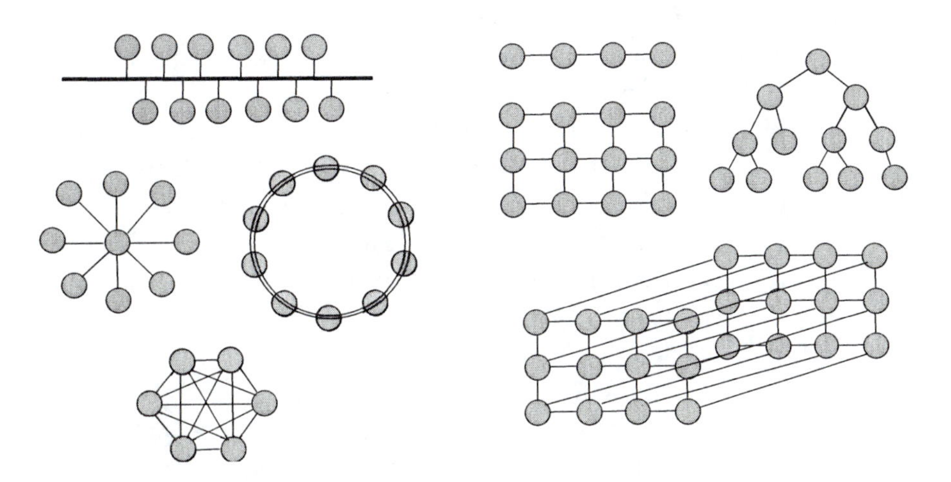

FIGURE 12.8 Various network topologies, nodes are computer resources, edges are connections.

shown in Figure 12.1, this could be twisted wire pair, coaxial cable, fiber optic cable, or some combination of these options). The various topologies are a bus topology (upper left), star topology (middle left), ring topology (to the right of the star), full mesh (lower left), and nearest neighbor topologies on the right side of the figure [one dimensional (1-D), 2-D, tree, 3-D]. Not shown in the figure is a 4-D topology known as a hypercube.

The simplest form of a network is a point-to-point network where there is a dedicated link between two resources. As a network topology, point-to-point is not often used because it limits how devices can communicate with each other. We do see point-to-point connections within a computer such as the bus connecting the ALU to registers in the CPU, or the connection between the monitor and the motherboard). We may also see point-to-point networks in the form of long distance connections such as dedicated telephone lines between two locations (e.g., the famous red phone that connects the White House and the Kremlin).

The *bus network* topology is the next simplest form and is very common. In this topology, every resource is connected to a single cable. The cable *is* the network, and it carries one message at a time. Devices connect to the bus through "T" connections. Each device would plug into a "T" connector that would plug into two cables, each leading to the next device and "T" connector. Each end of the network would be connected to a terminator. A cable will have a limited number of "T" connections however, so that the size of the network is restricted. Figure 12.9a shows a "T" connector and Figure 12.9b shows how computers connect to "T" connectors through network cards. A network card is plugged into an expansion slot on the motherboard of the computer. The "T" connector is slotted into a port so that it sticks out of the back of the system unit.

Another means to connect devices to the single network is through a hub or switch. In this case, devices plug into the single connection through the back of the hub or switch. If a network needs expansion, hubs or switches can be daisy chained together.

In the bus network, all devices listen to any communications over the single cable and ignore all messages that are not intended for them. The bus provides a dynamic network in that devices can be added and removed from the network at any time as long as there are still "T" connectors available.

The bus network is the cheapest of all network topologies. It is a reliable form of network and it does not degrade if a resource is either removed from the network or crashes. We will see that other forms of networks can degrade when a resource crashes. In spite of this, the bus network topology has a large drawback. All messages travel along the single cable and so the network's efficiency degrades as more devices are connected to it. The greater the number of devices connected to it, the greater the demand will be on the network. This, in turn, creates greater *message traffic*. So, the likelihood of two devices needing to use the network at the same time increases as we add devices to the network.

What happens if two devices try to communicate at the same time? This is known as message contention, and the result is that the messages will interfere with each other. Therefore, after contention is detected, at least one of the devices must wait while another device reattempts the communication. A strategy for handling message contention is discussed later in this chapter, developed for Ethernet technology.

(a)

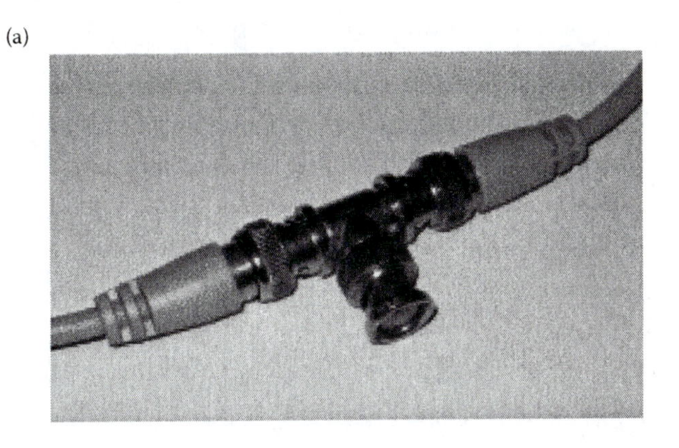

(b)

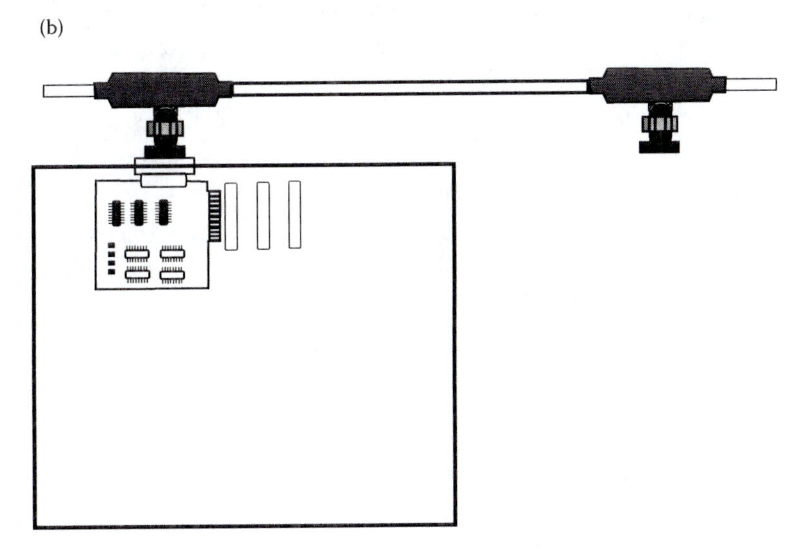

FIGURE 12.9 A "T" connection (a) and "T" connections to network cards (b).

The *star network* is in some ways the antithesis of the bus network. In the star network, all devices have a single point-to-point connection with a central server. This server is a device dedicated to act as a communication hub, routing messages from one machine to the destination. The advantages of the star network are its simplicity, ease of adding (or removing) devices, and the efficiency of message transferal. The star network does not have to contend with message traffic like the bus network (although the hub can quickly become a bottleneck if there are a lot of simultaneous messages). The number of links that it takes for any message to reach its destination is always two (or one if the message is intended for the hub device).

The star network has two detractors. First, it does require a dedicated device, making it a more expensive network than the bus. The hub of the star network may be a hub or switch as discussed in Networks at a Hardware Level. The hub of the star network may also be a server, which would make the star network more expensive. Second, although losing any single device would not degrade the network, if the hub is lost, all devices lose connectivity.

A common approach to building local area networks today is by connecting star networks together. This is done by daisy chaining hubs together. So, for instance, a hub is used to connect the resources of one network together and the hub is connected to another hub, which itself connects the resources of a second network together. This creates a larger single network in which all devices can communicate with each other. In this case, there is an additional transmission required between some of the resources if they are connected to different hubs. As hubs (as well as switches, routers, and gateways) will have a limited number of connections, this approach allows a network to easily grow in size. All that is needed is additional hubs. Figure 12.10 illustrates this concept. The "crossover cable" is used to connect the two hubs together. In this particular figure, the devices on the network are all connected to hubs by twisted wire pair.

In the *ring network*, devices have point-to-point connections with their neighbors. Communication between devices requires sending a message to a neighbor and having that neighbor forward the message along the ring until it reaches its destination. The ring is easily expanded by adding new devices between any two nodes. However, the larger the ring becomes, the greater the potential distance that a message might have to travel. This, in turn, can create lengthier transmission times. Rings are cheap, like buses, in that there is no dedicated resource that serves as a central point. A ring network could be unidirectional or bidirectional in terms of the direction that messages travel. In a unidirectional ring, if a device were to go offline, it would detach the network into two such that some devices could no longer communicate with others. Assume, for instance, that the ring topology shown in Figure 12.8 were unidirectional and all communication traveled

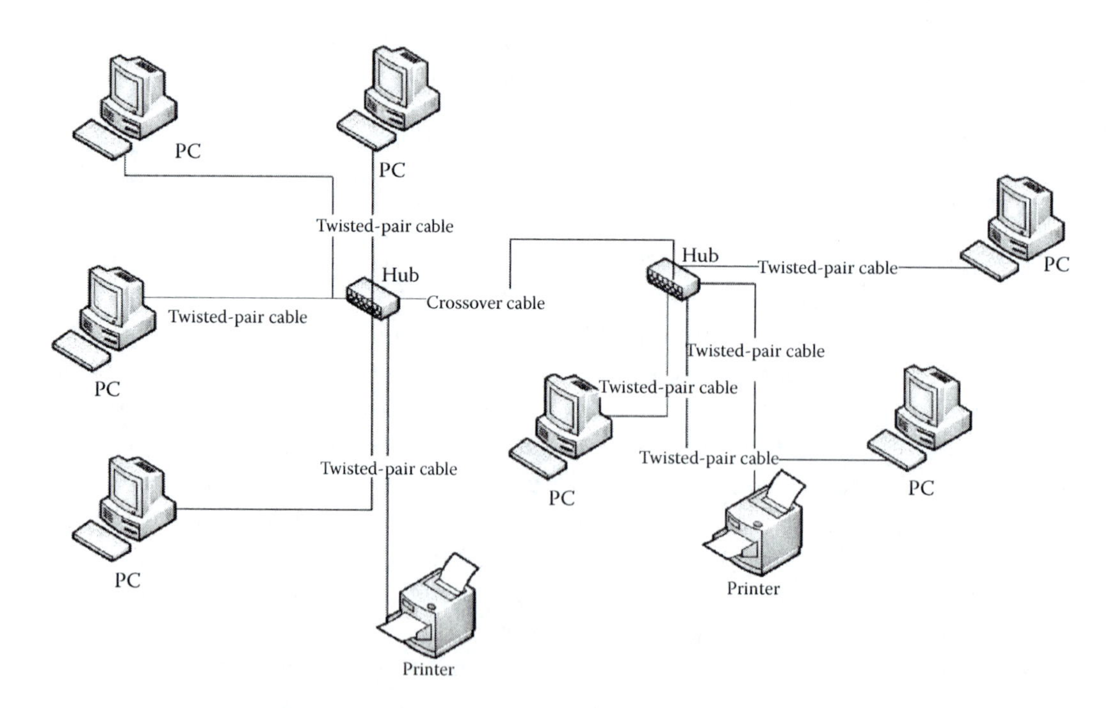

FIGURE 12.10 Two bus networks connected by hubs.

commercial release of Ethernet was produced by 3Com in 1980. Ethernet was standardized in 1982 and was a competitor of the Token Bus and Token Ring forms of networks. Although originally implemented using a bus topology over coaxial cable (with T connectors), modern Ethernet networks can use twisted wire pair and fiber optic cable with devices connecting to hubs or switches. By plugging resources directly into a hub or switch, not only is it more efficient but it also reduces installation costs and improves network management.

Ethernet technology has improved over the years, from a theoretical bandwidth of 10 Mbps to a current upper bandwidth of 100 Gbps. Ethernet introduced a number of networking concepts that are commonly found today. These include collision detection mechanisms, Ethernet repeaters, a 48-bit MAC (media access control) addressing scheme for source and destination addresses, the Ethernet frame format, error handling mechanisms, and a variety of adapters so that many different types of computers could connect to an Ethernet. Here, we briefly look at just the collision detection scheme.

The form of collision detection introduced by Ethernet is called Carrier Sense Multiple Access with Collision Detection (CSMA/CD). First, a device prepares a message to transmit. It attempts to sense whether the network media is busy. If so, the devices waits until the media is no longer busy. Then, it places its message onto the media. However, if another device is waiting, both devices could place messages on the media at the same, or nearly the same, moment. Therefore, even though it is transmitting, the device also attempts to sense if any other message is coming across the media. If so, a collision is detected. The device immediately stops transmitting the data and sends out a *jam signal*. The jam signal is used to alert other resources not to transmit at that moment. The jam signal also, since it is being transmitted over the same media that contains an actual message, overrides the message. Any receiving device will pick up the jam signal and know that the message it was receiving was corrupted. Now both sending devices wait a random amount of time before retrying their transmissions. When switches were added to Ethernet over hubs and bus connections, collisions were reduced, but CSMA/CD continued to be used.

We can also classify networks by the role that computers play within the network. Specifically, we refer to networks as either *peer-to-peer* or *client–server* networks. A *peer* means that each computer is roughly equal to every other computer. This differentiates a network from that of a client–server model. The *client* is a computer that will request information from another computer. The *server* is a computer (or device) that takes requests and responds with the requested information. The client–server network then is a network that contains one or more servers. Peer-to-peer networks are cheaper, although the need for servers forces most networks to follow the client–server model. Servers, in many cases, are more expensive computers as they typically require greater hard disk storage and faster response time than the other computers in the network. Servers often look like large system units, perhaps with multiple hard disk drives. Large file servers are often a collection of smaller units mounted into a cabinet, as shown in Figure 12.11.

There are a variety of types of servers based on the type of service desired. One type of server is the *file server*. In this case, the server's role is to send files over the network at the request of clients. The typical file server responds to requests over the LAN and services

FIGURE 12.11 Rack mounted file server.

only computers on the LAN. File servers may be used to support both application software and data files. That is, clients will store only their own operating systems. A user wishing to run software may have to load that software over the network on demand. Or, the file server may be limited to only a few, shared applications software and/or data files.

Another form of server is the web server. This is a special type of file server in that it still stores data files (web pages) and programs (scripts). However, the web server responds to client requests from anywhere on the Internet rather than just the LAN, and the requests are specifically http (hypertext transfer protocol) requests. Responses may be html files, documents stored on the file server, or web pages that were dynamically generated through server CGI (common gateway interface) scripts. The web server has several duties that the file server does not, including running server side scripts, logging requests and errors, and handling security.

Yet another form of server is the database server. It responds to client database queries with responses pulled from the database. Additionally, the server may generate reports from data obtained from the database management system. Like the file server, the database server typically responds only to local clients. Both the file server and the database server could respond to authorized clients from remote locations if the server was accessible over a wider area network. But like the web server, the duties of the database server go far beyond file transfer.

Other servers include print servers, mail servers, and ftp servers. Unlike the other servers, the print server does not return a file, but instead monitors print jobs and replies with an acknowledgment that the print job has completed, or an error message. In the case of a mail server, one submits e-mail to the server to be sent to another e-mail server. The two e-mail servers communicate with each other, and once the e-mail is received at its destination, the recipient server informs the recipient user that new e-mail has arrived. An ftp server is much like a web server or file server in that requests are for files and responses are the files.

NETWORK PROTOCOLS

A protocol is the set of rules established to govern how people behave and interact with each other. This might be considered a form of diplomacy or etiquette, or the means by which a researcher will report results to other researchers. A computer network exists at several different layers. We might think of how the network works at the hardware layer when discussing the media and the physical form that communication will take. We might discuss how the network will ensure reliability through error handling information. Or we might focus on how application software prepares messages for broadcast. In fact, each one of these steps is necessary and we must specify not only how they work, but how the steps work together. That is, we will consider network communication as a series of layers and a message must be translated from one form to another as it moves from layer to layer.

A *network protocol* provides the rules by which the layers of a network communicate with each other. This, in turn, provides rules by which different networks can communicate with each other. This also informs network programmers and administrators how to implement, manage, and maintain a network.

The most common protocol at the physical level is the Ethernet protocol, which specifies such aspects of the network as the types of cable that can be used, the collision processing mechanism (CSMA/CD), and the allowable types of topologies: bus, star, tree. Other existing popular protocols are LocalTalk, developed by Apple computers, Token Ring (developed by IBM), and ATM (asynchronous transfer mode), which directly supports audio and video transfer.

Perhaps the most commonly used network protocol is TCP/IP (Transmission Control Protocol/Internet Protocol),* which is a requirement of all computers that communicate over the Internet. So, although computers may use other network protocols, they must also run TCP/IP. As an alternative, the Open Systems Interconnection (OSI) model was a *proposed* standard for all network communication. So, although TCP/IP is a concrete protocol, OSI is often used as a target for new network developers. Today, both TCP/IP and OSI are commonly cited models for network protocols, so we will examine them both here.

The OSI model consists of seven layers, as shown in Table 12.1. Each layer is numbered, with the topmost layer numbered as 7 and the bottommost layer as 1. The top four layers are known as the *host* layers because the activities of these levels take place on a host computer. These layers package together the message to transmit, or receive the message and analyze it. The bottom three layers are known as the *media* layers because it is at these layers that messages are addressed, routed, and physically transmitted. Although OSI is not a specific protocol, it has been used to create network protocols, including Common Management Information Protocol, X.400 electronic mail exchange, X.500 directory services, and the IS–IS (Intermediate System to Intermediate System) routing protocol.

Layer 1, the lowest layer, is the physical layer. This layer dictates how the device that wishes to communicate (e.g., a computer or a hub) will carry out the communication over the transmission medium (coaxial cable, fiber optical cable, etc.). This layer requires such details as the voltage required for transmission, how to modulate the signal (if needed),

* TCP/IP is really a collection of protocols and is sometimes referred to as a protocol stack.

TABLE 12.1 OSI Model Layers

Data Unit	Layer Number and Name	Function
Data	7. Application	User interaction with application software
	6. Presentation	Data representation, encryption/decryption
	5. Session	Host-level communication, session management
Segments	4. Transport	Reliability and flow control
Packet/Datagram	3. Network	Logical addressing, routing
Frame	2. Data link	Addressing
Bit	1. Physical	Physical media, signal transmission in binary

how to establish and how to terminate the connection to the communication medium. When establishing a connection, the layer requires the capability of detecting message traffic (contention) and a means of resolution when there is traffic. When transmitting data, this layer receives the data from layer 2 and must convert it into a form suitable for transmission. When receiving data, this layer obtains the data from the transmission medium (electrical current, sound waves, light pulses) and converts the data into a binary format to be shared with layer 2. Thus, this layer deals with the message as a sequence of bits. There are numerous implementations of layer 1 including IEEE 802.3, IEEE 802.11, Bluetooth, USB, and hubs.

WHAT IS 802?

You may have seen notations such as IEEE 802.xx. What does this mean? IEEE is the Institute of Electrical and Electronics Engineers. Among the various efforts of the organization are a number of standards that they have established. Although standards are by no means laws or requirements, most implementers attempt to meet the established standards to guarantee that their efforts will be used and usable.

Most telecommunications standards are put forth by the IEEE and they are labeled as IEEE 802.xx, where the 802 stands for "February 1980" for the first month that IEEE met to discuss telecommunications standards (it was also the first freely available number).

What are some of the standards that go under IEEE 802?

- IEEE 802.3—Ethernet
- IEEE 802.7—Broadband LAN
- IEEE 802.10—LAN security
- IEEE 802.11—Wireless LAN (a number of variations have been produced, each given a letter such as 802.11 b, 802.11 g, and 802.11 n)
- IEEE 802.15.1—Bluetooth certification
- IEEE 802.22—Wireless area network
- IEEE 803.23—Emergency services workgroup

Aside from IEEE 802, the IEEE has established a number of other standards ranging from floating point representations in computers (IEEE 754) to the POSIX standard (portable operating system interface) that Unix and Linux systems meet (IEEE 1003) to standards in the field of software engineering (IEEE 610).

Layer 2 is the data link layer. If a message is intended for a device on the same network (e.g., one that shares the same hub or switch), communication will occur at this level and will not require layer 1. At this level, data are formed into units called *frames*. The frame must indicate frame synchronization, which is a sequence of bits at the start of the message. This layer contains two sublayers. The upper sublayer is the Logical Link Control (LLC) sublayer. This layer provides *multiplexing*, which is the ability to carry of several overlapping messages at a time. Through multiplexing, it is possible that multiple messages are of different network protocols, coexisting at the same time, sharing the same network media. The lower sublayer is the MAC. MAC addresses are used to denote a device's location within a given network. The MAC sublayer allows devices in the same network to communicate together by using only their MAC addresses. This sublayer serves as an interface between the LLC sublayer and the physical layer. Layer 2 implementations include IEEE 802.2, IEEE 802.3, PPP (Point-to-Point Protocol), X-25 packet switch exchange, and ATM. Ethernet is an implementation for both layers 1 and 2.

Layer 3 is the network layer. At this layer, the physical characteristics of the network are not a concern. Instead, this layer views data as variable length sequences with host and destination addresses. This layer handles *routing* operations. Messages that arrive at a device whose job is to route the message onward will examine the destination address with entries in its own routing table to determine which network to place the message onto. Therefore, routers operate at this level. At this layer, message components are formed into individual *packets*. The message being transmitted by the sending device will likely consist of multiple packets, perhaps dozens or hundreds depending on the message's size. Layer 3 implementations include IP (see TCP/IP below), AppleTalk, IPX (Internet Packet Exchange), ICMP (Internet Control Message Protocol), and ARP (Address Resolution Protocol).

Layer 4 is the transport layer. Its primary responsibility is to provide transparency between the upper levels of the protocol and the physical transfer of individual packets. At this level, messages are represented as segments to be divided into smaller units (e.g., packets). Among the services provided at this layer are reliability and control flow. For reliability, this layer must ensure that the packet is received, and received with no error. Details for handling reliability are described below. Control flow occurs when the two devices communicating with each other over the media are communicating at different rates (speeds), such as when two computers have different MODEM speeds.

To handle reliability, layer 4 must ensure that packets lost en route are replaced. There are a variety of mechanisms for handling this. A simple approach is to stamp every packet by its sequence number within the overall message, such as 4 of 7. The receiving device expects seven packets to arrive. If all packets arrive other than number 4, the sending device must resend it. Additionally, every packet will contain error detection information, such as a *checksum*. The checksum is a computation based on the binary values that makes up the message.

One simple computation for a checksum is to add up the number of 1 bits in the message. For instance, if a binary message consists of 256 bits, and 104 of those are 1s and the remaining 152 are 0s, then the checksum would be 104. However, typically, the checksum should be fixed in size. To accomplish this, a checksum function might add up the number of 1 bits and then divide this by a preselected value. The checksum then becomes the remainder of the division (this is the mod, or modulo, operator). The idea of using mod is known as a *hash function*. There are a number of different checksum algorithms including fingerprints, randomization functions and cryptographic functions. In any event, upon receipt of a packet, layer 4 determines if any of the data in the packet is erroneous by comparing the checksum value with the data in the packet. If an error is detected, layer 4 sends out a request so that the packet can be resent. Layer 4 implementations include TCP, UDP (both of these are discussed along with TCP/IP below), and SCTP (Stream Control Transmission Protocol).

Layer 5 is the session layer. This layer maintains a connection between two devices. When two devices communicate, they first establish a session. The session remains open until the devices terminate the connection. In between establishing and terminating the session, the session must be maintained. Additionally, a session that is prematurely terminated can be restored at this layer. It is this layer that handles these tasks (establishing, maintaining, restoring, terminating). Layer 5 implementations include NetBIOS, SAP (Session Announcement Protocol), PPTP (Point-to-Point Tunneling Protocol), and SOCKS (SOCKet Secure).

Layer 6 is the presentation layer. This layer is responsible for translating messages from the given application, which is generating the message, into the form of syntax required by the lower layers. Because of this, layer 6 is sometimes referred to as the syntax layer. It is at this layer that the original representation of the message is converted into a uniform representation. For example, in the C programming language, strings are terminated by a special character, \0. At this layer, any \0 characters can be stripped from the message. Another example is for hierarchically structured data, such as data in XML notation. Such data must be converted into a *flat* format. Encryption (for outgoing messages) and decryption (for incoming messages) takes place at this level. Layer 6 implementations include SSL (Secure Sockets Layer), TLS (Transport Layer Security), and MIME (Multipurpose Internet Media Extensions).

Layer 7, the highest layer, is the application layer. At this level, the end user or application program creates the message to be transmitted, or at this level, the application program presents a received message to the end user. This layer includes various network communication programs such as telnet; ftp; electronic mail protocols such as POP, SMTP, and IMAP; and network support services such as domain name system (DNS).

The OSI Model works as follows. A message is created through some application software at layer 7. This initial message is considered the data to be transmitted. Layer 7 affixes a header to the front of the message and maps the message to layer 6. At layer 6, an appropriate header is added to the message. The message, now consisting of the original data and two headers, is mapped into layer 5, and another header is affixed to the front of the message. The top three layers operate on the message as a whole rather than the segmented units operated on starting at layer 4. At layer 4, the message is divided up into packets and then into smaller units until, at layer 1, the message is treated as individual bits. Layers 4 and 3 also affix header information and layer 2 adds both header and footer information. Finally, at layer 1, the message is transmitted. A received message is similarly mapped upward through each layer, removing the header (and footer) as needed until the message arrives at the topmost layer and delivered to the application software. Figure 12.12 illustrates the mapping process in OSI.

TCP/IP is the other, commonly used, transmission protocol. TCP/IP comprises two separate protocols that were united early in the development of the Internet. TCP handles reliability and ordered delivery of packets. It operates at higher levels of the protocol stack, saving the details for how packets are transmitted to the IP portion. Although TCP is used extensively for Internet communication, some applications use User Datagram Protocol (UDP) in its place. Both TCP and UDP sit on top of IP, the Internet Protocol. Here, we briefly examine the layers of TCP/IP and compare them to the layers found in OSI. Unlike OSI, the layers of TCP/IP are not as proscriptive. That is, they describe what takes place at those layers, but they do not offer details on how those actions should take place. Therefore, there are many different ways to implement the TCP/IP stack.

The lowest layer of IP is called the Link Layer. The Link Layer corresponds directly to the lowest two levels of OSI. This layer performs all of the services in the physical and data link layers. The second layer of IP is the Internet Layer. This layer is responsible for sending packets across one or more networks. At this level, addressing and routing take place. It is at this level that IP addresses are utilized. TCP/IP layer 2 is similar to OSI layer 3. However, OSI permits communication between devices on the same network in its layer 2, whereas all routing takes place at layer 2 of TCP/IP. There are other substantial differences between OSI Layer 3 and TCP/IP layer 2, but these are not covered here.

It should be noted that the addressing and routing referred to here are from IP version 4 (IPv4), the current and more popular version of IP. However, because of a limitation on the number of IP addresses, and because of an interest in improving the Internet, IP version 6 has become available. IPv6 will make significant differences in Internet usage as demand increases over the next decade. One of the biggest differences between IPv4 and IPv6 is the form of addresses. In IPv4, machine addresses consist of 32 bits divided up into 4 octets. Each octet is 8 bits (1 byte) and thus can store a number between 0 and 255. The octets of the IP address are separated by periods. For instance, an IP address might be 127.31.49.6.

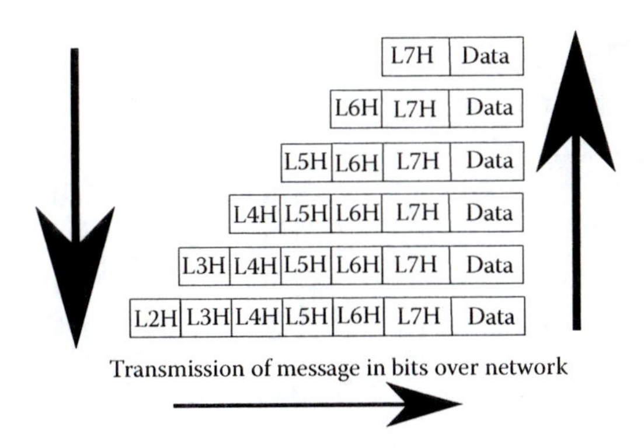

FIGURE 12.12 OSI model mapping layer by layer.

There are three classes of networks with respect to IP addressing (the practice of assigning classes to networks was discontinued in 1993 although network addresses are often still assigned this way). The class of network dictates which bits of the four octets specify the network on which the computer is housed and the remaining bits denote the machine's address on its network. Class A networks only use the first 7 bits of the first octet to denote the network address. This limits the number of class A networks to 128, but leaves 25 bits for addresses within the network, and thus class A networks can have as many as 16 M (more than 16 million) internal addresses. Class B networks are denoted with first octet addresses of 128–191 and use the first two octets to identify the network. This leaves 16 bits for network addresses, or 65,536. Class C networks, the most common, use the first three octets for the address allowing for millions of class C networks. However, the class C network then has only a single octet for each device's address on the network. Therefore, a class C network can only contain up to 256 addressable devices. See Table 12.2 for details. Notice that two classes have reserved first octet address values but have not been used to date.

IPv4 addresses are 32 bits long. This provides for 2^{32} different (distinct) addresses, or approximately 4 billion unique IP addresses. Although this looks like a large number, the limitation has created a problem in that we have reached this limit because of handheld devices (e.g., cell phones) connecting to the Internet. Furthermore, many of the IP addresses that would normally be available have not been utilized. For instance, class A networks may have enough addresses for more than 16 million internal devices, but this does not mean that every class A network uses all of the available addresses. On the other hand, an organization might be granted a class C network address but may require more than 256 distinct addresses.

In IPv6, addresses are 128 bits in length and often displayed using hexadecimal notation. The advantage of the longer address is that it provides as many as 2^{128} distinct IP addresses. This gives us plenty of addresses for a very, very long time, even if we provide a different address for every processor on the planet (including those in cell phones, sensors, and other electronic devices). The 128-bit address is usually composed of two parts, a 64-bit network address prefix used for routing across the Internet, and a 64-bit interface identifier to denote the host within the network. An example address might be 1234:5678:90ab:cdef:1234:5678:90ab:cdef. Notice the use of hexadecimal in the address rather than binary or decimal as we typically view IPv4 addresses.

TABLE 12.2 Internet Network Classes

Class	Octets for Network Address	Octets for Address in Network	Number of Addresses for Network	Legal Addresses (1st Octet)	Comments
A	1	3	16,777,216	0–127	Many addresses have gone unused
B	2	2	65,536	128–191	
C	3	1	256	192–223	
D	Not defined	Not defined	Not defined	224–239	Multicast addresses only
E	Not defined	Not defined	Not defined	240–255	Reserved (future use)

TCP, the upper layers of TCP/IP, also consists of two layers: the Transport Layer and the Application Layer. The Transport Layer is similar to OSI's Transport Layer (OSI layer 4). However, TCP allows for two different forms of data streams, those using TCP and those using UDP.

UDP does not provide the reliable form of communication that TCP does. When a UDP packet is dropped during transmission, there is no effort to resend it. Although this sounds like a negative, it can be advantageous when real-time communication is more important than data reliability. As an example, real-time audio transmission might use UDP instead of TCP. The rationale behind this is that a recipient would not want the audio signal interrupted while a packet is resent. The omission of one (or even a few packets) would almost certainly not interfere with the recipient's ability to understand the audio signal.

Figure 12.13 illustrates the contents of a TCP packet versus a UDP packet. The UDP packet is far more concise because it lacks such information as a sequence number, acknowledgment information, control flags to specify how the packet should be handled, and a data offset, indicating the size of the data field. The urgent pointer is optional in the TCP packet but can be useful for denoting the last *urgent* data byte in the data field.

The highest layer of TCP/IP is the Application Layer, which is roughly synonymous with OSI's top three layers. In the case of the Session Layer of OSI, where a connection

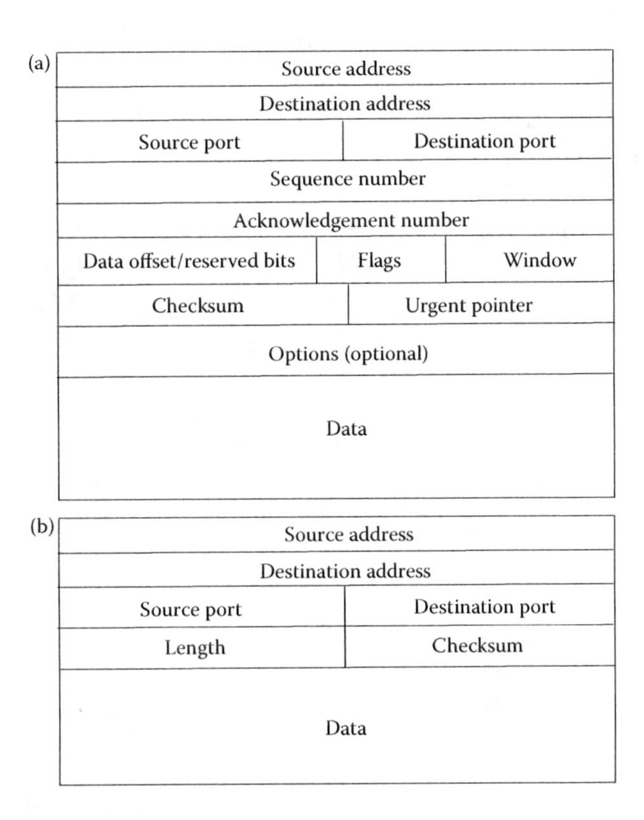

FIGURE 12.13 Standard TCP packet (a) and UDP packet (b).

is retained until termination, there is nothing precisely equivalent in TCP/IP. However, similar capabilities are handled in TCP/IP's Transport Layer.

Putting the four Layers of TCP/IP together, we see much of the same functionality as OSI. As with OSI, the message being transmitted or received is converted between layers. Figure 12.14 provides an illustration of how a packet is converted layer by layer in TCP/IP, where each lower layer adds its own header to the previous layer, and the Link Layer adds a footer as well.

Recall that both OSI and TCP/IP are protocols. They dictate several aspects of network communication although they leave other details to implementers. Both require that the protocols be used to translate a message from one format (as generated by some application software) to a format that can be transmitted, and then translated back from the transmitted form to a format that the destination application software can handle. Refer back to Figures 12.12 and 12.14 for the OSI and TCP/IP mappings, respectively. As you might notice, mappings take place from top down and from bottom up. As the message moves down the protocol, more and more details are added such as headers, error checking, and addressing information. As a message moves up the protocol, those added details are stripped off of the message.

One other aspect of note for TCP/IP is that of *network handshaking*. In fact, whenever two computers communicate over network, before any communication takes place, the two devices must perform a network handshake. In essence, this means that the first machine contacts the second with a message that indicates "I wish to communicate with you" and then it waits. When the second machine is ready, it responds with "I am ready". Once the handshake is done, the machines can freely communicate until communications are terminated at one end. TCP/IP introduced the notion of a three-way handshake. Here, the first machine sends a synchronization packet (SYN), to the second machine. The second machine, when available, responds back with a synchronization and acknowledgment packet (SYN/ACK). Finally, the first machine responds to the second with an acknowledgment packet of its own (ACK). At this point, the two machines are synchronized and ready to communicate, until communication is terminated. Referring back to Figure 12.13, you can see an "acknowledgement number" in the TCP packet. This is used to support the three-way handshake.

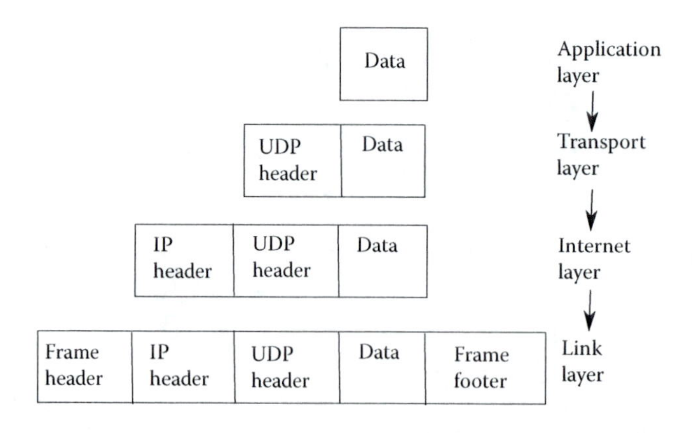

FIGURE 12.14 TCP/IP packet formation.

One concept not discussed in our examination of TCP/IP is that of a *port*. TCP and UDP messages not only require destination addresses but also port addresses. The port is merely an indication of the intended destination software for the message. Thus, when a message is received by a destination machine, the operating system can determine how to handle the message by the port address provided. TCP/IP dictates how that message should be decomposed, but not the application software that should handle it. Ports are dedicated for many different purposes. For instance, the common port used for the SMPT e-mail is 25, and the common ports for http used for web browsing are 80 and 8080. There are port numbers dedicated to ftp (20), telnet (23), ssh (22), and https (443).

Port addresses not only serve the OS, but can also be used by protection software. For instance, a firewall may disallow certain port addresses such as 23 (since telnet is not secure). Similarly a web server may ignore messages that come in from ports other than those expected (e.g., 80, 8080, and 443). Although we limit our discussion of ports here, you will no doubt gain experience with them as you further explore IT.

We finish off this section with one other TCP/IP-related topic: *network address translation* (NAT). NAT is the conversion of one IP address to another. Basic NAT simply uses one IP address as an alias for another. That is, there is a one-to-one mapping of one address to another. This form of translation might be useful if two networks with incompatible network addresses are attempting to communicate with each other. In basic NAT, the translation process must modify the message's IP address, IP header, and any checksum of the message that was computed by both the message content and the header information.

However, it is far more common to use NAT to hide an entire series of IP addresses. For instance, an organization might not have enough IP addresses for all of its internal devices. In this case, only some of the machines in the LAN are given actual IP addresses that can be "seen" from outside. The other devices are hidden from view externally. When a message arrives, the destination address must be converted into an internal, private IP address. This form of NAT is called many-to-one, or NAT overload or IP masquerading. A server must be able to determine which private IP address should be used in place of the external address. The translation to the private IP address requires additional modifications because the packet must be routed properly to the correct internal router. Among the modifications, it is possible or even likely that port addresses will be modified as well as IP addresses. The advantages of many-to-one NAT are that they offer a degree of protection because of the privacy of the internal addresses, and permit LANs to have additional IP addresses. The main disadvantage of many-to-one NAT is the complexity it introduces at the server that handles the Internet connection.

NETWORK SOFTWARE

In this section, we look at common forms of network software. These are programs that we might use to communicate between devices over a network. These are all used in TCP/IP intranets, so these are not necessarily limited to just Internet usage but can also provide service within an LAN.

Telnet is a program that allows an individual to log into another computer over the network. The person performing the telnet operation needs to have an account on the remote (host) computer. A variation of telnet is called rlogin, available in Linux. With rlogin, the user does not have to actually log in to the destination computer because the user's account information (name, password) will be the same on all networks computers. Linux actually has several "r-utilities", where "r" stands for "remote." Aside from rlogin, there is also rfile, rsh, rstatus, and rwho to name a few. The rfile program allows you to perform a variety of file management operations remotely. The rsh program opens a shell to another machine remotely. The rstatus program provides basic receiver status information. The rwho program performs the Linux who command on the remote machine, showing you who is logged in. The idea behind the r-utilities is that they run on a LAN of computers that share the same log in server. The r-utilities do not require that you log in to the remote computer as long as you have logged into your local computer. However, the r-utilities must be set up explicitly for use by a system administrator.

Telnet and rlogin share a problem; neither program is *secure*. Messages passed between local and remote computers are in clear text. This includes, in telnet's case, any passwords sent from the local to remote computer. It is therefore possible that messages can be intercepted by a third party who might then view your password or other sensitive material. The ssh utility is a Linux program that performs encryption and decryption automatically on any message sent, thus providing a secure form of communication, unlike telnet and rlogin. Although ssh is part of Linux, it is also available via the putty program for Windows.

MORRIS' INTERNET WORM

On November 2, 1988, a graduate student at Cornell University, Robert Morris, unleashed a program on the Internet intended to demonstrate security holes in Unix. Now known as Morris' Internet Worm, the program infiltrated about 6000 Unix machines over the course of 3 days, bringing a sizable portion of the Internet to a crawl, or down entirely (it was estimated that there were about 60,000 total computers on the Internet, so the worm impacted 10% of the Internet). The worm specifically attempted to gain entrance to Unix systems through four approaches:

- Guessing weak passwords (including no password at all)
- Exploiting known flaws in the Unix sendmail program such as a buffer overflow
- Exploiting known flaws in the Unix finger command
- Using r-utilities to propagate to other network machines

Once the worm reached a machine, it would upload its full version of itself and make copies to send to other network machines, machines listed in the current machine's /etc/host table, and other machines through the same four approaches as listed above.

In essence, the worm acted as a denial of service program in that it made so many copies of itself that the infected computers could do little else than run copies of the worm.

For his efforts, Morris paid $10,000 in fines and served 400 hours of community service. The cost of the damage of the worm was estimated at between $100,000 and $10 million. The worm prompted DARPA (Defense Advanced Research Projects Agency) to establish a group that would deal with Internet emergencies and threats in the future.

Ping is a program that will generate and send packets repeatedly to the destination device, outputting acknowledgment messages. This lets you know (1) if your machine can reach the network and (2) if the remote machine is accessible. You can also use ping to see how long the message and response take to determine if you have some abnormal network latency.

A similar program to ping is traceroute. This program outputs the route taken between your computer and the destination computer. You may use traceroute to see how a message (packet) makes it across the Internet, how much time it takes, and if there are routers along the way that are causing problems.

Both ping and traceroute are often considered insecure programs. Clever hackers might use either or both to investigate the computers on a LAN. In this way, the hackers can obtain IP addresses that might not otherwise be public knowledge, and use these IP addresses to stage an attack. Therefore, it is possible that the system or network administrator might disable both of these programs.

HTTP is the basis for web browser communication. An http request sent to a web server results in a page being returned and displayed in the browser. However, there may be times when you want to download the file quickly without displaying it in a browser. You can accomplish this from the command line in Linux using the program wget. In wget, you specify the complete URL (server name, directory path on the server, file name). The file is then downloaded to the current directory.

You can also download files directly using an FTP program. However, in FTP, a session is created between you and the server, which remains open until either you terminate the session, or the session times out. Although most FTP programs are graphical, there is also a command-line version. Graphical versions of ftp include Filezilla, Win-Ftp, and Win-SCP (which includes a secure form of FTP, similar to how ssh is a secure form of telnet). In using ftp to access files on a remote computer, you must either have an account on the remote computer or you must log in as an *anonymous user*. Anonymous users often only have access to public files on the ftp server.

One large difference between http and ftp is the URL. You commonly submit an http request by clicking on a link in an html document in a web browser. The link encodes the location of the document that you want to retrieve so that you do not have to memorize such information. In ftp, however, you must know the server's address, and then you must either know the directory path and the file name, or you have to search around on the server to locate what you are interested in. In both http and ftp, the server's address should be the machine's IP address. However, since IP addresses are hard to memorize, we have invented a shortcut whereby you can use the machine's IP alias instead. The alias is an English-like description (or abbreviation) of the destination machine. For instance, we have such aliases as www.google.com for businesses, www.nku.edu for universities and www.usa.gov for government sites.

Routers can only handle IP addresses, not IP aliases, so we have to provide a form of translation from alias to address. This is handled by DNSs. A DNS server is merely a computer running a DNS program and data files that provide mapping information. Additionally, the DNS has pointers to one or more DNSs on the Internet so that, if the local DNS does not have the mapping information, it can pass the request on to another DNS. DNSs are arranged hierarchically across the Internet so that a request might be passed along from one DNS to another to another to another before the mapping can finally be obtained.

The nslookup program provides a translation from IP alias to IP address for you. This is convenient if you want to make sure that a particular alias has been established in a DNS table. It is also useful if you want to know the physical IP address of a computer, knowing only its alias. If you want to learn your own computer's IP address, use ifconfig (in Linux) or ipconfig (in Windows).

Three other Linux network commands are arp, used to determine Ethernet connectivity and network card response; netstat, used to determine the status of network connectivity and socket status; and route, which lists the routing tables (including such things as network gateway IP address and local network IP address and masks). Also of note in Linux is the network service command. You can start or stop the network service or check its status. The command looks like this:

```
/sbin/service network command
```

where *command* is down, start, restart, status. The network service is a daemon that must be running for you to be able to use the network (send out messages or receive messages).

Linux has many network-related files worth exploring. A few are listed below.

- /etc/hosts: IP alias to address mapping for machines that your machine will often communicate with. By placing the IP addresses in this file, your machine does not have to communicate with the DNS first, saving time.
- /etc/resolv.conf: stores the address of your DNS. There may be several entries.
- /etc/sysconfig/network-scripts/: stores configuration files that are run during the Linux system initialization process to set up network connections.
- /etc/xinetd.d/: contains services that rely on the network daemon, xinetd.
- /etc/hosts.allow and/etc/hosts.deny: to permit or restrict access to your computer.
- /etc/hosts.equiv: contains IP addresses of "trusted" machines so that r-utilities will work for all computers defined as equiv (equivalent).

Placing your computer on a network invites hackers to attack your computer. This is perhaps one of the most vulnerable aspects to modern-day computing. The ultimate security is to not connect your computer to a network. Naturally, this drastically limits your computer's capabilities and your own. Therefore, network security software is available. Such software is not just useful these days but essential for anyone who wishes to access the Internet safely. Two of the most common network security software are *firewalls* and antiviral software.

A firewall can either be software or hardware (in the latter case, it is a dedicated server that runs firewall software). The firewall software contains a list of rules that describe the types of messages that should either be permitted to make it through the firewall and to your computer, and those that should be blocked. Rules can be based on the port number of the message, the type of message, the IP address of the originator of the message, whether the message is in TCP or UDP (or other) format, and protocol (e.g., http, https, ftp). We briefly examined the Linux firewall in Chapter 11.

Antiviral software attempts to identify if a file contains a virus, or more generally, some form of malware. Antiviral software can be run on demand, or you can set it up so that all incoming messages are scanned. Unfortunately, antiviral software will not necessarily catch every piece of malware. As programmers become more ingenious in how to attack an unsuspecting user, antiviral software must become more sophisticated. But the antiviral software always lags behind the innovations of the attackers. Furthermore, you must be diligent in updating your antiviral software often (new releases may come out daily or weekly).

Networks may have their own forms of security, such as intrusion detection and intrusion prevention software. There is no generic solution to preventing intrusions at this point as all operating systems have security holes to be exploited. In fact, hackers often attempt to probe a system's defenses before they mount an attack. Some of the openings that hackers may find include leaving gaps in the firewall (e.g., ports that are not inspected), leaving wireless connections unprotected, foregoing encryption, not requiring strong passwords, a lack of physical security around network access points, and permitting operations such as ping and traceroute.

Among the more common forms of attacks found today are denial of service attacks on web servers, IP spoofing to intercept messages, ARP poisoning to change IP addresses in a router to be of the attacker's machine rather than a local machine, buffer overflow that allows a hacker to invade computer memory with their own code, and SQL injections that permit a hacker to invade a database with their own SQL commands. Although these are all known forms of attacks, it does not necessarily mean that a system is protected against them. And as these attacks are known, newer attacks are being thought of. Any organization that values data integrity and the functionality of its computer systems will need operating system level and network security. It is common for mid-sized and large organizations to hire network security administrators in addition to system administrators. We examine data integrity and security in Chapter 15.

THE INTERNET

IP addresses are often difficult to remember, so we allow users to reference servers by aliases, such as www.google.com. In order to translate from an alias to the actual IP address, you need access to a DNS server. Every network either has a DNS or knows where to find a DNS. Most of this is hidden from you as a user, but if you are a network administrator, you will have to either set up a DNS or know where to find one. There are further aliases to restrict what a user might have to remember. For instance, it is common to further alias a web server so that the user does not require the www portion, and therefore although www.google.com is the alias of the Google search engine web servers, so is google.com. The DNS is a computer that has a table that contains IP addresses and aliases. When provided with an Internet alias, the DNS looks the entry up in its own DNS table and returns the corresponding IP address. The IP address is then sent back to your computer.

Computer networks have existed for decades. In the past, most computer networks were local area networks (LAN), and they were isolated—you could communicate within the network, but not outside of the network. The Internet changed all of this. To communicate over the Internet, a machine needs to run the TCP/IP protocol. With the popularity of the Internet, all computers today run TCP/IP. A computer connected to a LAN can also communicate over the Internet if the LAN has a connection to the Internet, and most do. We have blurred the lines of where one network stops and the next starts. A network connecting computers in a single room is connected by router to a network connecting computers in another room. The network that exists on one floor is connected to the network on another floor. The networks of one building are connected to the networks in other buildings. A LAN might be considered any one of these networks, or the network that covers the entire campus. With VPNs, the network can expand to include those logged in remotely from other sites. The network for the organization is most likely connected to the Internet as well. With Internet Service Providers (ISPs), computers in people's homes also connect to the Internet making the Internet a network of networks. So you might think of the Internet as an extension to the LAN of your current computer. There are three technologies "gluing" these computers together:

1. Packet switching. The telephone network uses circuit switching—when you dial a number, a pathway is set up between the source phone and the destination phone and that pathway remains fixed and established until one of the phones hangs up. Cell phone networks, carrying voice (but not data), are similar although the circuit can change as you move because you are routed to the nearest available cell phone tower. Packet switching, on the other hand, breaks the communication down into small packets of data and each is sent across the network separately. Each packet could potentially take a different pathway to reach the destination. Packet switching was pioneered in the 1960s when the Internet was first being constructed.

2. Routers. The router is the device that permits packet switching so that a message can be routed to the next link in the network "on the fly." The router is a device that receives a message, examines its destination IP address, and sends the message on to the next location in the network, another router. At the destination network, routers route messages until they reach a LAN, and finally the LANs router or switch sends the message to that one destination machine.

3. IP addresses. Part of the TCP/IP protocol is addressing rules. An IP address consists of four numbers, where each number is between 0 and 255. For instance, 10.11.241.105 is an IP address. Each number is actually stored in 1 byte of memory (1 byte is 8 bits, and 8 bits can store any number from 0 to 255). This makes an IP address 4 bytes

or 32 bits in size. Since IP addresses are difficult to remember, we often reference machines by an IP alias, an English-like name (for instance, www.google.com). In order to translate from an IP alias to an IP address, we use a DNS. As mentioned earlier, because of the limitation now being felt by the 32-bit addresses, IPv6 uses 128 bits for IP addresses.

The three classes of IP addresses currently in use dictate how to interpret the 4 octets of the address. In class A IP addresses, the first octet is the network ID and the remainder of the octets constitute the computer number. Class B addresses split the address into network ID for the first two octets and the machine number in the last two octets. Class C addresses use the first three octets for the network address leaving one octet (256) for the machine number. The class identification of the network part of the address can be subdivided even further through a process called subnetting.

You may have seen the term *subnet mask*. What is a mask (or netmask)? We often compute values in a computer using the AND or the OR operation on binary bits. For instance, the two bytes 10101010 and 00001111 when ANDed together yield 00001010 and when ORed together yield 10101111. These operations work bit-wise, that is, we perform the AND or OR operation on the first bit of both numbers to yield a new first bit (1 AND 0 is 0, 1 OR 0 is 1). We do this for all eight pairs of bits. The AND operation returns a 1 if both bits are a 1 (0 otherwise) and the OR operation returns a 1 if either bit are a 1 (0 otherwise).

Now, why do we use these? Imagine that you have the following IP address in binary:

```
00001010.00001011.11110001.01101001 (10.11.241.105)
```

If we AND this with 11111111.11111111.11111111.00000000, what do we get? We get

```
00001010.00001011.11110001.00000000
```

What does this value represent? We started with 10.11.241.105 and the result of the AND operation gives us 10.11.241.0. For a class C network, what we have obtained is the IP address of the network that houses our computer. On the other hand, if we AND the number with 00000000.00000000.00000000.11111111, we get

```
00000000.00000000.00000000.01101001
```

This value, 0.0.0.105, is our computer's specific address on the network.

The bit patterns 00000000 and 11111111 are the integer numbers 0 and 255, respectively. We use the values 255.255.255.0 and 255.0.0.0 as netmasks. Two combinations that we would not use are 0.0.0.0 and 255.255.255.255 because such netmasks would either return 0.0.0.0 or the entire original IP address, so neither would provide us with useful information. The subnet mask is used to easily obtain the network number and the subnet number is used to obtain the device's address on the network. You may also see an IP address referenced like this: 10.11.241.0/24 or 10.0.0.0/8. This is simply another technique for defining the subnet mask. The number after the / refers to the number of binary 1's in the mask. Therefore, a mask of 255.255.0.0 can also be referred to as /16.

How the Internet Works

Briefly, the Internet works as follows. You want to send a message from your computer to another computer (e.g., e-mail message, http request for a web page, ftp request, ssh, or telnet communication). Your computer takes your message and packages it up into one or more packets. Packets are relatively small. If the message is short, it can fit in one packet. As an example, the request http://www.google.com/ will be short, but an e-mail message could be thousands to millions of bytes long so it might be placed into multiple packets. Each packet is given the destination address (an IP alias), a checksum (error detection information base, which often is computed by summing up the byte values of several bytes and then doing some compression type operations on that sum and then summing the sums), and a return address (your machine's IP address so that the destination computer can send a response).

As the packet is assembled, the destination IP alias is translated into an IP address. Your computer stores the address of your site's DNS (in Linux, this is stored in /etc/resolv.conf). Your computer sends a message to the site's DNS to translate the IP alias to an IP address. If the local DNS does not recognize the IP alias, then a request is passed on to another DNS. The DNSs on the Internet are organized as a hierarchy. Ultimately, a DNS will recognize the alias and respond to those below it with the address. Each receiving DNS will update its own tables (if necessary). However, if the DNS at the highest level of the hierarchy does not recognize the alias, an error is returned and your message will never make it out of your computer.

Once the IP address has been provided, the packets are sent out. They are sent from your computer to your site's Internet point of presence (or gateway). The message is routed from the current LAN to this gateway. From there, the server sends each packet along one of its pathways that connect it to the Internet. Each packet finds its own way across the Internet by going from one router to another. Each router selects the next path based on the destination address and the current message traffic along the various paths available. The packet might traverse a single link, two links, or dozens of links. Use the traceroute command to explore how many hops a message might require.

Upon receiving a message at the destination site, the destination IP address may or may not be the actual IP address of the destination host. If NAT is being used at the remote site, the IP address is translated from an "external public" address to an "internal private" address. And now with the internal private address, the message continues to be routed internally to the destination network and finally, by switch to the destination computer. Usually there are fewer pathway choices internally and in fact a message may have only one route to take. Once received by the destination computer, that machine's operating system saves the packet. The packet is mapped up the protocol stack, and all packets of the same message are combined together as they arrive. They are sequenced based on sequencing numbers (such as packet 3 of 5). Once all packets have been received, the final message is presented to the appropriate application software (or stored to disk in such a case as an ftp or wget communication). Note that the ordering of packets is not done by IP, but rather by the higher-level TCP protocol. The destination machine packages together a response to the sending device to alert it that the message was received in full and correctly (or if any packets were dropped or arrived with errors). The entire process takes seconds or less (usually) even though the distance traveled may be thousands of miles and the number of hops may be in the dozens.

THE INTERNET VERSUS THE WORLD WIDE WEB

Although most people use the terms interchangeably, they are not the same. The Internet, as discussed in this chapter, is a computer network. It is a worldwide network but it is still a network—the computers and computer resources that we use to communicate with each other, the media connecting these devices together (whether cable, radio, microwave, or satellite-based), and the broadcast devices such as switches and routers. The Internet is a physical entity; it contains parts you can touch and point to.

The World Wide Web is a collection of documents, linked together by hyperlinks. These documents are stored on computers that we call servers. And these servers are among the various computers connected to the Internet.

The World Wide Web then sits on top of the Internet. We could continue to have an Internet without the web, but without the Internet, there would be no web.

Another point of distinction worth noting is of people who comment that they have lost their Internet, or that the Internet is down. What has happened is that they have lost their ability to connect to or communicate over the Internet. But the Internet should never be down. It was built to survive a nuclear war and it would take a lot of destruction for the Internet to be lost.

Brief History of the Internet

In 1968, four research organizations were funded by the Department of Defense (DOD) to create a computer network for long-distance communication. These organizations were the University of Utah, the University of California at Los Angeles (UCLA), the University of California at Santa Barbara, and Stanford Research Institute (SRI). Their intention was to build an "electronic think-tank" by having a computer network that could accommodate the transfer of files and allow remote access to each other's computers. They enhanced the fairly recently proposed packet switching technology ideas into practice and used telephone networks to connect these computers together. This network was dubbed the ARPANET. The first message was sent between UCLA and SRI on October 29, 1969.

The original protocol used was called 1822 protocol. This protocol was not very efficient and would eventually be replaced by TCP/IP in 1983. The first e-mail message was sent over the ARPANET in 1971. FTP was added in 1973.

As time went on, more computers were added to the network. By June 1970, there were nine computers on the network. By December 1970, there were 13 computers, and by September 1971, there were 18. Between 1972 and 1973, 29 and then 40 computers were placed on the ARPANET.

Two satellites made available to support ARPANET communication and computers in both Hawaii and Norway were added in 1973, making the network international. By

1975, the network had reached 57 computers. Also in 1975, the DOD handed control of the ARPANET from ARPA to the Defense Communications Agency (another part of the DOD). In 1983, the military portion of the network (MILNET) was separated. Before this split, there were 113 computers and afterward, 68.

In 1974, the UK developed its own form of packet switched network, X.25. X.25 became publicly available. Other networks were created in the late 1970s and early 1980s. These, along with ARPANET, would eventually become components of the Internet.

In 1979, two Duke students created UUCP for sharing messages via an electronic bulletin board. This became known as Usenet news. Usenet was made available over a number of networks and people could access the network using dial-up over a MODEM from their home computer or from a networked machine.

In 1983, with the creation of TCP/IP, the ARPANET was renamed the Internet. The various previous networks that were connected together became the NSF Backbone, which would become the Internet's backbone as the Internet continued to grow. In the 1980s and early 1990s, NSF sponsored an initiative to provide Internet access to as many U.S. universities as possible. In the meantime, pay networks (AOL, Genie, Compuserve, etc.) began connecting to the Internet.

From the late 1980s through the mid 1990s, there were several other developments that would change the Internet. First, access was being made available to the public through pay networks, libraries, schools, and universities. Second, hypertext documents were being pioneered. Third, the U.S. government passed legislation to make it easier for companies to offer Internet services and therefore provide more access to home computer users. However, many people were still uncomfortable trying to access the Internet.

Around 1994, the idea of a web browser was developed. The first one was known as Mosaic. The idea behind the web browser was that the browser would load and display a document automatically. The document would include hypertext information (links). Clicking on a hypertext link would activate a command for the browser to send out an http request for a new document to be sent from a server. The document would be sent back and displayed in the browser. Until this time, most users had to understand how to send messages (whether ftp, telnet, e-mail, or other) including knowing how to specify IP aliases or addresses. But with web browsers, the user only had to know how to click on links. Over time, the web browser has replaced most other methods of Internet communication except for e-mail for the typical user.

Where are we today? Hundreds of millions of computers make up the Internet. The exact number is difficult to tell because home computer users do not leave the computers on all the time. When we factor in handheld devices, the number exceeds 4 billion. The number of actual Internet users is also an estimate, but between home computers and handheld devices, it is estimated that (as of 2012) there are more than 2 billion users (approximately one-third of the population of the planet). Figure 12.15 indicates the growth that we have seen in Internet users since 1995.

And the web? There are perhaps trillions of documents available over the Internet. Many of these documents are hypertext (html) documents with hyperlinks to other documents.

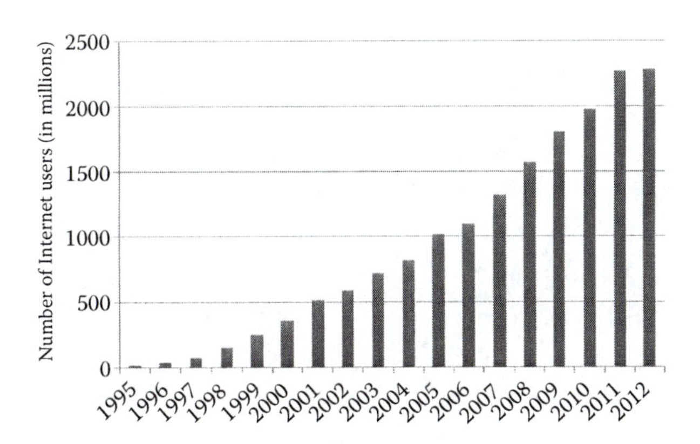

FIGURE 12.15 Number of Internet users from 1995 to 2012.

However, a large number of documents are data files (word documents, pdf documents, powerpoint documents, etc.).

Today, control of the Internet is largely in the hands of corporations that provide Internet access. These include many telecommunications companies thanks to the US 1996 Telecommunications Act. This act, among other things, allowed telephone companies, cable companies, cell phone companies, and others to serve as ISPs. Up until this point, Internet access was largely made available either through employers and universities or through pay WANs such as CompuServe, America On-line, and Prodigy.

So telecommunications companies control access to the Internet. But who *controls* the Internet itself? In some regard, all of the users control the Internet in that we are all responsible for the content being placed onto the Internet—through our web pages, through e-mails that we send and files that we upload and download, and through our participation on message boards, blog sites, and so forth. Some aspects of the Internet, however, have been handed over to a variety of international consortia.

One such organization is known as the Internet Corporation for Assigned Names and Numbers (ICANN). This organization provides IP addresses and handles domain name (IP alias) disputes. A department of ICANN is the Internet Assigned Numbers Authority, who as their name implies, hand out IP addresses. These addresses are based on geographical location, so, for instance, one set of IP addresses is issued to sites in North American and another in Europe and western Asia.

Another organization is the World Wide Web Consortium (W3C). This organization provides standards for web site construction including the next generation of the html markup language, the use of cascaded style sheets, the use of XML, and concepts related to the semantic web.

Future of the Internet

Obviously, the Internet has had a tremendous impact on our society. The advantages that our society has gained because of the Internet are perhaps too numerous to list. However, there are also significant problems with the current state of the Internet today. Solutions to these problems are currently being researched.

1. Dial-up access. Many people have used the telephone network to obtain access to the Internet through an ISP. The problem is that the telephone network consists primarily of twisted wire pair, and this medium can only support a limited bandwidth of message traffic (approximately 56,000 bps). Downloading a 1 MB file (a small picture for instance) would take 18 seconds at this bandwidth. Although other technologies are now available such as digital cable and direct digital subscriber lines (DSL) to people's households, there are still millions of households forced to use the slower technology. This has been called *last mile technology* because most of the telephone network uses fiber optic cable except for the last stretch between the subdivision and the home itself. To replace this last mile or two of twisted wire is cost prohibitive. The telephone companies are only slowly replacing it. Although it seems unlikely that last mile technology impacts people in larger cities, many people who live on the outskirts of a city or in rural areas largely must rely on the telephone lines for access to the Internet. One solution to this problem is through a satellite antenna. Downloading information is done very rapidly as the satellite has a much larger bandwidth than the telephone lines. However, uploading must still be done through the telephone lines, so this only offers a partial solution. Another solution to this problem is the use of cell phones. However, cell phone networks often are overcrowded and so their bandwidth will ultimately be limited as well. In fact, 3G networks are set up to drop packets when there are too many messages going across the network. The 4G networks are promising to resolve this, but 4G coverage is spotty.

2. TCP/IP seemingly offers 4 billion unique IP addresses (32 bits gives you 2^{32} or roughly 4 billion combinations of addresses). However, not all of the addresses are being used. Consider for instance a small organization that is given a Class C network address. They are given 256 distinct addresses but do not necessarily use them all. Another organization might require two Class C networks because they have more than 256 addressable devices but fewer than 512. Through NAT, organizations can provide a greater number of IP addresses for their devices, but this solution cannot extend to, for instance, all of the handheld devices that are now using the Internet. Considering that there are 7 billion people on the planet, we have run out of addresses for everyone.

3. The Internet backbone was not set up to support billions of users especially when the users are requesting large documents such as images, movies, and music. The result is that the Internet is sluggish.

To resolve some of the problems, IPv6 (IP version 6) is being used by some networks. However, IPv6 cannot solve all the problems and so researchers are investigating technologies for Internet2, a new form of Internet. Whether this will utilize the current backbone or require brand new hardware will be decided in the upcoming years.

Along with the upgrades to the Internet infrastructure and protocols, two additional concepts are present when discussing "Web 2.0". These are cloud computing and the semantic web. We wrap up this chapter with brief discussions of both of these ideas.

Cloud computing combines several different technologies: distributed computing, computer networks, storage area networks. The idea is to offer, as a service, the ability to offload computation and storage from your local computer to the cloud. Through the cloud, users can access resources that they may not normally have access to, such as increased processing power to support search engines, software that they do not have licenses for on their local machine, and storage in excess of their local file server or computer. Conceptually, a cloud is a step toward *ubiquitous computing*—the ability to have computation anywhere at any time. In effect, a cloud is the ultimate in a client–server network. The clients are the users who access the cloud remotely and the server is the cloud itself, although unlike a traditional client–server model, the cloud represents service on a very large scale.

To access a cloud, commonly the interface is made through a web portal. In this way, the client computers do not need any specific software downloaded onto their computers. In some cases, clients download specific applications for access. They are also provided a secure account (i.e., an account that requires a log in to gain access). Services offered by cloud computing range from simple storage to database and web services, e-mail, games, and load balancing for large-scale computational processing.

To support a cloud, a company needs a sophisticated network infrastructure. This infrastructure must handle very high bandwidth to support the hundreds, thousands, or millions of users. Data storage capacity might need to run on the order of Petabytes (the next magnitude of storage up from the Terabyte). If the cloud is to support computing as well as storage, dozens to hundreds of high-end processors may also be required. Companies that are making cloud services available (e.g., Amazon, Google, Oracle) are investing billions of dollars in the infrastructure. The result, for a small fee, is that businesses and individuals can have secure data storage (secure both in terms of secure access and the security of mind in that their data are backed up regularly) and remote access to powerful processing. Cloud computing may reach a point eventually where individually, we do not need our own desktop units but instead just interfaces (e.g., handheld devices or wearables), and all storage and computing take place elsewhere.

It should be noted that cloud computing does not have to done by the Internet, but for commercial and practical purposes, most cloud computing exists as a part of the Internet. If a company were to build their own private cloud, its use would be restricted to those who could access the company's network. In fact, we commonly see this with virtual private networks and extranets. The idea is that a company provides processing and storage, but through the VPN, users of the company's computing resources can remotely access them. Thus, they carry their computing with them no matter where they go.

Finally, we consider the *semantic web*. The World Wide Web is a collection of data files, hyperlinked together, and stored on servers on the Internet. The WWW contains a mass of information perhaps reaching the sum total of all human knowledge now that libraries are accessible via the web. However, the information is largely poorly indexed and organized.

It is up to an intelligence (human) to sift through the information and make sense of it. The semantic web is a pursuit led by researchers in artificial intelligence to organize the knowledge and provide automated reasoners to not only search through the information but to find useful pieces to present to the human users.

As a simple example, imagine that you want to take a vacation to Hawaii. Through various websites and search engines, you can book a flight, book a hotel, rent a car, and discover events and locations to explore in Hawaii. How much of this process could be automated? Let us assume that you have a representative-bot (a piece of software to represent you). You have described to your bot your interests, likes, and dislikes. You provide your representative-bot a calendar of important dates that limit your travel dates. You might also specify other criteria such as price restrictions. The representative-bot will use this information and communicate with the various websites to book the entire vacation for you, right down to obtaining maps to the locations you should see during your visit. Further, the bot would reason over the events to coordinate them so that proximity, travel time between locations, times of day/evening, and other factors are taken into account.

The semantic web consists of several different components. Website data are now being organized into *ontologies*: knowledge bases that not only contain information but how that information can be interpreted and used. Small programs, *intelligent agents*, are being built to handle different tasks such as scheduling, communication, and commerce decision making. New types of query languages are being created to facilitate communication between agents and between agents and ontologies. Combined, it is hoped that these new technologies will move the World Wide Web closer to the vision of a semantic web, an Internet whose resources are easily accessible and whose knowledge can be used without excessive human interaction. The semantic web, whose ideas date back to 2001, is only now being used in such areas as intelligence gathering and medical research. It will probably be another decade or more before we see practical, every day uses.

FURTHER READING

There are many texts that cover topics introduced in this chapter from computer science texts on network programming and network protocols to engineering books on the transmission media and protocols of networks to IT texts that detail how to set up a computer network and troubleshoot it. Additionally, there are numerous books that describe the development and use of the Internet. Here, we cite some of the more significant texts from an implementation perspective (the first list) and from an administration and troubleshooting perspective (the next list).

- Blum, R. *C# Network Programming*. New Jersey: Sybex, 2002.
- Forouzan, B. *Data Communications and Networking*. New York: McGraw Hill, 2006.
- Harold, E. *Java Network Programming*. Massachusetts: O'Reilly, 2004.
- Kurose, J. and Ross, K. *Computer Networking: A Top-Down Approach*. Reading, MA: Addison Wesley, 2009.

- Lin, Y., Hwang, R. and Baker, F. *Computer Networks: An Open Source Approach.* New York: McGraw Hill, 2011.
- Peterson, L and Davie, B. *Computer Networks: A Systems Approach.* San Francisco, CA: Morgan Kaufmann, 2011.
- Reese, G. *Cloud Application Architectures.* Massachusetts: O'Reilly, 2009.
- Rhoton, J. *Cloud Computing Explained: Implementation Handbook for Enterprises.* USA: Recursive Press, 2009.
- Segaran, T., Evans, C., and Taylor, J. *Programming the Semantic Web.* Cambridge, MA: O'Reilly, 2009.
- Sosinsky, B. *Cloud Computing Bible.* New Jersey: Wiley and Sons, 2011.
- Stallings, W. *Data and Computer Communications.* Upper Saddle River, NJ: Prentice Hall 2010.
- Stevens, W. *UNIX Network Programming: The Sockets Networking API.* Englewood Cliffs, NJ: Prentice Hall, 1998.
- Tanenbaum, A. *Computer Networks.* Upper Saddle River, NJ: Prentice Hall, 2010.
- Yu, L. *A Developer's Guide to the Semantic Web.* New Jersey: Springer, 2011.
- Allen, N. *Network Maintenance and Troubleshooting Guide: Field Tested Solutions for Everyday Problems.* Reading, MA: Addison Wesley, 2009.
- Corner, D. *Computer Networks and Internets.* Upper Saddle River, NJ: Prentice Hall, 2008.
- Dean, T. *Network+ Guide to Networks.* Boston, MA: Thomson Course Technology, 2009.
- Donahue, G. *Network Warrior.* Massachusetts: O'Reilly, 2011.
- Gast, M. *802.11 Wireless Networks: The Definitive Guide.* Massachusetts: O'Reilly, 2005.
- Hunt, C. *TCP/IP Network Administration.* Massachusetts: O'Reilly, 2002.
- Limoncelli, T., Hogan, C., and Chalup, S. *The Practice of System and Network Administration.* Reading, MA: Addison Wesley, 2007.
- Mansfield, K. and Antonakos, J. *Computer Networking for LANs to WANs: Hardware, Software and Security.* Boston, MA: Thomson Course Technology, 2009.
- Matthews, J. *Computer Networking: Internet Protocols in Action.* Hoboken, NJ: Wiley and Sons, 2005.
- Odom, W. *CCNA Official Exam Certification Library.* Indianapolis, IN: Cisco Press, 2007.

- Odom, W. *Computer Networking First-Step*. Indianapolis, IN: Cisco Press, 2004.
- Rusen, C. *Networking Your Computers & Devices Step By Step*. Redmond, WA: Microsoft Press, 2011.
- Sloan, J. *Network Troubleshooting Tools*. Massachusetts: O'Reilly, 2001.

In addition to the above texts, there are users' guides for various types of networks and study guides for certifications such as for the Cisco Certified Network Associate.

The original article that describes the semantic web was written by Tim Berners-Lee, who was one of the inventors of the World Wide Web. His article can be found on line at http://www.scientificamerican.com/article.cfm?id=the-semantic-web.

You can find a great many interesting statistics on Internet usage at http://www.internetworldstats.com/.

REVIEW TERMS

The following terms were introduced in this chapter:

Anonymous user	Gateway
ARPANET	Host
Bandwidth	Hub
Bus network	Ifconfig
Checksum	Ipconfig
Circuit switching	Internet
Client	Intranet
Client–server network	IP address
Cloud computing	IP alias
Coaxial cable	IPv4
Collision detection	IPv6
DNS	Last mile technology
Ethernet	Local area network
Extranet	Local computer
Fiber optic cable	MAC address
Firewall	Mesh network
FTP	MODEM

Nearest neighbor
Network address translation
Network handshake
Netmask
Network
Network topology
Nslookup
OSI
Packet
Packet switching
Peer-to-peer network
Ping
Point-to-point
Port
Protocol
Remote access
Remote computer
Ring network
Router
Routing
Semantic web
Server
Ssh
Star network
Subnet
Switch
T-connection
Telnet
TCP/IP
Traceroute
Transmission media
Tree topology
Twisted wire pair
UDP
Virtual private network
Wide area network

REVIEW QUESTIONS

1. What is the physical layer of a network composed of?
2. How does a switch differ from a hub?
3. How does a router differ from a switch?
4. What is a bus network?
5. How does a ring network differ from a bus network?
6. What are the advantages and disadvantages of a star network?
7. Why would you not expect to find a mesh network used in large networks?
8. What does message traffic mean? Of the bus, star and ring networks, which is most impacted by message traffic?

9. How does Ethernet handle message traffic?
10. How does a VPN differ from an extranet?
11. How does a MAN differ from a LAN?
12. In which layer of the OSI protocol do packets get formed?
13. In which layer of the OSI protocol is encryption handled?
14. In which layer of the OSI protocol does error detection get handled?
15. In which layer of the OSI protocol can messages be transmitted between devices on the same network?
16. In which layer of the OSI protocol are messages routed?
17. In which layer of the OSI protocol are applications software data converted into a uniform representation?
18. What is the difference between TCP and IP?
19. Which layer(s) of OSI is (are) similar to the application layer of TCP/IP?
20. Which layer(s) of OSI is (are) similar to the transport layer of TCP/IP?
21. Which layer(s) of OSI is (are) similar to the Internet layer of TCP/IP?
22. Which layer(s) of OSI is (are) similar to the link layer of TCP/IP?
23. How many bits is an IPv4 address? How many bits is an IPv6 address? Why are we finding it necessary to move to IPv6 addresses?
24. What is an octet?
25. Given an IPv4 address, what happens when you apply the netmask 255.255.255.0 to it? What about 0.0.0.255?
26. Given the IP address 201.53.12.251, apply the netmask 255.255.128.0 to it.
27. How does a TCP network handshake differ from other network handshaking?
28. Why might an organization use many-to-one network address translation?
29. Who funded the original incarnation of the Internet?
30. How does the ARPAnet differ from the Internet?
31. What significant event happened in 1994 that changed how we use the Internet today?

DISCUSSION QUESTIONS

1. To what extent should a user understand computer networks? To what extent should a system administrator?

2. The bus topology is the simplest and cheapest topology of network, but not very common today. Why not?

3. The primary alternative to the bus topology used to be the ring topology. Research these topologies and compare them in terms of their advantages and disadvantages.

4. What is the value of a metropolitan area network when just about all computers in that area would already be connected to the Internet?

5. What are the advantages for an organization to use an intranet? What are the advantages for that organization to enhance their network into an extranet?

6. In what way(s) does a virtual private network provide security for an organization?

7. IEEE 802.x are a series of standards produced by the IEEE organization. Why are standards important for computer network design and implementation? What is the significance behind the 802.x standards? That is, what have they all contributed toward?

8. The OSI model improves over TCP/IP in several ways. First, it has a built-in encryption component lacking from IPv4. Second, it is more concrete in its description. Why then is TCP/IP far more commonly used to implement networks?

9. Why are we running out of internet addresses? What are some of the solutions to this problem?

10. What is the difference between UDP and TCP? Why might you use UDP over TCP? Why might you use TCP over UDP?

11. Two network programs that might be useful are ping and traceroute, yet these programs are often disabled by system or network administrators. Explore why and explain in your own words what harm they could do.

12. Research Morris' Internet Worm. Why did Robert Morris write and unleash his worm? Why was it a significant event? What, if anything did we learn from the event?

13. You most likely have a firewall running on your home computer. Examine its rules and see if you can make sense of it. What types of messages does it prevent your computer from receiving? In Windows, to view the Firewall rules, go to your control panel, select Network and Sharing Center, and from there, Windows Firewall. From the pop-up window, click on Inbound and Outbound Rules and you will see all of the software that has rules. Click on any item and select properties to view the specific rules. From Linux, you can view your Firewall rules either by looking at the iptables file or from the GUI tool found under the System menu selection Security level and Firewall.

14. Figure 12.14 demonstrates the growth of the Internet in terms of hosts (computers). Research the growth of the Internet in terms of *users* and plot your own graph. How does it compare to the graph in Figure 12.14?

15. Explore the various companies that offer cloud computing and storage services. Make a list of the types of services available. Who should be a client for cloud computing and storage?

16. The original article describing the vision behind the semantic web was published here:

    Berners-Lee, T., Hendler, J., and Lassila, O. (May 17, 2001). The semantic web. *Scientific American Magazine.* http://www.sciam.com/article.cfm?id=the-semantic-web&print=true. Retrieved March 26, 2008. Read this article and summarize the idea behind it. How plausible is the vision?

CHAPTER 13

Software

This chapter begins with an introduction to software classifications and terminology. However, the emphasis of this chapter is on software management, specifically software installation in both Windows and Linux. In Linux, the discussion concentrates on the use of package managers and the installation of open source software. The chapter ends with an examination of server software with particular attention paid to installing and configuring the Apache web server at an introductory level.

The learning objectives of this chapter are to

- Discuss types of software and classification of proprietary and free software.
- Describe the process of software installation from an installation wizard and package manager.
- Illustrate through example how to install open source software in Linux using configure, make, and make install.
- Describe the function of popular server software.
- Introduce Apache web server installation and configuration.

Here, we look at the various types of software and consider what an IT specialist might have to know about. Aside from introducing software concepts and terminology, we will focus on software maintenance and installation. We end the chapter with an examination of servers, server installation, and maintenance.

TYPES OF SOFTWARE

The term software refers to the programs that we run on a computer. We use the term software because programs do not exist in any physical, tangible form. When stored in memory, they exist as electrical current, and when stored on disk, they exist as magnetic charges. Thus, we differentiate software from hardware, whose components we can point to, or pick up and touch.

This distinction may seem straightforward; however, it is actually not as simple as it sounds. Our computers are general purpose, meaning that they can run any program. The earliest computers from the 1930s and 1940s were not general purpose but instead only could run specific types of programs, such as integral calculus computations or code breaking processes. We see today that some of our devices, although having processors and memory, are also special purpose (e.g., navigation systems, mp3 players, game consoles). But aside from these exceptions, computers can run any program that can be compiled for that platform. On the other hand, anything that we can write as software can also be implemented using circuits. Recall that at its lowest level, a computer is merely a collection of AND, OR, NOT, and XOR circuits (along with wires to move data around). We could therefore bypass the rest of the computer and implement the given program at the circuit level. Capturing a program directly in hardware is often referred to as *firmware* (or a hardwired implementation). Similarly, any program that can be built directly into circuits can also be written in some programming language and run on a general purpose computer.

This concept is known as the *equivalence of hardware and software*. Thus, by calling one entity software and another hardware, we are differentiating how they were implemented, but not how they *must* be implemented. We typically capture our programs in software form rather than implement them as hardware because software is flexible. We can alter it at a later point. In addition, the creation of firmware is often far more expensive than the equivalent software. On the other hand, the fetch–execute process of the CPU adds time onto the execution of a program. A program implemented into firmware will execute faster, possibly a good deal faster, than one written as software. Therefore, any program could conceivably be implemented as software or firmware. The decision comes down to whether there is a great need for speed or a desire for flexibility and a cheaper production cost. Today, most programs exist as software. Those captured in firmware include computer boot programs and programs found inside of other devices, such as the antilock brake system or the fuel-injection system in a car.

We prefer to store our programs as software because software can be altered. Most software today is released and then revised and revised again. We have developed a nomenclature to reflect *software releases*. Most software titles have two sets of numbers listed after the title. The first number is the major release version and the second is the minor release version. For instance, Mozilla Firefox 5.2 would be the fifth major version of Firefox and the second minor release of version 5. A minor release typically means that errors were fixed and security holes were plugged. A major release usually means that the software contains many new features likely including a new graphical user interface (GUI). It is also possible that a major release is rewritten with entirely new code. Minor releases might appear every month or two, whereas major releases might appear every year or two. Service packs are another means of performing minor releases; however, service packs are primarily used to release operating system updates.

We generally categorize software as either *system software* or *application software*. Application software consists of programs that end users run to accomplish some task(s). The types of application software are as varied as there are careers because each career has

A BRIEF HISTORY OF PRODUCTIVITY SOFTWARE

1963—IBM developed GUAM, the first database management system, for IBM mainframes intended for use by the National Aeronautics and Space Administration.

1963—Based on a doctoral thesis, an MIT student released Sketchpad, a real-time computer drawing system using a light pen for input.

1964—IBM released MT/ST, the first word processor; I/O was performed on a Selectric keyboard, not directly with the computer.

Early 1970s—INGRES, the first relational database software, released by Relational Technology, followed shortly thereafter by Oracle.

1972—Lexitron and Linolex both introduced machines capable of displaying word process text on a screen (rather than operated on remotely).

1976—Electric Pencil, the first PC word processor, was released.

1979—VisiCalc, the first spreadsheet program, was released for the Apple II.

1979—The WordStar word processor was developed and would become very popular.

1982—Lotus 1-2-3 was released, which combined spreadsheets, graphics, and data retrieval in one software.

1984—Apple released Appleworks, the first office suite.

1985—PageMaker, for Macintosh, was released, a GUI-based word processor making desktop publishing available for personal computers.

its own support software. *Productivity* software consists of the applications that are useful to just about everyone. Productivity software includes the word processor, presentation software (e.g., PowerPoint), spreadsheet program, database management systems, calendar program, address book, and data organizer. Drawing software may also be grouped here although drawing software is more specific toward artists and graphic designers. Another type of application software is that based around Internet usage. Software in this class includes the e-mail client, web browser, and FTP client. There are also computer games and entertainment software (e.g., video players, DVD players, CD players); although not considered productivity software, they are used by just about everyone.

Productivity software is sometimes referred to as *horizontal software* because the software can be used across all divisions in an organization. *Vertical software* applies instead to software used by select groups with an organization or to a specific discipline. For instance, there are classes of software used by musicians such as music sequencers, samplers, and digital recorders. Programmers use development platforms that provide not only a language compiler but also programming support in the form of debugging assistance, code libraries, tracking changes, and code visualization. Filmmakers and artists use video editing, photographic editing and manipulation, and sound editing software. There is a large variety of accounting software available from specialized tax software to digital ledgers.

System software consists of programs that make up the operating system. System software is software that directly supports the computer system itself. Such software is often started automatically rather than by request of the end user. Some pieces of the operating system run in the background all of the time. These are often known as daemons or services. They wait for an event to arise before they take action. Other pieces of the operating system are executed based on a scheduler. For instance, your antiviral software might run every 24 hours, and a program that checks a website for any updates might run once per week. Yet other pieces, often called *utilities*, are run on demand of the user. For instance, the user might run a disk defragmentation utility to improve hard disk performance. Antiviral software is another utility, although as stated above, it may be executed based on a scheduler.

The various layers of the operating system, and their relationship with the hardware, the application software and the user, are shown in Figure 13.1. The figure conveys the layering that takes place in a computer system where the hardware operates at the bottommost level and the user at the top. Each action of the user is decomposed into more and more primitive operations as you move down the layer.

The core components of the operating system are referred to as the *kernel*. The kernel is loaded when the computer is first booted. Included in the kernel are the components that handle process management, resource management, and memory management. Without these, you would not be able to start a new program or have that program run efficiently. The kernel sits on top of the hardware and is the interface between hardware and software.

Another piece of the operating system is the collection of *device drivers*. A device driver is a critical piece of software. The user and applications software will view hardware devices generically—for instance, an input device might respond to commands such as provide input, check status, reboot. But every device requires unique commands. Therefore, the operating system translates from the more generic commands to more specific commands. The translation is performed by the device driver software. Although a few device drivers are captured in the ROM BIOS as firmware, most device drivers are software. Many popular device drivers are preloaded into the operating system, whereas others must be loaded off of the Internet or CD-ROM when a new hardware device has been purchased. In either case, the device driver must be installed before the new hardware device can be used. There are literally hundreds of thousands of drivers available (according to driverguide.com, more than 1.5 million), and there is no need to fill a user's hard disk drive with all of them when most will never be used.

Other system software components sit "on top" of the kernel (again, refer to Figure 13.1). Components that are on top of the kernel include the tailored user environment (shells or the user desktop), system utility programs, and services/daemons. As discussed in Chapter 9, Linux shells permit users to establish their own aliases and variables. Desktop elements include shortcut icons, the theme of the windows, menus and background, the appearance of toolbars and programs on the start menu, and with Windows 7, desktop gadgets.

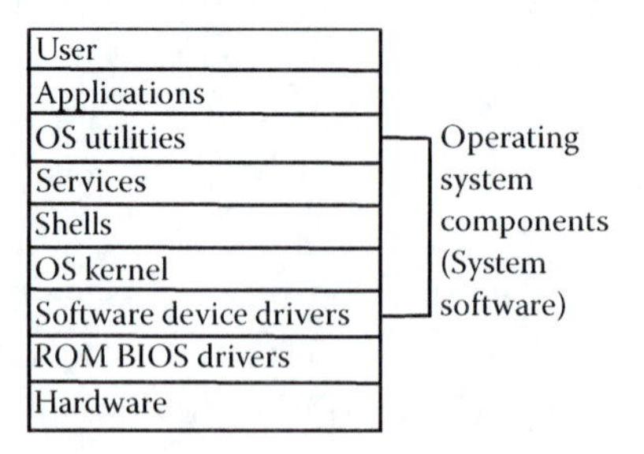

FIGURE 13.1 Layers of a computer system.

System *utilities* are programs that allow the user to monitor and improve system performance. We discussed some of these when we talked about processes. To improve system performance, utilities exist to defragment the hard disk and to scan files for viruses and other forms of malware. Usually, system utilities must be installed separately as they may not come with the operating system (often, these require purchasing and so many users may ignore their availability).

Services, or daemons, are operating system programs that are usually started when the operating system is loaded and initialized, but run in the background. This means that the program, while it is active, does not take up any CPU time until it is called upon. The operating system will, when a particular event arises, invoke the needed service or daemon to handle the event. There are numerous services in any operating system. We covered services in Chapter 11, Services, Configuring Sevices, and Establishing Services at Boot Time.

Aside from the system software and the application software, there is also server software. Servers include web servers, e-mail servers, and database servers, to name a few. We will examine servers in Services and Servers, with an emphasis on the Apache web server.

OLE!

One of the key components found in many forms of Windows-based productivity software today is the ability to link and embed objects created from one piece of software into another. OLE stands for Object Linking and Embedding. It was developed in 1990 as a successor to dynamic data exchange, made available in Windows 2.0, so that data could be transferred between two running applications. This allowed, for instance, for a Microsoft Access record to be copied or embedded into a Microsoft Excel spreadsheet. Modifying the record in Access would automatically update the values stored in Excel. This capability was implemented through tables of pointers whereby one application's data would point at data stored in other applications.

However, data exchange was limited. With OLE, improvements include the ability to capture an object as a bitmap to be transferred pictorially into another application through the clipboard. Also, the table of pointers was replaced through the Component Object Model (COM). In addition, the user's means to copy or embed objects was simplified through simple mouse operations.

OLE is a proprietary piece of technology owned by Microsoft. Today, Microsoft requires that any application software be able to utilize OLE if that piece of software is to be certified as compatible with the Windows operating system.

SOFTWARE-RELATED TERMINOLOGY

In this section, we briefly examine additional terms that help us understand software.

Compatibility describes the platforms that a piece of software can execute on. Most software that our computers run requires that the software be compiled for that computer platform. There are three issues here. First, the software must be compiled for the class of processor of the given computer. Intel Pentium processors, for instance, have different instruction sets than MIPS processors. If someone were to write a program and compile it for the Pentium processor, the compiled program would not run on a MIPS processor.

Because of the close nature between the hardware and the operating system, the compilation process must also be for a given operating system platform. A computer with a Pentium processor running Linux rather than Windows could not execute programs compiled for Windows in spite of having the same processor. Therefore, software must be compiled for both the processor and the operating system. Finally, the software might have additional requirements to run, or at least to run effectively, such as a minimum amount of RAM or hard disk space or a specific type of video card.

Backward compatibility means that a piece of software is capable of reading and using data files from older versions of the software. Most software vendors ensure backward compatibility so that a customer is able to upgrade the software without fear of losing older data. Without maintaining backward compatibility, a customer may not wish to upgrade at all. Maintaining backward compatibility can be a significant challenge though because many features in the older versions may be outdated and therefore not worth retaining, or may conflict with newer features. Backward compatibility also refers to newer processors that can run older software. Recall that software is compiled for a particular processor. If a new version of the processor is released, does it maintain backward compatibility? If not, a computer user who owns an older computer may have to purchase new software for the newer processor. Intel has maintained backward compatibility in their x86 line of processors, starting with the 8086 and up through the most recent Pentium. Apple Macintosh has not always maintained backward compatibility with their processors.

Upgrades occur when the company produces a new major or minor release of the software. Today, most upgrades will happen automatically when your software queries the company's website for a new release. If one is found, it is downloaded and installed without requiring user authorization. You can set this feature in just about any software so that the software first asks you permission to perform the upgrade so that you could disallow the upgrade if desired. However, upgrades often fix errors and security holes. It is to your advantage to upgrade whenever the company recommends it.

Patches are releases of code that will help fix immediate problems with a piece of software. These are typically not considered upgrades, and may be numbered themselves separately from the version number of the software. Patches may be released at any time and may be released often or infrequently depending on the stability of the software.

Beta-release (or beta-test) is an expression used to describe a new version of software that is being released before the next major release. The beta-release is often sent to a select list of users who are trusted to test out the software and report errors and provide feedback on the features of the software. The company collates the errors and feedback and uses this to fix problems and enhance the software before the next major release. A beta-release might appear a few months or a full year before the intended full release.

Installation is the process of obtaining new software and placing it on your computer. The installation process is not merely a matter of copying code onto the hard disk. Typically, installation requires testing files for errors, testing the operating system for shared files, placing files into a variety of directories, and modifying system variables such as the path variable. Installation is made easy today with the use of an installation wizard. However, some software installation requires more effort. This will be addressed in

Software Management. A user might install software from a prepackaged CD-ROM* or off of the Internet. In the latter case, the install is often referred to as a *download*. The term download also refers to the action of copying files from some server on the Internet, such as a music download. Installation from the Internet is quite common today because most people have fast Internet access.

There are several categories to describe the availability of software. *Proprietary*, or commercial, software is purchased from a vendor. Such software may be produced by a company that commercially markets software (usually referred to as a software house), or by a company that produces its own in-house software, or by a consultant (or consultant company) hired to produce software. Purchasing software provides you with two or three things. First, you receive the software itself, in executable form. If you have purchased the software from a store or over the Internet, the software is installed via an installation program. If you have purchased it from a consultant, they may perform the installation themselves. Second, you are provided a license to *use* the software. It is often the case that the software purchase merely gives you the right to use the software but it is not yours to own. Third, it is likely that the purchase gives you access to helpful resources, whether they exist as user manuals, technical support, online help, or some combination.

The other categories of software all describe free software, that is, software that you can obtain for free. Free does not necessarily mean that you can use the software indefinitely or any way you like, but means that you can use it for now without paying for it. One of the more common forms of software made available on the Internet is under a category called *shareware*. Shareware usually provides you a trial version of the software. Companies provide shareware as a way to entice users to purchase the software. The trial version will usually have some limitation(s) such as a set number of uses, a set number of days, or restrictions on the features. If you install shareware, you can often easily upgrade it to the full version by purchasing a registration license. This is usually some combination of characters that serves to "unlock" the software. The process is that you download the shareware. You use it. You go to the website and purchase it (usually through credit card purchase), and you receive the license. You enter the license and now the software moves from shareware to commercial software.

On the other hand, some software is truly free. These fall into a few categories. First, there is *freeware*. Such software is free from purchase, but not necessarily free in how you use it. Freeware is usually software that has become obsolete because a newer software has replaced it, or is software that the original producer (whether an individual or a company) no longer wishes to maintain and provide support for. You are bound by some agreements when you install it, but otherwise you are free to use it as if you purchased it. *Public domain* software, on the other hand, is software that has been moved into the public domain. Like freeware, this is software that no one wishes to make money off of or support. Anything found in the public domain can be used however you feel; no one is claiming any rights for

* Prior to optical disks, installation was often performed from floppy disks. Since floppy disks could only store about 1.5 MB, software installation may have required the use of several individual disks, perhaps as many as a dozen. The user would be asked to insert disk 1, followed by disk 2, followed by disk 3, etc.; however, it was also common to be asked to reinsert a previous disk. This was often called disk juggling.

it. Both freeware and public domain software are made available as executables. Finally, there is *open source* software. As discussed in Chapter 8, this software was created in the Open Source Community and made freely available as source code. You may obtain the software for free; you may enhance the software; you may distribute the software. However, you have to abide by the copyright provided with the software. Most open source software is made available using the GNUs Public License, and some are made available with what is known as a copyleft instead of a copyright.

SOFTWARE MANAGEMENT

Software management is the process of installing, updating, maintaining, troubleshooting, and removing software. Most software today comprises a number of distinct programs including the main executable, data files, help files, shared library files, and configuration files. Installation involves obtaining the software as a bundle or package. Once downloaded or saved to disk, the process requires uncompressing the bundled files, creating directories, moving files, getting proper paths set up, and cleaning up after the installation process. In the case of installing from source code, an added step in the process is compiling the code. Today, the installation process has been greatly simplified thanks to installation wizards and management packages. In Linux, for instance, users often install from yum or rpm. Here, we look at the details of the installation process, upgrading the software, and removing the software.

A software installation package is a single file that contains all of the files needed to install a given software title. The package has been created by a software developer and includes the proper instructions on how to install and place the various files within the package. You might perform the installation from an optical disk, or just as likely, by downloading the software from the Internet. Installation wizards require little of the user other than perhaps specifying the directory for the software, agreeing to the license, and answering whether shortcuts should be created.

Because the Windows end user is often not as knowledgeable about the operating system as the typical Linux end user, Windows installs have been simplified through the use of the Windows Installer program, an installation wizard. The Installer program is typically stored as an .msi file (MicroSoft Installer) and is bundled with the files necessary for installation (or is programmed to download the files from a website). The installation files are structured using the same linkages that make OLE possible (see the sidebar earlier in this chapter), COM files. The installer, although typically very easy to use, will require some user interaction. At a minimum, the user is asked to accept the software's licensing agreement. In addition, the installer will usually ask where the software should be installed. Typically, a default location such as C:\Program Files (x86)*softwarename* is chosen, where softwarename is the name of the software. The user may also be asked whether a custom or standard installation should be performed. Most installers limit the number of questions in the standard install in order to keep matters as simple as possible. Figure 13.2 shows four installation windows that may appear during a standard installation. Note that these four windows are from four different software installations, in this case all open source.

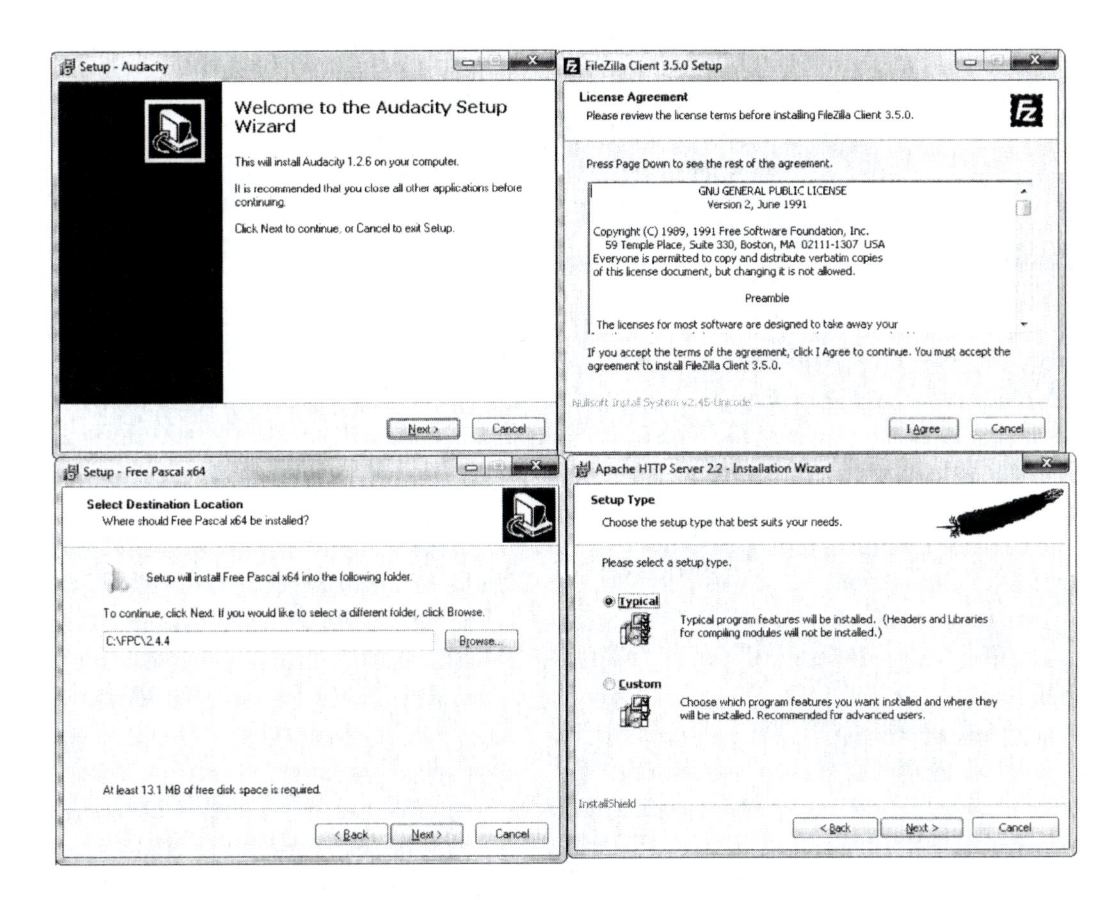

FIGURE 13.2 Various windows from Windows installers.

In spite of the ease of installation through Windows Installers, there are still things the user should know. The most important concern is whether you have the resources needed to run the software. The most critical resources are sufficient hard disk space, fast enough processor, and sufficient RAM. Before installing the software, you may want to check out the installation requirements. You also must make sure that you are installing the right version of the software—that is, the version compiled for your particular hardware and operating system platform. You will also need to make sure that you have sufficient access rights to install the software. Are you an administrator? Have you logged in as an administrator to perform the installation? Or, as a user, has your administrator permitted you access to install new software?

In Linux, there are several different ways to install software. First, there is the simplified approach using an installation wizard from downloaded software, just as you do in Windows. A second approach, also like Windows, is to install software from CD-ROM (or DVD). One difference here is that much software is available on the Linux operating system installation CD. To install from the CD, you must retrieve your installation CD and use the Add or Remove Programs feature (found under the Applications menu).

Another option in Linux is to install software packages from the command line. There are multiple approaches that one can take. The easiest command-line approach is through a *package manager program*. In Red Hat, there are two, yum (Yellow dog Updater, Modified) and rpm (Red Hat Package manager). The rpm program operates on rpm files. An rpm file is a package: an archive of compressed files that include the executable program(s) and related files as well as a description of the dependencies needed to run the files. Dependencies are the files that these files rely on. The rpm program will test these dependencies and alert the user of any missing files. The yum program uses rpm but handles dependencies for you, making it far easier.

To install software using rpm, the instruction is rpm –i packagename.rpm. To uninstall software, use –e (for erase) as in rpm –e packagename. To upgrade a package, use –u. Finally, you can also use –q to query the rpm database to find more information on either already installed packages, or packages that have yet to be installed (as long as the rpm files are present). You can combine the –q and –l options to get a listing of the files in a given package or the –q and –a options to get a list of all already installed packages. There are rpm repositories available at many locations on the Internet, such as http://centos.karan.org/el4/extras/stable/i386/RPMS/. Aside from using rpm from the command line prompt, you can also run GUI package managers.

Yum is a product from Duke University for Linux users. Yum calls upon rpm to perform the installation; however, it has the ability to track down dependencies and install any dependent packages as needed. As with rpm, yum can install or update software, including the Linux kernel itself. It makes it easier to maintain groups of machines without having to manually update each one using rpm. Using yum is very simple; it does most of the work by itself (although it can be time consuming to update packages and install new ones). A simple yum command is yum install *title*, where *title* is the name of the software package. Other yum commands (in place of install) include list (which lists packages that have a given name in them as in yum list "foobar"), chkconfig yum on (which schedules yum to perform an upgrade every night), yum remove *title* (to uninstall the titled package, removing all dependencies), and yum –y install *title* (which does the install, answering "yes" to any yes/no question so that the install can be done without human intervention).

The more challenging form of installation in Linux is installation of software from source code. Source code is the program code written in a high level language (e.g., C++ or Java). Such a program cannot be executed directly on a computer and therefore must be compiled first (we discuss compilation in the next chapter). Because so much of the software available in Linux is open source software, it is likely that a Linux user will need to understand how to install software from source code. Before attempting to install software from the source code, you might still look to see if the software is available in an executable format, either through an installation wizard or as an rpm package.

Installation from source code first requires manually unpackaging the software bundle. Most open source software is bundled into an archive. There are several mechanisms for archiving files, but the most common approach is through the Unix-based *tape archive* program called tar. The tar program was originally intended for archiving files to be saved onto tape. However, it is convenient and fairly easy to use, so many programmers use it to bundle their software together.

A tar file will have .tar as an extension. In addition, once tarred together, the bundle may be compressed to save space and allow faster transmission of the bundle across the Internet. Today, GNU's Zip program, gzip, is often used. This will add a .gz extension to the filename. For instance, a bundled and zipped archive might appear as bundle.tar.gz. The user must first unzip the compressed file using gunzip. This restores the archive, or bundle. To unbundle it, the next step is to untar the bundle, using tar.

When creating a tar file, you must specify both the source file(s) and the name of the tar file along with the options that indicate that you are creating a tar file (c) and that the destination is a file (f) instead of tape. This command could look like: tar –cf bundle1.tar *.txt or it might look like: tar –cf bundle2.tar /temp. The former case would take all.txt files and tar them into an archive, whereas the latter example takes all files in the directory /temp, including the directory itself, and places them in the archive. To untar the file, the command is tar –xf bundle2.tar. The x parameter stands for *extract*, for file extraction. If you were to untar bundle2.tar, it would create the directory /temp first and then place the contents that were stored in the original /temp directory into this new /temp directory.

Once the software has been unbundled, you must compile it. Most open source programs are written in either C or C++. The compilers most commonly used in Linux for C and C++ are gcc (GNU's C/C++ compiler) and g++ (gcc running specific c++ settings). Unfortunately, in the case of compiling C or C++ programs, it is likely that there are numerous files to deal with including header files (.h), source code files (.c or .cpp), and object files (.o). Without a good deal of instruction, a user trying to compile and install software may be utterly lost. Linux, however, has three very useful commands that make the process simple. These are the commands *configure*, *make*, and *make install*. These commands run shell scripts, written by the programmer, to perform the variety of tasks necessary to successfully compile and install the software.

The configure command is used to create or modify a Makefile script. The Makefile script is the programmer's instructions on how to compile the source code. The configure command might, for instance, be used by the system administrator to specify the directories to house the various software files and to specify modules that should be compiled and included in the software. If no Makefile exists, the configure command *must* be executed. If a Makefile already exists, the configure command is sometimes optional. The configure command is actually a script, also set up by the programmer. To run the configure script, one enters ./configure (rather than configure). The ./ means "execute this shell script". Once completed, there is now a Makefile.

The command *make* executes the Makefile script (NOTE: make will execute either Makefile or makefile, whichever it finds). The Makefile script contains all of the commands to compile the software. The Makefile script may also contain other sections aside from compilation instructions. These include an install section to perform the installation steps (if the software is to be installed into one or more directories), a clean section to clean up any temporarily created files that are no longer needed, and a tar section so that the combined files can be archived. If everything goes smoothly, all the user has to do is enter the commands ./configure, make, make install. A good Makefile will call make clean itself after performing the installation steps. These steps may take from a few seconds to several minutes depending on the amount of code that requires compilation.

Before trying to perform the compilation and installation of software, the administrator should first read the README text file that comes with the distribution. Like the Makefile script, README is written by the programmer to explain installation instructions. Reading the README file will help an administrator perform the compilation and installation steps especially if there are specific steps that the programmer expects the administrator to carry out in advance because the programmer has made certain assumptions of the target Linux system. For instance, the programmer might assume that gcc is installed in a specific location or that /usr/bin is available. The README file may also include hints to the administrator for correcting errors in case the Makefile script does not work.

Figure 13.3 provides an example Makefile script. This is a short example. Many Makefile scripts are quite large. The example Makefile script from Figure 13.3, although short, shows that the contents of a Makefile are not necessarily easy to understand. The first four lines of this script define variables to be used in the Makefile file. In this example, there are three program files: file1.c, file2.c, and file3.c. These are stored in the variable FILES. The next three lines define variables for use in the compilation steps. These variables define the name of the compiler (gcc, the GNU's C compiler), compiler flags for use during compilation, and linker flags for use during the linking stage of compilation. Note that the items on the right-hand side of the file are comments (indicated by being preceded by the // characters). Comments are ignored by the bash interpreter when the script executes.

The next six lines represent the commands to execute when the make command is issued. The idea behind these six lines is that each first defines a label, possibly with arguments (e.g., $(FILES)), and on the next line, indented, a command to carry out. The all: label has no arguments, and its sole command is to execute the command under the files: label. This is often used in a Makefile for the command *make all*. The files label iterates through all of the files in the variable $(FILES) (i.e., the three C files) and issues the gcc linking command on each. Beneath this is the line .c, which is used to actually compile each C program. The compilation and linking steps can be done together, but are often

```
FILES = file1.c file2.c file3.c    // define program files to be compiled
CC = gcc                           // C compiler
CFLAGS = -c -Wall                  // C compiler options
LFLAGS = -Wall                     // C compiler options for linking
all:                               // start here - call files
     files
files: $(FILES)                    // for each item in the FILES variable
     $(CC) $(LFLAGS) $(FILES) -o   // perform linking
.c:                                // for each .c file, compile it
     $(CC) $(CFLAGS) -o
clean:                             // if make is called with clean, do this
     \rm *.o myprog                // step (remove all files except the c files)
tar:                               // if make called with tar, tar all files
     tar cfv myprogram.tar file1.c file2.c file3.c myprogram.h
```

FIGURE 13.3 Sample Makefile script.

separated for convenience. The last four lines pertain to the instructions that will execute if the user issues either make clean or make tar. The former cleans up any temporary files from the current directory, and the latter creates a tar file of all relevant program files. Because this was such a simple compilation process, there is no need for an install section.

The programmer(s) writes the Makefile file and packages it with all source code and any other necessary files (such as documentation). The administrator then installs the software by relying on the Makefile script. It is not uncommon that you might have to examine the script and make changes. For instance, imagine that the Makefile moved the resulting files to the directory /usr/local/bin but you wanted to move it to /usr/bin instead, requiring that you make a minor adjustment in the script, or if you wanted to change the settings for the compiler from –g and –Wall to something else.

With the use of package management facilities such as rpm and yum, why are configure, make, and make install important? They may not be if you are happy using the executable code available through the rpm repository websites. However, there are two situations where they become essential. First, some open source software does not come in an executable form and therefore you have no choice but to compile it yourself. Second, if you or someone in your organization wanted to modify the source code, you would still have to install it using the configure, make, and make install commands. We will briefly look at the use of configure in the next section.

SERVICES AND SERVERS

The term *server* may be used to refer to three different things. The first is a dedicated computer used to provide a service. Servers are typically high-performance computers with large hard disk drives. In some cases, servers are built into cabinets so that several or even dozens of hard disk drives can be inserted. The server is commonly a stand-alone computer. That is, it is not someone's desktop machine, nor will users typically log directly into the server unless they happen to be the administrator of the server.

The second use of the term server is of the software that is run on the machine. Server software includes print service, ftp server, webserver, e-mail server, database server, and file server. The software server must be installed, configured, and managed. These steps may be involved. For instance, configuring and managing the Apache web server could potentially be a job by itself rather than one left up to either a system administrator or web developer.

The third usage for server is the collection of hardware, software, and data that together constitutes the server. For instance, we might refer to a web server as hardware, software, and the documents stored there.

In this section, we look at a few types of software servers. We emphasize the Apache web server with a look at installation and minimal configuration. Below is a list of the most common types of servers.

Web Server

The role of a web server is to respond to HTTP requests. HTTP requests most typically come from web browsers, but other software and users can also submit requests. HTTP is the hypertext transfer protocol. Most HTTP requests are for HTML documents (or variants of html), but could include just about any web-accessible resource. HTTP is similar to FTP in that a file is transferred using the service, but unlike FTP, the typical request is made by clicking on a link in a web browser—and thus, the user does not have to know the name of the server or the name or path of the resource. These are all combined into an entity known as a URL. The open source Apache server is the most popular web server used today. Aside from servicing requests, Apache can execute scripts to generate dynamic pages, use security mechanisms to ensure that requests are not forms of attacks, log requests so that analysts can find trends in the public's browser behavior of the website, and numerous other tasks. We look at Apache in more detail below. Internet Information Server (also called Internet Information Services), or IIS, is a Microsoft web server, first made available for Windows XP operating systems and also comes with Windows Vista and Windows 7.

Proxy Server

A proxy server is used in an organization to act as a giant cache of web pages that anyone in the organization has recently retrieved. The idea is that if many people within the organization tend to view the same pages, caching these pages locally permits future accesses to obtain the page locally rather than remotely. Another function of the proxy server is to provide a degree of anonymity since the IP address recorded by the web server in response to the request is that of the proxy server and not the individual client. Proxy servers can also be used to block certain content from being returned, for instance, rejecting requests going out to specific servers (e.g., Facebook) and reject responses that contain certain content (e.g., the word "porn"). Squid is the most commonly used proxy server although Apache can also serve as a proxy server. Squid, like Apache, is open source. Among the types of things a system administrator or web administrator might configure with Squid are the number and size of the caches, and the Squid firewall to permit or prohibit access of various types.

Database Server

A database is a structured data repository. A database management system is software that responds to user queries to create, manipulate, and retrieve records from the database. There are numerous database management system packages available. A database server permits database access across a network. The database server will perform such tasks as data analysis, data manipulation, security, and archiving. The database server may or may not store the database itself. By separating the server from the database, one can use different mechanisms to store the database, for instance a storage area network to support load balancing. MySQL is a very popular, open source database server. MySQL is actually an umbrella name for a number of different database products. MySQL Community is the database server. There is also a proxy server for a MySQL database, a cluster server for high-speed transactional database interaction, and a tool to support GUI construction.

FTP Server

An FTP server is like a webserver in that it hosts files and allows clients to access those files. However, with FTP, access is in the form of uploading files and downloading files. Any files downloaded are saved to disk unlike the web service where most files are loaded directly into the client's web browser. FTP access either requires that the user have an account (unlike HTTP), or requires that the user log in as an *anonymous* user. The anonymous user has access to public files (often in a special directory called /pub). FTP is an older protocol than HTTP and has largely been replaced by HTTP with the exception of file uploading. An FTP server is available in Linux, ftpd. This service is text-based. An extension to FTP is FTPS, a secure form of FTP. Although SFTP is another secure way to handle file transfer, it is not based on the FTP protocol. One can also run ssh and then use ftp from inside of ssh. Popular FTP client tools for performing FTP include WS-FTP and FileZilla, both of which are GUI programs that send commands to an FTP server. FileZilla can also operate as an FTP server, as can Apache.

File Server

A file server is in essence a computer with a large hard disk storing files that any or many users of the network may wish to access. The more complex file servers are used to not only store data files but also to store software to be run over the network. File servers used to be very prominent in computer networks because hard disk storage was prohibitively expensive. Today, with 1 TB of hard disk space costing $100, many organizations have forgone the use of file servers. The file server is still advantageous for most organizations because it supports file sharing, permits easy backup of user files, allows for encryption of data storage, and allows remote access to files. There are many different file servers available with NFS (the Network File System) being the most popular in the Unix/Linux world.

E-mail Server

An e-mail server provides e-mail service. Its job is to accept e-mail requests from clients and send messages out to other e-mail servers, and to receive e-mails from other servers and alert the user that e-mail has arrived. The Linux and Unix operating systems have a built-in e-mail service, sendmail. Clients are free to use any number of different programs to access their e-mail. In Windows, there are a large variety of e-mail servers available including Eudora, CleanMail, and Microsoft Windows Server. Unlike FTP and Web servers, which have a limited number of protocols to handle (primarily ftp, ftps, http, and https), there are more e-mail protocols that servers have to handle: IMAP, POP3, SMTP, HTTP, MAPI, and MIME.

Domain Name System

For convenience, people tend to use IP aliases when accessing web servers, FTP servers, and e-mail servers. The aliases are easier to remember than actual IP addresses. But IP aliases cannot be used by routers, so a translation is required from IP alias to IP address. This is typically performed by a domain name system (DNS). The DNS consists of mapping information as well as pointers to other DNSs so that, if a request cannot be successfully mapped, the request can be forwarded to another DNS. In Linux, a common DNS program is bind. The dnsmasq program is a forwarder program often used on small Linux (and MacOS) networks. For Windows, Microsoft DNS is a component of Windows Server. Cisco also offers a DNS server, Cisco Network Registrar (CNR).

We conclude this chapter by looking at how to install and configure the Apache webserver in Linux. Because much software in Linux is open source, we will look at how to install Apache using the configure, make, and make install steps.

The first step is to download Apache. You can obtain Apache from httpd.apache.org. Although Apache is available for both Linux/Unix and Windows, we will only look at installing it in Linux because the Windows version has several limitations. From the above-mentioned website, you would select Download from the most recent stable version (there is no need to download and install a Beta version because Apache receives regular updates and so any new features will be woven into a stable version before too long). Apache is available in both source code and binary (executable) format for Windows, but source code only in Linux. Select one of the Unix Source selections. There are encrypted and nonencrypted versions available. For instance, selecting http-2.4.1.tar.gz selects Apache version 2.4.1, tarred and zipped. Once downloaded, you must unzip and untar it. You can accomplish this with a single command:

```
tar -xzf httpd-2.4.1.tar.gz
```

The options x, z, and f stand for "extract", "unzip", and "the file given as an argument", respectively. The result of this command will be a new directory, httpd-2.4.1, in the current directory. If you perform an ls on this directory, you will see the listing as shown in Figure 13.4.

The directory contains several subdirectories. The most important of these subdirectories are server, which contains all of the C source code; include, which contains the various C header files; and modules, which contains the code required to build the various modules that come with Apache. There are several scripts in this directory as well. The buildconf and configure scripts help configure Apache by altering the Makefile script to fit the user specifications. The Makefile.in and Makefile.win scripts are used to actually perform the compilation steps.

The next step for the system administrator is to perform the configure step. The usage of this command is ./configure [OPTION]... [VAR = VALUE]... If you do ./configure –h, you will receive help on what the options, variables, and values are. There are few options available and largely will be unused except possibly –h to obtain the help file and –q to operate in "quiet mode" to avoid lengthy printouts of messages. The more common arguments for configure are establishing a number of environment variables. Many of these are used to

```
ABOUT_APACHE      BuildBin.dsp    httpd.dsp        Makefile.in       ROADMAP
acinclude.m4      buildconf       httpd.spec       Makefile.win      server
Apache-apr2.dsw   CHANGES         include          modules           srclib
Apache.dsw        config.layout   INSTALL          NOTICE            support
apache_probes.d   configure       InstallBin.dsp   NWGNUmakefile     test
ap.d              configure.in    LAYOUT           os                VERSIONING
build             docs            libhttpd.dsp     README
BuildAll.dsp      emacs-style     LICENSE          README.platforms
```

FIGURE 13.4 Listing of the Apache directory.

either alter the default directories or to specify modules that should be installed. Here are some examples:

```
--bindir=DIR –replace the default executable directory with DIR
--sbindir=DIR –replace the default system administration
  executable directory with DIR
--sysconfidir=DIR –replace the default configuration directory
  with DIR
--datadir=DIR –replace the default data directory (usually/var/
  www/html) with DIR
--prefix=DIR –replace all default directories with DIR
--enable-load-all-modules –load all available modules
--enable-modules=MODULE-LIST –load all modules listed here
--enable-authn-file –load the module dealing with authentication
  control
--enable-include –enable server side includes module (to run CGI
  script)
--enable-proxy –enable proxy server module
--enable-ssl –enable SSL/TLS support module
```

A typical command at this point might be:

```
./configure --prefix=/usr/local/ --enable-modules=…
```

Where you will fill in the desired modules in place of…, or omit –enable-modules= entirely if there are no initial modules that you want to install. The configure command will most likely take several minutes to execute. Once completed, your Makefile is available. Type the command make, which itself will take several more minutes to complete. Finally, when done, type make install. You will now have an installed version of Apache, but it is not yet running.

There are other means of installing Apache. For instance, an executable version is available via the CentOS installation disk. You can install this using the Add/Remove Software command under the Applications menu. This does require that you have the installation disk available. An even simpler approach is to use yum by entering yum install httpd. However, these two approaches restrict you to default options and the version of Apache made available in their respective repositories (the one on the disk and the one in the RPM website) and not the most recent version as found on the Apache website.

Now that Apache is installed, we can look at how to use it. To run Apache, you must start the Apache service. Find the binary directory, in this case it would be /usr/local/apache2/bin. In this subdirectory is the script apachectl. You will want to start this by executing ./apachectl start. Your server should now be functional.

Using the above configuration, all of your configuration files will be stored under /usr/local/apache2/conf. The main conf file is httpd.conf. You edit this file to further configure and tailor your server to your specific needs. This file is not the only configuration file. Instead, it is common to break up configuration commands into multiple files. The Include directive is used to load other configuration files. You might find in the httpd.conf file that many of the Include statements have been commented out.

Let us consider what you might do to configure your server. You might want to establish additional IP addresses or ports that Apache will respond to. You might want to alter the location of log files and what is being logged. You might want to utilize a variety of modules that you specified in the ./configure command. You might want to alter the number of children that Apache will spawn and keep running in the background to handle HTTP requests. Additionally, you can add a variety of containers. Containers describe how to specifically treat a subdirectory of the file space, a group of files that share the same name, or a specific URL. Any time you change your configuration file, you must restart apache to have those changes take effect. You can either issue the commands ./apachectl stop and ./apachectl start or ./apachectl restart (the latter command will only work if Apache is currently running).

The location of your "web space", as stored in your Linux computer, should be /usr/local/apache2/web. This is where you will place most of your web documents. However, you might wish to create subdirectories to better organize the file space. You can also better control access to individual subdirectories so that you could, for instance, place certain restricted files under https access that requires an authorized log in, by using directory containers.

The Apache configuration file is composed of comments and directives. Comments are English descriptions of the directives in the file along with helpful suggestions for making modifications. The directives are broken into three types. First, server directives impact the entire server. Second, container directives impact how the server treats specific directories and files. Finally, Include statements are used to load other configuration files.

The server directives impact the entire server. The following are examples of seven server directives.

```
ServerName www.myserver.com
User apache
Group apache
TimeOut 120
MaxKeepAliveRequests 100
Listen 80
DocumentRoot "/usr/local/apache2/web"
```

ServerName is the IP alias by which this machine will respond. User and Group define the name by which this process will appear when running in Linux (e.g., when you do a ps command). TimeOut denotes the number of seconds that the server should try to communicate with a client before issuing a time out command. MaxKeepAliveRequests establishes the number of requests that a single connection can send before the connection is closed and a new connection must be established between a client and the server. Listen identifies the port number(s) that the server will listen to. Listen also permits IP addresses if the server runs on a computer that has multiple IP addresses. DocumentRoot stores the directory location on the server of the web pages. The default for this address typically /var/www/html; however, we overrode this during the ./configure step when we specified the location in the file system for all of Apache.

The Include directive allows you to specify other configuration files. For instance,

```
Include conf.d/*.conf
```

would load all other.conf files found in the conf.d subdirectory at the time Apache is started. This allows you to separate directives into numerous files, keeping each file short and concise. This also allows the web server administrator to group directives into categories and decide which directives should be loaded and which should not be loaded by simply commenting out some of the Include statements. In the CentOS version of Linux, there are two .conf files found in this subdirectory: one to configure a proxy service and a welcome .conf file.

The last type of directive is actually a class of directives called containers. A container describes a location for which the directives should be applied. There are multiple types of containers. The most common type are <Directory> to specify directives for a particular directory (and its subdirectories), <Location> to specify directives for a URL, and <Files> to specify directives for all files of the given name no matter where those files are stored (e.g., index.html).

The container allows the website administrator to fine-tune how Apache will perform for different directories and files. The directives used in containers can control who can access files specified by the container. For instance, a pay website will have some free pages and some pages that can only be viewed if the user has paid for admittance. By placing all of the "pay" pages into one directory, a directory container can be applied to that directory that specifies that authentication is required. Files outside of the directory are not impacted and therefore authentication is not required for those.

Aside from authentication, numerous other controls are available for containers. For instance, encryption can also be specified so that any file within the directory must be transmitted using some form of encryption. Symbolic links out of the directory can be allowed or disallowed. Another form of control is whether scripts can be executed from a given directory. The administrator can also control access to the contents of a container based on the client's IP address.

What follows are two very different directory containers. In the first, the container defines access capabilities for the entire website. Everything under the web directory will obey these rules unless overridden. These rules first permit two options, Indexes and

FollowSymLinks. Indexes means that if a URL does not include a filename, use index.html or one of its variants like index.php or index.cgi. FollowSymLinks allows a developer to place a symbolic link in the directory to another location in the Linux file system. AllowOverride is used to explicitly list if an option can be overridden lower in the file space through a means known as htaccess. We will skip that topic. Finally, the last two lines are a pair: the first states that access will be defined first by the Allow rule and then the Deny rule; however, because there is no Deny rule, "Allow from all" is used, and so everyone is allowed access.

```
<Directory "/usr/local/apache2/web">
      Options Indexes FollowSymLinks
      AllowOverride None
      Order allow,deny
      Allow from all
</Directory>
```

The following directory container is far more complex and might be used to define a subdirectory whose contents should only be accessible for users who have an authorized access, that is, who have logged in through some login mechanism. Notice that this is a subdirectory of the directory described above. Presumably, any files that were intended for an authorized audience would be placed here and not in the parent (or other) directory. If a user specifies a URL of a file in this subdirectory, first basic authentication is used. In the pop-up log in window, the user is shown that they are logging into the domain called "my store". They must enter an account name and password that match an entry in the file /usr/localapache2/accounts/file1.passwd. Access is given to anyone who satisfies one of two constraints. First, they are a valid user as authenticated by the password file, or second, they are accessing the website from IP address 10.11.12.13. If neither of those is true, the user is given a 401 error message (unauthorized access). Finally, we alter the options for this subdirectory so that only "ExecCGI" is available, which allows server-side CGI scripts stored in this directory to execute on the server. Notice that Indexes and FollowSymLinks are not available.

```
<Directory "/usr/local/apache2/web/payfiles">
      AuthType Basic
      AuthName "my store"
      AuthUserFile/usr/local/apache2/accounts/file1.passwd
      Require valid-user
      Order deny,allow
      Deny from all
      Allow from 10.11.12.13
      Satisfy any
      Options ExecCGI
</Directory>
```

There is much more to explore in Apache, but we will save that for another textbook!

FURTHER READING

As with previous chapters, the best references for further reading are those detailing specific operating systems (see Further Reading section in Chapter 4, particularly texts pertaining to Linux and Windows). Understanding software categories and licenses can be found in the following texts:

- Classen, H. *Practical Guide to Software Licensing: For Licensees and Licensors.* Chicago, IL: American Bar Association, 2012.
- Gorden, J. *Software Licensing Handbook.* North Carolina: Lulu.com, 2008.
- Laurent, A. *Understanding Open Source and Free Software Licensing.* Cambridge, MA: O'Reilly, 2004.
- Overly, M. and Kalyvas, J. *Software Agreements Line by Line: A Detailed Look at Software Contracts and Licenses & How to Change Them to Fit Your Needs.* Massachusetts: Aspatore Books, 2004.
- Tollen, D. *The Tech Contracts Handbook: Software Licenses and Technology Services Agreement for Lawyers and Businesspeople.* Chicago, IL: American Bar Association, 2011.

Installing software varies from software title to software title and operating system to operating system. The best references are found on websites that detail installation steps. These websites are worth investigating when you face "how to" questions or when you need to troubleshoot problems:

- http://centos.karan.org/
- http://support.microsoft.com/ph/14019
- http://wiki.debian.org/Apt
- http://www.yellowdoglinux.com/

The following texts provide details on various servers that you might wish to explore:

- Cabral, S. and Murphy, K. *MySQL Administrator's Bible.* Hoboken, NJ: Wiley and Sons, 2009.
- Costales, B., Assmann, C., Jansen, G., and Shapiro, G. *sendmail.* Massachusetts: O'Reilly, 2007.
- Eisley, M., Labiaga, R., and Stern, H. *Managing NFS and NIS.* Massachusetts: O'Reilly, 2001.
- McBee, J. *Microsoft Exchange Server 2003 Advanced Administration.* New Jersey: Sybex, 2006.

- Mueller, J. *Microsoft IIS 7 Implementation and Administration.* New Jersey: Sybex, 2007.
- Powell, G., and McCullough-Dieter, C. *Oracle 10g Database Administrator: Implementation and Administration.* Boston, MA: Thomson, 2007.
- Sawicki, E. *Guide to Apache.* Boston, MA: Thomson, 2008.
- Schneller, D. and Schwedt, Ul., *MySQL Admin Cookbook.* Birmingham: Packt Publishing, 2010.
- Silva, S. *Web Server Administration.* Boston, MA: Thomson Course Technology, 2008.
- Wessels, D. *Squid: The Definitive Guide.* Massachusetts: O'Reilly, 2005.

Finally, the following two-page article contains a mathematical proof of the equivalence of hardware and software:

- Tan, R. Hardware and software equivalence. *International Journal of Electronics,* 47(6), 621–622, 1979.

REVIEW TERMS

Terminology introduced in this chapter:

Apache	E-mail server
Application software	Equivalence of hardware and software
Background	File extraction
Backward compatibility	File server
Beta-release software	Firmware
Compatibility	Freeware
Compilation	FTP server
Configure (Linux)	Horizontal software
Container (Apache)	Installation
Daemon	Kernel
Database server	Make (Linux)
Device driver	Makefile (Linux)
Directive (Apache)	Make install (Linux)
Download	Package manager program

Patch

Productivity software

Proprietary software

Proxy server

Public domain software

Server

Service Shareware

Shell

Software

Software release

Source code

System software

Tape archive (tar)

Upgrade

Utility

Yum (Linux)

Version

Vertical software

Web server

Windows installer

REVIEW QUESTIONS

1. What are the advantages of firmware over software?
2. What are the advantages of software over firmware?
3. Is it true that anything implemented in software can also be implemented in hardware?
4. What is the difference between application software and system software?
5. Why are device drivers not necessarily a permanent part of the operating system?
6. When do you need to install device drivers?
7. What is the difference between the kernel and an operating system utility?
8. What types of things can a user do to tailor the operating system in a shell?
9. A service (daemon) is said to run in the background. What does this mean?
10. Why might a software company work to maintain backward compatibility when releasing new software versions?
11. What is a patch and why might one be needed?
12. When looking at the name of a software title, you might see a notation like Title 5.3. What do the 5 and 3 represent?
13. What does it mean to manage software?
14. What is rpm? What can you use it for? Why is yum simpler to use than rpm?
15. What is the Linux tar program used for? What does tar –xzf mean?

16. What does the Linux command configure do? What does the Linux command make do? What does the Linux command make install do?

17. What are gcc and g++?

18. What is the difference between a web server and a proxy server? Between a web server and an FTP server?

DISCUSSION QUESTIONS

1. We receive automated updates of software over the Internet. How important is it to keep up with the upgrades? What might be some consequences of either turning this feature off or of ignoring update announcements?

2. Intel has always maintained backward compatibility of their processors whenever a new generation is released, but this has not always been the case with Apple Macintosh. From a user's perspective, what is the significance of maintaining (or not maintaining) backward compatibility? From a computer architect's perspective, what is the significance of maintaining backward compatibility?

3. Explore the software on your computer. Can you identify which software is commercial, which is free but proprietary under some type of license, and which is free without license?

4. Have you read the software licensing agreement that comes with installed software? If so, paraphrase what it says. If not, why not?

5. What are some ways that you could violate a software licensing agreement? What are some of the consequences of violating the agreement?

6. Assume that you are a system administrator for a small organization that uses Linux on at least some of their computers. You are asked to install some open source software. Provide arguments for and against installing the software using an installation program, using the yum package manager, using the rpm package manager, and using configure/make/make install.

CHAPTER 14

Programming

IT administrators will often have to write shell scripts to support the systems that they maintain. This chapter introduces the programming task. For those unfamiliar with programming, this chapter covers more than just shell scripting. First, the chapter reviews the types of programming languages, which were first introduced in Chapter 7. Next, the chapter describes, through numerous examples, the types of programming language instructions found in most programming languages: input and output instructions, assignment statements, selection statements, iteration statements, subroutines. With this introduction, the chapter then turns to scripting in both the Bash interpreter language and DOS, accompanied by numerous examples.

The learning objectives of this chapter are to

- Describe the differences between high level language and low level programming.
- Discuss the evolution of programming languages with an examination of innovations in programming.
- Introduce the types of programming language instructions.
- Introduce through examples programming in both the Bash shell and DOS languages.

In this chapter, we look at programming and will examine how to write shell scripts in both the Bash shell and DOS.

TYPES OF LANGUAGES

The only types of programs that a computer can execute are those written in the computer's machine language. Machine language consists of instructions, written in binary, that directly reference the computer processor's instruction set. Writing programs in machine language is extremely challenging. Fortunately though, programmers have long avoided writing programs directly in machine language because of the availability of *language translators*. A language translator is a program that takes one program as input, and creates a machine language program as output. Each language translator is tailored to translate from one specific language to the machine language of a given processor. For instance, one translator would be used to translate a C program for a Windows computer and another would be used to translate a C program for a Sun workstation. Similarly, different translator programs would be required to translate programs written in COBOL (COmmon Business-Oriented Language), Ada, or FORTRAN.

There are three types of language translators, each for a different class of programming language. These classes of languages are *assembly* languages, high level *compiled* languages, and high level *interpreted* languages. The translators for these three types of languages are called *assemblers*, *compilers*, and *interpreters*, respectively.

Assembly languages were developed in the 1950s as a means to avoid using machine language. Although it is easier to write a program in assembly language than in a machine language, most programmers avoid assembly languages as they are nearly as challenging as machine language programming. Consider a simple C instruction

```
for(i=0;i<n;i++)
        a[i]=a[i]+1;
```

which iterates through *n* array locations, incrementing each one. In assembly (or machine) language, this simple instruction would comprise several to a few dozen individual instructions. Once better languages arose, the use of assembly language programming was limited to programmers who were developing programs that required the most efficient executable code possible (such as with early computer games). Today, almost no one writes assembly language programs because of the superiority of high level programming languages.

High level programming languages arose because of the awkwardness of writing in machine or assembly language. To illustrate the difference, see Figure 14.1. In this figure, a program is represented in three languages: machine language (for a fictitious computer

```
1111 0000000000001          Load #1         int a = 1, b = 5, c;
0010 1010100110000          Store a         while(a < b)
1111 0000000000101          Load #5         {
0010 1010100110001          Store b                 c = c + 2;
0001 1010100110000   Top:   Load a                  b = b - 1;
0100 1010100110001          Subt b          }
1000 0000000000000          Skipcond 00     printf("%d", c);
1001 0001000101010          Jump exit
0001 1010100110011          Load c
1110 0000000000010          Add #2
0010 1010100110011          Store c
0001 1010100110001          Load b
1101 0000000000001          Subt #1
0010 1010100110001          Store b
1001 0001000100010          Jump top
0001 1010100110011   Exit:  Load c
0101 0000000000000          Output
```

FIGURE 14.1 A simple program written in machine language (leftmost column), assembly language (center column), and C (rightmost column).

called MARIE), assembly language (again, in MARIE), and the C high level programming language. Notice that the C code is far more concise. This is because many tasks, such as computing a value (e.g., a = b – 1) require several steps in machine and assembly languages but only one in a high level language. Also notice that every assembly language instruction has an equivalent machine language instruction, which is why those two programs are of equal size.

High level programming languages come in one of two types, compiled languages and interpreted languages. In both cases, the language translator must translate the program from the high level language to machine language. In a compiled language, the language translator, called a compiler, translates the entire program at one time, creating an executable program, which can be run at a later time. In an interpreted language, the program code is written in an interpreted *environment*. The interpreter takes the most recently entered instruction, translates it into machine language, and executes it. This allows the programmer to experiment while developing code. For example, if the programmer is unsure what a particular instruction might do, the programmer can enter the instruction and test it. If it does not do what was desired, the programmer then tries again. The interpreted programming process is quite different from the compiled programming process as it is more experimental in nature and developing a program can be done in a piecemeal manner. Compiled programming often requires that the entire program be written before any of it can be compiled and tested.

Most of the high level languages fell into the compiled category until recently. Scripting languages are interpreted, and more and more programmers are using scripting languages. Although interpreted programming sounds like the easier and better approach, there is a large disadvantage to interpreted programming. Once a program is compiled, the chore of language translation is over with and therefore, executing the program only requires execution time, not additional translation time. To execute an interpreted program, every instruction is translated first and then executed. This causes execution time to be lengthier than over a similarly compiled program. However, many modern interpreted languages have optional compilers so that, once complete, the program can be compiled if desired.

Another issue that arises between compiled and interpreted programming is that of software testing. Most language compilers will find many types of errors at the time a program is being compiled. These are known as *syntax* errors. As a programmer (particularly, a software engineer), one would not want to release software that contains errors. One type of error that would be caught by a compiler is a *type mismatch* error. In this type of error, the program attempts to store a type of value in a variable that should not be able to store that type, or an instruction might try to perform an operation on a variable that does not permit that type of operation. Two simple examples are trying to store a fractional value with a decimal point into an integer variable and trying to perform multiplication on a string variable. These errors are caught by the compiler at the time the program is compiled. In an interpreted program, the program is not translated until it is executed and so the interpreter would only catch the error at run-time (a *run-time error*), when it is too late to avoid the problem. The user, running the interpreted program, may be left to wonder why the program did not run correctly.

A run-time error is one that arises during program execution. The run-time error occurs because the computer could literally not accomplish the task at hand. A common form of run-time error is an arithmetic overload. This happens when a computation results in a value too large to place in to the designated memory location. Another type of run-time error arises when the user inputs an inappropriate value, for instance, entering a name when a number is asked for.

The type mismatch error described above is one that can be caught by a compiler. But if the language does not require compilation, the error is only caught when the particular instruction is executed, and so it happens at run time.

A third class of error is a *logical error.* This type of error arises because the programmer's logic is incorrect. This might be something as trivial as subtracting two numbers when the programmer intended to add two numbers. It can also arise as a result of a very complicated code because the programmer was unable to track all of the possible outcomes.

Let us consider an example that illustrates how challenging it is to identify a logical error from the code. The following code excerpt is from C (or C++ or Java).

```
int x = 5;
float y;
y = 1/x;
```

The code first declares two variables, an integer x and a floating point (real) y. The variable x is initialized with the number 5. The third instruction computes 1/x and stores the result in y. The result should be the value 1/5 (0.2) stored in y. This should not be a problem for the computer since y is a float, which means it can store a value with a decimal point (unlike x). However, the problem with this instruction is that, in C, C++, and Java, if the numerator and the denominator are both integers, the division is an integer division. This means that the quotient is an integer as well. Dividing 1 by 5 yields the value 0 with a remainder of 1/5. The integer quotient is 0. So y = 1/x; results in y = 0. Since y is a float, the value is converted from 0 to 0.0. Thus, y becomes 0.0 instead of 0.2! Logical errors plague nearly every programmer.

A BRIEF EXAMINATION OF HIGH LEVEL PROGRAMMING LANGUAGES

Programming languages, like computer hardware and operating systems, have their own history. Here, we take a brief look at the evolution of programming languages, concentrating primarily on the more popular high level languages of the past 50 years.

The earliest languages were native machine languages. Such a language would look utterly alien to us because the instructions would comprise only 1s and 0s of binary, or perhaps the instructions and data would be converted into octal or hexadecimal representations for somewhat easier readability. In any event, no one has used machine language for decades, with the exception of students studying computer science or computer engineering. By the mid 1950s, assembly languages were being introduced. As with machine languages, assembly languages are very challenging.

Starting in the late 1950s, programmers began developing compilers. Compilers are language translators that translate high level programming languages into machine language.

The development of compilers was hand in hand with the development of the first high level programming languages. The earliest high level language, FORTRAN (FORmula TRANslator) was an attempt to allow programmers who were dealing with mathematical and scientific computation a means to express a program as a series of mathematical formulas (thus the name of the language) along with input and output statements. Most programmers were skeptical that a compiler could produce efficient code, but once it was shown that compilers could generate machine language code that was as efficient as, or more efficient than, that produced by humans, FORTRAN became very popular.

FORTRAN, being the first high level language, contained many features found in assembly language and did not contain many features that would be found in later languages. One example is FORTRAN's reliance on implicit variable typing. Rather than requiring that the programmer declare variables by type, the type of a variable would be based on the variable's name. A variable with a name starting with any letter from I to N would be an integer and any other variable would be a real (a number with a decimal point). Additionally, FORTRAN did not contain a character type so you could not store strings, only numbers. Variable names were limited to six characters in length. You could not, therefore, name a variable income_tax.

FORTRAN did not have a useful if-statement, relying instead on the use of GO TO statements (a type of unconditional branching instruction). Unconditional branches are used extensively in machine and assembly language programming. In FORTRAN, the GO TO permits the programmer to transfer control of the program from any location to another. We will examine an example of this later in this section. Because of the GO TO rather than a useful if-statement, FORTRAN code would contain logic that was far more complex (or convoluted) than was necessary. In spite of its drawbacks, FORTRAN became extremely popular for scientific programming because it was a vast improvement over assembly and machine language.

As FORTRAN was being developed, the Department of Defense began working with business programmers to produce a business-oriented programming language, COBOL (COmmon Business-Oriented Language). Whereas FORTRAN often looked much like mathematical notation, COBOL was expressed in English sentences and paragraphs, although the words that made up the sentences had to be legal COBOL statements, written in legal COBOL syntax. The idea behind a COBOL program was to write small routines, each would be its own paragraph. Routines would invoke other routines. The program's data would be described in detail, assuming that data would come from a database and output would be placed in another database. Thus, much of COBOL programming was the movement and processing of data, including simple mathematical calculations and sorting. COBOL was released a couple of years after FORTRAN and became as successful as FORTRAN, although in COBOL's case, it was successful in business programming.

COBOL also lacked many features found in later languages. One major drawback was that all variables were *global* variables. Programs are generally divided into a number of smaller routines, sometimes referred to as procedures, functions, or methods. Each routine has its own memory space and variables. In this way, a program can manipulate its local variables without concern that other routines can alter them. Global variables violate this idea because a variable known in one location of the program, or one routine, is known throughout the program. Making a change to the variable in one routine may have unexpected consequences in other routines.

COBOL suffered from other problems. Early COBOL did not include the notion of an array. There was no parameter passing in early COBOL (in part because all variables were global variables). Like FORTRAN, COBOL had a reliance on unconditional branching instructions. Additionally, COBOL was limited with respect to its computational abilities. COBOL did, however, introduce two very important concepts. The first was the notion of structured data, now referred to as *data structures*, and the second was the ability to store *character strings*. Data structures allow the programmer to define variables that are composed of different types of data. For instance, we might define a Student record that consists of a student's name, major, GPA, total credit hours earned, and current courses being taken. The different types of data would be represented using a variety of data types: strings, reals, integers.

Researchers in artificial intelligence began developing their own languages in the late 1950s and early 1960s. The researchers called for language features unavailable in either FORTRAN or COBOL: handling lists of symbols, recursion and dynamic memory allocation. Although several languages were initially developed, it was LISP—the LISt Processing language—that caught on. In addition to the above-mentioned features that made LISP appealing, LISP was an interpreted language. By being an interpreted language, programmers could experiment and test out code while developing programs and therefore programs could be written in a piecemeal fashion. LISP was very innovative and continued to be used, in different forms, through the 1990s.

Following on from these three languages, the language ALGOL (ALGOrithmic Language) was an attempt to merge the best features of FORTRAN, COBOL, and LISP into a new language. ALGOL ultimately gained more popularity in Europe than it did in the United States, and its use was limited compared to the other languages, eventually leading to ALGOL's demise. However, ALGOL would play a significant role as newer languages would be developed.

In the 1960s, programming languages were being developed for many different application areas. New languages were also developed to replace older languages or to provide facilities that older languages did not include. Some notable languages of the early 1960s were PL/I, a language developed by IBM to contain the best features of FORTRAN, COBOL, and ALGOL while also introducing new concepts such as exception handling; SIMULA, a language to develop simulations; SNOBOL (StriNg Oriented and SymBOlic Language), a string matching language; and BCPL, an early version of C.

By the late 1960s, programmers realized that these early languages relied too heavily on unconditional branching statements. In many languages, the statement was called a GO TO (or goto). The *unconditional branch* would allow the programmer to specify that the program could jump from its current location to anywhere else in the program.

As an example, consider a program that happens to be 10 pages in length. On page 1, there are four instructions followed by a GO TO that branches to an instruction on page 3. At that point, there are five instructions followed by an if–then–else statement. The then clause has a GO TO statement that branches to an instruction on page 6, and the else clause has a GO TO statement that branches to an instruction on page 8. On page 6, there are three instructions followed by a GO TO statement to page 2. On page 8, there are five instructions followed by an if–then statement. The then clause has a GO TO statement that branches to page 1. And so forth. Attempting to understand the program requires tracing through it. Tracing through a program with GO TO statements leads to confusion. The trace might begin to resemble a pile of spaghetti. Thus, unconditional branching instructions can create what has been dubbed *spaghetti code*.

Early FORTRAN used an if-statement that was based around GO TO statements. The basic form of the statement is *If (arithmetic expression) line number1, line number2, line number3*. If the arithmetic expression evaluates to a negative value, GO TO *line number1*, else if the arithmetic expression equals zero, GO TO *line number2*, otherwise GO TO *line number3*. See Figure 14.2, which illustrates spaghetti code that is all too easily produced in FORTRAN. The second instruction is an example of FORTRAN's if-statement, which computes I–J and if negative, branches to line 10, if zero, branches to line 20, otherwise branches to line 30.

Assuming I = 5, J = 15, and K = –1, can you trace through the code in Figure 14.2 to see what instructions are executed and in what order? If your answer is no, this may be because of the awkwardness of tracing through code that uses GO TO statements to such an extent.

Programmers who were developing newer languages decided that the unconditional branch was too unstructured. Their solution was to develop more structured *control statements*. This led to the development of conditional iteration statements (e.g., while loops), counting loops (for-loops), and nested if–then–else statements. Two languages that embraced the concept of *structured programming* were developed as an offshoot of ALGOL: C and Pascal. For more than 15 years, these two languages would be used extensively in programming (C) and education (Pascal). Developed in the early 1980s, Ada would take on features of both languages, incorporate features found in a variety of other languages, introduce new features, and become the language used exclusively by the U.S. government for decades.

```
          READ (*,*) I, J, K
          If (I-J), 10, 20, 30
10        If(K) 30, 40, 50
20        I=J*K
          If(I-25) 40, 50, 60
30        J=K*3
          GO TO 20
40        K=K+1
          GO TO 10
50        WRITE (*,*) I, J, K
          GO TO 70
60        WRITE (*, *) J, K
70        END
```

FIGURE 14.2 Spaghetti code in FORTRAN.

In 1980, a new concept in programming came about, *object-oriented programming*. Developed originally in one of the variants of Simula, the first object-oriented programming language was Smalltalk, which was the result of a student's dissertation research. The idea was that data structures could model real-world objects (whether physical objects such as a car or abstract objects such as an operating system window). A program could consist of objects that would communicate with each other. For example, if one were to program a desktop object and several window objects, the desktop might send messages to a window to close itself, to move itself, to shrink itself, or to resize itself. The object would be specified in a *class* definition that described both the data structure of the object and all of the code needed to handle the messages that other objects might pass to it. Programming changed from the idea of program subroutines calling each other to program objects sending messages to each other.

In the late 1980s, C was upgraded to C++ and LISP was upgraded to Common Lisp, the two newer languages being object-oriented. Ada was later enhanced to Ada 95. And then, based on C++, Java was developed. Java, at the time, was a unique language in that it was both compiled and interpreted. This is discussed in more detail later. With the immense popularity of both C++ and Java, nearly all languages since the 1990s have been object-oriented. Another concept introduced during this period was that of *visual programming*. In a visual programming language, GUI programs could be created easily by using a drag-and-drop method to develop the GUI itself, and then program the "behind-the-scenes" code to handle operations when the user interacts with the GUI components. For example, a GUI might contain two buttons and a text bar. The programmer creates the GUI by simply inserting the buttons and text bar onto a blank panel. Then, the programmer must implement the actions that should occur when the buttons are clicked on or when text is entered into the text bar.

The 1990s saw the beginning of the privatization of the Internet. Companies were allowed to sell individuals the ability to access the Internet from their homes or offices. With this and the advent of the World Wide Web, a new concept in programming was pioneered. Previously, to run a program on your computer, you would have to obtain a compiled version. The programmer might have to find a compiler for every platform so that the software could be made available on every platform. If a programmer wrote a C++ program, the programmer would have to compile the program for Windows, for Macintosh, for Sun workstations, for Intel-based Linux, for IBM mainframes, and so forth. Not only would this require the programmer to obtain several different compilers, it might also require that the programmer modify the code before using each compiler because compilers might expect slightly different syntax.

When developing the Java programming language, the inventors tried something new. A Java compiler would translate the Java program into an intermediate language that they called *byte code*. Byte code would not run on a computer because it was not machine language itself. But byte code would be independent of each platform. The compiler would produce one byte code program no matter which platform the program was intended to run on. Next, the inventors of Java implemented a number of Java Virtual Machines (JVM), one per platform. The JVM's responsibility would be to take, one by one, each byte code instruction of the program, convert it to the platform's machine language, and execute it. Thus, the JVM is an interpreter. Earlier, it was stated that interpreting a program is far less efficient than compiling a program, which is true. However, interpreting byte code can be done nearly as efficiently as running a compiled program because byte code is already a partially translated program instruction. Thus, this combination of compilation and interpretation of a program is nearly as efficient as running an already compiled program. The advantage of the Java approach is that of *portability*; a program compiled by the Java compiler can run on any platform that has a JVM. JVMs have been implemented in nearly all web browsers, so Java is a language commonly used to implement programs that run within web browsers (such programs are often called applets).

The desire for platform-independent programs has continued to increase over time with the increased popularity of the Internet. Today, Microsoft's .net ("dot net") programming platform permits a similar approach to Java's byte codes in that code is compiled into byte code and then can be interpreted. .Net programming languages include C# (a variation of C++ and Java), ASP (active server pages), Visual Basic (a visual programming language), C++, and a variant of Java called J++. The .Net platform goes beyond the platform-independent nature of Java in that code written in one language of the platform can be used by code written in another one of the languages. In this way, a program no longer has to be written in a single language but instead could be composed of individual classes written in a number of these languages.

It is unknown what new features will be added to languages in the future, or where the future of programming will take us with respect to new languages. But to many in computer science, a long-term goal is to program computers to understand *natural languages*. A natural language is a language that humans use to communicate with each other. If a computer could understand English, then a programmer would not have to worry about writing code within any single programming language, nor would a computer user have to learn the syntax behind operating system commands such as Linux and DOS. Natural languages, however, are rife with ambiguity, so programming a computer to understand a natural language remains a very challenging problem.

TYPES OF INSTRUCTIONS

Program code, no matter what language, comprises the same *types* of instructions. These types of instructions are described in this section along with examples in a variety of languages.

Input and Output Instructions

Input is the process of obtaining values from input devices (or files) and storing the values in variables in memory. Output is the process of sending literal values and values from variables to output devices (or to files).

Some languages have two separate sets of I/O statements: those that use standard input (keyboard) and output (monitor) and those that use files. Other languages require that the input or output statement specify the source/destination device. In C, C++, and Java, for example, there are different statements for standard input, standard output, file input, and file output. In FORTRAN and Pascal, the input and output statements default to standard input and output but can be overridden to specify a file. In addition, some languages use different I/O statements for different data types. In Ada, there are different input statements for strings, for integers, for reals, and so forth. Most languages allow multiple variables to be input or output in a single statement.

Figure 14.3 demonstrates file input instructions from four languages: FORTRAN, C, Java, and Pascal. In each case, the code inputs three data—x, y, and z (an integer and two real or floating point numbers)—from the text file input.dat. Each set of code requires more than just the input instruction itself.

The FORTRAN code first requires an OPEN statement to open the text file for input. The file is given the designator 1, which is used in the input statement, READ. The READ statement is fairly straightforward although the (1, 50) refer to the file designator and the FORMAT line, respectively. The FORMAT statement describes how the input should be interpreted. In this case, the first two characters are treated as an integer (I) and the next six characters are treated as a floating point (real) number with two digits to the right of the decimal point (F6.2) followed by another float with the same format.

In the case of C, the first statement declares a variable to be of type FILE. Next, the file is opened with the fopen statement. The "r" indicates that the file should be read from but not written to. The notation "%d %f %f" indicates that the input should be treated as a decimal value (integer) and two floating point values. The & symbol used in C before the variables in the scanf statement is required because of how C passes parameters to its functions.

```
   OPEN(1,FILE='input.dat', ACCESS='DIRECT', STATUS='OLD')
   READ(1,50) X, Y, Z
50 FORMAT(I2, F6.2, F6.2)


FILE *file;
file = fopen("input.dat", "r");
fscanf("%d %f %f", &x , &y, &z);


var input : text;
...
Assign(input, 'input.dat');
Reset(input);
Readln(input, x, y, z);


try {
        BufferedReader input = new BufferedReader(new FileReader("input.dat"));
        data=input.readLine( );
        StringTokenzier token = new StringTokenizer(data, " ");
        x=Integer.parseInt(token.nextToken( ));
        y=Float.parseFloat(token.nextToken( ));
        z=Float.parseFloat(token.nextToken( ));
}
catch(IOException e) { }
```

FIGURE 14.3 Example file input in four programming languages.

In Pascal, the textfile must be declared as type text. Next, the text variable is assigned the name of the file and opened. Finally, the values are input. Pascal, the least powerful of these four languages, is the simplest. In fact, Pascal was created with simplicity in mind.

In Java, any input must be handled by what is known as an exception handler. This is done through the try-catch block (although Java gives you other mechanisms for handling this). Next, a variable of type BufferedReader is needed to perform the input. Input from BufferedReader is handled as one lengthy string. The string must then be broken up into the component parts (the three numbers). The string tokenizer does this. Additionally, the results of the tokenizer are strings, which must be converted into numeric values through the parse statements. If an exception arises while attempting to perform input, the catch block is executed, but here, the catch block does nothing.

Assignment Statements

The assignment statement stores a value in a variable. The variable is on the left-hand side of the statement, and the value is on the right-hand side of the statement. Separating the two parts is an assignment operator. The typical form of assignment operator is either = or : = (Pascal and Ada). The right-hand side value does not have to be a literal value but is often some form of expression. There are three common forms of expressions: arithmetic, relational, and Boolean. The arithmetic expression comprises numeric values, variables, and arithmetic operators, such as a * b – 3. In this case, a and b are variables presumably storing numeric values. Arithmetic operators are +, –, *, /, and modulo (division yielding the remainder), often denoted using %. Relational expressions use relational operators (less than, equal to, greater than, greater than or equal to, equal to, etc.) to compare values. Boolean expressions use AND, OR, NOT, XOR. There can also be string expressions using such string functions as concatenation, and function calls that return values, such as sqrt(y) to obtain the square root of y. Function calls are discussed in Subroutines and Subroutine Calls.

Since the equal sign (=) is often used for assignment, what symbol do languages use to test for equality (i.e., "does a equal b")? We usually use the equal sign in mathematics. Should a programming language use the same symbol for two purposes? If so, this not only might confuse the programmer, but it will confuse the compiler as well. Instead, languages use two different sets of symbols for assignment and equality. Table 14.1 provides a

TABLE 14.1 Assignment and Equality Operators in Several Languages

Language	Assignment Operator	Equality Operator	Not Equal Operator
Ada	: =	=	/=
C/C++/Java, Python, Ruby	=	= = (2 equal signs)	! =
COBOL	assign	Equals (or =)	Is not equal to (or NOT =)
FORTRAN	=	.EQ.	.NE.
Pascal	: =	=	<>
Perl	=	eq (strings), = = (numbers)	ne (strings), ! = (numbers)
PL/I	=	=	<>

comparison in several languages. It also shows what the "not equal" operator is. Notice that PL/I is the only language that uses = for both assignment and equality.

The assignment statement in most languages assigns a single variable a value. Some languages, however, allow multiple assignments in one instruction. In C, for instance, x = y = z = 0; sets all three variables to 0. In Python, you can assign multiple variables by separating them on the left-hand side with commas, as in x, y = 5, x + 1, which assigns x to the value 5, and then assigns y to the value of x+1 (or 6). This shortcut can reduce the number of instructions in a program but may make the program less readable. C also has shortcut operators. These include the use of ++ to increment a variable, as in x++, which is the same as x = x + 1, and + =, which allows us to write x+ = y instead of x = x + y. C also permits assignments within the expression part of an assignment statement. The statement a = (b = 5) * c; is two assignment statements in one. First, b is set equal to 5, and then a is set equal to b * c using b's new value.

If the language is compiled and variables are declared, then the compiler can test for type mismatches. This type of error will arise if the right-hand side of the assignment statement generates a value that is not compatible with the variable on the left-hand side. For example, if x is an integer variable, then the assignment statement x = 3.1415 * y will yield a type mismatch error because the right-hand side involves a real (floating point) number. This is true no matter what type y is.

Selection Statements

Conditions are expressions that evaluate to true or false. Conditions are used in programs to determine what the program should do next. They are used in two types of statements, selection statements and iteration (or loop) statements (covered in Iteration Statements). The idea behind a condition is that the value stored in one or more variables is compared. Comparisons can be based on relational operators and can also use Boolean operators and arithmetic operators. For instance, a condition might test to see if x is greater than y, or x > y. Another condition might test if x equals y or x equals z. Different languages use different symbols for the relational operators (less than, greater than, equal to, etc.) and the Boolean operators (and, or, not).

A *selection* statement decides whether a series of instructions should be executed or skipped based on the evaluation of a condition. This type of statement is often called an if–then statement (or just an if statement). In the if–then statement, if the condition is true, the then portion of the statement (usually called the *then clause*) is executed. If the condition is false, the then clause is skipped.

In an if–then–else statement, if the condition is true the then clause is executed, and if the condition is false the *else clause* is executed. You would use an if–then statement if you only wanted code to execute when a condition is true. This is sometimes called a one-way selection. You would use an if–then–else statement if you wanted code to execute no matter what the condition turned out to be. This is sometimes called a two-way selection.

Figure 14.4 provides example code in three languages: Pascal, C, and FORTRAN IV (an early version of FORTRAN). Each example is of the same logical problem, adding 1 to x if x is equal to y or x is equal to z, and subtracting 1 from x otherwise. The Pascal code is the easiest to understand. In C, two equal signs make up the "equality" operator and two vertical

```
If x = y or x = z then
x := x + 1
Else
x := x - 1;


if ( x = = y || x = = z)
        x = x + 1;
else
        x = x - 1;


        IF ( X .EQ. Y .OR. X .EQ. Z) GO TO 10
        X = X - 1
        GO TO 20
10      X = X + 1
20 ....
```

FIGURE 14.4 Three examples of if–then–else statements.

bars (||) make up the OR operator. The FORTRAN code, which relies on GO TO statements, is less intuitive. Read through the code and see if you can understand it. You might notice that in C, the word "then" is omitted and the condition is placed inside of parentheses.

The if–then–else statement selects between two sets of code, a "one thing or the other" situation. What if there were more than just the two possible sets of code to select between? Most languages allow the then-clause and the else-clause to contain if–then and if–then–else statements. This creates what is called a *nested* if–then–else statement. Figure 14.5 shows an example in Pascal, demonstrating how to determine which letter grade to assign based on the student's numeric grade. We use the typical 90/80/70/60 breakdown. Based on the condition that matches, the code assigns the variable *letter* one of the following values: 'A', 'B', 'C', 'D', or 'F'.

Notice that the last statement does not require an if. The idea is that if the first condition is true, then the first then clause executes and the entire nested statement ends; otherwise, the first else clause executes. That else clause contains an if–then–else statement. For that second condition, if it is true, then the second then clause executes and the statement ends; otherwise, the second else clause executes. That else clause also contains an if–then–else statement. If that clause's condition (the third condition) is true, then the third then clause executes and the statement ends; otherwise, the third else clause executes. The third else clause contains yet another if–then–else statement. If that fourth condition is true, then

```
If grade >= 90 then letter := 'A'
          Else if grade >= 80 then letter := 'B'
                    Else if grade >= 70 then letter := 'C'
                              Else if grade >= 60 then letter := 'D'
                                        Else letter := 'F';
```

FIGURE 14.5 Nested if–then–else in Pascal.

the fourth then clause executes and the statement ends; otherwise, the final else clause executes.

A couple of other comments are worth mentioning here. The indentation provided above helps make the code more readable, but does not impact the code's accuracy or performance. That is, all of the tabs inserted are ignored by the computer. Also, capitalization is unimportant. Finally, you might notice that the entire set of code contains a single semicolon at the end. One difference between Pascal and C, C++, and Java is the use of semicolons to end statements. In Pascal, only one is used, to end the entire structure.

Notice in the above examples that the then and else clauses consisted of single operations. For instance, in the example shown in Figure 14.4, each clause did one thing: increment x or decrement x. It is just as often the case that the clause will contain multiple instructions. In Pascal, for instance, you might have code like this:

```
If x > = y
        Then x: = z;
             y: = 0;
q : = x + y;
```

Indentation and other white space is ignored, so how does the compiler know when the then clause ends? That is, is q : = x + y; part of the then clause? In fact, it is not, and neither is y : = 0; even though the spacing implies that it is. In languages such as Pascal and C, the compiler only expects a clause to have a single instruction. If you want to specify multiple instructions, you must place them into a *block*. A block denotes a collection of instructions that should be treated as one. This allows you to have multiple instructions in a then or else clause. In Pascal, a block is denoted using the words *begin* and *end*, whereas in C, the block is denoted using the symbols { and }. The correct Pascal and C versions of the above if–then statement are shown below. You might notice that y : = 0 does not end with a semicolon. Pascal semicolon rules are somewhat complicated.

```
If x > = y Then
        Begin
              x : = z;
              y : = 0
        End;
q : = x + y;
```

```
if (x > = y) {
        x = z;
        y = 0;
}
q = x + y;
```

In some languages such as Ada, an explicit endif ends the statement. In early FORTRAN instead, you denoted the end of clauses through GO TO statements. We will see the block return when we get to iteration statements and in Subroutines and Subroutine Calls. Below is the same code from above but this time written in Python. Python is a unique language in that indentation *does* matter. Notice how we do not need the block indicators (begin and end or { and }) but instead, the indentation indicates that both instructions after if x > = y: are part of the then clause. Without the indentation for the last statement, it means that it is not a part of the then clause.

```
If x > = y:
      x = z
      y = 0
q = x + y
```

Let us look at another example of a nested if–else statement, this time in C. You will see that there are three different clauses that can execute based on the values of x, y, and z. Given values for x, y, and z, you should be able to determine which of the three clauses executes. For instance, if x = 5, y = 3, and z = 0, which clause would execute?

```
if(x > 0)
      if(y > 0)
            …//clause 1
      else if(z > 0)
            …//clause 2
      else …//clause 3
```

Clause 1 will execute if x > 0 and y > 0. Clause 2 will execute if x > 0, y < = 0, and z > 0 (if x < = 0, we never reach the second if-statement and therefore we never reach the else if-statement, and if y > 0, we never reach the else if-statement). When does clause 3 execute? That is, is the last else clause paired with the condition (z > 0) or the condition (x > 0)? It is not paired with (y > 0) because the else if statement contains the else clause for that condition.

In this example code, it is unclear if the last else is associated with the first if or the third if. This is a situation known as a *dangling else*. If we were to use block notation, we could easily resolve this as one of the following:

```
if(x > 0) {
      if(y > 0)
            …//clause 1
      }
else if(z > 0)
      …//clause 2
else …//clause 3
```

```
if(x > 0)
      if(y > 0) {
            …//clause 1
      else if(z > 0)
            …//clause 2
      }
else …//clause 3
```

In the code on the left, the last else goes with the second if statement (y > 0) and in the code on the right, the last else goes with the first if statement (x > 0). However, in C, we do not need to use the block notation because, by default, any dangling else clause is always associated with the *most recent* condition. So clause 3 executes if x > 0 and y < = 0 and z < = 0. If we did not want the default to apply (the case on the right above), we would then use the block notation. Different languages have different rules regarding dangling else statements. By the way, if x = 5, y = 3, and z = 0, clause 3 will execute.

There are other forms of multiple selection statements. In Pascal, they are called case statements. In C and Java, they are called switch statements. We will not cover them here although you would see them in any introductory programming course.

LANGUAGE BARRIERS

As described in Types of Instructions, all programming languages have the same types of programming language instructions: input, output, assignment, selection, iteration, subroutines. If this is true, then should you not be able to solve a given problem in any language? And if so, then why are there so many different programming languages? It seems unreasonable to have dozens, hundreds or thousands of languages when one will do!

Early in the history of computers, languages offered different features. COBOL, for instance, allowed you to structure your data very precisely, whereas ALGOL introduced useful control statements and LISP provided for both symbolic processing and recursion. But if you could write a program to solve a problem in any language, what exactly were the differences? The differences were a matter of convenience. For instance, you could not write recursive code in FORTRAN or COBOL. If the problem were to call for recursion, you could still solve it in FORTRAN or COBOL, but it would be far more challenging because your program would have to mimic the aspects of recursion that were required.

Even today, we see differences in the advantages and disadvantages of the various programming languages that lead some people to use one and other people to use another. These features include how safe a language is, whether the language can be interpreted, if the language produces byte code, and even what the language looks like. To a C++ or Java programmer, for instance, Python is very unusual because it uses indentation rather than blocks denoted with { } symbols.

Iteration Statements

There are times when a set of code should be performed multiple times. Let us consider as an example that you want to write a payroll program that will perform payroll operations. The program needs to input an employee's data, compute the employee's pay and taxes, output the results, and maintain totals throughout. You might think to write the program as a straight line of code that performs the input steps, the computations, and the outputs, and then ends. This would compute one employee's pay. You would not want to run the program one time per employee because it would be annoying to keep rerunning the program. Additionally, you would not be able to compute the totals of all employees. Instead, you would write the "straight line code" and place it inside of an iteration statement. The *iteration* statement is used to repeat a set of code.

There are two general forms of iteration statement. The *conditional*, or *logical*, loop repeats the code while the condition evaluates to true. The *counting* loop (often known as a for-loop) iterates a number of times based on either a starting and ending point (for instance, 1 to 10) or the values stored in a list. The reason that counting loops are sometimes called *for-loops* is that most languages use the reserved word *for* to indicate that it is a counting loop. The following examples are of Pascal code. First, we see a logical loop.

```
While x > y
      begin
            ...
      end;
```

In the preceding loop, the condition is x > y. This means that the loop will continue to execute while x remains greater than y. Notice the use of the begin…end block. As with if–then and if–then–else statements, if the while loop's code (known as the *loop body*) consists of more than one instruction, the use of a block is required. In the above loop, once y is no longer less than x, the loop terminates and the program continues with the next instruction after the end statement. In C, the same loop would look like this:

```
while (x > y) {
        ...
}
```

The for-loop can be used in one of two ways. First, the loop iterates through a series of values denoted as the starting value and the terminating value by units called the *step size.* For instance, we might specify a loop from 1 to 11 by 2s (the loop would iterate by counting 1, 3, 5, 7, 9, 11). Second, the loop iterates through a list. For instance, if a variable stored a list of values, say x = {1, 10, 25, 39, 44}, the loop would iterate five times, once per value in the list. In either case, the loop contains a *loop index.* This variable stores the value that is currently being iterated over. In the counting loop, the index would take on each value of 1, 3, 5, 7, 9, 11, and in the list loop, it would take on each value of the list (1, 10, 25, 39, 44).

The loop below is a counting loop. The loop index I takes on the value of 1 during the first iteration, 2 during the second iteration, 3 during the third, and so forth until it reaches 10, its final iteration.

```
for I : = 1 to 10 do
        begin
                ...
        end;
```

The loop index can be referenced in the loop body. You might use this value as part of a computation. The following code not only iterates from 1 to 10, but adds the current loop index to a running sum. This code then sums up the values from 1 to 10 (i.e., it computes 1 + 2 + 3 + … + 10). This is known as a summation loop.

```
Sum := 0;
For I := 1 to 10 do
        Sum := Sum + I;
```

Notice that we did not need to place the loop body inside of begin…end statements because the loop body consists of a single instruction. The same code is given below in C. The C for-loop looks odd in comparison to the relatively simple Pascal for loop. In C, the first clause in the parentheses initializes any loop index(es), the second contains the terminating condition, and the third clause performs the step increment or decrement of the index.

```
sum = 0;
for(i=0;i<=10;i=i+1)
        sum = sum + i;
```

C actually provides a number of shortcuts. One of which is the statement i++, which can be used in place of i = i + 1. Another is sum +=i, which can be used in place of sum = sum + i. The following is a for-loop (just the loop statement, not the loop body) that will iterate downward instead of upward. for (i = 10; i > 0; i--). One difference between C and Pascal is that C can vary its step size (for instance, iterating by 2s or 5s), whereas Pascal can only iterate upward or downward by ones.

Some languages offer multiple forms of both conditional and counting loops. C and Java have both while loops and do–while loops. The difference between them is where the condition is tested. In the while loop, the condition is tested before entering the loop body, whereas in the do–while loop, the condition is tested after executing the loop body. Although this may seem like the same thing, the do–while loop tests the condition after executing the loop body so that the loop body is executed at least one time. If a while loop's condition is initially false, then the loop body does not execute at all. Here is an example that compares two loops in C.

```
while(x > 0)
      x = x/2;

do {
      x = x/2;
} while (x > 0);
```

Assume in both of these loops above that x is an integer initially equal to −1. In the first loop, the condition is initially false and so the loop body (x = x/2) never executes. Thus, x remains −1. In the second loop, the loop body executes one time because the condition is not checked until after the loop body executes, so x is changed from −1 to 0 (the value −1/2 cannot be stored in an integer, so instead x gets the value 0). The loop then terminates because x is not greater than 0.

Newer languages often forego the counting loop in favor of an *iterator* loop. This version of a counting loop iterates over a list rather than over a sequence of values. For example, if you have a list, 1, 5, 10, 19, then the loop will iterate four times, once for each of the four values in the list. Python is an example where the for-loop iterates over a list. You can use this type of loop to simulate a counting loop by iterating over a range from 1 to the upper limit by stating range(1, 100). C++ and Java originally only had the counting form of for loop but today have both counting and iterator loops.

One last comment regarding conditional loops is that the loop body should contain instructions that manipulate at least one of the variables used in the loop's condition. Consider the following code in C.

```
while(x > y)
      x = x + 1;
```

If x is initially greater than y, the loop body will be executed. However, because x starts at a value larger than y, repeatedly adding to x will never make x become less than or equal to y, and the result is that the loop will never terminate. This is known as an *infinite loop.*

In order to ensure that you do not have an infinite loop, you have to make sure that there is code in the loop body that changes some value in the condition so that, eventually, the condition can become false. Consider the following C code.

```
while(x > 0)
      printf("%d\n", x);          //print out x
      x = x - 1;
```

This loop looks like it will count down from whatever x starts at to 0, printing out each value. But because the loop body contains two instructions, we need to place them inside a block using the { } symbols. Unfortunately, the compiler examines this code and decides that the statement x = x – 1; exists outside of (or after) the loop. Therefore, the while loop only performs the printf output statement. The result is that, if x is greater than 0 to begin with, the printf executes and the condition is tested again. Since the printf did not alter x, it is still greater than 0 and the printf executes again. This repeats, over and over again and never stops (until the user forces the program to abort). The reason for this infinite loop is a logical error, the programmer's logic was not correct. To fix this problem, the loop should be as follows.

```
while(x > y) {
      printf("%d\n", x);          //print out x
      x = x - 1;
}
```

Infinite loops are often a problem for programmers, even those with extensive experience. This type of logical error is often difficult to detect, and the programmer only discovers the error when testing the program and finding that it seems to "hang". In this case, a loop never terminates so the program does not advance on to the next step.

Recall from earlier that Python does not use special block designators such as begin… end and {…}. Instead, indentation is used. So in this case, the infinite loop from earlier would probably not occur because the indentation would resolve it.

Subroutines and Subroutine Calls

A *subroutine* is a set of code separate from the code that might invoke it. A *subroutine call* invokes the subroutine. The subroutine might be part of your program, or it might be part of another program or some independent piece of code made available through a library of subroutines. Most programmers will write programs as a series of subroutines. In this way, designing, writing, and debugging a program is simplified because the programmer need only concentrate on one subroutine at a time. The concept is known as *modular* programming where the modules are relatively small and independent chunks of code.

What follows is a brief example of two subroutines written in C. In C, subroutines are called *functions*. The function main is always the first to execute in a program. Here, main will execute its first printf (output) statement and then call the function foo. The function foo will output its own statement and terminate. When foo terminates, main resumes from the point immediately after the call to foo, which is the last printf statement. The function main then terminates and the program is over.

```
void foo( ) {
      printf("are we in foo? Yes!");
}

void main( ) {
      printf("Hello world. We are about to enter foo.");
      foo( );
      printf("We have now returned from foo.");
}
```

The output of this program is as follows:

```
Hello world. We are about to enter foo.
are we in foo? Yes!
We have now returned from foo.
```

Obviously, there is no reason to write this program in two functions; we could have instead written all three printf statements in main. However, most programs are too complicated to write in a single function.

In the preceding example, the function call is foo( ). The parentheses provide a means of communicating between the calling function (main) and the called function (foo). This communication comes in the form of variables and values. We call them *parameters*. For instance, if a function will compute the square root of a value, we must tell the function what value we want the square root of. If the function is called sqrt, we might call the function using notation such as sqrt(5) to compute the square root of 5, or sqrt(x) to compute the square root of the value stored in variable x. What follows is an example of passing a parameter in a function in C. The function computes a value (the reciprocal of the parameter) and outputs the result.

```
void determineReciprocal(float x) {
      if(x==0)
            printf("There is no reciprocal of 0");
      else printf("The reciprocal is %f", 1.0/x);
}
```

Here, the function is passed a float value. The function will output a different message for each different parameter. For instance, what does it output if x = 1.0? What if x = 10.0? What if x = 0?

Special Purpose Instructions

These might include invocations of the operating system, string operations, computer graphics routines, random number operations, and error handling routines.

Aside from the list of executable instructions above, some programming languages also require explicit *variable declarations*. In languages such as Pascal, C, and Java, variables

```
Var x, y : integer;
Var z : real;

int x, y;
float z;

var x, y, z;
```

FIGURE 14.6 Three forms of variable declarations.

require being typed, that is, declared a specific *type* of value, which they then remain as during the course of the program. In JavaScript, although variables must be declared by name, you do not type them. In Figure 14.6, three variables, x, y, and z, are declared in three languages: Pascal (top), C (center) and JavaScript (bottom). In C, int means integer and float is the same as Pascal's real (a number with a decimal point).

Some languages allow you to define your own types. In Ada, you can declare a type to be a subset of another type. For example, we might define the type Grades to be a subset of integers with an allowable range of 0..100. In this way, a variable of type Integer is any integer where a variable of type Grade is an integer within the range 0 to 100. In Java, types are known as classes.

WHAT'S IN A NAME?

Most early programming languages were named by acronyms that explained the language (e.g., FORTRAN, COBOL, ALGOL, LISP). Today, language names are more creative. Here are some of the more intriguing ones.

- Axum
- BeanShell
- Blue
- Boo
- Boomerang
- Caml
- Chef
- Cola
- EASY
- Fancy
- Go
- Groovy
- Joy
- Lua
- Nice
- Oxygene
- Pure
- Rust
- Scratch
- Shakespeare
- Squeak
- ZZ T-oop

SCRIPTING LANGUAGES

As a system administrator, it is important that you understand programming because you will often have to write scripts. A *script* is an interpreted program. Scripts are often fairly short (as compared to applications or systems software). But in spite of not having to write code for large-scale applications, a system administrator must still understand the commands of a programming language as presented.

Types of Instructions

Popular scripting languages today include shell scripting languages (like the Bash scripting language and DOS) as well as more complete and complex programming languages such as Python, Ruby, Perl, and PHP. Here, we will concentrate on the Bash Shell and DOS languages only. We will specifically examine the language features described in Types of Instructions, and see numerous examples. As this is not a programming text, these sections only introduce shell scripting.

Some of the code that you can place into a shell script is the same as the commands that you use at the command line. In Linux, for example, a script can contain cd, ls, rm, echo, and so forth. And similarly, commands that you will see in the next two subsections can also be typed at the command line. However, there are some differences. A script can receive parameters making the script more powerful.

Bash Shell Language

Shell scripts are stored in text files. These files must be both readable *and* executable. It is best to set such files with 745 or 755 permission. To execute a script named foo, you would type ./foo. This implies that your working directory stores the file foo. All Bash shell scripts should start with the following line, which alerts the Bash interpreter to execute: #!/bin/bash.

Variables in Bash scripts are handled the same way as from the command line. You can assign a variable a value using an assignment statement of the form `VAR=VALUE`.* The value on the right-hand side must be a string, a literal number, or the value computed from an integer-based arithmetic expression. If the string has spaces, enclose the entire string in "" marks. We look at integer-based arithmetic expressions below. To access the value in a variable, precede the variable's name with a $. Some examples follow.

```
X=0
Y="Hi there"
NAME=Frank
Z=$X
MESSAGE="$Y $NAME"
```

X contains the value 0, Y the string "`Hi there`", and NAME the string "`Frank`" even though no quote marks were used in the assignment statement. Z obtains the value

* Note that variable names do not need to be written in upper-case letters. I will use upper-case letters to easily distinguish a variable from an instruction.

currently stored in X, 0. `MESSAGE` obtains the string of the value stored in Y (`Hi there`), a space, and the value stored in NAME (Frank), or "Hi there Frank".

An integer-based operation is tricky because, if it has variable values, it must be enclosed in peculiar syntax: $(( )). For example, to set Z to be X + 5, you would use Z = $((X + 5)), and to increment X, you would use X = $((X + 1)). The legal operators for arithmetic expressions are +, –, *, /, and % (% performs the modulo operation, i.e., it provides the remainder of a division). The division operation (/) operates only on integers and returns an integer quotient. For example, if X = 5 and Y = 2, then Z = $((X/Y)) results in Z being set to 2, not 2.5. The operation Z = $((X%Y)) results in Z being set to 1 because 5/2 is 2 with a remainder of 1.

The output statement is *echo*. You saw this earlier in the textbook. The echo statement contains a list of items to be output. This list can contain variables, literal values, and Linux commands. If you place this list '', the output is provided literally (i.e., variable values are not returned, instead you get the names of the variables). For example, if NAME = Frank and AGE = 53, then

```
echo $NAME is $AGE
```

provides the output

```
Frank is 53
```

but the instruction

```
echo '$NAME is $AGE'
```

provides the output

```
$NAME is $AGE.
```

You can also place your output within "", which provides the same result as having no quote marks. In most cases, the "" can be omitted in an echo statement. However, consider the following two echo statements:

```
echo $NAME is $AGE
echo "$NAME is $AGE"
```

The first echo statement contains three arguments (parameters), whereas the second echo statement only contains one. In addition, the use of "" suppresses Bash' file name expansion. For instance, echo * will result in Bash first expanding * to be all items in the local directory, and then echo will output them, so echo * is the same as ls or ls *. However, echo "*" will literally output * as the file name expansion does not occur.

You can use echo to combine the output of a literal message with the output of another Linux command. For instance, if you wished to output "the date and time are" along with the current date/time, you can combine the echo statement and the date operation. However, to keep echo from outputting the word "date" literally, you must indicate that Bash should execute the date command. This is accomplished using the back quote marks, ` `. For instance, the statement

```
echo the date is date
```

will literally output "the date is date". However, the statement

```
echo the date is `date`
```

causes Linux to execute the date command and then insert its output into the echo statement. The result would look like this: "the date is Thu Mar 29 10:35:32 EDT 2012".

Input is accomplished through the *read* statement. This statement must always be followed by a variable name. For instance, "read NAME" would be used to obtain an input from the user and place the entered value into the variable NAME. When a script with an input statement is executed, a read statement will result in the cursor blinking on a blank line. Since the user may not know what is going on, or what the program is expecting, it is important to precede any read statement with a prompting statement. A prompt is simply an output statement that instructs the user on what to do. The following example would prompt the user for their name, input the value, and then provide a personalized output statement. Notice that the variable NAME in the input statement is not preceded with $, whereas the $ is used in the output statement.

```
echo Enter your name
read NAME
echo Hello $NAME, nice to meet you!
```

The user may enter anything in the read statement. For instance, they might enter their name, or another string entirely, or even a number. Nothing in the code above prevents this. If the input has a space in it, no special mechanism is required in the code, unlike an assignment statement such as NAME="Frank Zappa", which requires that the string be placed in quote marks. If the user were to enter Frank Zappa for the above code, the output would be as expected, "Hello Frank Zappa, nice to meet you!".

At this point, we will put together everything we have seen. The following script will compute the average of four values. Notice the use of parentheses in the two assignment statements. Recall that to perform an arithmetic operation that contains variables, you use the notation $((…)). However, in this case, since we want to sum num1, num2, num3, and num4 before performing the division or mod operation, we place that summation in another layer of parens. Without the parens, the division (num4/4) is performed before the additions.

```
#!/bin/bash
echo Please input your first number
read num1
echo Please input your second number
read num2
echo Please input your third number
read num3
echo Please input your fourth number
read num4
quotient = $(((num1+num2+num3+num4)/4))
remainder = $(((num1+num2+num3+num4)%4))
echo The average is $quotient with a remainder of $remainder/4
```

The conditional statement in Bash has very peculiar syntax. Because of this, it is often challenging for introductory scripters to get their scripts running correctly. Make sure you stick to the syntax. The form of the statement is "if [condition]; then action(s); fi". The semi-colons must be positioned where they are shown here. Furthermore, there must be spaces between the various components of the if statement as shown. Finally, the word "fi" must end your if statement. The actions are individual statements, separated by semicolons. For instance, you might have assignment statements, input statements, output statements, and even nested if-statements.

The conditional statement requires a condition. There are two common forms of the condition:

- variable comparison value
- filetest filename

In the former case, the variable's name is preceded by a $. The comparisons are one of == or ! = for string comparisons, or one of –eq, -ne, -gt, -lt, -ge, -lt for numeric comparisons (equal to, not equal to, greater than, less than, greater than or equal to, less than or equal to). The value maybe a literal value, another variable (whose name is again preceded by $), or an arithmetic expression.

The second case is a filetest condition followed by a filename. A filetest is one of –d (item is a directory), –e (file exists), –f (file exists and is a regular file), –h (item is a symbolic link), –r (file exists and is readable), –w (file exists and is writable), and –x (file exists and is executable). The filetest may be preceded by the symbol ! to indicate that the condition should be negated. That is, the test –r foo.txt asks if foo.txt is readable, whereas ! –r foo.txt asks if foo.txt is not readable. Some example conditions follow. Notice the blank spaces between each part of the condition. Without these blank spaces, your script will generate an error and not execute.

```
[ $NAME == "Frank" ]
[ $X -gt $Y ]
[ $X -eq 0 ]
[ $STUDENT1 ! = $STUDENT2 ]
[ -x $FILENAME ]
[ ! -r $FILE2 ]
```

A complete example is shown below. In this if–then statement, the variable $NAME is tested against the value "Frank", and if they are equal, the instruction both outputs the message "Hi Frank" and adds 1 to the variable X. The fi at the end states that this ends the if-statement.

```
if [ $NAME == "Frank" ]; then echo Hi Frank; $X=$((X+1)); fi
```

In some cases, a condition requires more than one comparison, combined with a Boolean AND or OR. For example, to test to see if a variable's value lies between two other values, you would have to test the variable against both values. If we want to see if the value stored in the variable X is between 0 and 100, we would test $X –ge 0 AND $X –le 100. The syntax becomes even more peculiar in this case. Such a conditional is called a *compound* conditional and requires an additional set of [ ] around the entire condition. The symbols used for AND and OR are the same as are used in C and Java, && for AND, and || for OR. To test X between 0 and 100, we would use the following: [[$X –ge 0 && $X –le 100]].

An else clause can be added to the if-statement, in which case the word "else" appears after the action(s) in the then clause, followed by the else clause action(s). The "fi" appears only after the else clause. Below are three examples. Notice that the second example really makes no sense because there are no floating point computations in Linux, so the operation $((1/X)) would not provide a true reciprocal.

```
if [ $X -gt $Y ]; then Y=0;
      else X=0;
fi
if [ $X -ne 0 ]; then Y=$((1/X)); echo The reciprocal of $X is $Y;
      else echo Cannot compute the reciprocal of 0;
fi
if [ $NAME == "Frank" ]; then echo "Hi Frank"; X=$((X+1));
      else echo Who?; X=0;
fi
```

Note that you do not have to place the else clause or the fi statement on separate lines, they are written this way here for readability purposes.

There is also a nested if–then–else statement, where the "else if" is written as elif. The structure looks like this:

```
if [ condition ];
      then
            action(s);
      elif [ condition ]
      then
            action(s);
      elif [ condition ]
      then
            action(s);
      …
      else
            action(s);
fi
```

If you refer back to Figure 14.5, there is an if–then–else statement in Pascal. The equivalent statement in Bash is shown in Figure 14.7.

There are two types of loops available in the Bash scripting language. The for-loop is an iterator loop that will iterate over a list. The list can consist of one of several things. First, it can be an enumerated list of values. Second, it can be a list that was supplied to the script as a list of parameters (this is discussed later in this section). Third, it can be a list generated by globbing, that is, file name expansion.

The basic form of the Bash for loop is:

```
for variable in list; do statement(s); done
```

Here, *variable* is the loop index. The loop index takes on each value in the list, one at a time. The loop index can then be referenced in the statements in your loop body. For instance,

```
for I in 1 2 3 4 5; do echo $I; done
```

will output each of 1, 2, 3, 4, and 5 on separate lines, one per loop iteration.

```
if [ $grade -ge 90 ];
      then letter=A;
elif [ $grade -ge 80 ];
      then letter=B;
elif [ $grade -ge 70 ];
      then letter=C;
elif [ $grade -ge 60 ];
      then letter=D;
else letter=F;
fi
```

FIGURE 14.7 Nested if–then–else in Bash.

The for loop is primarily used for two purposes. The first is to iterate through the items in a directory, taking advantage of globbing. The second is to iterate through parameters in order to process them. Again, we will hold off on looking at parameters until later in this section. The following loop is an example of using globbing. In this case, the loop is merely performing a long listing on each item in the current directory. This could more easily be accomplished using ls –l *.*.

```
for file in *.*; do ls -l $file; done
```

Globbing in a for loop becomes more useful when we place an if–then statement in the loop body, where the if–then statement's condition tests the file using one of the filetest conditions. The following for loop example examines every text file in the directory and outputs those that are executable.

```
for file in *.txt; do
      if [ -x $file ];
            then echo $file;
      fi
done
```

The indentation is not necessary, but adds readability. This loop iterates through all .txt files in the current directory and displays the file names of those files that are executable.

The other type of loop is a conditional loop. This loop is similar to the while loop of C. You specify a condition, and while that condition remains true, the loop body is executed. The syntax for the Bash while loop is:

```
while [ condition ]; do statement(s); done
```

Recall that a while loop can be an infinite loop. Therefore, it is critical that the loop body contain code that will alter the condition from being true to false at some point. One common use of a while loop is to provide user interaction that controls the number of loop iterations. For instance, we might want to perform some process on a list of filenames. We could ask the user to input the file name. In order to control whether the loop continues or not, we could ask the user to enter a special keyword to exit, such as "quit". In this case, "quit" would be called a *sentinel value*. The following script repeatedly asks the user for file names, outputs those files that are both readable and writable, and exits the loop once the user enters "quit". The if–then statement counts the number of times the condition is true for a summary output at the end of the code.

```
#!/bin/bash
COUNT=0
echo Enter a filename, quit to exit
read FILENAME
```

```
while [ $FILENAME ! = quit ]; do
      if [[ -r $FILENAME && -w $FILENAME ]];
            then ls -l $FILENAME; COUNT=$((COUNT+1));
      fi
      echo Enter a filename, quit to exit
      read FILENAME
done
echo There were $COUNT files that were both readable and writable
```

Notice in the previous example that two instructions appeared both before the loop and at the end of the loop. Why do you suppose the echo and read statements had to appear in both locations? Consider what would happen if we omitted the two statements from before the loop. When the loop condition is reached, $FILENAME returns no value, so the loop is never entered. Therefore, we must provide an initial value to filename. If we omitted the echo and read from inside of the loop, then whatever file name you entered initially remains in the variable FILENAME. Thus, the while loop's condition is always true and we have an infinite loop!

You would tend to use a while loop when either you wanted to perform some task on a number of user inputs, waiting for the user to enter a value that indicates an exiting condition (like the word quit), or when you wanted to do a series of computations that should halt when you reached a limit. The former case is illustrated in the previous code. The following example computes the powers of 2 less than 1000, and outputs the first one found to be greater than or equal to 1000.

```
#/bin/bash
VALUE=1
while [ $VALUE -lt 1000 ];
      do
            VALUE=$((VALUE*2));
      done
echo The first power of two greater than 1000 is $VALUE
```

Bash scripts can be passed parameters. These are entered at the command line prompt after the script name. A script is executed using the notation ./script <enter>. Parameters can be added before the <enter>, as in ./script 5 10 15 <enter>. The parameters can then be used in your script as if they were variables, initialized with the values provided by the user. There are three different ways to reference parameters. First, $# stores the number of parameters provided. This can be used in an if statement to determine if the user has provided the proper number of parameters. Second, each individual parameter is accessed using $1, $2, $3, and so forth. If you are expecting two parameters, you would access them as $1 and $2. Finally, $@ returns the entire list of parameters. You would use $@ in a for loop as the list. The following script outputs the larger of two parameters. If the user does not specify exactly two parameters, an error message is provided instead.

```
#!/bin/bash
if [ $# -ne 2 ];
      then echo This script requires two parameters;
      elif [ $1 -gt $2 ];
            then echo Your first parameter is larger;
      else echo Your second parameter is larger;
fi
```

The following script expects a list of file names. If no such list is provided, that is, if $# is 0, it outputs an error message. Otherwise, it iterates through the list and tests each file for executable status and provides a long listing of all of those files that are executable. Notice that this script has an if–then–else statement where the else clause has a for loop and the for loop has an if–then statement. Because of the two if–then statements in the script, there is a need for two fi statements.

```
#/!bin/bash
if [ $# -eq 0 ]; then
      echo No parameters provided, cannot continue;
else
      for FILE in $@; do
            if [ -x $FILE ]; then
                  ls -l $FILE;
            fi
      done
fi
```

Let us conclude with a script that will expect a list of numbers provided as parameters. The script will compute and output the average of the list. Assuming that the script is called avg, you might invoke it from the command line by typing

```
./avg 10 381 56 18 266 531
```

```
#!/bin/bash
if [ $# -eq 0 ]; then echo No parameters, cannot compute average;
else
      SUM=0;
      for NUMBER in $@; do
            SUM=$((SUM+NUMBER));
      done
      AVERAGE=$((SUM/$#));
      echo The average of the $# values is $AVERAGE;
fi
```

Notice in this example that all of the numbers input were specified at the time the user executed the script. What if we want to input the values from the keyboard (using the read command)? How would it differ? This problem is left to an exercise in the review section. However, here are some hints. You would use a while loop. Let us assume all numbers entered will be positive. The condition for the while loop would be [$NUMBER –gt 0] so that, once the user input a 0 or negative number, the while loop would terminate. You would have to add both an echo statement (to prompt the user) and read statement before and in the while loop to get the first input, and to get each successive input.

One last comment on Bash shell scripting. The use of the semicolon can be very confusing. In general, you do not need it if all of your instructions are placed on separate lines. This requires moving the words such as "then", "do", and "else" onto separate lines. The above script could also be rewritten as follows:

```
#!/bin/bash
if [ $# -eq 0 ]
      then
            echo No parameters, cannot compute average
      else
            SUM = 0
            for NUMBER in $@
                  do
                        SUM=$((SUM+NUMBER))
                  done
            AVERAGE=$((SUM/$#))
            echo The average of the $# values is $AVERAGE
      fi
```

As an introductory shell scripter though, it is always safe to add the semicolons.

MS-DOS

In this section, we examine the commands and syntax specific to MS-DOS. You should read The Bash Shell Language before reading this section as some of the concepts such as loops will not be repeated here. Instead, it is assumed that you will already know the concepts so that we can limit our discussion to the commands, syntax, and examples.

As with the Bash language, DOS allows variables, which can store strings and integer numbers. There are three types of variables: parameters that are provided from the command line when the script is executed, environment variables (defined by the operating system or the user), and for-loop index variables. Parameters are denoted using %n, where n is the number of the parameter. That is, the first parameter is %1, the second is %2, up through %9 for the ninth parameter. Although you are not limited to nine parameters, using more parameters is tricky (it is not as simple as referencing %10, %11, and so on). Environment variables appear with the notation %name% (with the exception of numeric values, see the examples that follow). For-loop variables appear with the notation %name.

To assign a variable a value, the assignment statement uses the word set, as in "set x = hello". To assign a variable a value stored in another variable, use the notation "set x =%y%". The percent signs around the variable name tell the interpreter to return the variable's value; otherwise, the interpreter literally uses the name, for example, "set x = y" would set x to store the value "y", not the value stored in the variable y. However, this differs if we are dealing with arithmetic values as described in the next paragraph.

By default, values in variables are treated as strings, even if the values contain digits. Consider the following two instructions

```
set a=5
set b=%a%+1
```

seem to store the value 6 in the variable b. But instead, the instructions store the string 5 + 1 in b. This is because the items on the right-hand side of the equal sign are treated as strings by default and therefore the notation %a%+1 is considered to be a string concatenation operation by combining the value stored in a with the string "+1". To override this default behavior, you have to specify /a in the set statement to treat the expression as an arithmetic operation rather than a string concatenation. Thus, the second statement should be "set /a b =%a%+1", which would properly set b to 6. For arithmetic statements like this, you can omit the percent signs from around variables, so "set /a b = a+1" will accomplish the same thing.

Input uses the set command as well, with two slight variations. First, the parameter /p is used to indicate that the set statement will prompt the user. Second, the right-hand side of the equal sign will no longer be a value or expression, but instead will be a prompting message that will be output before the input. If you omit the right-hand side, the user only sees a blinking cursor. The following example illustrates how to use input.

```
set /p first=Enter your first name
set /p middle=Enter your middle initial
set /p last=Enter your last name
set full=%first% %middle% %last%
```

Notice that spaces are inserted between each variable in the last set statement so that the concatenation of the three strings has spaces between them, as in Frank V Zappa.

Output statements in DOS use the echo command, similar to Linux. Output statements will combine literal values and variables. To differentiate whether you want a literal value, say Hello, or the value stored in a variable, you must place the variable's name inside of percent signs. Assuming that the variable name stores Frank Zappa, the statement "echo hello name" will literally output "hello name", whereas "echo hello %name%" will output "hello Frank Zappa".

A DOS script is set up to output each executable statement as it executes. In order to "turn this feature off", you need to add @echo off. To send output to a file rather than the screen, add "> filename" in your echo statement to write to the file and ">> filename" to append to the file. For example, assuming again that name stores Frank Zappa. The following three output statements show how to output a greeting message to the screen, to the file greetings.txt and to append to the file greetings.txt.

```
echo hello %name%
echo hello %name% > greetings.txt
echo hello %name% >> greetings.txt
```

The if-statement permits three types of conditions. You can test the error level as provided by the last executed program, compare two variables/values, and test if a file exists. For each of these types of conditions, you can precede the condition with the word not. Comparison are equ (=), neq (not equal), lss (<), leq (< =), gtr (>), and geq (>=) for numeric comparisons and = = for strings. The if-statement has an optional else clause. The entire if-statement (including the optional else clause) must appear on a single line of the script, thus it limits the amount of actions that the then clause or else clause can perform. If you use the optional else, place both the then clause and the else clause inside of ( ). The word "then" is omitted, as it is in C. Here are some examples of if-statements.

```
If errorlevel 0 echo Previous program executed correctly
If %x% gtr 5 (echo X is greater than 5) else (echo X is not greater than 5)
If %x% lss 0 set/a x = 0-%x%
If not %name% = = Frank (echo I do not know you)
If Exist c:\users\foxr\foo.txt del c:\users\foxr\foo.txt
```

DOS contains a for-loop but no while loop. The for-loop is an iterator loop like Linux's for-loop. It only iterates over a list of values. The format is as follows:

```
for options %var in (list) do command
```

The following example will sum up the values in the list. It uses no options in the for loop.

```
for %i in (1 2 3) do set /a sum=sum+%i
```

Aside from providing a list to iterate over, you can specify a command such as dir so that the for-loop iterates over the result from the dir command. To do this, you must include the option /f in the for-loop. For example, the following will output the items found when performing a dir command, one at a time.

```
for /f %file in ('dir') do echo %file
```

Unfortunately, combining the for-loop with the ('dir') command may not work as expected because the dir command provides several columns' worth of items, and the echo statement only picks up on the first item (column) of each line (which is the date of the last modification). The command dir /b gives you just the item names themselves. To modify the above instruction, you would specify

```
for /f %file in ('dir /b') do echo %file
```

As with Linux, DOS commands can be placed in DOS shell scripts. This includes such commands as del (delete), move, copy, ren (rename), dir, type (similar to Linux' cat), print (send to a printer), rmdir, mkdir, cd, and cls (clear screen). So you might write scripts, for instance, to test files in the file system, and delete, rename, or move them if they exist or have certain attributes.

FURTHER READING

- Blum, R. and Bresnahan, C. *Linux Command Line and Shell Scripting Bible.* New Jersey: Wiley and Sons, 2011.
- Dietel, P. and Dietel, H. *C++: How to Program.* Upper Saddle River, NJ: Prentice Hall, 2011.
- Glass, R. *In the Beginning: Personal Recollections of Software Pioneers.* New Jersey: Wiley and Sons, 1997.
- Kelley, A. and Pohl, I. *A Book on C: Programming in C.* Reading, MA: Addison Wesley, 1998.
- Kernighan, B. and Ritchie, D. *The C Programming Language.* Englewood Cliffs, NJ: Prentice Hall, 1988.
- Liang, Y. *Introduction to Java Programming.* Boston, MA: Pearson, 2013.
- Mitchell, J. *Concepts in Programming Languages.* UK: Cambridge, Cambridge University Press, 2003.
- Newham, C. *Learning the bash Shell: Unix Shell Programming.* Massachusetts: O'Reilly, 2005.
- Reddy, R., and Ziegler, C. *FORTRAN 77 with 90: Applications for Scientists and Engineers*, St. Paul, MN: West Publishing, 1994.
- Robbins, A. and Beebe, N. *Classic Shell Scripting.* Massachusetts: O'Reilly, 2005.
- Sebesta, R. *Concepts of Programming Languages.* Boston, MA: Pearson, 2012.
- Tucker, A. and Noonan, R. *Programming Languages: Principles and Paradigms.* New York: McGraw Hill, 2002.
- Venit, S., and Subramanian, P. *Spotlight on Structured Programming with Turbo Pascal.* Eagan, MN: West Publishing, 1992.
- Webber, A. *Modern Programming Languages: A Practical Introduction.* Wilsonville, OR: Franklin, Beedle and Associates, 2003.
- Weitzer, M. and Ruff, L. *Understanding and Using MS-DOS/PC DOS.* St. Paul, MN: West Publishing, 1986.

REVIEW TERMS

Terms introduced in this chapter:

- Ada
- ALGOL
- Assembler
- Assembly language
- Assignment statement
- Byte code
- COBOL
- Compiler
- Condition
- Conditional loop
- Counting loop
- Done (Linux)
- Dot net (.net)
- Echo (Linux and DOS)
- Elif (Linux)
- Else clause
- Fi (Linux)
- FORTRAN
- GO TO
- High-level language
- Infinite loop
- Input statement
- Interpreter
- Iteration
- Iterator loop
- Java virtual machine
- Language translator
- LISP
- Logical error
- Loop body
- Loop index
- Machine language
- Modular programming
- Object-oriented programming
- Output statement
- Parameter
- Pascal
- PL/I
- Portability
- Read (Linux)
- Run-time error
- Scripting language
- Selection statement
- Sentinel value
- Set (DOS)
- Spaghetti code
- Syntax error
- Subroutine
- Subroutine call
- Then clause
- Type mismatch
- Unconditional branch
- Variable declaration
- Visual programming

REVIEW QUESTIONS

1. In what ways are high level programming languages easier to use than machine language and assembly language?
2. What is the difference between a compiled language and an interpreted language?
3. Name three improvements to high level programming languages between the first language, FORTRAN, and today.
4. What is spaghetti code and what type of programming language instruction creates spaghetti code?
5. What is the difference between a conditional loop and a counting loop?
6. What is the difference between a counting loop and an iterator loop?
7. You have a high level language program written in the language C. You want to be able to run the program on three different platforms. What do you need to do? If your program were written in Java, what would differ in how you would get the program to run?
8. For each of these languages, explain why it was created (what its purpose was): FORTRAN, COBOL, LISP, PL/I, Simula, SNOBOL.
9. What significant change occurred with the creation of Pascal and C?
10. What significant change occurred with the creation of Smalltalk?
11. What is the .Net platform?

REVIEW PROBLEMS

1. Write a Linux script that will receive two parameters and output whether the two parameters are the same or not.
2. Rewrite #1 so that, in addition, the script will output an error message if exactly two parameters were not provided.
3. Rewrite #1 as a DOS script assuming the two parameters are strings.
4. Rewrite #3 assuming the two parameters are numbers.
5. Write a DOS script that will test the last program for error level 0 and output an ok message if the error level was 0 and an error message if the error level was not 0.
6. Write a Linux script that will receive a directory name as a parameter and will output all of the subdirectories found in that directory.

7. Write a DOS script that will receive a directory name as a parameter and will output all items found in that directory.

8. Write a Linux script that will compute the average of a list of values that are input, one at a time from the keyboard, until the user ends the input with a 0 or negative number. Hint: use a while loop and the read statement.

9. If you have successfully solved #8, what would happen if you did not have the read statement before the while loop? What would happen if you did not have the read statement inside of the while loop?

10. Recall that you can use regular Linux commands in a Linux script. Write a Linux script that will iterate through the current directory and delete any regular file that is not readable. This requires the use of rm –f and the filename. If your loop index is called file, the command might look like rm –f $file. Rewrite your program to use rm –i instead of –f. What is the difference?

DISCUSSION QUESTIONS

1. In your own words, explain the problem with unconditional branches (e.g., the GO TO statement in FORTRAN). How did structured programming solve this problem?

2. Examine various code segments provided in Types of Instructions. Just in looking at the code, which language is easiest to understand, FORTRAN, C, Pascal, or Java? Which language is most difficult to understand? Explain why you feel this way.

3. As a computer user, is there any value in understanding how to write computer programs? If so, explain why.

4. As a system administrator, would you prefer to write code in the bash scripting language, Python, Ruby, C, Java, or other language? Explain.

CHAPTER 15

Information

The textbook has primarily emphasized technology. In this chapter, the focus turns to the information side of IT. Specifically, the chapter examines the types of information that organizations store and utilize along with the software that supports information processing. The chapter then examines information security and assurance: the ability of an organization to ensure that their information is protected. Toward that end, the chapter describes numerous forms of attacks that an organization must protect against, along with technological methods of security information. The chapter examines hardware solutions such as RAID storage and software solutions including encryption. The chapter ends with a look at legislation created to help protect the information technology and individuals' privacy.

The learning objectives of this chapter are to

- Discuss the various forms that information comes in with an introduction to the Data–Information–Knowledge–Wisdom hierarchy.
- Describe the use of databases, database management systems, and data warehouses.
- Introduce the field of information security and assurance and the processes used in risk analysis.
- Discuss the threats to information and solutions in safeguarding information.
- Describe encryption technologies.
- Introduce U.S. legislation that facilitates the use of information and protects individuals privacy.

Up to this point of the textbook, we have concentrated mostly on the technology side of information technology (IT). We have viewed computer hardware, software, operating systems, programming, and some of the underlying theories behind their uses and functions. In this chapter, we look at the other side of IT, information.

WHAT IS INFORMATION?

Information is often defined as processed data. That is, raw data are the input and information is the output of some process. In fact, more generically, we should think of information as any form of *interpreted* data. The data can be organized, manipulated, filtered, sorted, and used for computation. The information is then presented to humans in support of decision making. We might view data and information as a spectrum that ranges from raw data, which corresponds to accumulated, but unorganized findings, to intellectual property, which includes such human produced artifacts as formulas and recipes, books and other forms of literature, plans and designs, as well as computer programs and strategies.

From an IT perspective, however, although information is often the end result of IT processes, we need to consider more concretely the following issues:

- How the information should be stored
- How (or if) the information should be transmitted
- What processes we want to use on the data to transform it into information
- How the information should be viewed/visualized
- What the requirements for assuring the accuracy of the data and information are
- What the requirements for assuring the accessibility and security of the data and information are

Information itself is not the desired end result of information processing. Researchers in Information Science have developed the "DIKW Hierarchy". This hierarchy defines the transition of data to information to knowledge to wisdom. Each level of the hierarchy is defined in terms of the previous level.

Data are the input directly received by the human (or computer). Data typically are thought of as signals. For a human, this might be the signals received by the senses (sight, sound, smell, taste, feel). For a computer, data might be the values entered by the user when running an application, although in today's computing, data might also be received through camera, microphone, bar code reader, sensor, or pen tablet, to name but a few. The input is generally not usable until it has been converted into a relevant form. For a computer, that would be a binary representation.

In the hierarchy, information is the next level. Information is often defined in the hierarchy as having been inferred from data. The term inferred (or inference) means that one or more processes have been applied to the data to transform it into a more useful form. Implicit in any process applied is that the resulting information is more structured and more relevant than the data by itself. The idea here is that data tend to be too low level for humans to make use of directly. A list of numbers, for instance, may not itself be useful in decision making, whereas the same list of numbers processed through statistical analysis may provide for us the mean, median, standard deviation, and variance, which could then tell us something more significant.

Knowledge is a more vague concept than either information or data. Knowledge is sometimes defined as information that has been put to use. In other cases, knowledge is a synthesis of several different sources of information. One way to think of knowledge is that it is information placed into a context, perhaps the result of experience gained from using information. Additionally, we might think of knowledge as being refined information such that the user of the knowledge is able to call forth only relevant portions of information when needed. As an example, one may learn how to solve algebraic problems by studying algebraic laws. The laws represent information, whereas the practice of selecting the proper law to apply in a given situation is knowledge.

Finally, wisdom provides a social setting to knowledge. Some refer to wisdom as an understanding of "why"—whether this is "why things happened the way they did" or "why people do what they do". Wisdom can only come by having both knowledge *and* experience. Knowledge, on the other hand, can be learned from others who have it. Wisdom is the ultimate goal for a human in that it improves the person's ability to function in the modern world and to make informed decisions that take into account beliefs and values.

In some cases, the term understanding is added to the hierarchy. Typically, understanding, if inserted, will be placed between knowledge and wisdom. Others believe that understanding is a part of wisdom (or is required for both the knowledge and wisdom levels).

In the following section, we focus on data and one of its most common storage forms, the database. We also examine data warehouses and the applications utilized on data warehouses. We then focus on two primary areas of concern in IT: information assurance and information security. Among our solutions to the various concerns will be risk management, disaster planning, computer security, and encryption. We end this chapter with a look at relevant legislation that deals with the use of IT.

DATA AND DATABASES

Information takes on many different forms. It can be information written in texts. It can be oral communications between people or groups. It can also be thoughts that have yet to be conveyed to others. However, in IT, information is stored in a computer and so must be represented in a way that a computer can store and process. Aside from requiring a binary form of storage, we also want to organize the information in a useful manner. That is, we want the information stored to model some real-world equivalent. Commonly, information is organized as records and stored in databases.

A *database* is an organized collection of data. We create, manipulate, and access the data in a database through some form of *database management system* (DBMS). DBMSs include Microsoft Access, SQL and its variations (MySQL, Microsoft SQL, PostgreSQL, etc.), and Oracle, to name a few.

The typical organizing principle of a database is that of the relation. A *relation* describes the relationship between individual records, where each record is divided into attribute values. For instance, a relation might consist of student records. For each student, there is a first name, a last name, a student ID, a major, a minor (optional), and a GPA (Grade Point Average).

The relation itself is often presented as a table (see Table 15.1). Rows of the table represent records: individuals of interest whether they are people, objects, or concepts (for instance, a record might describe a customer, an automobile, or a college course). Formally, the records (rows) are called tuples. All records of the relation share the same attributes. These are called *fields*, and they make up the columns of the relation. Typically, a field will store a type of data such as a number, a string, a date, a yes/no (or true/false) value, and within the type there may be subtypes, for instance, a number might be a long integer, short integer, real number, or dollar amount.

In Table 15.1, we see a student academic relation. In this relation, student records contain six fields: student ID, first name, last name, major, minor, and GPA. Each of the entries is a string (text that can combine letters, punctuation marks, and digits) except for GPA, which is a real number. Student ID could be a number, but we tend to use numbers for data that could be used in some arithmetic operation. Since we will not be adding or multiplying student IDs, we will store them as strings.

The fields of a relation might have restrictions placed on the values that can be stored. GPA, for instance, should be within the range 0.000–4.000. We might restrict the values under Major and Minor to correspond to a list of legal majors and minors for the university. Notice for one record in the relation in Table 15.1 that there is no value for Minor. We might require that all records have a Major but not a Minor.

Given a relation, we might wish to query the database for some information such as "show me all students who are CIT majors", "show me all students who are CIT majors with a GPA > = 3.0", "show me all students who are CIT majors or CSC major with a GPA > = 3.0". These types of queries are known as *restrictions*. The result of such a query is a list of those records in the relation that fit the given criteria. That is, we restrict the relation to a subset based on the given criteria. We can also refer to this type of operation as filtering or searching.

Queries return data from the database although they do not have to be restrictions. That is, they do not have to restrict the records returned. A *projection* is another form of query that returns all of the records from the relation, but only select attributes or fields. For instance, we could project the relation from Table 15.1 to provide just the first and last names.

Another operation on a relation is to *sort* the records based on some field(s). For instance, the records in Table 15.1 are sorted in ascending order by last name followed by first name (thus, Ian Underwood precedes Ruth Underwood).

TABLE 15.1 Example Database Relation

Student ID	First Name	Last Name	Major	Minor	GPA
11151631	George	Duke	MUS	PHY	3.676
10857134	Mike	Keneally	MUS	CHE	2.131
19756311	Ian	Underwood	MUS	HIS	3.801
18566131	Ruth	Underwood	MUS	MAT	3.516
18371513	Frank	Zappa	MUS		2.571

We could also perform *insert* operations (add a record), *update* operations (modify one or more attributes of one or more records), and *delete* operations (delete one or more records). By using some filtering criteria with the update or delete, we are able to modify or remove all records that fit some criteria. For instance, being a generous teacher, I might decide that all CIT majors should have a higher GPA, so an update operation may specify that GPA = GPA * 1.1 where Major = "CIT". That is, for each record whose Major is "CIT", update the GPA field to be GPA * 1.1.

A database will consist of a group of relations, many of which have overlapping parts. In this way, information can be drawn from multiple relations at a time. For instance, we might have another student relation that consists of student contact information. So, the relation presented in Table 15.1 might be called the student scholastic information, whereas the student contact information might look like that shown in Table 15.2.

Given the two relations, we can now use another database query called a *join.* A join withdraws information from multiple relations. For instance, we can join the two relations from Tables 15.1 and 15.2 to obtain all student records, providing their student ID, first and last names, majors, minors, GPA addresses, and phone numbers.

We can combine joins with restrictions and/or projections. For instance, a join and projection could provide the phone numbers and first and last names of all students. A join and a restriction could provide all records of students who are CIT majors who live in OH. Combining all three operations could, for instance, yield the first and last names of all CIT majors who live in OH.

There are additional operations available on a database. These include set operations of union, intersection, and difference. Union combines all records from the relations specified. Intersection retrieves only records that exist in both (or all) relations specified. Difference returns those records that are not in both (or all) relations specified.

Notice that data in the various relations can have repeated attribute values. For instance, there are three students from OH, two students with the last name of Underwood, and five students who are Music majors. For a database to work, at least one of the attributes (field) has to contain unique values. This field is known as the *unique identifier* or the *primary key.* In the case of both relations in Tables 15.1 and 15.2, the unique identifier is the student ID.

The DBMS will use the unique identifier to match records in different relations when using a join operation. For instance, if our query asks for the phone numbers of students with a GPA $\geq$ 3.0, the DBMS must first search the scholastic information for records whose GPA matches the criteria, and then withdraw the phone numbers from the contact relation. To identify the students between the relations, the unique identifier is used. Also

TABLE 15.2 Another Database Relation

Student ID	Address	City	State	Zip	Phone
10857134	8511 N. Pine St	Erlanger	KY	41011	(859) 555-1234
11151631	315 Sycamore Dr	Cincinnati	OH	45215	(513) 555-2341
18371513	32 East 21st Apt C	Columbus	OH	43212	(614) 555-5511
18566131	191 Canyon Lane	Los Angeles	CA	91315	(413) 555-1111
19756311	32 East 21st Apt C	Columbus	OH	43212	(614) 555-5511

notice that the records in Table 15.2 are ordered by the student ID (as opposed to last name as the records are ordered in the relation from Table 15.1).

The relational database is only one format for database organization, although it is by far the most common form. Since the 1980s, research has explored other forms of databases. A few of the more interesting formats are listed here.

- Active database—responds to events that arise either within or external to the database, as opposed to a static relational database that only responds to queries and other database operations.
- Cloud database—as described in Chapter 12, the cloud represents a network-based storage and processing facility, so the cloud database is simply a database that exists (is stored) within a cloud and thus is accessible remotely.
- Distributed database—a database that is not stored solely within one location. This form of database will overlap the cloud and network databases and quite likely the parallel database.
- Document database—a database whose records are documents and which performs information retrieval based on perhaps less structured queries (such as keyword queries as entered in a search engine). Document databases are often found in libraries, and search engines such as Google could be considered a document database.
- Embedded database—the database and the DBMS are part of a larger application that uses the database. One example is a medical diagnostic expert system that contains several components: a natural language interface, a knowledge base, a reasoner, and a patient records database.
- Hierarchical database—a database in which data are modeled using parent–child relationships. Records are broken up so that one part of a record is in one location in the database, and another part is in a different branch. For instance, an employee database might, at the highest level, list all of the employees. A child relation could then contain for a given employee information about that employee's position (e.g., pay level, responsibilities) and another child relation might include information about that employee's projects.
- Hypermedia database—the database comprises records connected together by hypermedia links. The World Wide Web can be thought of as a hypermedia database where the "records" are documents. This is not a traditional database in that the records are not organized in any particular fashion, and the relations are not an organized collection of records.
- Multidimensional database—a database relation is usually thought of as a table: a two-dimensional representation. The multidimensional database stores relations that consist of more than two dimensions. Such a relation might organize data so that the third dimension represents a change in time. For instance, we might have a group of

relations like that of Table 15.1, where each relation represents the students enrolled in a particular semester. Although we could combine all of these relations together and add a semester field, the three-dimensional relation provides us with a more intelligent way to view the data because it contains a more sensible organization.

- Network database—this is not the same as a distributed database; instead, the term network conveys a collection of data that are linked together like a directed graph (using the mathematical notion of a graph). The network database is somewhat similar to the hierarchical database.
- Object-oriented database—rather than defining objects as records placed into varying relations, an object's data are collected into a single entity. Additionally, following on from object-oriented programming, the object is encapsulated with operations defined to access and manipulate the data stored within the object. In essence, this is a combination of databases and object-oriented programming.
- Parallel database—this is a database operated upon by multiple processors in parallel. To ensure that data are not corrupted (for instance, changed by one processor while another processor is using the same data), synchronization must be implemented on the data.
- Spatial or temporal database—a database that stores spatial information, temporal information, or both. Consider, for instance, medical records that contain patient records as sequences of events. Specialized queries can be used to view the sequences. Did event 1 occur before event 2? Or did they overlap or correspond to the same period? The spatial database includes the ability to query about two- and three-dimensional interactions.

A database is a collection of records organized in some fashion. The relational database uses relations. As listed above, other organizations include object-oriented, hierarchical, hypermedia, network, spatial, and temporal. A database can consist of a few records or millions. The larger the collection of data, the more critical the organization be clearly understood and represented. As databases grow in their size and complexity, we tend to view the database as not merely a collection of data, but a collection of databases. A collection of organized databases is called a *data warehouse*. The typical data warehouse uses an ETL process.

- Extract data—data comes from various sources, these sources must be identified, understood and tapped.
- Transform data—given that the data come from different sources, it is likely that the data are not organized using the same collection of attributes. Therefore, the data must be transformed to fit the model(s) of the data warehouse. This will include altering data to relations or objects (or whichever format is preferred), recognizing the unique identifier(s), and selecting the appropriate attributes (fields).

- Load data—into the storage facility, that is, the transformed data must be stored in the database using the appropriate representation format.

For a business, the data for the data warehouse might come from any number of sources. These include:

- Enterprise resource planning systems
- Supply chain management systems
- Marketing and public relations reports
- Sales records
- Purchasing and inventory records
- Direct input from customers (e.g., input from a web form or a survey)
- Human relations records
- Budget planning
- Ongoing project reports
- General accounting data

Once collected and transformed, the data are stored in the data warehouse. Now, the data can be utilized. There are any number of methods that can be applied, and the processed results can be stored back into the warehouse to further advance the knowledge of the organization as described below. These methods are sometimes called *data marts*—the means by which users obtain data or information out of the warehouse.

Data warehousing operations range from simple database management operations to more sophisticated analysis and statistical algorithms. In terms of the traditional DBMS, a user might examine purchasing, inventory, sales, and budget data to generate predicted manpower requirements. Or, sales and public relations records might help determine which products should be emphasized through future marketing strategies.

In OLAP (Online Analytical Processing), data are processed through a suite of analysis software tools. You might think of OLAP as sitting on top of the DBMS so that the user can retrieve data from the database and analyze the data without having to separate the DBMS operations from the more advanced analysis operations. OLAP operations include slicing, dicing, drilling down (or up), rolling-up, and pivoting the data.

Slicing creates a subset of the data by reducing the data from multiple dimensions to one dimension. For instance, if we think of the data in our database as being in three dimensions, slicing would create a one-dimensional view of the data. Dicing is the same as slicing except that the result can be in multiple dimensions, but still obtaining a subset of the data. For instance, a dice might limit a three-dimensional collection of data into a smaller three dimensional collection of data by discarding certain records and fields.

Drilling up and down merely shifts the view of the data. Drilling down provides more detail, drilling up provides summarized data. Rolling up is similar to drilling up in that it summarizes data, but in doing so, it collapses the data from multiple items (possible over more than one dimension) into a single value. As an example, all HR records might be collapsed into a single datum such as the number of current employees, or a single vector that represents the number of employees in each position (e.g., management, technical, support).

Finally, a pivot rotates data to view the data from a different perspective. Consider a database that contains product sales information by year and by country. If the typical view of the data is by product, we might instead want to pivot all of the data so that our view first shows us each year. The pivot then reorganizes the data from a new perspective. We could also pivot these data by country of sale instead.

Figure 15.1 illustrates, abstractly, the ideas behind slicing, dicing, and pivoting. Here, we have a collection of data. Perhaps each layer (or plane) of the original data represents a database relation from a different year. For example, the top layer might be customer records from 2012, the next layer might be customer records from 2011, and the bottom two layers are from 2010 and 2009, respectively. A slice might be an examination of one full layer, say that of 2010. A dice might be a subset in multiple dimensions, for instance, restricting the data to years 2012, 2011, and 2010, records of those customers from Ohio, and fields of only last name, amount spent, and total number of visits. The pivot might reorganize the data so that, rather than first breaking the data year by year, the data are first broken down state by state, and then year by year.

OLAP analysis has existed since the 1960s. Today, however, OLAP becomes more important than ever because the data warehouse is far too large to obtain useful responses from mere database operations.

Another suite of tools that can be applied to a data warehouse is data mining. Unlike OLAP, *data mining* is a fairly recent idea, dating back to the 1990s as an offshoot of artificial intelligence research. In data mining, the idea is to use a number of various statistical

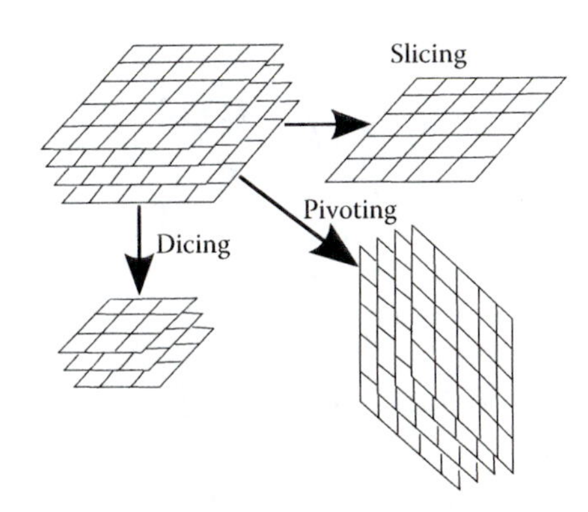

FIGURE 15.1 Demonstrating OLAP (Online Analytical Processing) analysis techniques.

operations on a collection of data and see if the results are meaningful. Unlike OLAP, which is driven by the human, data mining attempts somewhat random explorations of the data. There is no way to know in advance if the results will be worthwhile. There are a number of different algorithms applied in data mining. Here, we look at a few of the most common techniques.

In *clustering*, data are grouped together along some, but not all, of the dimensions (fields). If you select fields wisely, you might find that the data clump together into a few different groups. If you can then identify the meaning of each clump, you have learned something about your data. In Figure 15.2, you can see that data have been clustered using two dimensions. For instance, if these data consist of patient records, we might have organized them by age and weight. Based on proximity, we have identified three clusters, with two outlying data. One of the outlying data might belong to cluster 2, the other might belong to cluster 3, or perhaps the two data belong to their own cluster. Perhaps cluster 2 indicates those patients, based on age and weight, who are more susceptible to a specific disease.

In order to identify clusters automatically, the computer determines distances between individual records. It finds a center among groups and then adds other data elements one by one by determining which cluster it comes closest to.

In forming clusters, we might find that our original selection of fields did not give us any useful groupings. For instance, if we chose weight and height instead of weight and age, we might not have identified any useful characteristics. Imagine that we want to identify patients who might be particularly susceptible to type 2 diabetes. Fields such as age, height, and income level would probably be useless. However, weight (or at least body mass index), ethnicity, and degree of job stress might find more meaningful clusters.

This leads to a problem. Whereas age, height, weight, and even income are easy enough to graph, how do we graph job stress and ethnicity? We must find a way to convert data from categories into numeric values.

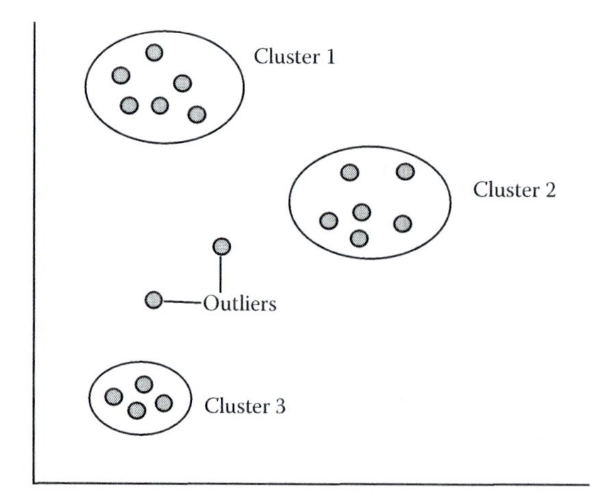

FIGURE 15.2 Results of clustering data.

Clustering is a very common data analysis technique, but as implied here, there are problems. Computationally, we might need to try to cluster over a variety of different combinations of fields before we find any useful clusters. As the number of fields increase, the number of combinations increases exponentially. For instance, with 10 fields, there are 1024 different combinations that we might try, but with 20 fields, the number of combinations increases to more than 1 million!

A variation of clustering is to identify *nearest neighbors*. A nearest neighbor is the datum that lies closest in proximity as computed by some mathematical equation (such as the Euclidean distance formula).* The nearest neighbor algorithm can be used in clustering to position records and identify clusters by those whose distance is less than some preset amount. The *k*-nearest neighbor algorithm computes distances along *k* of the fields. The selection of which k is another computationally intensive problem. For instance, if there are 10 fields and $k = 6$, we might want to try using all combinations of six fields out of 10 to determine a nearest neighbor.

Association rule learning searches data for relationships that might be of interest, and provides them as if–then types of rules. For instance, in analyzing 10,000 grocery store receipts, we find that 4000 of the receipts show that someone bought peanut butter, and 3600 of those 4000 receipts show that the person also bought bread. This becomes a rule:

if customer buys peanut butter then they buy bread

The rule also has a frequency of 90% (3600 out of 4000). The frequency describes how often the rule was true. The frequency, however, did not tell us how useful the rule might be. Consider that we find 12 people (out of 10,000) bought sushi and all of those 12 people also bought white wine. Even though the frequency is very high (100%), the rule is not very useful.

We can use the association rule to help us make decisions such as in marketing and sales. We might, for instance, decide to move the peanut butter into the bread aisle, and we might decide to put on a sale to promote the two products by saying "buy bread and get a jar of peanut butter for 25% off". The result of association rule learning, like clustering, may or may not be of value. The sushi rule would probably not convince us to do anything special about sushi and wine. Or, imagine another rule that tells us that people who buy wine do not typically also buy beer. Is there any value in knowing this correlation? The advantage of association rule learning is that we can provide the receipt data to the data mining tool and let it find rules for us. We might tell the tool to only return rules whose frequency is greater than 80%. Then, it is up to management to decide how to use the rules.

Another product of data mining is a *decision tree*. A decision tree is a tree structure that contains database fields as nodes such as age, sex, and income. The branches of the tree represent different possible values of the fields. For instance, if a node has the field sex, it will

---

* The Euclidean distance formula between two data, 1 and 2, is $\sqrt{(a_1 - a_2)^2 + (b_1 - b_2)^2 + (c_1 - c_2)^2}$ assuming that our data has three attribute values, *a*, *b*, and *c*.

DATA MINING TO THE RESCUE

Data mining is now a tool of business in that it can help management make decisions and predictions. However, data mining has been found to be useful far beyond profits. Here are a few interesting uses of data mining.

The Minnesota Intrusion Detection System analyzes massive amounts of data pertaining to network traffic to find anomalies that could be intrusions of various types into computer systems. Experiments on more than 40,000 computers at the University of Minnesota have uncovered numerous break-in attempts and worms.

In Bioinformatics, data mining is regularly used to help sift through literally billions of pieces of data pertaining to the human genome. Recent applications have led to improved disease diagnoses and treatment optimizations and better gene interaction modeling. In addition, BioStorm combines data mining and web information to track possible pandemic and bioterrorist incidents.

The National Security Agency has used data mining on more than approximately 1.9 trillion telephone records looking for links between known terrorists and citizens. Other data mining techniques have been used to create the Dark Web, a group of tools and links to websites of suspected terrorist organizations.

Law enforcement agencies are using data mining to identify possible crime hot spots in their cities, to predict where future crimes are likely to occur, and to track down criminals from previous crimes.

Search engines and recommender sites regularly use data mining to find links of interest based on your input queries. Google and Amazon are both pioneers in data mining.

have two branches, one for male and one for female (a third branch is possible if the datum for sex is not known or available). For a field such as age, rather than providing a branch for every possible value, values are grouped together. So, for instance, there might be a branch for adolescent ages (0 to 12 years old), a branch for teenage years, a branch for young adults (e.g., 20 to 32), a branch for middle age (e.g., 33 to 58), and a branch for retirement age (e.g., 59 and older). The leaf nodes of the decision tree represent the decisions that we want to

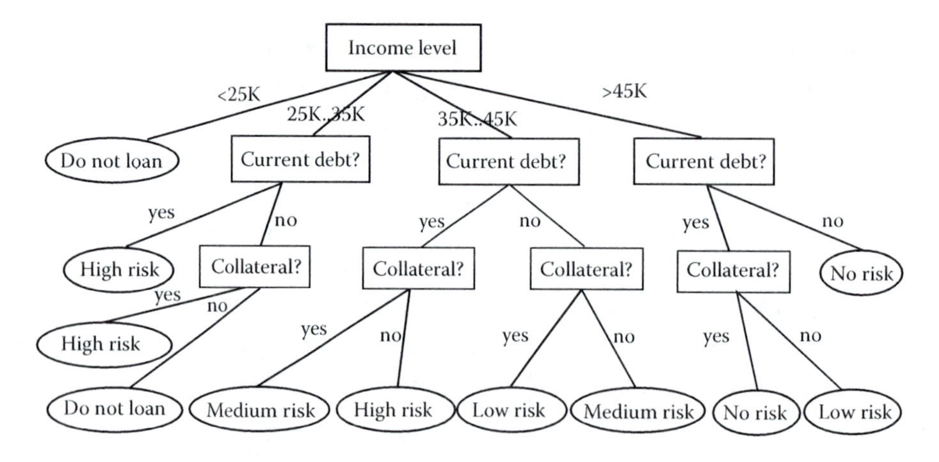

FIGURE 15.3 Decision tree generated from bank loan data.

make. The decision tree then represents a process of making a decision by following the proper branches of the tree.

Figure 15.3 provides a common example for a decision tree. Given banking decisions of previous loans granted or not granted, we want to automate the process. In this case, we construct the decision tree from records of customers given their income level, whether they have current debt or not, and if they have collateral. For our tree, these three fields represent our nodes (indicated in the figure with boxes). Branches are annotated with the possible values for each field. We have broken down income level into four possible values, "< 25K", "25K...35K", "35K...45K", and ">45K". The other two fields, current debt and collateral, have yes/no responses. The leaf nodes represent loan decisions made in the past. These nodes are indicated with ovals. For example, a person who has no current debt and an income of more than $45,000 has no risk, whereas a person whose income is less than $25,000 should not be given a loan under any circumstances. A person whose income is between $25,000 and $35,000 and has no debt will still need collateral to receive a loan, but even then is considered a high risk.

Decision trees can be automatically generated from database data. However, as implied in the previous example, if a field has more than just a few values (e.g., age, income), then the data should first be segmented. This can be done by a human or the data mining algorithm. In either event, decision tree generating algorithms are also computationally intensive because the algorithm must decide which fields to use in the decision tree (sex and age were not used in the tree in Figure 15.3) and the order that the fields are applied. Given a database relation of 10 fields, there are 1024 different combinations that the fields can be used, not to mention millions of potential orderings of fields (from top to bottom in the tree).

Today, data warehouses are a very common way to organize and utilize data for large organizations. Data warehousing is utilized by corporations, financial institutions, universities, hospitals and health care providers, telecommunications companies, the airline industry, the intelligence community, and many government agencies. Along with cloud computing (introduced in Chapter 12), data warehousing denotes a shift in the scale of IT in support of what many are now calling "Big Data".

This discussion of databases and data warehouses lacks one important characteristic. Although databases and data warehouses are stored using IT, they actually are classified as forms of information systems. What is the difference? An information system is a system that stores and permits the processing of information (and data). However, the idea of an information system is neutral to technology. Although it is true that nearly all information systems today are stored by computer and utilize software to process the information, it does not have to be this way. An information system might be a card catalog system of a library stored on 3 × 5 index cards, or a database of a company's inventory stored by paper in a filing cabinet. In fact, the term information system has existed for a lot longer than computers. Aside from databases and data warehouses, people often view expert systems and decision support systems, geographic information systems, and enterprise systems as forms of information system. The term information technology describes technology infrastructure, which itself might support information systems but could also support many other types of systems such as telecommunications systems, sensor networks, and distributed processing to name but a few, which are distinct from information systems. We briefly examine information systems as a career option in Chapter 16.

INFORMATION ASSURANCE AND SECURITY

Information assurance and information security are often described together as information assurance and security (IAS). In this section, we introduce the concepts behind IAS. We leave some of the details of information security to the section Threats and Solutions. The IAS model combines the three components of IT—communications, hardware, and software—with the information utilized by the three components. Focusing on how information is utilized, IAS proscribes three goals, commonly referred to as CIA: confidentiality, integrity, and availability.

Confidentiality requires that data be kept secure so that they are not accidentally provided to unauthorized individuals and cannot be obtained by unauthorized users. Confidentiality goes beyond security mechanisms as it deals with policy issues of who is an authorized user of the data, how data should be kept and stored, and privacy policies. As an example, we discussed earlier in the chapter the need for a unique identifier. An organization may have a policy that employee records will use social security numbers as unique identifiers. However, such a policy violates the employee's freedom to protect their social security number (which should only be made available for payroll purposes).

Additionally, security mechanisms must go far beyond those that protect against unauthorized access over computer network. For instance, confidentiality must ensure that either sensitive data are not available on laptop computers, or that those laptop computers should not be accessible to unauthorized individuals. A theft of a laptop should not violate the confidentiality of the organization.

Integrity requires that data are correct. This requires at a minimum three different efforts. First, data gathering must include a component that ensures the accuracy of the collected data. Second, data must be entered into the system accurately. But most importantly, data modification must be tracked. That is, once data are in a database, changes to that data must leave behind a record or trace to indicate who made the change, when the change was made, and what the change was. This permits auditing of the data and the ability to roll the data back if necessary. Furthermore, if a datum is being used by one individual, the datum should be secure from being altered by another. A lack of proper data synchronization could easily lead to corrupt data, as has been discussed previously in this text.

Finally, *availability* requires that information is available when needed. This seems like an obvious requirement, but it conflicts to some extent with both confidentiality and integrity. We would assume that most organizations will make their information available in a networked fashion so that users can obtain the data from computers positioned all over the organization, including remote access from off-site locations. To maintain confidentiality, proper security measures must then be taken including secure forms of login, encryption/decryption, and proper access controls (e.g., file permissions). Additionally, if one user is accessing some data, to maintain integrity, other users are shut out of accessing that data, thus limiting their availability. To promote integrity, timely backups of data need to be made, but typically during the time that files are being backed up, they are not available. An additional complication for availability is any computer downtime, caused when doing computer upgrades, power outages, hardware failures, or denial of service attacks.

Y2K

Data integrity is a worthy goal, but what about the integrity of the software? There have been numerous software blunders but probably none more well known than Y2K itself. What is Y2K, and why did it happen?

Back in the 1960s and 1970s, it was common for programmers to try to save as much memory space as possible in their programs because computers had very limited main memory sizes. One way to save space was to limit the space used to store a year. Rather than storing a four-digit number, they would usually store years as two-digit numbers. So, for instance, January 31, 1969 would be stored as 01, 31, and 69. No one realized that this would be a problem until around 1997.

With the year 2000 approaching, programmers began to realize that any logic that involved comparing the current year to some target year could suddenly be incorrect. Consider the following piece of code that would determine whether a person was old enough to vote:

```
If (currentyear - birthyear >= 18) …
```

Now let us see what happens if you were born in 1993 and it is currently 2013. Using four-digit years, we have:

```
If (2013 - 1993 >= 18) …
```

This is true. So the code works correctly. But now let us see what happens with two-digit years:

```
If (13 - 93 >= 18) …
```

This is false. Even though you are old enough to vote, you cannot!

Y2K was a significant threat to our society because it could potentially prevent people from receiving their social security checks, cause banking software to compute the wrong interest, and even prevent students from graduating! But no one knew the true extent of the problem and we would not know until January 2000. Would missile silos suddenly launch their missiles? Would power companies automatically shut down electricity to their customers? Would medical equipment malfunction? Would planes fall out of the sky or air traffic control equipment shut down?

So programmers had to modify all of the code to change two-digit years to four. In the United States alone, this effort cost more than $100 billion!

Two additional areas that are proscribed by some, but not all, in IAS are authenticity and non-repudiation. These are particularly important in e-commerce situations. Authenticity ensures that the data, information, or resource is genuine. This allows a valid transaction to take place (whether in person or over the Internet). Authenticity is most commonly implemented by security certificates or other form of digital signature. We examine these in more detail in Threats and Solutions. Non-repudiation is a legal obligation to follow through on a contract between two parties. In e-commerce, this would mean that the customer is obligated to pay for the transaction and the business is obligated to perform the service or provide the object purchased. As these two areas only affect businesses, they are not acknowledged as core IAS principles by everyone.

IAS is concerned primarily with the protection of IT. IAS combines a number of practices that define an organization's information assets, the vulnerabilities of those assets, the threats that can damage those assets, and the policies to protect the assets. The end result of this analysis, known as *strategic risk analysis*, is security policies that are translated into mechanisms to support information security. We will focus on the risk management process.

To perform risk management, the organization must first have goals. These goals are utilized in several phases of risk management. Goals are often only stated at a very high level such as "the organization will provide high-quality service to members of the community with integrity and responsiveness". Goals may be prioritized as some goals may conflict with others. For instance, a goal of being responsive to customers may conflict with a goal to ensure the integrity of data as the former requires a quick response, whereas the latter requires that data be scrutinized before use.

Now the organization must perform a risk assessment. The first step in a risk assessment is to identify the organization's information assets. These are physical assets (e.g., computers, computer network, people), intellectual property (ideas, products), and information (gathered and processed data). Aside from identifying assets, these can be categorized by type (e.g., hardware, process, personnel) and prioritized by their importance to the organization.

In the case of information, security classifications are sometimes applied. A business might use such categories as "public", "sensitive", "private", and "confidential" to describe the data that they have gathered on their clients. Public information might include names and addresses (since this information is available through the phone book). Sensitive information might include telephone numbers and e-mail addresses. Although this is not public information, it is information that will not be considered a threat to a person's privacy if others were to learn of it. Private information is information that could be a threat if disclosed to others such as social security and credit card numbers, or health and education information. This information is often protected from disclosure by federal legislation. Finally, confidential information consists of information that an organization will keep secret, such as patentable information and business plans. The government goes beyond these four categories with a group of classified tags such as confidential, secret, and top secret.

The next step in risk assessment is to identify vulnerabilities of each asset. Vulnerabilities will vary between assets, but many assets will share vulnerabilities based on their categorized type. For instance, imagine in risk assessment, that the three types of assets are managers, technical staff, and clerical staff. These are all people, and some of their vulnerabilities will be the same. Any person might be recruited away to another organization by a higher paying salary, and any person might be vulnerable to social engineering attacks (covered in Threats and Solutions). Similarly, most hardware items will share certain vulnerabilities such as being vulnerable to power outage and damage from power surge, but not all hardware will have vulnerabilities from unauthorized access because some hardware may not be networked.

Once vulnerabilities are identified, risk assessment continues with threats. That is, given a vulnerability, what are the possible ways that vulnerability can be exploited? As stated in the previous paragraph, a person may be vulnerable to a social engineering attack.

However, such an attack on someone with access to confidential files is a far greater threat than an attack on someone without such access. Similarly, unauthorized access to a file server storing confidential data is a far greater threat than unauthorized access to a printer or a laptop computer that is only used for presentation graphics.

With the risk assessment completed, the organization moves forward by determining how to handle each risk. Risks, like goals, must be prioritized. Here, the organizational goals can help determine the priorities of the risks. For instance, one organizational goal might be to "recruit the best and brightest employees", but a goal of "high quality service" is of greater importance. Thus, risks that threaten the "service" goal would have a higher priority than risks that threaten the "employee" goal.

In prioritizing risks, risk management may identify that some risks are acceptable. It might, for instance, be an acceptable risk that employees be recruited away to other organizations. However, some risks may be identified that can critically damage the organization such as attacks to the organization's web portal (e.g., denial of services) that result in a loss of business for some length of time. The risks, in turn, require the creation of policies that will reduce the threats. Policies must then be enacted. Policies might involve specific areas of the organization such a hiring practices for human resources, management practices for mid-level management, and technical solutions implemented by IT personnel.

Once the risk management plan is implemented, it must be monitored. For instance, after 6 months have elapsed, various personnel may examine the results. Such an examination or self-study of the organization could result in identifying areas of the plan that are not working adequately. Thus, the risk management process is iterative and ongoing. The time between examinations will be based on several criteria. Initially, iterations may last only a few months until the risk management plan is working well. Iterations may then last months to years depending on how often new assets, vulnerabilities, and threats may arise.

A risk management plan may include a wide variety of responses. Responses can vary from physical actions to IT operations to the establishment of new procedures. Physical actions might include, for instance, installing cameras, sensors, and fire alarms; locking mechanisms on computers; and hiring additional staff for monitor or guard duty. IT operations are discussed in the next section such as firewalls. New procedures might include hiring processes, policies on IT usage and access control, management of information assets, maintenance, and upgrade schedules.

One other component of a risk management plan is a *disaster recovery plan*. A disaster recovery plan is a plan of action for the organization in response to a disaster that impacts the organization to the point that it has lost functionality. There are several different types of disasters, and a recovery plan might cover all (or multiple) types of disasters, or there may be a unique recovery plan for each type of disaster. Disasters can be natural, based on severe weather such as tornados, hurricanes, earthquakes, and floods. Disasters can be man-made but accidental such as a fire, an infrastructure collapse (e.g., a floor that collapses within the building), a power outage that lasts days, or a chemical spill. Disasters can also be man-made but purposeful such as a terrorist attack or act of war. Here are examples of the latter two categories. The power outage that hit the East Coast in 2008 was deemed accidental, yet the outage lasted several days for some communities. And, of course, on September 11, 2001, terrorists crashed four planes in New York City, Washington, DC, and Pennsylvania, causing a great deal of confusion and leading to airline shutdowns for several days.

A disaster recovery plan should address three things. First, preventive measures are needed in an attempt to avoid a disaster. For instance, a fire alarm and sprinkler system might help prevent fires from spreading and damaging IT and infrastructure. Proper procedures when dealing with chemical spills might include how to handle an evacuation including how to quickly shut down any IT resources or how to hand off IT processing capabilities to another site. Simple preventative measures should include the use of uninterrupted power supplies and surge protectors to ensure that computers can be shut down properly when the power has gone out, and to protect against surges of electricity that could otherwise damage computers. Second, there needs to be some thought to how to detect a disaster situation in progress. Obviously, a tornado or a hurricane will be noticeable, but a power surge may not be easily detected. Finally, the disaster recovery plan must address how to *recover* from the disaster itself.

Focusing on IT, the best approaches for preparing for and handling disasters are these. First, regular and timely backups of all data should be made. Furthermore, the backed up data should be held off-site. Redundancy should be used in disk storage as much as possible. This is discussed in more detail in the next section. Simple measures such as uninterrupted power supplies and surge protectors on the hardware side, and strong password policies and antiviral software should always be used. Monitoring of equipment (for instance, with cameras) and logging personnel who have access to equipment that stores sensitive data are also useful actions. And, of course, fire prevention in the form of fire alarms, fire extinguishers, fire evacuation plans, and even fire suppression foam are all possible.

Preparation is only one portion of the disaster recovery plan. Recovery is the other half. Actions include restoring backed up data, having spare IT equipment (or the ability to quickly replace damaged or destroyed equipment) and having backup services ready to go are just some of the possible options. In many large organizations, IT is distributed across more than one site so that a disaster at one site does not damage the entire organization. During a disaster, the other site(s) pick up the load. Recovery requires bringing the original site back up and transferring data and processes back to that site.

The risk management plan is often put together by management. But it is critical for any organization that relies on IT to include IT personnel in the process. It is most likely the case that management will not have an understanding of all of the vulnerabilities and threats to the technology itself. Additionally, as new threats arise fairly often, it is important to get timely input. Do not be surprised if you are asked to be involved in risk management at times of your career.

THREATS AND SOLUTIONS

Information security must protect at its core the data/information of the organization. However, surrounding the data are several layers. These are the operating system, the applications software, the computer and its resources, the network, and the users. Each of these layers has different threats, and each will have its own forms of security. As some threats occur at multiple levels, we will address the threats rather than the layers. However, information security is not complete without protecting every layer.

Social engineering is a threat that targets users. The idea is that a user is a weak link in that he or she can be tricked, and often fairly easily. A simple example of a social engineering attack works like this. You receive a phone call at home one evening. The voice identifies itself as IT and says that because of a server failure, they need your password to recreate your account. Without your password, all of your data may be lost. You tell them your password. Now they can break into your account because they are not IT but in fact someone with malicious intent.

Social engineering has been used to successfully obtain people's passwords, bank account numbers, credit card numbers, social security numbers, PIN (personal identification number) values, and other confidential information. Social engineering can be much more subtle than a phone call. In a social setting, you are far more likely to divulge information that a clever hacker could then use to break your password. For instance, knowing that you love your pet cats, someone may try to obtain your cats' names to see if you are using any of them as your password. Many people will use a loved one's name followed by a digit for a password.

A variation on social engineering is to trick a user by faking information electronically. *Phishing* involves e-mails to people to redirect them to a website to perform some operation. The website, however, is not what it seems. For instance, someone might mock up a website to make it look like a credit card company's site. Now an e-mail is sent to some of the credit card company customers informing them that they need to log into their accounts or else the accounts will be closed. The link enclosed in the e-mail, however, directs them to the mocked up website. The user clicks on the link and is taken to the phony website. There, the user enters secure information (passwords, credit card number, etc.) but unknowingly, this information is made available to the wrong person.

Another class of threat attacks the computer system itself whether the attack targets the network, application software, or operating system. This class includes *protocol attacks, software exploits, intrusion*, and insider attacks. In a protocol attack, one attempts to obtain access to a computer system by exploiting a weakness or flaw in a protocol. There are, for instance, known security problems in TCP/IP (Transmission Control Protocol/Internet Protocol). One approach is called TCP Hijacking, in which an attacker spoofs a host computer in a network using the host computer's IP address, essentially cutting that host off from its network.

Many forms of protocol attacks are used as a form of reconnaissance in order to obtain information about a computer network, as a prelude to the actual attack. An ICMP (Internet Control Message Protocol) attack might use the ping program to find out the IP addresses of various hosts in a computer network. Once an attacker has discovered the IP addresses, other forms of attack might be launched. A smurf attack combines IP spoofing and an ICMP (ping) attack where the attacker spoofs another device's IP address to appear to be a part of the network. Thus, the attacker is able to get around some of the security mechanisms that might defeat a normal ICMP attack.

Software exploits vary depending on the software in question. Two very popular forms of exploits are *SQL injections* and *buffer overflows*. In the SQL injection, an attacker issues an SQL command to a web server as part of the URL. The web server, which can accept queries as part of the URL, is not expecting an SQL command. A query in a URL follows a "?" and includes a field and a value, such as www.mysite.com/products.php?productid = 1. In this case, the web page products.php most likely accesses a database to retrieve the entry productid = 1. An SQL injection can follow the query to operate on that database. For instance, the modified URL www.mysite.com/products.php?product = 1; DROP TABLE products would issue the SQL command DROP TABLE products, which would delete the relation from the database. If not protected against, the web server might pass the SQL command onto the database. This SQL command could potentially do anything to the database from returning secure records to deleting records to changing the values in the records.

The buffer overflow is perhaps one of the oldest forms of software exploit and is well known so that software engineers should be able to protect against this when they write software. However, that is not always the case, and many pieces of software are still susceptible to this attack. A buffer is merely a variable (typically an array) that stores a collection of values. The buffer is of limited size. If the software does not ensure that insertions into the buffer are limited to its size, then it is possible to insert into the buffer a sufficient amount so that the memory locations after the array are filled as well. Since memory stores both data and code, one could attempt to overflow a buffer with malicious code. Once

WHITE HAT VERSUS BLACK HAT

The term hacker conveys three different meanings:

- Someone who hacks code, that is, a programmer
- A computer hobbyist
- Someone who attempts to break into computer systems

Historically, the hacker has been a person who creates software as a means of protest. For instance, early hackers often broke into computer-operated telephone systems to place free long distance calls. They were sometimes called phreakers.

In order to differentiate the more traditional use of hacker, a programmer, with the derogatory use of someone who tries to break into computer systems, the term cracker has been coined. The cracker attempts to crack computer security. But even with this definition, we need to differentiate between those who do this for malicious purposes from those who do it either as a challenge or as an attempt to discover security flaws so that the security can be improved.

The former case is now referred to as a black hat hacker (or just a black hat). Such a person violates security in order to commit crime or terrorism.

The latter case of an individual breaking security systems without malicious intent is referred to as a white hat hacker (or just a white hat). However, even for a white hat, the action of breaking computer security is still unethical.

As a case in point, the organization of crackers who call themselves Anonymous purport to violate various organization's computer systems as a means of protest. They have attacked the Vatican in protest over the Catholic church's lack of response to child abuse claims. They have attacked sites run by the Justice Department, the Recording Industry Association of America, and Motion Picture Association of America to protest antipiracy legislature. And they attacked the sites of PayPal, MasterCard, and Visa when those companies froze assets of Wikileaks. Although touting that these attacks are a form of protest, they can also be viewed as vigilante operations violating numerous international laws.

stored in memory, the processor could potentially execute this code and thus perform the operations inserted by the attacker.

Intrusion and other forms of active attacks commonly revolve around first gaining unauthorized access into the computer system. To gain entrance, the attacker must attempt to find a security hole in the operating system or network, or obtain access by using someone else's account. To do so, the attacker will have to know a user's account name and password.

As stated above, there are social engineering and phishing means of obtaining passwords. Other ways to obtain passwords include writing a program that continually attempts to log in to a user's account by trying every word of the dictionary. Another approach is to simply spy on a person to learn the person's password, perhaps by watching the person type it in. Or, if a person is known to write passwords down, then you can look around their computer for the password if you have access to that person's office. You might even find the password written on a post-it note stuck to the computer monitor!

Another means of obtaining a password is through *packet sniffing*. Here, the attacker examines message traffic coming from your computer, for instance e-mail messages. It is possible (although hopefully unlikely) that a user might e-mail a password to someone else or to him/herself.

Aside from guessing people's passwords, there are other weaknesses in operating systems that can be exploited to gain entrance to the system. The Unix operating system used to have a flaw with the telnet program that could allow someone to log into the Unix system without using a password at all. Once inside, the intruder then can unleash their attack. The active attack could potentially do anything from deleting data files, copying data files, and altering data files to leaving behind malicious code of some kind or creating a backdoor account (a hidden account that allows the attacker to log in at any time).

An even simpler approach to breaking through the security of an IT system is through an inside job. If you know someone who has authorized access and either can be bribed or influenced, then it is possible that the attacker can delete, copy, or alter files, insert malware, or otherwise learn about the nature of the computer system through the person. This is perhaps one of the weakest links in any computer system because the people are granted access in part because they are being trusted. That trust, if violated, can cause more significant problems than any form of intrusion.

Malware is one of the worst types of attacks perpetrated on individual users. The original form of malware was called a Trojan horse. The Trojan horse pretends to be one piece of software but is in fact another. Imagine that you download an application that you think will be very useful to you. However, the software, while pretending to be that application, actually performs malicious operations on your file system. A variation of the Trojan horse is the computer virus. The main differences are that the virus hides inside another, executable, file, and has the ability to replicate itself so that it can copy itself from one computer to another through a floppy disk (back when we used them), flash drive, or e-mail attachment.

Still other forms of malware are network worms that attack computer networks (see, e.g., Morris' Internet worm discussed in Chapter 12) and spyware. Spyware is often downloaded unknown to the user when accessing websites. The spyware might spy on your browsing behavior at a minimum, or report back to a website sensitive information such as a credit card number that you entered into a web form. Still another form of malware will hijack some of your software. For instance, it might redirect your DNS information to go to a different DNS, which rather than responding with correct IP addresses provides phony addresses that always take your web browser to the wrong location(s).

One final form of attack that is common today, particularly to websites, is the denial of service attack. In the denial of service attack, one or more attackers attempts to flood a server with so many incoming messages that the server is unable to handle normal business. One of the simplest ways to perform a denial of service attack is to submit thousands or millions (or more) HTTP requests. However, this only increases the traffic; it does not necessarily restrict the server from responding to all requests over time. A UDP attack can replace the content of a UDP packet (defined in Chapter 12) with other content, inserted by the attacker. The new content might require that the server perform some time-consuming operation. By performing UDP flooding, large servers can essentially be shut down. Other forms of denial of service utilize the TCP/IP handshaking protocol, these are known as TCP SYN and TCP ACK flood attacks.

The above discussion is by no means a complete list of the types of attacks that have been tried. And, of course, new types of attacks are being thought of every year. What we need, to promote information security, are protection mechanisms to limit these threats to acceptable risks. Solutions are brought in from several different approaches.

First, the organization's users must be educated. By learning about social engineering, phishing, and forms of spying, the users can learn how to protect their passwords. Additionally, IT policies must ensure that users only use strong passwords, and change their passwords often.

In some cases, organizations use a different approach than the password, which is sometimes referred to as "what you know". Instead, two other approaches are "what you have" and "who you are". In the former case, the access process includes possession of some kind of key. The most common form of key is a key card (swipe card). Perhaps this can be used along with a password so that you must physically possess the key and know the password to log in. In the latter case, the "who you are" constitutes some physical aspect that cannot be reproduced. Biometrics are used here; whether in the form of a fingerprint, voice identification match, or even a lip print, the metric cannot be duplicated.

Next, we need to ensure that the user gains access only to the resources that the user should be able to access. This involves access control. Part of the IT policy or the risk management plan must include a mechanism whereby a user is given access rights. Those rights should include the files that the user is expected to access while not including any files that the user should not be accessing. Access rights are usually implemented in the operating system by some access control list mechanism (see Chapter 6 for a discussion on this). In Windows and Linux, access control is restricted to file and directory access per user using either lists (Windows, Security Enhanced Linux) or the 9-bit rwxrwxrwx scheme (Linux).

Many advanced DBMS use role-based access control. In such a system, roles are defined that include a list of access methods. For instance, the supervisor may be given full access to all data, whereas managers are given access to their department's data. A data analyst may only be given read access to specific relations. Once roles are defined, individuals are assigned to roles. One person may have different roles allowing the person to have a collection of access rights.

Most of the other forms of attack target the computer system. Protection will be a combination of technologies that protect against unauthorized access and other forms of intrusion, denial of service attacks, software exploits, and malware. Solutions include the firewall to prevent certain types of messages from coming into or out of the network, antiviral software to seek out malware, and intrusion detection software. Resolving software exploits require first identifying the exploits and then modification of the source code of the software. If the software is a commercial product or open source, usually once the exploit is found, an update is released to fix it within a few days to a few weeks. Denial of service attacks and intrusions are greater challenges, and the IT staff must be continually on guard against these.

Two types of threats arise when communicating secure information over the Internet whether this involves filling out a web form or sending information by e-mail. First, Internet communications are handled with regular text. If someone can intercept the messages, the secure information is open to read. Second, the sender must be assured that the recipient is who they say they are, that is, the recipient should be authenticated. To resolve the former problem, encryption can be used. To resolve the second problem, we might use a third party to verify the authenticity of the recipient. To handle both of these solutions, we turn to digital signatures and certificates.

The idea behind a certificate or digital signature is that a user (organization, website) creates a sign to indicate who they are. This is just data and might include for instance an e-mail signature, a picture, or a name. Next, the data are put together into a file. Encryption technology (covered in the next section) is used to generate a key. The key and the data are sent to a certificate authority. This is a company whose job is to ensure that the user is who they claim to be and to place their stamp of approval on the data file. The result is a signed certificate or digital signature. The signed document is both the authentication proof and the encryption information needed for the sender to send secure information without concern of that information being intercepted and accessed.

If a user receives a certificate or signature that is out of date, not signed, or whose data do not match the user's claimed identity, then the recipient is warned that the person may not be whom they claim. Figure 15.4 shows two certificates. The one on the left is from a university and is *self-signed*. You might notice near the top of this certificate the statement "Could not verify this certificate because the issuer is not trusted." This indicates that the site that uses the certificate did not bother to have a certificate authority sign their certificate. Any user who receives the certificate is warned that doing business with that site may be risky because they are not authenticated. Fortunately, the certificate is not used to obtain secure information so that a user would not risk anything substantial. The certificate on the right side of the figure is from Visa and is signed by SSL Certificate Authority, thus the site can be trusted.

Information security attempts to ensure the confidentiality, integrity, and availability of information when it is stored, processed, and transmitted (across a network). The approaches listed above primarily target the transmission of information by network

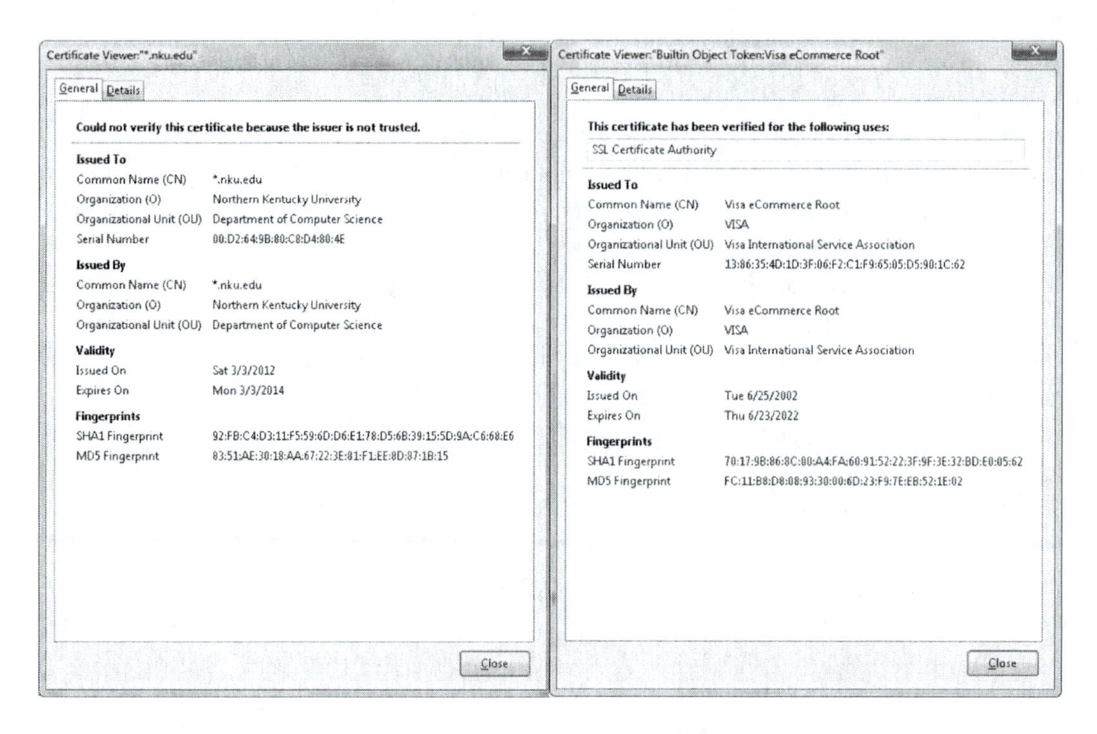

FIGURE 15.4 Two certificates, one self-signed (left) and one signed (right).

and processing of information. We have to use another solution to ensure proper storage: redundancy. Redundancy provides a means so that data are available even when storage devices are damaged or offline. There are two common approaches to promote redundancy. The first is through backups of the file system. The second is through redundant storage, particularly RAID.

Backups date back to the earliest computer systems. A backup is merely a copy of a collection of data. Backups used to be stored onto magnetic tape. Backup tapes would be stored in separate rooms to ensure that if anything damaged the computer system, the tapes should be protected. Thus, the data would be easily available for restoration if needed. Ironically, many of the tapes used for data storage dating back to the 1950s have long degraded, and the data have been lost.

Floppy disks and optical disks eventually replaced tape for backup purposes. However, with the low cost of hard disk storage today, many users will back up their file system to an external hard disk, or use cloud storage. Backups can either be of the entire file system, or of those portions of the file system that have been modified since the last backup. In the latter case, this is known as an *incremental backup.* Operating systems can be set up to perform backups automatically. As part of the risk management plan, a backup policy is required.

RAID stands for redundant array of independent disks (alternatively, the "I" can also stand for inexpensive). The idea behind RAID is to provide a single storage device (usually in the form of a cabinet) that contains multiple coordinated hard disks. Disk files are divided into blocks as usual, but a block is itself dividing into stripes. A disk block is then stored to one or more disk surfaces depending on the size of the stripe. A large stripe will use few disks simultaneously whereas small stripes will use many disks simultaneously. Thus, the stripe size determines how independent the disks are. For instance, if a disk block is spread across every disk surface, then access to the disk block is much faster because all disk drives are accessed simultaneously and thus each surface requires less time to access the bits stored there. On the other hand, if a RAID cabinet were to use say eight separate disk drives and a stripe is spread across just two drives, then the RAID cabinet could potentially be used by four users simultaneously.

Although the "independent" aspect of RAID helps improve disk access time, the real reason for using RAID is the "redundancy". By having multiple drives available, data can be copied so that, if a disk sector were to go bad on one disk drive, the information might be duplicated elsewhere in the cabinet. That is, the redundant data could be used to restore data lost due to bad sectors. Redundancy typically takes on one of three forms. First, there is the mirror format in which one set of disks is duplicated across another set of disks. In such a case, half of the disk space is used for redundancy. Although this provides the greatest amount of redundancy, it also is the most expensive because you lose half of the total disk space, serving as a backup. Two other approaches are to use parity bits (covered in Chapter 3) and Hamming codes for redundancy. Both parity bits and Hamming codes provide a degree of redundancy so that recovery is possible from minimal damage but not from large-scale damage. The advantage of these forms of redundancy is that the amount of storage required for the redundancy information is far less than half of the disk space, which makes these forms cheaper than the mirror solution.

RAID technology has now been in existence since 1987 and is commonplace for large organizations. It is less likely that individual users will invest in RAID cabinets because they are more expensive. RAID itself consists of different levels. Each RAID level offers different advantages, disadvantages, and costs. RAID levels are identified by numbers where a larger number does not necessarily mean a better level of RAID. Here is a brief look at the RAID levels.

RAID 0 offers no redundancy, only the independent access by using multiple drives and stripes. RAID 1, on the other hand, is the mirror format where exactly half of the disks are used as a mirror. Thus, RAID 0 and RAID 1 offer completely opposite spectrums of redundancy and cost: RAID 0, which uses all disk space for the files, is the cheapest but offers no redundancy. RAID 1, which uses half of the disk space for redundancy, is the most expensive but offers the most redundancy. RAID 2 uses Hamming codes for redundancy and uses the smallest possible stripe sizes. RAID 3 uses parity bits for redundancy and the smallest possible stripes sizes. Hamming codes are complex and time consuming to compute. In addition, RAID 2 requires a large number of disks to be successful. For these reasons, RAID 2 is not used.

RAID 4, like RAID 3, uses parity bits. But unlike RAID 3, RAID 4 uses large stripe sizes. In RAID 4, all of the parity bits are stored on the same drive. Because of this, RAID 4 does not permit *independent* access. For instance, if two processes attempted to access the disk drive at the same time, although the two accesses might be two separate stripes on two different disk drives, both accesses would collide when also accessing the single drive containing the parity information. For this reason, RAID 4 is unused. RAID 5 is the same as RAID 4 except that parity information is distributed across drives. In this way, two processes could potentially access two different disk blocks simultaneously, assuming that the blocks and their parity information were all on different disk drives. RAID 6 is the same as RAID 5 except that the parity information is duplicated and distributed across the disks for additional redundancy and independence.

In Figure 15.5, RAID 4, 5, and 6 are illustrated. In RAID 4, all parity data are on one and only one drive. This drive would become a bottleneck for independent accesses. In RAID 5, the parity data are equally distributed among all drives so that it is likely that multiple disk operations can occur simultaneously. In RAID 6, parity data are duplicated and distributed so that not only is the redundancy increased, but the chance of simultaneous access is increased. You will notice that RAID 6 has at least one additional disk drive and so is more expensive than either of RAID 4 or RAID 5 but offers more redundancy and possibly improved access.

More recently, hybrid RAID levels have come out that combine levels 0 and 1, known as RAID 0+1 or RAID 10, and those that combine 3 and 5, called RAID 53. With backups and/or RAID cabinets, we can ensure the redundancy of our files and therefore ensure availability and integrity of storage. However, this does not resolve the issue of confidentiality. In order to support this requirement, we want to add one more technology to our solution: encryption. Through encryption, not only can we ensure that data are held confidential, but we can also transmit it over an open communication medium such as the Internet without fear of it being intercepted and understood. Since encryption is a complex topic, we will present it in the next section.

CRYPTOGRAPHY

Cryptography is the science behind placing information into a code so that it can be securely communicated. Encryption is the process of converting the original message into a code, and decryption is the process of converting the encrypted message back into the

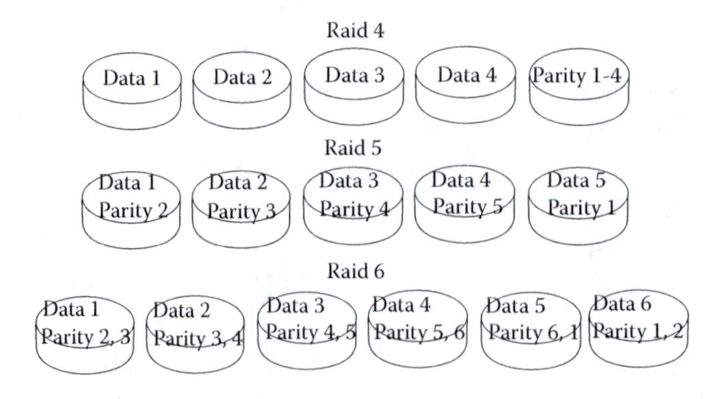

FIGURE 15.5 Parity data stored in RAID 4, 5, and 6.

original message. The key to encryption is to make sure that the encrypted message, even if intercepted, cannot be understood.

The use of codes has existed for centuries. In the past, codes, or what is sometimes called classic encryption, were largely performed by pencil and paper. A message was written in code, sent by courier, and the recipient would decode it. If the courier was intercepted (or was not trustworthy), the message could not be read because it was in code.

A simple code merely replaces letters of the alphabet with other letters. For instance, a rotation code moves each letter down in the alphabet by some distance. In rotation+1, an 'a' becomes a 'b', a 'b' becomes a 'c', and so forth, and a 'z' wraps around to an 'a'. During World War II, both German and Allied scientists constructed computing devices in an attempt to break the other sides' codes. Today, strong encryption techniques are used because any other form of encryption can be easily broken by modern computers.

There are two general forms of cryptography used today: symmetric key and asymmetric key. Both forms of encryption are based on the idea that there is a key, a mathematical function, that will manipulate the message from an unencrypted form into an encrypted form, and from an encrypted form back into the unencrypted form. In symmetric-key encryption, the same key is used for both encryption and decryption. This form of encryption is useful only if the message (or file) is not intended for others, or if the recipients can be given the key. So, for instance, you might use this form of encryption to store a file such that no one else can view the file's contents. You would not share the key with others because that would let others view the file.

Consider the case of e-commerce on the Internet. You want to transmit your credit card number from your web browser to a company's web server so that they can charge you for a purchase (well, you may not "want" to transmit the number, but you do want to make the purchase). The company you are doing business with wants you to trust them, so they use an encryption algorithm to encrypt your credit card number before it is transmitted. To do this, your web browser must have the key. However, if they provide your web browser with the key, you can also obtain the key. If you have the key, you could potentially intercept other messages to their server and break the encryption, thus obtaining other customers' credit card numbers.

Therefore, for e-commerce, we do not want to use symmetric-key encryption. In its place, we move on to asymmetric-key encryption. In this form of encryption, the original organization generates a key. This is the *private key*. Now, using the private key, they can create a *public key*. The public key can be freely shared with anyone. Given the public key, you can encrypt a message. However, you are unable to use the public key to decrypt a message. The organization retains the private key to themselves so that they can decrypt incoming messages. We usually refer to symmetric-key encryption as *private key encryption* because it only uses one key, a private (or secret) key. Asymmetric-key encryption is often called *public key encryption*, using two keys, a private key and a public key. Both of these forms of encryption are illustrated in Figure 15.6.

Both public and private key encryption use the notation of a block. The block is merely a sequence of some number of characters—in the case of computers, we will use a sequence of bits. So, we take some n bits of the message that we want to transmit and convert those

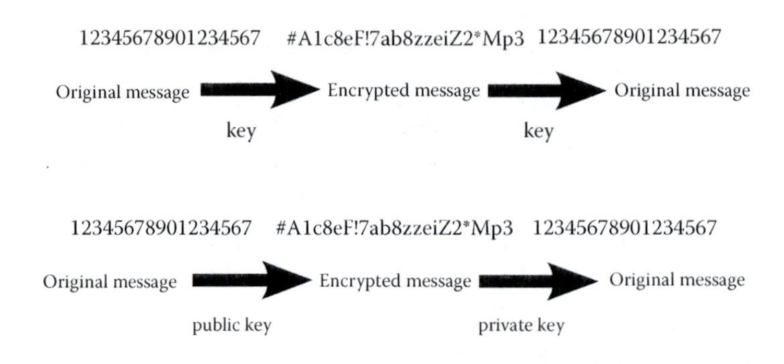

FIGURE 15.6 Private key encryption (top) versus public key encryption (bottom).

n bits using the key. We do this for each n-bit sequence of the block, putting together a new message. The recipient similarly decomposes the received message into n-bit blocks, decrypting each block. As a simple example, let us assume the message is "Hello". In ASCII, this message would consist of five 8-bit sequences (we add a 0 to the 7-bit ASCII character to make each character take up 1 byte). This gives us: 01001000 01100101 01101100 01101100 01101111. Now we take these 40 bits and divide them up into n-bit sequences. For our example, we will use n = 13. This gives us instead four sequences of bits: 0100100001100 1010110110001 1011000110111 1. Because the last sequence is only a single bit, we will pad it with 0s to become 1000000000000.

The key will treat each block as a binary number. It will apply some mathematical operation(s) to the binary number to transform it into a different number. One approach is to use a list of integer numbers, for instance, 1, 2, 5, 9, 19, 41, 88, 181, 370, 743, 1501, 3003, 6081. Since there are 13 integer numbers, we will add each of these integer numbers together if the corresponding bit in the 13-bit sequence is a 1. For instance, for 0100100001100, we will add 2, 19, 743, and 1501 together (because the only 1 bits in the block are the second, fifth, 10th and 11th, and 2, 19, 743, and 1501 are the second, fifth, 10th, and 11th in the sequence). The sum is 2265. So, our key has transformed 0100100001100 into 2265, which we convert back into binary to transmit.

Notice that our sequence of 13 integer numbers is not only in increasing order, but for any number, it is greater than the sum of all numbers that precede it. For instance, 88 > 41 + 19 + 9 + 5 + 2 + 1. This is crucial to being able to easily decrypt our number. To decrypt a number requires only finding the largest number in our key less than or equal to the number, subtracting that value, and then repeating until we reach 0. So, to decrypt 2265, we do the following:

Largest value < = 2265: 1501. 2265 – 1501 = 764.

Largest value < = 764: 743. 764 – 743 = 21.

Largest value < = 21: 19. 21 – 19 = 2.

Largest value < = 2: 2. 2 – 2 = 0. Done.

So we have decrypted 2265 into 1501, 743, 19, and 2, or the binary value 0100100001100.

For public key encryption, we have to apply some additional mathematics. First, we use the original list of numbers (our private key) to generate a public key. The public key numbers will not have the same property of increasingly additive. For instance, the public key might be 642, 1284, 899, 1156, 643, 901, 1032, 652, 1818, 940, 2266, 552, 723. The numbers in our public key are our private key numbers having been transformed using two additional values, some large prime numbers.

We can hand out our public key to anyone. Given the public key, the same process is used to compute a sum given a binary block. For our previous binary block (0100100001100), we add up the corresponding numbers in the public key rather than the private key. These numbers, again corresponding to the second, fifth, 10th, and 11th numbers, are 1284, 643, 940, and 2266. This sum is 5133. Instead of transmitting 2265, we transmit 5133. Using the public key, 5133 is very difficult to decrypt because our public key numbers do not have that same increasing, additive property. In fact, since the numbers are not in this order, to decrypt a number requires trying all possible combinations of values in our public key. There are 2^{13} different combinations of these 13 numbers. For instance, we might try the first public key number by itself, 642. That is not equal to 5133, so we can now try the first two numbers, 642 + 1284. That is not equal to 5133 so we next try the first and third numbers, 642 + 899. This is still not 5133, so now we try the first three numbers, 642 + 1284 + 899. Ultimately, we may have to try all 2^{13} different combinations before we hit on the proper solution.

You might recall from Chapter 3 that 2^{13} equals 8192. For a human, trying all of these combinations to decrypt the one number, 5133, would take quite a while. A computer, however, would be able to try all 8192 combinations in far less than a second. For this reason, our strongest encryption technologies today use blocks of at least 200 bits instead of something as short at 13 bits. With 200 bits, it would require testing at least 2^{200} different combinations in order to decrypt our number. This value, 2^{200}, is an enormous number. It is greater than 1,000,000,000,000,000,000,000,000,000,000,000,000,000,000,000,000,000,000,000,000,000!

If our encrypted value is so difficult to decrypt, how does the recipient decrypt it? Remember that the recipient has the private key. We have to transform the received number first by applying the two prime numbers. This would convert 5133 into 2265. And now we use our private key to easily and quickly decrypt the number. Whether the key uses block sizes of 13 or 200, decryption takes almost no time at all.

Not explained above is the use of the prime numbers to create the public key and to transform the received numbers. We apply a mathematical operation known as a *hash function*. A hash function maps a large series of values into a smaller range. Hash functions have long been used to create hash tables, a form of data storage in memory for quick searching, and are also used to build caches, filters, and error correction information (checksums). The idea is to take the original value and substitute it with an integer. This is already done for us using public key encryption, but if we were to convert say a person's name, we might add up the ASCII values of the characters in the name. Next, we use the modulo operator to divide the integer number by a prime number. Recall modulo (mod) performs division and gives us the integer remainder. For instance, if our prime number is 1301, then 5133 mod 1301 = 1230. The result from our mod operation will be a new integer between 0 and the prime number minus one (in this case, a number between 0 and 1300).

There are many popular encryption strategies used today. Here, we touch on a few of them. The following are all private key encryption algorithms.

- Data Encryption Standard (DES)—developed in the 1970s, this algorithm uses 56-bit block sizes. It was used by the national security agency but is now considered to be insecure.
- Advanced Encryption Standard (AES)—the follow-up from DES, it uses a substitution-permutation network that can be easily implemented in software and hardware for fast encryption and decryption. It uses a 128-bit block size but keys can take on sizes from 128 bits to 256 bits, in 32-bit increments.
- Triple DES—although DES is considered insecure because it can be broken in just hours, Triple DES uses three DES keys consecutively to encrypt and decrypt a message. For instance, if the keys are K1, K2, and K3, encrypting a message is done by applying $K3(K2^{-1}(K1(message))$, where $K2^{-1}$ means to use K2 to decrypt the message. Decryption is then performed by using $K3^{-1}(K2(K1^{-1}(encrypted\ message)))$. Each key is 56 bits in size and operates on 64-bit blocks. However, the application of three keys as shown here greatly increases the complexity of the code, making it far more challenging to break. In spite of the challenge, the result is roughly equivalent to a 112 bit key, which can also be broken with today's computers.
- Message-Digest Algorithm (MDn)—the most recent version is MD5. This 128-bit encryption algorithm uses a hash function applying a 32-digit hexadecimal hash number. However, between the limited size of the key and flaws discovered starting in 1996, MD5 is no longer in use and has been superseded by SHA-1.
- SHA-0 and SHA-1 are similar algorithms that are based on 160-bit blocks. As with MD5, they are message digest algorithms that apply a hash function. And although they do not have the flaws found in MD5, they still have their own weaknesses. Today, SHA-2 is used by the federal government, improving on SHA-1 by combining four different hash functions whose sizes are upward of 224 bits apiece. SHA-3 is being introduced in 2012.

Two of the most commonly used public key encryption algorithms are RSA and DSA (digital signature algorithm). RSA, named after the three men who wrote the algorithm in 1978, uses two large prime numbers to generate the public key from the private key. These two prime numbers, along with the private key, must be kept secure. DSA is a more complex algorithm than RSA. DSA is the standard algorithm used by the U.S. government for public key encryption.

These algorithms are used in a number of different software and protocols. Here, we take a brief look at them.

- WEP (Wired Equivalent Privacy) was released in 1999 as a means to permit encrypted wireless transmission equivalent to what can be found in a wired network. Using twenty-six 10-digit values as a key, it has been found to have numerous flaws and has been replaced first by WPA and then WPA2.
- WPA and WPA2 (Wi-Fi Protected Access) are protocols used to encrypt communication between resources in a wireless network. WPA2 replaced WPA around 2004. WPA2 is based on the AES algorithm. WPA and WPA2 were designed to work with wireless hardware produced before the releases of the protocols so that the protocols are backward compatible with older hardware. There are known security flaws with both WPA and WPA2, particularly in using a feature called Wi-Fi Protected Setup. By avoiding this feature, both WPA and WPA2 are far more secure.
- SSH is the secure shell in Linux, a replacement of the insecure telnet, rlogin, and rsh communication protocols. It uses public key encryption. Several versions have existed, generally referred to as SSH (or SSH-1), SSH-2, and OpenSSH. SSH uses port 22 whereas telnet uses port 23. SSH is also used to provide secure file transfer with SCP (secure copy). An alternate form of SSH is SFTP, secure file transfer protocol. This differs from using FTP within an SSH session.
- SSL and TLS are Secure Socket Layers and Transport Layer Security, respectively. TLS has largely replaced SSL. These communication protocols are used to enhance the transport layer of TCP/IP by providing encryption/decryption capabilities that are missing from TCP/IP. Communication can be done by private key encryption or public key encryption. In the latter case, the keys are provided during the TCP/IP handshake and through a certificate sent by the server. SSL and TLS are used in a number of different protocols including HTTPS, SMTP (e-mail), Session Initiation Protocol (SIP) used in Voice over IP, and VPN communication.
- HTTPS is the secure form of HTTP using port 431 instead of port 80. HTTPS requires the transmission of a certificate from the server to the client (refer back to Figure 15.4). The certificate is used in two ways. First, it contains the public key so that the client can encrypt messages. Second, if signed by a certificate authority, it authenticates that the server is who it claims to be. An out-of-date certificate, a self-signed certificate, or no certificate at all will prompt the client's web browser to issue a warning.

LAWS

We wrap up this chapter by examining several important legislations enacted by the U.S. government worth noting because of their impact on information security. This is by no means a complete list, and a website that contains more complete listing is provided in the Further Reading section. This list does not include legislation from other countries, but you will find similar laws in Canada and the United Kingdom.

- Privacy Act of 1974—the earliest piece of legislature that pertains to the collection and usage of personal data, this act limits actions of the federal government. The act states that:

  > No agency shall disclose any record which is contained in a system of records by any means of communication to any person, or to another agency, except pursuant to a written request by, or with prior written consent of, the individual to whom the record pertains….*

  The act also states that federal agencies must publicly make available the data that they are collecting, the reason for the collection of the data, and the ability for any individual to consult the data collected on them to ensure its integrity. There are exemptions to the law such as for the Census Bureau and law enforcement agencies.

- The Family Educational Rights and Privacy Act of 1974 (FERPA)—gives students the right to access their own educational records, amend those records should information be missing or inaccurate, and prevent schools from disseminating student records without prior authorization. This is an important act because it prevents faculty from sharing student information (e.g., grades) with students' parents unless explicitly permitted by the students. In addition, it requires student authorization to release transcripts to, for instance, graduate schools or prospective employers.

- Foreign Intelligence Surveillance Act of 1978—established standards and procedures for the use of electronic surveillance when the surveillance takes place within the United States.

- Electronic Communication Privacy Act of 1986—similar to the Foreign Intelligence Surveillance Act, this one establishes regulations and requirements to perform electronic wiretapping over computer networks (among other forms of wiretapping).

- Computer Matching and Privacy Protect Act of 1988—amends the Privacy Act of 1974 by limiting the use of database and other matching programs to match data across different databases. Without this act, one agency could potentially access data from other agencies to build a profile on a particular individual.

- Drivers Privacy Protection Act of 1994—prohibits states from selling data gathered in the process of registering drivers with drivers' licenses (e.g., addresses, social security numbers, height, weight, eye color, photographs).

- Health Insurance Portability and Accountability Act of 1996 (HIPPA)—requires national standards for electronic health care records among other policies (such as limitations that a health care provider can place on an employee who has a preexisting condition). Similar to FERPA, HIPPA prohibits agencies from disseminating health care information of an individual without that individual's consent. This is

* You can find the full law on the Department of Justice's website at http://www.justice.gov/opcl/privstat.htm.

one reason why going to the doctor today requires first filling out paperwork dealing with privacy issues.

- Digital Millennium Copyright Act of 1998 (DMCA)—implements two worldwide copyright treaties that makes it illegal to violate copyrights by disseminating digitized material over computer. This law includes clauses that make it illegal to post on websites or download from websites copyrighted material whether that material originally existed as text, sound, video, or program code. Additionally, the act removes liability from Internet Service Providers who are providing Internet access for a criminal prosecuted under this act. As a result of this law, YouTube was not held liable for copyrighted content posted to the YouTube site by various users; however, as a result, YouTube often orders the removal of copyrighted material at the request of the copyright holder.
- Digital Signature and Electronic Authentication Law of 1998 (SEAL)—permits the use of authenticated digital signatures in financial transactions and requires compliance of the digital signature mechanisms. It also places standards on cryptographic algorithms used in digital signatures. This was followed 2 years later with the Electronic Signatures in Global and National Commerce Act.
- Security Breach Notification Laws—since 2002, most U.S. states have adopted laws that require any company (including nonprofit organizations and state institutions) notify all parties whose data records may have been compromised, lost, or stolen in the event that such data has been both unencrypted and compromised. For instance, if files were stored on a laptop computer and that computer has gone missing, anyone who potentially had a record stored on files on that computer must be notified of the action.
- Intelligence Reform and Terrorism Prevention Act of 2004—requires, as much as is possible, that intelligence agencies share information gathered and processed pertaining to potential terrorist threats. This act was a direct response to the failures of the intelligence community to prevent the 9/11 terrorist attacks.

In addition to the above acts and laws, the U.S. government has placed restrictions on the export of cryptographic technologies. For instance, an individual or company is not able to sell or trade certain cryptographic algorithms deemed too difficult to break by the U.S. government. The government has also passed legislature to update many laws to bring them up to the computer age. For instance, theft by computer now is considered theft and not some other category. There are also laws that cover censorship and obscenity with respect to the Internet and e-mail.

FURTHER READING

The analysis of the data, information, knowledge, and wisdom hierarchy is specifically an IT topic; you might find it useful background in order to understand the applications that IT regularly uses. A nice book that provides both practical and theoretical views of the application of the hierarchy is *Information Lifecycle Support: Wisdom, Knowledge, Information and Data Management*, by Johnson and Higgins (The Stationery Office, London, 2010). Jennifer Rowley provides a detailed examination of the hierarchy in her paper, "The wisdom hierarchy: representations of the DIKW hierarchy" (*Journal of Information Science*, 33(2), 63–180, 2007).

Information System texts vary widely just like Computer Science texts. Here, we focus on IS texts related directly to databases and DBMS. As an IT student, you will no doubt have to learn something about databases. Database design, DBMS usage, relational database, and DBMS administration are all possible topics that you will encounter. The list below spotlights a few texts in each of these topics along with a couple of texts on data mining and data warehouses.

- Cabral, S. and Murphy, K. *MySQL Administrator's Bible*. Hoboken, NJ: Wiley and Sons, 2009.
- Elmasri, R. and Navathe, S. *Fundamentals of Database Systems*. Reading, MA: Addison Wesley, 2010.
- Fernandez, I. *Beginning Oracle Database 11g Administration: From Novice to Professional*. New York: Apress, 2009.
- Golfarelli, M. and Rizzi, S. *Data Warehouse Design: Modern Principles and Methodologies*. New York: McGraw Hill, 2009.
- Halpin, T. and Morgan, T. *Information Modeling and Relational Databases*. San Francisco, CA: Morgan Kaufmann, 2008.
- Han, J., Kamber, M., and Pei, J. *Data Mining: Concepts and Techniques*. San Francisco, CA: Morgan Kaufmann, 2011.
- Hoffer, J., Venkataraman, R. and Topi, H. *Modern Database Management*. Upper Saddle River, NJ: Prentice Hall 2010.
- Janert, P. *Data Analysis with Open Source Tools*. Cambridge, MA: O'Reilly, 2010.
- Knight, B., Patel, K., Snyder, W., LoForte, R., and Wort, S. *Professional Microsoft SQL Server 2008 Administration*. New Jersey: Wrox, 2008.
- Lambert III, J. and Cox, J. *Microsoft Access 2010 Step by Step*. Redmond, WA: Microsoft, 2010.
- Rob, P. and Coronel, C. *Database Systems: Design, Implementation, and Management*. Boston, MA: Thomson Course Technologies, 2007.
- Shmeuli, G., Patel, N., and Bruce, P. *Data Mining for Business Intelligence: Concepts, Techniques and Applications in Microsoft Office Excel with XLMiner*. New Jersey: Wiley and Sons, 2010.
- Whitehorn, M. and Marklyn, B. *Inside Relational Databases with Examples in Access*. Secaucus, NJ: Springer, 2006.

IAS encompasses risk management, network and computer security, and encryption. As with databases, thousands of texts exist on this widespread set of topics. This list primarily emphasizes computer and network security as they are more important to the IT professional.

- Ciampa, M. *Security+ Guide to Network Security Fundamentals.* Boston, MA: Thomson Course Technologies, 2011.
- Cole, E. *Network Security Bible.* Hoboken, NJ: Wiley and Sons, 2009.
- Easttom, W. *Computer Security Fundamentals.* Indiana: Que, 2011.
- Gibson, D. *Management Risk in Information Systems.* Sudbury, MA: Jones and Bartlett, 2010.
- Jang, M. *Security Strategies in Linux Platforms and Applications,* Sudbury, MA: Jones and Bartlett, 2010.
- Oriyano, S, and Gregg, M. *Hacker Techniques, Tools and Incident Handling.* Sudbury, MA: Jones and Bartlett, 2010.
- Paar, C., Pelzl, J. and Preneel, B. *Understanding Cryptography: A Textbook for Students and Practitioners.* New Jersey: Springer, 2010.
- Qian, Y., Tipper, D., Krishnamurthy, P. and Joshi, J. *Information Assurance: Dependability and Security in Networked Systems.* San Francisco, CA: Morgan Kaufmann, 2007.
- Schneier, B. *Applied Cryptography: Protocols, Algorithms, and Source Code in C.* Hoboken, NJ: Wiley and Sons, 1996.
- Solomon, M. *Security Strategies in Windows Platforms and Applications.* Sudbury, MA: Jones and Bartlett, 2010.
- Stallings, W. and Brown, L. *Computer Security: Principles and Practices.* Upper Saddle River, NJ: Prentice Hall, 2011.
- Stewart, J. *Network Security, Firewalls, and VPNs.* Sudbury, MA: Jones and Bartlett, 2010.
- Vacca, J. *Computer and Information Security Handbook.* California: Morgan Kaufmann, 2009.

Many laws related to IT are described on different websites. Issues pertaining to privacy and civil liberty can be found on the Department of Justice website at http://www.justice.gov/opcl/ and issues and laws regarding cybercrime at http://www.justice.gov/criminal/cybercrime/. Other IT related laws are provided on the U.S. Government's cio.gov website. Your rights with respect to computer usage and free speech are always a confusing topic. Many people believe that the first amendment gives them the right to say anything that they want over the Internet. The book *Cyber Rights: Defending Free Speech in the Digital Age* (New York: Times Book, 2003) by M. Godwin provides a useful background and history to free speech on the Internet.

REVIEW TERMS

Terminology used in this chapter:

- Asymmetric key encryption
- Availability
- Backup
- Buffer overflow
- Certificate
- Certificate authority
- Clustering
- Confidentiality
- Database
- Data mining
- Data warehouse
- Decision tree
- Denial of service
- Disaster recovery
- Field (database)
- Information
- Information asset
- Integrity
- Intrusion
- Join (database)
- Malware
- OLAP
- Phishing
- Projection (database)
- Protocol attack
- Public key
- Public key encryption
- Private key
- Private key encryption
- Query (database)
- RAID
- Record (database)
- Redundancy
- Restriction (database)
- Relation
- Relational database
- Risk assessment
- Risk management
- Social engineering
- Software exploit
- SQL injection
- Symmetric key encryption
- Threat
- Vulnerability

REVIEW QUESTIONS

1. What is the difference between data and information?
2. Information can come in many different forms, list five forms.
3. How does a database differ from a relation?
4. Given the relations in Tables 15.1 and 15.2, answer the following:
   a. Show the output of a restriction on the relation in 15.1 with the criteria GPA > 3.0.
   b. Show the output of a restriction on the relation in 15.1 with the criteria Minor = "Physics" or Minor = "Chemistry".
   c. Show the output of a projection on the relation in 15.1 of First Name, Major, and Minor.
   d. Show the output of a join on the relations in 15.1 and 15.2 with a projection of First Name, Last Name, and State with the criteria of GPA > 3.0.
5. What is the difference between a database and a data warehouse?
6. Refer back to Figure 15.3. If someone has an income of $40,000, no debt, and no collateral, what conclusion should we draw regarding the risk of a loan? What if the person has an income of $30,000, no debt, and no collateral?
7. What type of vulnerability might an employee of a bank have toward the bank's risk management? What type of vulnerability might a bank's database files and computers have?
8. What is the best defense from social engineering?
9. Would a buffer overflow be considered a phishing attack, software exploit, protocol attack, or intrusion attack?
10. Would packet sniffing be considered a phishing attack, software exploit, protocol attack, or intrusion attack?
11. List four forms of malware.
12. What is a self-signed certificate? What happens in your web browser if you receive a self-signed certificate when using https?
13. How does RAID 0 differ from the other levels of RAID?
14. Which RAID level provides the greatest amount of redundancy?
15. How is RAID 5 an improvement over RAID 4? How is RAID 6 an improvement over RAID 5?

16. Rewrite the message "hello" using rotation+1.

17. How does symmetric-key encryption differ from asymmetric-key encryption?

18. Given the private key numbers from Threats and Solutions, encrypt the binary value 0001100100101.

19. Given the private key numbers from "Threats and Solutions", decrypt the value 6315.

20. Assume you have received the value 4261. Why would it be difficult to decrypt this value using the public key from Threats and Solutions?

21. Are your grades protected such that a teacher is not allowed to freely announce or give away that information? Is your medical history protected?

DISCUSSION QUESTIONS

1. Assume we have the following 10 pieces of data (here, we only examine two fields of the data, given as integer numbers). Each is listed by name (a letter) and the values of the two fields, such as a(10,5). Graph the 10 pieces of data and see if you can identify reasonable clusters. How many clusters did you find? Are there any outliers?

   a(10,5), b(3,6), c(9,7), d(8,4), e(9,3), f(1,8), g(3,7), h(2,4), i(2,8), j(10,2)

2. Consider your own home. Define some of your information assets. What vulnerabilities and threats do you find for those assets? For instance, one information asset is your checking account number and the amount of money there. Vulnerabilities include that the account information should be kept secret and that you must be aware of the amount of money in the account. Threats include someone forging your signature on a check or overdrawing on your account. Attempt to define five other assets and their vulnerabilities and threats.

3. List three man-made disasters that have occurred in your area of the country in the past year. List three natural disasters that have occurred in your area of the country in the past year. From your list, which ones might impact a nearby company's IT?

4. As a computer user, how might you use encryption with respect to storing data? How might you use encryption with respect to Internet access?

5. Why does the U.S. government have laws against exporting certain types of encryption technology?

6. Stop Online Piracy Act (SOPA) was a proposed piece of legislation that was eventually removed from consideration. Explore what SOPA attempted to do and why it was abandoned. Provide a brief explanation.

7. SOPA is not the only piece of legislation proposed to stop online piracy. Research this topic and come up with other pieces of legislature. Of those you have discovered, which ones were similarly removed from consideration and which remain under consideration?

8. Research the Anonymous cracker group. Do you find their actions to be justified or not? Explain.

CHAPTER 16

Careers in Information Technology

IT CAREERS

The textbook ends where it began: by examining IT careers and careers related to the IT field. In this chapter, the various areas of IT are examined with an emphasis on the specific roles that each area requires. Among the details shared are forms of certifications that an IT person might seek and recent salaries. As students of IT may have not given a great deal of thought to their careers as they start college, the text describes a variety of topics that may extend the reader's interest. These include topics such as IT ethics as well as social concerns in the digital world such as the digital divide, Internet censorship, and Internet addiction. The chapter ends by discussing various forms of continuing education for the IT person.

The learning objectives of this chapter are to

- Discuss careers in IT.
- Describe forms of continuing education in IT.
- Introduce social concerns related to IT.
- Discuss ethical conduct in the IT field.

Are you interested in IT because of the money? Perhaps. More likely, you have grown up as a technology user. Your interest may have started because of social media or from schoolwork or through computer games. Or perhaps you got interested because of friends and family. Whatever interests led you into technology, you are most likely interested in IT because you have fallen in love with using technology. You are probably an analytical person, a problem solver. It is a common path that leads to an IT career. Fortunately for you, careers in IT are not only readily available, but can lead to high-paying and satisfying careers!

Here, we focus on IT career paths. Later in the chapter, we examine some of the aspects of an IT career that you may not have thought of. Beyond the technology, beyond the need to continually upgrade your knowledge and improve your skills, there are issues that you might face: legal, ethical, social.

Careers in IT are high paying, and the outlook for job growth is tremendous for many years to come. Table 16.1 illustrates job growth projections and salary information for a number of different computer-related careers. Notice that computer programmer positions are expected to decrease as the software development field moves from a demand for programmers to a demand for software engineers.

Of the careers listed in Table 16.1, the latter three are careers that an IT major might consider. Database administrators, network administrators, and system administrators are all projected for high growth through 2018, and there is no reason to expect that this trend will not continue further into the future. Other careers not listed here include web developer, website administrator, network architect, and computer security specialist.

All of these careers require a high degree of technical skill. These skills include an understanding of computer systems, operating systems, software, and hardware. Additionally, it is expected that IT personnel be problem solvers. They should also be self-starters and lifelong learners. It is also very important for all IT professionals to be proficient in written and verbal communication, team collaboration, and leadership, have a firm understanding of IT ethics, and the desire to continue to develop their own skills and education. A foundation in business practices can help further an IT specialist's career.

Today, most organizations looking to hire IT personnel prefer graduates of 4-year accredited university programs. Graduates of these programs have received a much broader education than those who have only focused on technical skills (as found in 2-year technical, vocational, or community colleges). However, 4-year IT degrees are still somewhat rare, so organizations looking to hire IT personnel will sometimes look to related disciplines such

TABLE 16.1 U.S. Computer Career Job Growth (Numbers in Thousands)

Title	**2008**	**2018 (Est.)**	**Inc.**	**Inc. %**	**Median 2008 Salary**
Computer programmers	426.7	414.4	−12.3	−2.87	$69,620
Computer software engineers, applications	518.4	689.9	175.1	34.01	$85,430
Computer software engineers, systems software	394.8	515.0	120.2	30.44	$92,430
Computer systems analysts	532.2	640.3	108.1	20.31	$75,500
Database administrators	120.4	144.7	24.4	20.26	$69,740
Network and computer systems administrators	339.5	418.4	78.9	23.23	$66,310
Network systems and data communications analysts	292.0	447.8	155.8	53.36	$71,100

Source: U.S. Bureau of Labor Statistics, www.bls.gov/emp/tables.htm.

as computer science, business administration or management information systems (IS), and information science. Students whose degrees do not fit the IT profile may still be desirable if they have taken appropriate coursework in computer systems, networks, databases, and so forth, or have picked up suitable experience.

The breadth of responsibility in these careers is often dependent on the size of the organization. In large organizations, you may find individuals whose sole responsibility is in an IT specialty area, for example, network security or web server administration. In smaller organizations, it is not uncommon for a single individual to have responsibility for several IT roles. Here, we focus on each of the types of positions that an IT specialist may have and the roles and responsibilities of each of those positions. Additionally, we will examine forms of certification (if any) and salary expectations.

Network Administration

Network administrators are responsible for the installation and maintenance of hardware and software that make up a computer network. Network hardware includes specialized network devices such as switches, routers, hubs, firewalls, adaptive security appliances, wireless access points, and voice-over-IP (VoIP) phones. Network software comprises the various operating systems installed on those hardware devices, the protocols to communicate over the network, and the applications. This includes routing protocols, access lists, trunking protocols, spanning tree protocols, virtual LANs, virtual private networks, IPv4, IPv6, and intrusion detection systems. Network administrators may be responsible for installing, configuring, and maintaining such software. Network administrators may also require securing the network; this task may be accomplished in conjunction with other IT personnel such as system administrators.

If the network slows or shuts down, a network administrator may occasionally find himself or herself in a high-pressure troubleshooting role. The administrator must use his or her skills and knowledge to diagnose the issues, identify potential solutions, and implement the selected solution. A solution should solve the given problem most effectively in terms of user convenience, efficient access, and reasonable cost. In addition, issues and solutions should be well documented.

Higher level network administration positions, sometimes referred to as *network architects* or network engineers, are often responsible for designing networks that meet the performance and capacity needs of an organization. This job includes choosing appropriate network hardware and planning the allocation of network addresses across an organization.

Network administrators all require a basic knowledge of the TCP/IP stack as well as other network protocols that follow the Open Systems Interconnect (OSI) model. However, network administrators often focus on specialties such as network security, wireless administration, VoIP, management of internet service providers, and network reliability and capacity planning.

From this list of duties, it is obvious that network administrators require a significant amount of training and experience to be successful in their roles. Because of the changing nature of technology, network administrators must constantly learn new technologies through trade journals, conferences, publications, and continuing education. Various career certifications in these specialty areas are a valuable addition to a network administrator's resume. Among the most sought-after certifications are the CCNA (Cisco Certified Network Associate), CCNP (Cisco Certified Network Professional), Novell CNE (Certified Novell Engineer), CWNA (Certified Wireless Network Administrator), and CompTia's Network +.

Salaries for network administrators (as of 2008) range from $41,000 to as high as $104,000 and a median average salary of $66,310. Management of company networks currently pays around $3000 more than working for telecommunications carriers, which pays substantially better than similar jobs working for schools of any level.

Systems Administration

Whereas network administrators are responsible for the performance of networking hardware and software, *systems administrators* are responsible for the installation and maintenance of the resources on that network. The system administrator's role covers hardware, software, and system configuration. Computers might include mainframe, minicomputers, and personal computers, as well as network servers. The system administrator should also be familiar with multiple operating systems as it is likely that the organization has computers of several platforms.

The responsibilities are varied and will vary by organization. The system administrator will, at a minimum, be required to install and configure operating systems on the organization's machines, install new hardware, perform software upgrades and maintenance, handle user administration, perform backup and recovery, and perform system monitoring, tuning, and troubleshooting. The system administrator may also be involved with application development and deployment, system security, and reliability, and have to work with networked file systems. The system administrator would most likely have to perform some programming through shell scripting so that repetitive tasks can be automated. A greater amount of programming might be required depending on the organization's size and staff. Policy creation and IT infrastructure planning may also be tasks assigned to system administrators.

The system administrator's role may overlap or completely subsume the requirements of a network administrator depending on the size of the organization and the need. Smaller organizations will hire fewer IT staff, and thus duties may be more diverse. On the other hand, smaller organizations will likely have lesser needs because of small networks, fewer computer resources, and perhaps fewer situations that require a system administrator's attention. The knowledgeable system administrator will turn to any number of web forums available whereby system administrators discuss recent problems and look for solutions.

As with a network administrator, the system administrator will require a great deal of training and experience to ensure that the computer systems run efficiently, effectively, and securely. The system administrator should be proactive in his or her continuing development. Like the network administrator, there are several chances for certification as well as trade journals, conferences, and publications. The certification opportunities include Microsoft's MCSA (Microsoft Certified Systems Administrator), MCSE (Microsoft Certified Solutions Expert), and MCITP (Microsoft Certified IT Professional), Red Hat's

RHCE (Red Hat Certified Engineer) and RHCSS (Red Hat Certified Security Specialist), and Novell's CNA (Certified Novell Administrator) as well as the network certifications mentioned earlier.

Salaries for system administrators are in the same basic range as those of network administrators. The two job categories combined promise to see increasing demand through at least 2018 with tens of thousands of new positions opening up.

Web Administration

The web administrator, also known as a webmaster, but more precisely referred to as a *web server administrator*, is responsible for maintaining websites. This differs from the development of websites (see Web Developer, in Related Careers). Specifically, the web administrator must install, configure, maintain, secure, and troubleshoot the web server. The web server is software that runs on one or more computers. The web administrator may or may not be responsible for the hardware. Typically, the web administrator is not a system administrator on that or those computers. Therefore, the web administrator will have to work with the system administrator on some aspects of installation and configuration.

Configuration involves setting up configuration files to specify such directives as where the web pages will be located, who can access them, whether there is any authorization requirements, what forms of security might be implemented, whether and what documents will be encoded, for instance, using encryption, and what scripting languages the server may execute.

Monitoring the web server's performance permits the administrator to fine-tune aspects of the server. This might include, for instance, installing larger capacity or more hard disks, adding file compression, and reducing the number of server side scripts. Additionally, monitoring via automatically generated log files provides the administrator with URL requests that could be data mined to discover client browsing behavior.

Securing the web server becomes critical for businesses. The web administrator may be required to set up password files and establish locations where server scripts can be stored and tested. Security may be implemented directly through the web server software, or added through the operating system, or some combination. Security issues may be discovered by analyzing the same log files that are used for data mining.

Web server software is made up of various lesser programs. These programs handle any number of server-related tasks from the simple retrieval of files of URL requests to logging to error handling to URL redirections. There are free web servers such as Apache as well as popular commercial web servers such as IBM WebSphere, Oracle WebLogic, and Microsoft IIS. The web administrator may or may not be involved in the selection of the web server software, but it is the web administrator's job to understand the software selected and make recommendations on how best to support the software with adequate hardware. Hardware decisions might involve whether the server should be distributed across numerous physical computer servers, whether the server requires multiple IP addresses, and what type of storage system should be used. In monitoring the web server's performance, the web administrator can make informed recommendations on upgrades and improvements to the hardware.

Web server administrators may be required to either perform application development themselves or to support the developers. Support might be in the form of installation of additional web server modules such as those that can execute Perl or PHP code. There may also be policy decisions that impact development, such as the requirement that all scripts execute outside the web server on another computer or all scripts being stored in a common cgi-bin directory.

Although the web server administration may work for a large organization that wishes to have a presence on the Internet, there are also opportunities to work for companies that host websites. In such a case, a single server might be used to host multiple websites of different companies. The administrator has the added challenges of ensuring the web server(s) performance and dealing with several groups of web developers.

Web master salaries can vary greatly from $40,000 to $90,000 with an average around $55,000. There are fewer certifications available for web administrators. The primary form is known as a Certified Internet Webmaster Server Administrator. This is one of many under the abbreviation CIW, but most of the CIWs involve web development.

Database Administration

Like web administrators and system administrators, *database administrators* have a long list of important responsibilities, but with a focus on the design, development, and support of database management systems (DBMSs). Tasks will include installation, maintenance, performance analysis, and troubleshooting as with the other administrative areas. There are specific applications, performance tuning parameters, account creation, and procedures associated with the DBMS that are separate from those of system administrators and web administrators. The database administrator may also be involved with the integration of data from older systems to new. As many or most database systems are now available over the Internet (or at least over local area networks), the administrator must also ensure proper security measures for remote access.

There are many different types of DBMSs. The most popular are relational databases such as Oracle, MySQL, or Microsoft SQL Server. But other types of database technology are on the rise including NoSQL databases and databases focused on cloud computing. As with all areas of technology, the continued growth of new database technology requires database administrators to be in a pattern of constant learning and self-improvement.

Unlike a web administrator who will probably not be involved in web development, a database administrator is often expected to play a role in application design and development. This may be by assisting the application developers with the design of the databases needed to effectively support those applications. This requires a unique set of skills that include database modeling, schema development, normalization, and performance tuning.

Because of the more specialized nature of the database administrator, salaries are higher than in the other IT roles listed above. Median average salaries are around $70,000 with the highest salaries being over $110,000. Certifications are primarily those of particular DBMS software such as the Microsoft Certified Database Administrator, Oracle Certified, and MySQL Certification.

Computer Support Specialist

Computer support specialist is somewhat of an umbrella term. Certainly, some of the roles of the support specialist could be a responsibility of a system administrator. The support specialist primarily operates as a trainer and troubleshooter. Training tasks may include the production of material such as technical manuals, documentation, and training scripts. Training may take on numerous forms from group training, individual training, training videos, and the production of training videos and software.

Troubleshooter might involve working the *help desk* (see below), training help desk personnel, and overseeing the help desk personnel. It might also involve helping out other IT personnel such as the system administrator(s) and network administrator(s). Tasks for the computer support specialist may be as mundane as explaining how to log in or how to start a program, stepping a user through some task such as connecting to the network file server or a printer, or as complex as discovering why organizational members are unable to access a resource.

At an entry level, the computer support specialist serves on the help desk. Such a person will answer phone calls and e-mails of questions from users. The types of questions posed to a help desk person will vary depending on the organization that the person works for. For instance, if you were on the help desk for an Internet Service Provider, you might receive phone calls from clients who cannot connect or have slow Internet connections. As a help desk person, you would have to be able to answer common questions and know how to locate answers for less common questions or be able to refer clients to others who have more specialized knowledge. You would also hear a lot of questions that might cause you to laugh. Be prepared to talk to some ignorant (technologically speaking) people! Here are some examples of (reportedly) true help desk conversations.*

- "What kind of computer do you have?" "A white one."
- Customer: "Hi, this is Rose. I can't get my diskette out." Helpdesk: "Have you tried pushing the button?" Customer: "Yes, sure, it's really stuck." Helpdesk: "That doesn't sound good; I'll make a note." Customer: "No. Wait a minute. I hadn't inserted it yet. It's still on my desk. Sorry."
- Helpdesk: "Click on the 'My Computer' icon on to the left of the screen." Customer: "Your left or my left?"
- Helpdesk: "Good day. How may I help you?" Male customer: "Hello, I can't print." Helpdesk: "Would you click on start for me." Customer: "Listen pal; don't start getting technical on me! I'm not Bill Gates, you know!"
- Customer: "Hi, good afternoon, this is Martha, I can't print. Every time I try, it says, 'Can't find printer'. I've even lifted the printer and placed it in front of the monitor, but the computer still says it can't find it."

* Taken from the website http://www.funny2.com/computer.htm.

- Helpdesk: "Your password is the small letter 'a' as in apple, a capital letter 'V' as in Victor, and the number '7'." Customer: "Is that '7' in capital letters?"
- Customer: "I have a huge problem. A friend has put a screensaver on my computer, but every time I move the mouse, it disappears!"
- Helpdesk: "How may I help you?" Customer: "I'm writing my first e-mail." Helpdesk: "Okay, and what seems to be the problem?" Customer: "Well, I have the letter 'a' in the address, but how do I get the circle around it?"

With remote desktop access, a computer support specialist is able to "take control" of a person's computer remotely. This allows the specialist to better understand, diagnose, and resolve problems. It also allows the specialist to work with little input from the person who is having difficulty. This can save time if the person is unknowledgeable about the computer or the problem.

Unlike the administrative careers, support specialists may not have a 4-year degree nor specialized skills other than those necessary to work at the help desk. However, those with a 4-year degree are more likely to receive promotions and move on to more challenging, and probably more enjoyable, tasks. Salaries vary especially since support specialists can be part-time or full-time employees. Salary ranges can be as low as $25,000 and as high as $70,750, although support specialists more commonly make in the $30,000 to $45,000 range.

IT Management

In addition to requiring strong and often specialized technical skills, all of the IT careers described here require communication and collaboration skills. These soft skills permit the IT specialist to work well with end users, application developers, and upper management. This is one of the reasons why employers are now seeking IT personnel among those potential employees who have a 4-year IT degree that provides a well-rounded education, beyond merely technical training.

Those IT personnel who can demonstrate interpersonal communication, leadership, and understanding of business practices may find themselves well positioned for future IT management positions. The IT Manager is one who understands both the technical side of the IT infrastructure in order to understand the problems that need resolving as well as the management side of the business in order to oversee technical staff and communicate readily with upper management. An IT Manager will be a project manager who will have to establish deadlines, work with external constituents, handle and possibly propose budgets, and make decisions regarding the project(s) being managed. Additionally, hiring and firing may be part of the IT Manager's responsibilities.

The management positions are typically at a higher pay rate than the technical fields described earlier. If your goal ultimately is to obtain such a management position, you would be best served by increasing your opportunities during your college education. These opportunities include joining college clubs and organizations, participating in student governance, and taking additional classes in business areas such as project management, economics, accounting, and leadership. Oftentimes, a student who is seeking further career advancement will return to school for an advanced business degree such as an MBA or MIS, or an advanced technical degree such as a Master's Degree in CIT. IT Managers could potentially make salaries twice that of technical employees.

SOFT SKILLS

So what are these soft skills being referred to? What follows are some of the cited soft skills demanded by recent employers:

- Articulate and observant
- A good listener
- Able to write and document work
- Time management skills
- Problem solving, critical thinking, and organizational skills
- Has common sense
- Strong work ethic, motivated, hardworking, has integrity
- Positive attitude
- Team player
- Confident, courteous, honest, reliable
- Flexible, adaptable, trainable
- Desire to learn, able to learn from criticism
- Self-supervising
- Can handle pressure and stress
- Makes eye contact with people, has good personal appearance
- A company person (cares about the company's success), politically sensitive
- Can relate to coworkers
- Able to network (in social settings)

RELATED CAREERS

There are other careers that overlap the content covered in a 4-year IT degree. The following list discusses a few of these.

Computer Forensics

Typically, a person who enters a career in computer forensics will have either a computer forensics degree itself, or will have a technical degree with additional background in criminal justice. The technical degree might be computer science, IT, or computer engineering. As computer forensics is a new and growing field, there are few undergraduate programs that offer a complete curriculum in the topic. However, many computer science and IT programs are adding computer forensics classes or minors. Additionally, graduate programs in computer forensics, or the related IS security, are on the rise.

Aside from technical skills learned in IT or computer science programs, one would need practical knowledge in operating systems, computer security, computer storage, software applications, cryptology, and possibly both programming languages and software engineering. Additionally, strong analytical skills are an essential.

A person who works in computer forensics is not just securing computer systems or looking for evidence of break-ins and tampering. Instead, one must provide evidence in computer crime cases to the law enforcement agencies that might hire you. Therefore, you must understand how to build a case and how to identify, collect, preserve, and present evidence in a legal manner.

Job opportunities in computer forensics abound today. Options include working directly for law enforcement, consulting with law enforcement, or working for organizations that have a vested interest in protecting their data and prosecuting violators, such as banks and corporations. Salaries are substantially higher presently than those of other IT careers, with annual salaries being in the $75,000–$115,000 range and an average of $85,000. However, without the specific degree in computer forensics, your options may be more limited than students who have obtained the full 4-year degree.

Web Development

There are many different roles that one might play in web development. An organization might hire a single individual to get their company "up on the web" or a company may hire a staff of web developers. Therefore, the specific role that a web developer might play could vary depending on the company and the situation. What is clear, however, is that the web developer has different skills than the website administrator discussed in Web Administration.

Web developer skills include the ability to use HTML, CSS (cascaded style sheets) and various scripting languages. Scripting might be done in JavaScript, Ruby, Python, Java Applets, Active Server Pages (ASP), Visual Studio .Net (including Visual Basic, C++, or C#), Perl, and PHP. Although a web developer would not be expected to know all of these languages, it is likely that the web developer would use several of them. While the programming aspect of the web developer is important, the web developer would not necessarily require a computer science degree (although we do tend to see a lot of computer scientists as web developers).

A web developer may need to know how to build aesthetically pleasing web pages with the use of proper color, layout, graphics, animation, and so forth. The web developer would also need to understand many of the technical and societal issues related to the Internet. For instance, understanding the nature of network communication, bandwidth speeds, disk caching, and the use of proxy servers would help a web developer build a more efficient and easier to download website. Understanding societal issues can be critical in securing a website from attacks, ensuring data integrity of the website, and ensuring that the website content is politically correct. There are also a number of accessibility concerns so that people with various types of handicaps can still visit and maneuver around the site without complication. It is now a federal law that websites of federal agencies and agencies funded by the government be accessible to people with handicap.

Web development itself includes several different tasks that incorporate graphic design and art, business, database access, telecommunications, human–computer interaction, and programming. Specifically, web application development involves writing server-side scripts and other server applications in support of dynamic web pages and interactions with the server and database.

Web developer positions are in high demand these days. On the other hand, they have a tremendous salary range, possibly reflecting the wide range of companies that are looking to create a web portal for their companies. Salaries can be as low as $30,000 and as high as the mid $70,000 range. There are also a number of certifications available to improve your status. These include the Microsoft Certified Professional Developer, several CIW Web Developer certificates (see the "Web Administration"), HTML Developer Certification, and Sun Certified Web Component Developer, to name a few.

Programming/Software Engineer

Historically, computer programmers have come from any number of disciplines including the sciences, mathematics, and business. The computer science degree became popular starting around 1980. Today, companies do not hire programmers, they hire graduates of computer science programs (or in some cases, computer engineering). With the increased emphasis on software engineering, some schools now have software engineering programs, which are similar in many ways to computer science programs.

The computer science discipline provides a different type of foundation than that of IT. For the computer scientist, a mathematical grounding is necessary to study topics in computability and solvability. These topics give the computer scientist the tools to measure how challenging a given problem is to solve (or whether it is solvable at all). The computer scientist also studies computer organization and operating systems to understand the underlying nature of the computer system. The computer scientist also studies a wide variety of algorithms and data structures and implements programs in a number of programming languages.

As with the IT individual, the computer scientist/software engineer must keep up with the field. However, whereas the IT specialist will largely study the latest operating systems, their features, tools, and security measures, the computer scientist will study new languages and programming techniques. There are certifications for programmers although with a 4-year degree in the field, the certifications may not be very necessary. Salaries, as shown in Table 16.1, tend to be higher in computer science/software engineering than they do in other IT areas, with the notable exception of IT management.

One particularly potent mixture of coursework is to combine the IT degree with computer science coursework. This provides the IT individual with an excellent programming foundation. Alternatively, the computer scientist is well served by taking additional IT courses to gain an improved understanding of operating systems, networks, and computer security.

Information Systems

As described in Chapter 15, IS have existed for longer than computers. An information system is in essence a collection of data and information used to support management of an organization. Most commonly, the organization is some sort of business although it does not have to be. The field of IS is somewhat related to computer science in that IS includes the study of computer hardware, software, and programming. However, the emphasis on software revolves around business software (e.g., DBMSs, spreadsheets, statistical software, transaction processing software), and the study of hardware often revolves around networks and to a lesser extent, storage. The IS discipline covers a number of business topics including introductory accounting and economics and quite often includes several courses in management as well as soft skills that are less emphasized in computer science.

The IS employee might take on any number of roles within an organization from project management to systems development to research. As a project manager, a person is in charge of planning, designing, executing, monitoring, and controlling the project. During the planning and design phases, a manager will be required to produce specifications, and time and cost estimations. The manager may be required to hire employees to support the project. During execution, monitoring, and control phases, the manager coordinates with each project implementation group to keep tabs on progress, determine problems, and resolve the problems by obtaining additional resources. Although the IS manager does not have to have the technical skills of the employees that he or she is managing, it is essential that the IS manager understand the underlying technology.

In IS development, the IS employee is part of the implementation team. Implementation may require programming in a high level language (e.g., Visual Basic, COBOL, or Java) but is just as likely or more likely to use tools that provide the IS person with a greater degree of abstraction over the specific details of the computer code. Many of these tools are referred to as 4th Generation Languages (4GLs), implying that these are programming languages developed during the fourth generation of computing. These languages include report generators, form generators, and data managers such as Excel, Access, SAS, SPSS, RPG, Mathematica, and CASE Tools, and more recently XBase++, Xquery, and Oracle Reports.

IS research examines IS modeling approaches and IS tools. In modeling, one attempts to identify a way to represent some organizational behavior, whether it is the transactions used in an organization's IS or the way that knowledge is transmitted between organizational units or some process in between. Given a model, the researcher defines methods to perform some task on the model. The model and methods must be specific enough so that the researcher is able to simulate the organization's processes sufficiently to demonstrate that the model and methods capture the expected behavior. Given the model, the researcher can then attempt to define improvements to the organization's processes and policies.

Computer Technician

The computer technician is often someone who will work directly on computer hardware installation, repair, and troubleshooting. Commonly, a 2-year technical degree is enough to train a person for such a position. Experience and skill may be gained in other ways such as by putting together computers from parts as a hobby. The computer technician will have to understand the role of the various pieces of hardware in a computer system and understand the electrical engineering technology behind it—such as testing whether a device is receiving power. Skills might include wiring/soldering, using various electrical tools, and troubleshooting. Computer technicians are often not paid very well as there is both less need for computer technicians today and because the skills are not highly specialized.

IT ETHICS

As computers play an increasingly important role in our society, IT specialists must become aware of the impacts that computers, and IT in general, have on society. The issues include maintaining proper ethical conduct, understanding legal issues, enforcing IT-related policies, maintaining data integrity and security, and identifying risks to data and employees. Not only will you have to understand these for yourself, but you may also be responsible for the development, implementation, and education of company policies.

Figure 16.1 lists the 10 Commandments of computer ethics, as published by the Computer Ethics Institute, Washington, DC.* Computer ethics is nothing new (the concept originated in 1950, although most of the emphasis on computer ethics has targeted proper ethics for programmers). But with IT as a growing field, the notion of ethical conduct must be expanded to include those in charge of computer systems rather than those who create computer systems. The primary concern regarding IT specialists is that they have greater privileges with respect to the computer systems—they have access to all or many resources. Therefore, they must act in an appropriate manner so that they do not violate the trust that the employers place in them to supervise these systems.

Aside from the 10 Commandments, as listed in Figure 16.1, other organizations have provided computer ethics codes of conduct. These include the Association for Computing Machinery (ACM), the British Computer Society, and the Institute of Electrical and Electronics

1. Thou shalt not use a computer to harm other people.
2. Thou shalt not interfere with other people's computer work.
3. Thou shalt not snoop around in other people's computer files.
4. Thou shalt not use a computer to steal.
5. Thou shalt not use a computer to bear false witness.
6. Thou shalt not copy or use proprietary software for which you have not paid.
7. Thou shalt not use other people's computer resources without authorization or proper compensation.
8. Thou shalt not appropriate other people's intellectual output.
9. Thou shalt think about the social consequences of the program you are writing or the system you are designing.
10. Thou shalt always use computers in ways that insure consideration and respect for fellow humans.

FIGURE 16.1 The 10 Commandments of computer ethics.

* The Computer Ethics Institute has a website at http://computerethicsinstitute.org/home.html, which includes a number of white papers that discuss cyberethics, the 10 Commandments listed in Figure 16.1 and numerous other issues.

Engineers (IEEE). These codes of conduct primarily relate to programmers and emphasize professional competence, understanding the laws, accepting professional review of their work, providing a thorough risk assessment of computer systems, honoring contracts, striving to improve one's understanding of computing and its consequences, and so forth. These codes combine notions of ethics as they pertain to technology as well as ethics of being a professional.

There are a number of issues that any organization must tackle with respect to proper computer ethics, and in particular, the role of administrators. These include notions of privacy, ownership, control, accuracy, and security.

- Privacy—under what circumstances should an administrator view other people's files? Under what circumstances should a company use the data accumulated about their customers?
- Ownership—who owns the data accumulated and the products produced by the organization?
- Control—to what degree will employee actions be monitored?
- Accuracy—to whom does the responsibility of accuracy fall, specific employees or all employees? To what extent does the company work to ensure that potential errors in data are eliminated?
- Security—to what extent does the company ensure the integrity of their data and systems?

In this section, we will focus first on ethical dilemmas that might arise for an IT professional. Afterward, we briefly examine some policy issues that an IT professional might be asked to provide in an organization.

For the following situations, imagine yourself in the given position. What would you do?

- Your boss has asked you to monitor employee web surfing to see if anyone is using company resources to look at or download pornography. Is it ethical for you to monitor the employees?
- Your boss has asked you to examine a specific employee's file space for illegally downloaded items. Is this ethical?
- You find yourself with some downtime in your job. You decide to brush up on your knowledge by visiting various technology-related websites. Among them, you read about hacking into computer systems. Is it ethical for you to be reading about hacking? Is it ethical for you to be reading about hacking during work hours?
- You have noticed that the e-mail server is running out of disk space. You take it upon yourself to search all employee e-mail to see if anyone is using company e-mail for personal use. Is this ethical?

- You have installed a proxy server for your organization. The server's primarily purpose is to cache web pages that other users may want to view, thus saving time. However, the proxy server also creates a log of all accesses. Is it ethical for you to view those accesses to see what sites and pages fellow employees have been viewing?
- You are a web administrator (not a web developer). One of the sites that your company is hosting contains content that you personally find offensive. What should you do? If the content is not obscene by the community's standards, would this alter your behavior? What if you feel the content would be deemed obscene by your community?
- You are a web developer. You have been asked by the company to produce some web pages using content that they provide you. You believe the content is copyrighted. What should you do?
- You work for an organization that is readying a product for delivery. Although you are not part of the implementation team, being an IT person, you know that the product is not yet ready but management is pressuring them to release the product anyway, bugs and all. What should you do?
- Your boss asks you to upgrade the operating system to a newer version that has known security holes. Is it ethical to do so?
- You fear that your boss is going to fire you because of a number of absences (all of which you feel were justified). You decide, as a precaution, to set up a secret account (a backdoor). Is this ethical? Is this legal?
- Your organization freely allows people to update their Facebook pages on site. You are worried that some of your employees are putting up too much information about themselves. Is it your place to get involved?
- Your organization has a policy prohibiting people from using Facebook at work. You notice one of your colleagues has modified his Facebook page during work hours. Should you report this?

As can be seen from these example situations, you may be asked to put yourself into a position that violates employees' privacy or confidence. When asked to monitor employees, you are being asked to spy on them. In order to be ethical, you need to consider many factors. These factors include the company's policy on employee usage of computers and facilities and the role of the IT specialists. Without stated policies, it is difficult to justify spying on employees. On the other hand, even if a company has policies, how well are those policies presented to the employees? If they are unaware of the policies, the policies do little good.

As an IT specialist, your role might include the development and implementation of IT policies (on the other hand, the policies may be specified solely by management). Policies should reflect the ethical stance of the company as well as the law. Policies should cover all aspects of IT usage. Specifically, policies should define:

- The role that IT plays in the company
- Proper employee usage of IT
- The employee's rights to privacy (if any)
- Proper usage of data
- Security and privacy of data
- Ownership of ideas developed through IT in the company (intellectual property)
- How the company handles copyright protection issues

Drafting policies may be a foreign concept to an IT specialist. It might be best handled by management. However, the IT specialist must be well versed in the organization's policy in order to make decisions that do not violate the policy. Without knowing the policy, your answers to the earlier questions could put you at risk.

Just as an IT specialist should understand both the policies of the organization as well as the proper ethical stance, understanding the laws is equally important. Consider a situation in which your boss has asked you to install software that he hands to you on a CD-R. Was this software purchased legally? Are there licenses with it? If not, do you go ahead with the installation? Just because your employer has asked you to perform a task does not mean that the task is either ethical or legal. Knowing the law can not only protect you but can also help you inform your employer about potentially illegal situations. Unfortunately, understanding the laws that regulate IT and software can be challenging. Even if you can identify the laws, they are often written vaguely and in legalese.

Consider as an example that your company has created a policy whereby all employees will be known by their social security number. That is, the social security number will serve as their unique identifier for database entries. As such, it is also used in their log in process. Because it is part of their log in, social security numbers are stored in world-readable files in the operating system (for instance, /etc/passwd). As the system administrator, what will you do? Is the use of social security numbers a violation of a company policy? No. Is it illegal? Not strictly speaking. Is it unethical? Probably.

You convince management that using the social security number for log in purposes is a bad idea, and therefore these secure data are removed from any world-readable file. Yet, the values are still available in company databases of which you and the database administrators and database managers still have access. Is this unethical? No, because those who have access are those that the company decrees as having a legitimate need to access such data. However, how can the company assure their employees that the information is secure? The people who have access to secure data need additional training in understanding the sensitivity of such data. The IT staff have the added responsibility of protecting devices that store this sensitive data, such as laptop computers.

FLAME AND STUXNET

As this book is being readied for publication, news has come out about a new virus called Flame. First, some background. In 2009, the Stuxnet virus (thought to be a product of a U.S.–Israeli partnership), attacked various Iranian computers, particularly those involved in Iran's uranium enrichment program. Stuxnet, and another similar virus, DuQu, were small viruses, some 500 KB in size. Both Stuxnet and DuQu have been heralded as groundbreaking malware-based sabotage.

Although Flame has similar targets, computers in Iran, as well as those in other Middle East and Eastern European countries, analysis indicates that the virus was written by a completely different organization.

Flame only came to light in 2012, but investigation of the virus has led some to believe that some computers have been infected since as far back as December, 2007.

The Flame virus is some 20 MB in size and of far greater complexity than Stuxnet. Its primary purpose seems to be to spy on users. It can steal data (documents, files), record conversations by surreptitiously turning on a computer's microphone, record keystrokes, obtain telephone numbers from nearby Bluetooth-enabled devices, create and store screenshots, and scan local Internet traffic to obtain usernames and passwords as they are passed across the network for authentication. If that were not enough, the virus can also create a backdoor account for people to log into so that they can upload more components of the virus. The virus is programmed to transmit accumulated information to one of more than 80 different servers on the Internet.

Flame itself is a collection of units; the initial units, when infecting a computer, are used to upload other units. The main unit can upload, extract, decompress, and decrypt other components. Furthermore, the virus seems capable of hiding itself from detection.

The Flame virus can spread over the network and through USB drives. Flame has been given several names by different organizations including Wiper, Viper, and Flamer.

OTHER SOCIAL CONSIDERATIONS

The IT professional must understand the consequences of IT decisions. Thus, beyond ethics and law, beyond organizational policy, the IT specialist should also understand key social concerns. Some of the concerns that our society faces are described in this section.

Digital divide. There are gaps in access to computers and the Internet. These gaps impact races, genders, income levels, and geographical locations. The gender gap has been greatly reduced in most societies, especially with the popularity of social networks. But the gaps that occur between technologically advanced and wealthy nations over poor nations continue to grow. Because the Internet has afforded those who have access a tremendous amount of information at their fingertips, those without access become even more disadvantaged with respect to an informed life, a healthy lifestyle, education, and employment. Governments and multinational corporations have addressed the digital divide, but it continues to be a concern. Although, as an IT specialist, you may never have to deal with the ramifications of the digital divide, understanding it might help your organization create reasonable policies.

Computer-related health problems. Heavy computer usage can cause a number of health problems. The primary problems are repetitive stress injuries (RSI) and eye strain. One of the more common forms of RSI is carpal tunnel syndrome, a muscular problem that impacts the wrist. This is often caused by extensive misuse of the keyboard and mouse. Aside from RSI, back strain, sore shoulders and elbows, and even hip and knee strain are often attributed to heavy computer usage. The list of problems also includes headaches and stress. These problems are well documented, and there are a number of solutions. *Ergonomics*, the study of designing equipment and devices that better fit the human body, and HCI (human–computer interaction) have given us better furniture, computer monitors that are easier on the eye, and less destructive forms of interaction with the computer (e.g., a track point instead of a mouse). In spite of ergonomics, people continue to have health problems because of poor posture or just simply because they are not remembering to take breaks from their computer usage. The IT specialist should understand ergonomic solutions and utilize them when requesting computing resources.

Maintaining privacy. Threats to an individual's privacy continue to increase. There is already a plethora of private information about you that is made public. This includes such facts as your date of birth, your residence, the car(s) you drive, and when and to whom you are married. Most of this information is made publicly available because, by itself, it does not pose a serious threat to your privacy or life. However, through social networking, websites, and blogs, people are regularly volunteering more information.

Consider a person who tweets that he and his family are leaving town for a weeklong trip. A clever thief could locate the person's home and, knowing that the house will be empty for a week, break in, and rob the person. Such publicizing of one's own private information has been called *digital litter.* Understanding the threats of such self-advertising is important to all computer users in this age of identity theft. Informing employees of the threats may fall on the IT personnel. However, the threats can be far more serious. Phishing attempts are used by scammers to obtain secure information from people directly. A common example is a phone call (or e-mail) claiming to be from an authorized IT person telling you that IT has lost pertinent data (such as your password or social security number) and that you must provide the caller with this information or else your account will be disabled.

Internet addiction. The number of hours that many people are now spending using computers has surpassed the number of hours that they view television. This is a remarkable change in our society. Part of the reason for this shift is that people have become obsessed with electronic forms of communication. These include e-mail, blogging, and social networking sites. This need to always be online has been dubbed Internet addiction, and some people suffer so greatly that they cannot keep their hands off of their smart phones when they are away from home. We see this in the extreme when people are at social events (e.g., movies, out to dinner) spending more time on their smart phones than interacting with people in person. This particular social concern may not directly impact your organization. However, many employers have become concerned with the diversions that impact their employees. Internet addiction can have a tangible impact on work performance.

Obscenity and censorship. The Internet has opened up the ability for people to read and view almost anything on the planet. Unfortunately, not everything presented on the Internet would pass people's moral compasses. Between pornography websites, violent video content (viewable at sites such as YouTube), and the diatribes posted by disturbed individuals, there is material on the Internet that people will label obscene. Obscenity has long been an issue in our society. Rating services, the Federal Communications Commission (FCC), and watchdog organizations attempt to police movies, television, and radio. However, policing the Internet is an entirely different matter because of its global nature and enormous content. Censorship might be implemented within an organization through a firewall and/or proxy server to block specific content, or content from specific sources. As with many of the issues discussed above, awareness

DIGITAL LITTER

People have gotten so used to posting their thoughts on Facebook or via Twitter or on various blogs and message boards that they are careless with the information that they supply. Consider the following, seemingly innocent update

"Going to a movie with the fam today, home around 3"

Imagine that this person has a "friend" who has no qualms with a little thievery, now the friend knows when you will be gone and for how long. Why advertise that your house will be empty? And yet people do, regularly (going on a vacation, taking the family out to dinner, will be working all night, etc.).

If you do not believe your innocent posts can get you into trouble, do a little research. There are a number of websites set up to obtain information about people via social media and public record. The site spokeo.com is pretty thorough. Check it out and enter your name. You might find that the following information has been obtained on you:

Age, marital status, financial worth, politics and religion, interests/hobbies, occupation, health, your family members' names, location of your home, value of your home, pictures of your home

If that were not enough to scare you, consider that there are dozens of sites that have similar information available about you:

- mylife.com
- whozat.com
- socialpulse.com
- peekyou.com

to list but a few!

and education are important. Policies on proper computer usage may also provide solutions. Web developers and administrators should have familiarity with the issues to ensure that their website complies with laws and regulations.

CONTINUING EDUCATION

We have already used the term "lifelong learner." What does this mean? The phrase implies that you will continue to study throughout your life. This is critical for an IT professional because IT continues to change. Recall the various histories presented in this text of computer hardware, software, operating systems, and programming languages. Each area (computer engineering, software engineering, systems engineering) has required that the employees working in that field continue to learn. Without this, as newer technologies emerge, the employee is not able to use them. The employee is now holding the company back from improving itself. It is likely in such circumstances that the employee is determined to be a liability and eventually laid off. Therefore, continuing education in any technology-oriented field is critical. For IT professionals, continuing education is not merely a matter of remaining valuable to your organization, but a matter of helping your organization make informed decisions on future technology directions.

There are numerous forms of continuing education available in IT. Most of these forms revolve around you learning on your own. There are many sources available to learn from, but it is up to you to take the initiative to find and use those sources. Among the forums available are trade magazines and journals, whether they are part of the popular press such as *Wired*, *New Scientist*, *Dr. Dobb's Journal*, or more academically oriented such as IEEE and ACM publications. Table 16.2 provides a list of some of the more relevant IT publications from IEEE and ACM.*

Two problems with reading the more academically focused journals are that they tend to emphasize cutting-edge research rather than the industry-available technology and they are presented at a level that might be more amenable to graduate students and faculty. Therefore, as an IT professional, you might prefer instead to research technology through the wide variety of books and websites.

There is a large market in technology books. In fact, just about every IT topic will have several dozen books published on it, from users' guides to "dummy guides for" style books. Among the more popular IT publishers are Thompson Course Technologies, O'Reilly, Elsevier, and Springer Verlag.

Many technology-oriented websites are created and maintained by IT professionals. They publish their knowledge as a service to others who can benefit by learning from them. There are, for instance, a large number of websites that include information about the Linux operating system, the Windows 7 operating system, the Internet, the Apache web server, and so on. These sites can not only provide you with "how to" knowledge, but also up-to-date tips on known problems and security patches.

* For more information on IEEE, see www.ieee.org, and for more information on ACM, visit www.acm.org.

TABLE 16.2 ACM and IEEE IT-Oriented Journals

Journal Name	IT-Related Topics
ACM Transactions of Information and System Security	Cryptography, authorization mechanisms, auditing, protocols, threats, data integrity, privacy issues
ACM Transactions on Internet Technology	Digital media and digital rights, electronic commerce, Internet crime and law, Internet performance, peer-to-peer networks, security and privacy
ACM Transactions on Storage	Storage systems and architecture, GRID storage, storage area networks, virtualization, disaster recovery, caching
ACM Transactions on the Web	Browser interfaces, e-commerce, web services, XML, accessibility, performance, security and privacy
Communications of the ACM	Communications/networking, computer systems, computers and society, data storage and retrieval, performance, personal computing, security
IEEE Internet Computing	IT services, WWW technologies
IEEE Networks	Network protocols and architectures, protocol design, communications software, network control and implementation
IEEE Technology and Society	Impacts on society, history of social aspects, professional responsibilities
IEEE Transactions on Computers	Operating systems, communication protocols, performance, security
IEEE Transactions on Dependable and Secure Computing	Evaluation of dependability and security of operating systems and networks, simulation techniques, security, performance
IEEE Transactions on Wireless Communications	Theoretical and practical topics in wireless communication including prototype systems and new applications
IT Professional	Organizing data, cross-functional systems, IT breakthroughs, capitalizing on IT advances, emerging technologies, e-commerce, groupware, broadband networks, security tools

If you are not very comfortable learning on your own, guided forms of continuing education are available. These include courses (whether online, through technical schools, through your own organization, or offered by universities) that can prepare you for certification or can broaden or deepen your knowledge. Many companies will support their IT professionals by paying for various forms of continuing education. For instance, the company might pay for your tuition in seeking either an advanced degree or a certificate. A variety of certifications are listed in IT Careers. Today, there exist a very large number of certifications in or related to the IT field. When you interview for IT jobs, you might ask what certifications the organization finds valuable.

There are also trade conferences taking place all over the world, several times per year. These conferences include those sponsored by Adobe, Oracle, VMware, Microsoft, and IBM. Others cover topics such as cloud computing, web services, web 2.0, databases, networking, virtualization, interoperability, emerging technologies, cryptography, and IT best practices to name but a few. As with continuing education courses, it is common for companies to pay for their IT professionals to attend such conferences.

In essence, to continue your IT education, you have to be willing to learn. Hopefully, you are already a self-learner and that continuing your education, whether because your employer finds it valuable or because you are curious, will be something that you will enjoy.

FURTHER READING

There are many books dealing with computer ethics. Here are a few of the more notable ones.

- Edgar, S. *Morality and Machines*. Sudbury, MA: Jones and Bartlett, 2003.
- Ermann, M., Williams, M., and Shauf, M. *Computers, Ethics, and Society*. New York: Oxford University Press, 1997.
- Johnson, D. *Computer Ethics*. Englewood Cliffs, NJ: Prentice Hall, 1985.
- Johnson, D. and Snapper, J. *Ethical Issues in the Use of Computers*. Belmont, CA: Wadsworth, 1985.
- Spinello, R. *Cyberethics: Morality and Law in Cyberspace*. Sudbury, MA: Jones and Bartlett, 2011.
- Spinello, R., and Tavani, H (ed.). *Readings in CyberEthics*. Sudbury, MA: Jones and Bartlett, 2004.

Information on IT careers can be found at a number of websites including careerbuilder .com, cio.com, dice.com, infoworld.com, itjobs.com, ncwit.org, techcareers.com, and ACM's SIGITE (Special Interest Group for Information Technology Education). IT certification information can be found through MC MCSE certification resources at www.mcmcse.com or directly at the various websites that offer the certifications such as through Microsoft and Cisco.

REVIEW TERMS

Terminology from this chapter

Censorship	Internet addiction
Certification	IT manager
Computer forensics	Network administrator
Computer scientist	Obscenity
Computer support specialist	Phishing
Computer technician	Repetitive stress injuries
Database administrator	Soft skills
Digital divide	Software engineering
Ergonomics	System administrator
Forensics	Website administrator
Help desk staff	Web developer

REVIEW QUESTIONS

1. Why is it important to earn a 4-year IT degree from an accredited institution if you want to work in the IT field?
2. What nontechnical skills might an IT person be required to have?
3. What are the skills necessary to be a system administrator?
4. What are the skills necessary to be a network administrator?
5. What are the skills necessary to be a database administrator?
6. What are the skills necessary to be a website administrator?
7. How does website administration differ from web development?
8. What are the skills necessary to be an IT manager?
9. What types of material will a person study for a career in computer forensics that differs from IT?
10. Why do help desk and computer technician careers require less education than other IT areas?
11. Which of the various IT related careers require an understanding of computer programming?
12. Of the various IT related careers, which one(s) would not necessarily require a 4-year degree?
13. What are the various forms of continuing education available to an IT professional?
14. What types of training would a computer support specialist be asked to provide in an organization?
15. As a system administrator, is it your responsibility to understand current laws as they pertain to IT? Would your answer change if you were a network administrator, database administrator or web administrator?
16. Under what circumstances might you be expected to write computer user policies for your organization?
17. As a system administrator, what kind of policy might you propose for disk quotas in an organization where all employees use a shared file server? Would this policy differ if you were the system administrator for a school/university?
18. As an IT professional, why would you have to understand health-related problems of computer usage?
19. As an IT professional, why would you have to know about privacy concerns?

20. For which of the various IT careers would an understanding of obscenity and FCC regulations be important?

DISCUSSION QUESTIONS

1. Imagine that you have been trained (either through a formal education, on your own, or through prior work experience) to be a system administrator. To what extent should you learn about the other IT areas (e.g., network administration, web server administration)?

2. Should an IT education cover all of the various areas of IT such as computer forensics, database administration, web developer, or should it concentrate on one area (presumably system administration)?

3. Your employer has asked you to do something that you believe is unethical but legal. What should you do about it?

4. Your employer has asked you to do something that is illegal. What should you do about it?

5. What is the difference between something being unethical and something being illegal? Why are they not the same?

6. Your best friend at the organization that you work for is doing something that violates your organization's policies. What should you do about it?

7. As a system administrator, under what circumstance(s) might it be ethical for you to examine employee files and e-mails? If you discover that an employee has illegally downloaded content stored on company file servers, under what circumstances should you report this? Are there circumstances when you should not report this?

8. To what extent should an organization seek input from the IT personnel regarding IT policies? If management creates such policies without involvement from the IT personnel, should the IT personnel provide a response?

9. Assume you work for a small organization of perhaps 15–20 individuals. You would like to have policies that address the digital divide. Try to come up with two to three policies that either promote education about the digital divide, or might help your organization (or your community) lessen the divide. If you worked for a large organization of hundreds of individuals, would your answers differ?

10. Explore one of the codes of conduct mentioned in IT Ethics. Do you find that the listings pertain to IT specialists as much as they do to computer programmers?

11. Your organization has a policy that says that work computers should never be used for personal use. Your boss has asked you, as system administrator, to collect all employees' Facebook passwords so that the boss can log into their Facebook accounts to

make sure that they are not accessing Facebook during work hours. Is the company's policy legal? What should you do?

12. It is not necessarily an IT person's responsibility to tackle issues covered in Other Social Considerations (e.g., digital divide). How important is it for an IT person to be aware of the issues?

13. How might an organization be impacted by obscenity, Internet addiction, and the digital divide?

14. As an IT person, should you provide input into your organization's IT policies in order to help resolve such issues as the digital divide and digital litter?

15. Continuing education is perhaps one of the most critical aspects to keeping up with the continual evolution of IT. Explain how you hope to continue your education.

16. Go to spokeo.com and enter your name and location. Find yourself and examine the content that has been made available. Do you feel that a website such as spokeo.com is doing a public service or disservice? How much of the information that you found there do you feel threatens your privacy? Do you feel that you should be able to do something about it like protest or have the information removed?

17. Given the privacy threats that a site like spokeo.com point out because of the Internet and social media, discuss the pros and cons of social media. Do the benefits outweigh the privacy concerns?

Appendix A: Glossary of Terms

Abacus—earliest computing device; invented thousands of years ago.

Absolute path—directory path starting from the root of the file system. In Linux, this path would start with a /; in Windows it would start with the partition letter such as C:\.

Access control list—means of controlling user permissions to access system resources, commonly, the list contains user names and access types such as read, read/write, execute; applied to files and directories.

Accessibility—the capacity of hardware, software, or website to be accessible by people with impairments.

Accumulator (AC)—a register in the CPU that stores data.

Ada—high-level programming language developed by the U.S. government named after Lady Ada Augusta Lovelace (a mathematician from the early nineteenth century considered to be the world's first programmer).

Address bus—a portion of the computer's bus used by the CPU to pass addresses to memory and I/O devices.

Administrator—a special user, the administrator is given access to all aspects of the computer system.

Alias—instance where two or more names point to the same entity such as hard links and soft links pointing to the same file.

ALU—arithmetic/logic unit of the CPU that performs all arithmetic and logic operations in the computer.

Analytical Engine—designed by Charles Babbage in the early 1830s to automate the computation of many mathematical equations. It was designed to be powered by a steam engine and perform all four parts of the IPOS cycle, making it a general-purpose computer. Babbage never completed it as he ran out of funds, but analytical engines have been constructed since then as proofs of the concept. Lady Ada Lovelace wrote programs for the Analytical Engine.

Anonymous user—a designation given to individuals who log into an FTP server with no account. They are given limited access.

Apache—open source web server that has become the most commonly used web server.

Application software—computer programs that users run to accomplish some task.

Archive—a bundled collection of files and directories.

ARPANET—Wide Area Network developed in 1968 connecting four computers in the western United States that eventually grew into the Internet.

ASCII—the American Standard Character Interchange Interface, a 7-bit representation for storing characters (letters of the alphabet, digits, punctuation marks) in binary.

Assembler—a computer program that translates an assembly language program into machine language.

Assembly language—a primitive programming language which uses mnemonics (abbreviations of operation names) and variable names to make it easier to use than machine language.

Assignment statement—a type of program instruction that assigns a value to a variable.

Asymmetric key encryption—a form of encryption where the key to encrypt the message differs from the key to decrypt the message.

Attribute—a field (column) in a database relation; for a student relation, examples might include "last name", "major" and "GPA".

Availability—a feature of information assurance and security that requires that data be available when that information is needed.

Background—a process that operates as the CPU has time for it, and does not require direct interaction with the user. Background processes essentially do not interfere with the user's interaction with the computer.

Backup—storing a copy of files in the file system (perhaps the entire file system) elsewhere for security purposes. Backups used to be performed on magnetic tape, but today it is common to use an external hard disk or disk storage space available over the Internet.

Backward compatibility—the ability for a newer computer to run older software or for a new piece of software to be able to access files created by an older version of the software. Backward compatibility allows users of newer Windows-based computers to still run software from old Windows and DOS-based computers.

Bandwidth—the amount of data that can be transmitted over a unit of time, such as bits per second. Bandwidth is used to measure the performance of a computer network, MODEM, or other form of telecommunications.

Base—the radix of the numbering system that determines the legal digits in that numbering system and how those digits are interpreted. Decimal (base 10) is the primary base used by people, whereas binary (base 2), octal (base 8), and hexadecimal (base 16) are all used by humans to represent information stored in a computer.

Batch—a form of process management that would restrict the operating system in executing only one program at a time until it completed. Furthermore, in batch processing, all program input would have to be specified at the time the program was submitted and all output would be saved to file so that there would be no interaction.

Beta-release software—a release of software before a major distribution so that specific users could test the software to identify flaws such as logical and run-time errors.

Binary—the base 2 numbering system using only 0s and 1s; computers represent all information in binary.

BIOS—the basic input/output system is a program that computers use to communicate to their basic input and output devices such as the keyboard. Today, the BIOS is almost always stored in ROM.

Bit—the smallest unit of storage in a computer, a bit is either a 0 or a 1.

Bitmap—a type of image file.

Block—the physical unit of disk storage; all files are broken into fixed-sized blocks.

Boot Loader—a program used to load an operating system into memory when the computer is booted up.

Booting—the process of starting up the computer; booting tests the CPU and makes sure it can communicate with the various devices like memory and disk drive, and then runs the boot loader to load and start the operating system.

Brace expansion—a feature of the Linux Bash shell in which files and directories listed in { } are expanded before the instruction executes so that the instruction executes on all of the items in the expanded list.

Buffer overflow—a flaw in some programs that causes values to be placed in memory outside of a buffer; if the overflow are program instructions, it could potentially allow the processor to run what is in the overflow area, allowing an attacker to take control of the processor.

Bug—a slang term for an error found in a program.

Bus—the device in the computer that connects the hardware components together so that they can communicate with each other; the bus is made up of wires over which electrical current travels.

Bus network—a form of computer network in which devices are connected to a single communication line through T-connectors; the bus network is the simplest and cheapest form of network but is often inefficient because of message contention.

Byte—8 bits used to store a single piece of information which can be a number from –128 to +127, a number from 0 to +255, or a character in ASCII or EBCDIC.

Byte code—an intermediate form of code that is produced by a compiler so that a later virtual machine can interpret it. The Java programming language compiles source code into byte code permitted it to be platform independent as long as the platform has a Java Virtual Machine. Microsoft's.net (dot net) platform can also compile into byte code.

Certificate—a means of identification used by websites as a digital signature to assure the client that the website is legitimate. The certificate may also include a public key for encryption.

Certificate authority—an organization that can sign certificates for authenticity.

Certification—an acknowledgement received by a person who successfully passes a class or examination in some IT-related area; often used as a means of advancing one's career in IT.

Checksum—a means of ensuring that data received from telecommunications is correct.

Child process—a process that was spawned by another process (the parent).

Circuit switching—a form of network in which the pathway from source to destination is established at the beginning of communication and retained until communication terminates; the public telephone network uses circuit switching.

Client—the name given to a computer (or user) requesting a service from another computer.

Client-server network—a network that consists of devices specifically designated as clients and servers.

Cloud computing—a recent development in computer networks where an organization offers services available over the Internet. The services include storage space and distributed processing.

Clustering—a form of data mining used to see how instances of data group together into possible categories.

Coaxial cable—a form of media used in telecommunications. Cable TV signals are commonly carried over coaxial cable. Using coaxial cable provides users with potentially larger bandwidth than twisted wire pair.

COBOL—Common Business Oriented Language, one of the earliest high-level languages, developed by the U.S. government for use in business. The language's most unique feature is that it reads like English.

Collision detection—the ability of a computer network to detect if two or more messages have been placed on the network at the same time.

Command line—in some operating systems, commands can be entered by keyboard as textual commands rather than by using pointing devices and a GUI interface. Linux, Unix, and DOS all have command line interfaces.

Command line editing—a feature of the Bash interpreter that allows the user to edit the current command through a series of keystrokes.

Compatibility—software that can run on multiple platforms.

Competitive multitasking—a form of concurrent process management in which a timer interrupts the CPU to force it to perform a context switch from one process to another.

Compilation—the process of translating a high-level language program into machine language so that it can be executed.

Compiler—a program that translates high-level language programs into machine language. A specific compiler is needed for each language and each platform.

Compression—a means of reducing file sizes. Two forms of compression are lossy compression, in which data is lost in order to reduce the file size, and lossless compression, in which the file is not accessible until it is uncompressed.

Computer—a programmable device that performs the IPOS cycle; computers come in a variety of sizes, capabilities, and costs from supercomputers to handheld devices.

Computer forensics—an area of study (and career) that combines legal knowledge of law enforcement, evidence gathering, and courtroom processes with technical knowledge of computer functionality to support prosecution of computer crimes.

Computer scientist—an area of study (and career) that revolves around software development and software engineering; the field combines programming, computer organization, data structures, programming languages, mathematical concepts related to solvability, project management, and the software life cycle.

Computer security—providing proper mechanisms, physical and software, to ensure the integrity and safety of computer systems.

Computer support—IT support for organization members through training, help desk, and documentation.

Computer technician—an area of study (and career) in which a person understands the electronic nature of computer hardware and can diagnose, repair, and replace such components and assemble and repair computers.

Concurrent processing—a form of process management where multiple processes are active at a time but where the CPU switches off between them.

Condition—a test, used in a program instruction, that evaluates to true or false; conditions are used in selection statements (e.g., if–then, if–then–else) and loops to control the behavior of the program.

Conditional loop—a programming instruction that uses a condition to determine whether to repeat a body of code or exit the loop.

Confidentiality—a feature of information assurance and security that specifies that the information must not be disclosed to anyone other than authorized and legitimate users of that information.

Configuration file—a file that dictates settings to an operating system service. Configuration files are used in Linux so that, when a service is started, it reads its configuration file first.

Context switch—an event that causes the CPU to switch from one process to another.

Control bus—part of the computer bus used to send control information (commands) from the control unit to other devices in the computer, and used for devices to send back status information to the CPU.

Control unit—the portion of the CPU that is in charge of controlling the components within the computer by performing the fetch–execute cycle.

Cooperative multitasking—a form of concurrent process management in which processes can voluntarily give up access to the CPU.

Core dump—a file created in Linux when a program terminates with an abnormal error; the file contains process status information that might be useful in debugging the program.

Counting loop—a programming language instruction that continues to repeat over a set of code based on a starting and ending count value, such as 1 to 10 or 100 down to 1.

CPU—central processing unit, or processor; the component in a computer that executes programs.

CPU cooling unit—a hardware device placed on top of the CPU to keep it from overheating due to power consumption.

Cylinder—a reference to a common location (the same track and sector) across all disk surfaces.

Daemon—a service program in Linux; see service.

Data bus—a part of the bus used to move data (and program instructions) between the CPU and memory and I/O subsystem.

Data mining—a form of data processing that attempts to sift through huge collections of data to discover meaningful information.

Data warehouse—a collection of many databases making up the sum total of data that an organization may have in its holdings.

Database—a way to store data as records and relations, created and manipulated by a database management system; operations include restrictions, projections and joins.

Database administrator—an IT person whose job is to manage the database management system and database(s), and whose duties include software installation, security of data, creation of accounts to access the database, and configuring the database to be accessible over network.

Database server—a program, usually running on a server, that provides clients remote access to a database.

Deadlock—a situation in a concurrent processing operating system where two or more processes are holding on to resources that the other processes need; the result is that none of the processes involved can progress.

Debian Linux—one of the early and most popular distributions of Linux.

Decimal—the base 10 numbering system that most people use, consisting of digits 0–9.

Decision tree—a product of data mining that creates models from a collection of data that people can use to make decisions.

Decode—a step in the fetch–execute cycle where the machine language instruction is decoded into microcode.

Defragmentation—a disk utility that takes disk blocks and moves them closer together on a disk so that disk access becomes more efficient.

Delayed start—a selection by administrators whereby services in the Windows operating systems are started automatically shortly after system initialization time when there is time to start them.

Denial of service—a type of attack on a server whereby so much message traffic is received that the server denies legitimate client requests.

Device driver—a program added to the operating system that lets the computer communicate with a new piece of hardware.

Difference Engine—designed by mathematician Charles Babbage in the early 1800s to perform automated computation of a certain class of math problems; it was never constructed because Babbage abandoned it to design the Analytical Engine.

Digital divide—a phenomenon in our society whereby many in the lower class have limited or no access to computers and the Internet.

Directory—a logical division in a file system used by people to organize and collect their files into related categories.

Disaster recovery plan—a plan to describe the efforts taken to prepare for and recover from a disaster, whether man-made or natural.

DNS—a domain name system translates IP aliases to IP addresses; needed so that users of the Internet do not have to remember IP addresses.

Dot net (.net)—a programming platform developed by Microsoft that permits programs of one .Net language to call upon components written in other .Net languages.

Download—the action of copying a file from a remote computer over a network to a local computer. Alternatively, the file itself is sometimes referred to as a download.

DRAM—dynamic random access memory used to create the main (or primary form of) memory of a computer; it is known as dynamic because the technology used can only retain a charge for a short amount of time before the current is discharged and so it must be continually refreshed. As DRAM is slower than SRAM, modern computers use both SRAM and DRAM.

Dumb terminal—used starting in the third generation, the dumb terminal is an input/output device with no memory or processor, allowing a user to connect to a computer (mainframe or minicomputer) over a network.

EBCDIC—a form of character representation used on IBM mainframe computers; an 8-bit alternative to ASCII.

Else clause—a programming language instruction used in if statements to offer an alternate action.

E-mail server—a computer (or a program) whose job is to collect e-mail messages and let individual users view their messages and send outgoing messages to other servers.

ENIAC—the first general-purpose, electronic computer; first started in 1946, it was a large, expensive, and slow computer by today's standards.

Environment variable—a variable storing a value that can be accessed by the user or running software, often used when writing Linux scripts.

Equivalence of hardware and software—the idea that any problem that can be solved using a computer program can also be solved by implementing that solution using computer hardware, and any problem that can be solved using computer hardware can also be solved by a computer program.

Ergonomics—the study of how to improve devices, whether computer, furniture, car, or other, to be easier for the human user and have less impact on the person's health.

Ethernet—a type of local area network technology, introduced in 1980, it became one of the most common forms of local area network.

Executable—a program stored in the computer's native machine language that can be executed on a computer.

Execute access—one of the levels of file/directory permissions; allows a user to execute the file or change into the directory.

Extranet—an intranet network that is extended to permit access from off-site.

FAT—the file allocation table is used in Windows operating systems to denote the physical location of all file blocks.

Fetch-execute cycle—the main execution cycle of the CPU, it fetches an instruction from memory, decodes the instruction, and executes it. The CPU repeatedly performs this cycle to execute any and all programs.

Fiber optic cable—a medium used for network communication, it carries data as light pulses and is thus faster than any other form of network cable.

Field—attributes that make up the database relation; these are the columns of a relation such as "first name", "address", "GPA", "major".

File—the logical unit of storage in the file system.

File extraction—removing files and directories from an archive.

File server—a computer dedicated to provide storage access over a network.

File system—the structure of storage for computers; commonly made up of one or more hard disk drives and possibly removable storage devices.

Filename expansion—prior to a command execution, any wildcards are replaced by all matching files; for instance, the * indicates "anything" whereas in Linux, the ? indicates "any 1 character".

Firewall—software (or a server running software) that prevents malicious attacks over a computer network; the firewall consists of rules to determine what messages should be permitted and which should be rejected.

Firmware—programs stored in hardware (chips); although firmware is more expensive to produce and inflexible as compared to software, it executes faster than software.

Floating point—a representation to store a number with a decimal point (e.g., 123.456) using integers to represent the mantissa and exponent.

Folder—another name for a directory, a logical division in a file system.

Foreground—running processes that can directly interact with the user.

FORTRAN—the earliest high-level programming language, developed in 1958 by IBM, primarily used for mathematical and scientific computation.

Fragment—a leftover piece formed when contiguous allocation is used; fragments can be formed in memory or on disk.

Free software movement—movement created by programmers working on Unix and Linux operating system software so that software being produced would be made freely available in source code form so that others could modify the code and make new code available.

Freeware—a category of software that, although it may have licensing restrictions, is free of charge.

FTP—file transfer protocol, used for file uploading and downloading across a network; predates HTTP and web servers.

FTP server—a computer that hosts files to be transferred using FTP.

Gateway—a broadcast device used to connect different types of networks together so that messages of one protocol can be translated to another protocol.

GHz—abbreviation for "gigahertz" that specifies the frequency of clock "pulses"; 1 GHz is a clock pulse every 1 billionth of a second (1 nanosecond).

GIF—Graphics Interchange Format, a lossless form of image file compression that uses a standard palette of 256 colors.

Globbing—a slang expression used to describe filename expansion in Linux using wildcards.

GNU GPL—the General Public License, a "copyleft" (rather than a copyright) applied to open source software as well as other works so that they can be shared among the open source community. This license not only permits free access to the item, it also permits others to manipulate the item (software, image file, etc.) and share the result for free.

GO TO—a type of program instruction, first used in FORTRAN, to permit a programmer to specify an unconditional branch from any location in the program to any other location; use of GO TO statements leads to spaghetti code.

Grounding strap—a tool used by computer technicians to prevent static charges from damaging circuitry due to a static charge released when a person touches the computer.

Group—a classification used in Linux and Windows whereby users can be given some permission to files.

GUI—graphical user interface is the term given to windowing-based operating systems and programs so that the user can control the software easily through a pointing device rather than command line input.

Hard disk drive—the primary form of storage; the hard disk is sealed within the disk drive, permitting far greater storage capacity over outdated floppy disk drives.

Hard link—a pointer in a file system from the file's name to the file's location.

Hardware—the physical components of a computer system which include the system unit and its components (motherboard, fan, disk drives, CPU, memory chips) and peripheral devices.

HCI—human–computer interaction studies how to improve our ability to use a computer—both to reduce strain that might be caused by poorly designed interface devices (such as the keyboard and mouse) and to make interaction more natural, particularly for people who have limited physical movement or disabilities.

Help desk—a call center staffed by IT personnel to help answer user questions.

Hexadecimal—the base 16 numbering system often used to group binary bits together to make it easier to read over binary.

High-level language—a class of computer programming language developed so that programmers would not have to write at the lower levels of machine and assembly languages; high-level languages often use English words and mathematical notation.

Hit rate—the percentage of accesses to a level of the memory hierarchy where the item sought is found there rather than having to move further down the hierarchy.

Horizontal software—software that is used throughout departments of an organization such as productivity software.

Host—a computer on a network that serves as a server or that can be logged into.

Hub—a type of network broadcast device that broadcasts any incoming message to all devices attached to it.

I/O queue—waiting lines, managed by the operating system, for processes that need servicing by an I/O device.

Infinite loop—a logical error in a program where a loop never exits because the loop's condition is always true.

Information—processed data.

Information asset—a piece of information that an organization values and wishes to protect.

Information Technology—the term used to describe both the computer systems that organizations use to process their information and the information itself; also used to refer to the collective group of people who manage the organization's IT infrastructure.

Initialization script—a set of instructions run after the operating system has been loaded to bring up services and prepare the computer for use by the user.

inode—the name given to a data structure that stores information about a Linux file; the inode also stores pointers to the file's blocks.

Input—the process of taking information or data from outside the computer and bringing it into the computer.

Input statement—a programming language instruction that causes the program to pause while input is performed.

Installation—the process of adding new hardware or software to a computer.

Instruction register (IR)—a register in the CPU that stores the current machine language instruction. The IR is used by the control unit to determine what instruction is to be executed and what the data for the instruction are.

Integrated circuit—also known as a chip or IC, a semiconducting electronic device that contains logic gates to perform operations when current flows through them; the IC is used for many parts of modern computers.

Integrity—a feature of information assurance and security that requires that information of the organization be accurate.

Interactivity—the ability of a computer system to have real-time input and output with the user.

Internet—a global wide area computer network connecting billions of people that supports telecommunications, commerce, social networks, and numerous other human endeavors.

Internet addiction—a recent phenomenon, in part brought on by social networks and mobile devices, whereby a person tends to access one or more Internet accounts often (e.g., their Facebook page, their e-mail).

Interpreter—a form of programming language translator that translates one instruction or command immediately upon input into machine language and executes it; the interpreter offers the user a session to work in, unlike a compiler, which translates an entire program into machine language so that the translated program can be used at a later time.

Interrupt—a situation in which a hardware device requires the attention of the CPU so that the fetch–execute cycle must be interrupted, or a situation in which the software itself generates an interrupt because the program requires the attention of the operating system.

Interrupt handler—a part of the operating system set up to handle interrupts; each interrupt handler is designed to handle one type of interrupt (e.g., mouse moved, key pressed, disk drive fault).

Intranet—a type of local area network that uses TCP/IP so that the networked devices can communicate over the Internet.

Intrusion—an attack on a computer system whereby an unauthorized user has been able to log in and/or launch processes; usually intrusions have malicious intent.

IP address—a unique identifier given to each device on the Internet so that messages can be routed to the device; IPv4 addresses are 32 bits (4 octets) and IPv6 addresses are 128 bits.

IP alias—an English-like description of an Internet machine, used because IP addresses are hard to memorize; IP aliases must first be converted into IP addresses before communication can take place.

IPOS Cycle—input, processing, output, and storage, the four activities that all computers perform.

IT manager—an IT person who has an understanding of both the management side of the organization and the IT infrastructure so that the person can manage the IT staff and projects; IT management often requires an advanced business degree.

Iteration—a looping behavior in a program to repeat some section of code.

Iterator loop—a type of programming language instruction that loops over a set of code one time for each item in stored in a list.

Java virtual machine (JVM)—an interpreter that can execute Java byte code; the JVM is built into all web browsers so that Java programs can be executed on any platform.

Join—a database operation that combines two or more database relations together to withdraw data.

JPG—the Joint Photographic Experts Group image format (jpeg or jpg) is a form of lossy image compression most commonly used to reduce the storage size of digitized photographs.

Kernel—the portion of the operating system responsible for managing the computer's resources (CPU, memory) and execute programs.

Kill (a process)—terminating a running process, usually because the process has stopped responding.

Language translator—a class of programs that translates a computer program from one format to another. The most common types are compilers, interpreters, and assemblers.

Last mile technology—a term that applies to the twisted wire cable that connects houses to the telephone network; as twisted wire offers lower bandwidth, unless households have other forms of media (e.g., coaxial cable, fiber optic cable, satellite), those households are limited to 56 Kbps Internet access.

Linux—an operating system that competes with Windows and MacOS; noteworthy because it (and much of the application software available in Linux) is available for free.

LISP—an early interpreted high-level programming language developed to support artificial intelligence research.

Live CD—an optical disc storing a bootable operating system; many operating systems can only be booted from hard disk.

Load—the operation of reading a datum from memory and moving it into the CPU.

Local area network—a network within one location/site, such as one room, one floor, one building.

Local computer—the computer that the user is on, to differentiate it from a remote computer.

Log file—a file of messages automatically generated from software to record meaningful events that arise during program execution.

Logging—the process of writing information to a log file.

Logical error—a type of error in a program where the logic is incorrect, leading to the program not functioning correctly.

Logon type—the specification of the owner of a Windows service so that it can acquire a certain level of access.

Loop body—in a programming language's loop, the loop body is the code to be repeated.

Loop index—for counting and iterator loops, a variable is often used to store the value of the iteration, for instance, the count (1 to 10) or the current value in the list.

Lossless compression—a form of compression where data/content is not lost when the file is compressed.

Lossy compression—a form of compression where data/content is purposefully discarded in order to reduce the file's storage size, resulting in a blurrier image or an audio file that has a more limited range.

MAC address—the Media Access Control address is assigned to network interfaces (e.g., network cards in computers) and used for communication at the lowest level(s) of the network (as opposed to the network addresses used at higher levels like IP addresses).

Machine language—the programming language native to the given processor; a computer can only execute a program written in machine language such that a program written in any other language must be translated first.

Macintosh OS—a GUI operating system; with MacOS X (version 10), the OS sits on top of a Unix-like operating system.

Magnetic core memory—form of main memory used in the second generation of computers, composed of small, iron rings (cores), each of which would store 1 bit; 1 MB of magnetic core memory could cost up to $1 million in the early 1960s.

Mainframe—primary form of commercially marketed computers from the 1950s through the 1980s, these computers are expensive but powerful, supporting hundreds to thousands of users; have largely been replaced by personal computers and servers since the 1990s.

Malware—a class of software that has malicious intent; includes Trojan horses, viruses, worms, and spyware.

Megaflops—millions of floating point operations per second, a rate used to express the performance of a processor on programs that contain floating point computations.

Memory—hardware used to store program code and data; there are many forms of memory that make up the memory hierarchy.

Memory chips—memory stored in integrated circuits that are plugged into the motherboard of a computer.

Memory hierarchy—the organization of the different types of memory (registers, cache, main memory, swap space, hard disk file system, removable storage) such that the higher forms of memory are faster but more expensive and so there is less of it in the typical computer system.

Memory management—an operating system task to handle memory accesses of running processes, typically composed of virtual memory mapping, swapping, and the prevention of memory violations.

Memory violation—a situation in which a process requests access to memory owned by another process; if not prevented, a memory violation could corrupt the memory owned by another process, and therefore memory violations usually lead to runtime errors and termination of the process that causes the violation.

Mesh network—a topology of computer network in which all resources have a direct connection with all other resources. Although this presents an optimal means of communication, it is prohibitively expensive.

Metacharacters—characters used in regular expressions to describe how to interpret literal characters; metacharacters can express such notions as "1 or more instances of" or "match any character in the given list".

Microcomputer—a class of computers whose CPU is built using a microprocessor so that the computer can be relatively small in size; all personal computers and laptop computers are of this category.

Microprocessor—a CPU placed entirely on one chip; the earliest microprocessor was the Intel 4004, released in 1971.

Minicomputer—a category of computer, the minicomputer is a scaled-back mainframe. It has less power and fewer resources to keep the cost down.

Minix—a Unix-like operating system developed by Andrew Tanenbaum to accompany an operating system textbook and used in college courses.

MIPS—millions of instructions per second, a rate used to express the performance of a processor when executing an integer-based program.

MODEM—a piece of hardware used to convert digital information to an analog form to be broadcast over the telephone network, and to convert received analog messages back to digital; MODEM stands for MOdulation and DEModulation.

Motherboard—the printed circuit board that houses the CPU, memory chips, and expansion cards in any computer.

Mount point—the physical location in a file system where a partition is added.

Mounting—the process of adding a partition to a file system.

MS-DOS—Microsoft Disk Operating System, the original IBM PC operating system; it is text-based and single tasking.

Multiprocessing—a form of processing in which there are multiple processors (CPUs) to execute multiple processes simultaneously.

Multiprogramming—a form of concurrent processing in which the CPU switches off between processes whenever one process surrenders the CPU either because it needs to perform time-consuming I/O, is waiting for an event to happen, or is of lower priority than other waiting processes; also known as cooperative multitasking.

Multitasking—a form of concurrent processing in which the CPU switches off between processes; the two forms are competitive multitasking and cooperative multitasking.

Multithreading—a form of concurrent processing in which the CPU switches off between both processes and threads.

Mutually exclusive—a restriction placed on processes when attempting to access a shared resource in a computer system; access of the resource must be wholly performed by one process at a time to ensure that the shared resource is not corrupted in case one process is interrupted mid-access and another process is then given access.

Nearest neighbor—a form of computer network topology in which each computer connects only to the nearest computers/resources near it; nearest neighbor can be one-, two-, three-, or four-dimensional.

Netmask—a sequence of binary bits applied to an IP address using the AND operation to return a portion of the IP address that either matches the resource's network address or the address of the resource on the network.

Network—a collection of computers and resources connected by various media so that they can communicate with each other.

Network address translation (NAT)—a process of converting an external address into an internal local area network address; the many-to-one variation of NAT permits an organization to have a single external IP address but multiple internal addresses.

Network administrator—a role in IT whereby a person (or people) is responsible for installation, configuration, maintenance, securing, and troubleshooting of the computer network.

Network handshake—in network communication, the sending device must establish a connection with the recipient through a two-way or three-way handshake.

Network topology—a way to categorize the structure of a network.

NFS—the network file system, often used in local area networks for Unix and Linux machines.

Non-volatile memory—a form of memory that does not require a constant power supply to retain its contents; both read-only memory (ROM) and solid state memory (e.g., flash drives) are of this form.

Obscenity—a piece of information (image, communication, video, etc.) found to be unacceptable by a community because it violates that community's acceptable standards; obscenity is a concern of IT in that content may need to be blocked to protect some members of the community (particularly children).

Octal—the base 8 numbering system which uses digits 0–7 and is often used in place of binary because it is somewhat easier to read.

Octet—a portion of an IPv4 address equivalent to 8 bits or a single number from 0 to 255.

Off-chip cache—added SRAM in a computer because the on-chip cache must be limited due to space restrictions on the CPU.

OLAP—online analytical processing, used in business to analyze large collections of data for decision making.

On-chip cache—small SRAM placed on the CPU to help support fast memory access.

Open architecture—publishing the details of a computer architecture to make it available to others; IBM did this with the IBM PC, which resulted in many compatible computers (clones).

Open source—software that is made available as source code.

Open source initiative—to support Unix and later Linux, a number of early programmers were willing to make their code available as open source and for free.

OSI—the open systems interconnections network model, a seven-layer model, often used when developing new computer networks.

Output—taking computer processed results and delivering them to the outside world; common forms of output are displayed via the monitor, printed output, and audio over speakers.

Output statement—an instruction that will deliver output to an output device (or stored to a file).

Owner—the user who has full access and authority to a resource.

Package manager program—a program that allows a user to easily install, uninstall, or upgrade software packages.

Packet—the unit of information to be transmitted over the Internet.

Packet switching—a type of network routing in which a packet sent over the network finds its way en route, rather than having a pre-established route.

Page—a piece of an executable program; used to support virtual memory; a page is loaded from swap space into a free frame in memory upon demand.

Page fault—a situation where the address generated by the CPU is of a page that is not currently in memory and so must be swapped into memory from swap space by the operating system.

Page table—mapping information so that a CPU address, which specifies locations by page number, can be converted into a physical memory address, specified by frame number.

Parameter—a piece of information passed from one program routine (procedure, function, method) to another.

Parent process—a process that starts (spawns) another process.

Parity—the evenness or oddness of a value, used for error detection and correction.

Parity bit—a simple form of error detection that is an added bit to a byte such that the 9 bits contain an even number of 1s.

Partition—divisions of the file system such that each partition can be handled independently of others; for instance, one partition can be removed to perform a backup without impacting the other partitions.

Password—a mechanism used in operating systems to authenticate users.

Patch—a software update released to fix bugs or security flaws.

PATH variable—an operating system environment variable that stores directories that should be searched whenever any instruction or file name is entered at the command line prompt.

Peer-to-peer network—a type of network in which all computers are roughly equivalent peers (i.e., there is no server in the network).

Peripheral—a hardware device connected to the computer outside of the system unit.

Permission—a means of allowing users to control who can access their resources (files, directories). Permissions are usually limited to read, write, and execute.

Phishing—a means of obtaining secure information from a computer user by tricking the user through a fraudulent e-mail or website.

PID—process ID, a unique identifier assigned to each executing process.

Ping—a network program that allows one to determine if a network resource is responding.

Pixel—picture element, the smallest part of an image as stored by a computer.

Platter—one disk within a hard disk drive.

Port—a portion of a network address used to specify the intended application that should handle the message.

Portability—the capability of software to be run on multiple platforms.

Priority—as used in scheduling processes, the higher the process' priority, the sooner the CPU will get to it.

Private key—the key used to decrypt messages and must be kept secure.

Private key encryption—symmetric encryption in which one key (the private key) is used to both encrypt and decrypt messages.

Process—a program that is in some state of execution.

Process management—the operating system task of running processes on the computer; two basic categories are single tasking (which includes batch processing) and concurrent tasking (which includes multiprogramming, multitasking, and multithreading).

Process status—the current execution status of a process, commonly one of waiting, ready, running, suspended, or terminated.

Process tree—the relationships between parent and child processes.

Processing—executing programs.

Processor—also known as the central processing unit or CPU, the hardware device responsible for processing (executing) programs.

Productivity software—the category of software that contains the types of software found useful for most individuals and companies: word processing, spreadsheet software, database management system software, presentation graphics software.

Program—a listing of instructions that the computer is to execute to accomplish a task.

Program counter (PC)—a register in the CPU storing the location in memory of the next program instruction.

Projection—a database operation to retrieve specified fields of all records in a relation.

Proprietary software—programs that are sold. Purchasing the software does not give you ownership, just the right to use the software.

Protection—an operating system task that ensures that a user or a user's process does not incorrectly use system resources.

Protocol—rules for communication, commonly referred to as how an operating system packages up (or unpackages) a message to be sent over a network.

Protocol attack—a type of network attack that exploits a weakness in the network protocol.

Proxy server—a type of program used by organizations so that recently accessed web pages are cached for efficiency. The proxy server can also be used to filter content.

Public domain software—a category of software that is freely available to use any way you wish.

Public key—the key used to encrypt messages in public key encryption; the public key can be made available to anyone as it cannot be used to decrypt messages.

Public key encryption—an asymmetric form of encryption using two keys, a public key for encryption and a private key for decryption.

Query—a command sent to a database to retrieve, insert, delete, or modify records.

Queue—a waiting line, used in operating systems to organize waiting processes.

RAID—redundant array of independent (or inexpensive) disks, a form of storage used to ensure that data are reliably stored; extra disks are used to provide the opportunity for multiple disk accesses at a time and for storing redundancy (error correction) information.

Read—a type of programming language instruction that accepts input from the user.

Read access—the ability to read (input) a file.

Read/write head—the mechanism in a disk drive that reads and writes magnetic information from and onto the surface of the disk.

Ready queue—the waiting line in the operating system that stores the processes that the CPU is currently switching off between.

Record—a row in a database relation pertaining to an individual (e.g., a customer, a student, a piece of inventory).

Recovery—the amount of time that the Windows operating system should wait before attempting to restart a service that has stopped working.

Red Hat Linux—a distribution or dialect of Linux, one of the most popular (along with Debian).

Redundancy—a means of ensuring data availability and integrity by adding extra data in the form of either an exact duplicate of the file (a mirror), or error detection and correction information through Hamming code bits or parity bits.

Register—temporary storage in the CPU used to store data or to support the fetch–execute cycle.

Regular expressions—a means of expressing a search pattern in strings.

Relation—a table in a database, consisting of records and fields.

Relational database—a database that stores relations; the most popular format of database.

Relative path—a specification of a file or directory as indicated from the current directory.

Remote access—the ability to access a computer over a network.

Remote computer—the computer being accessed over the network.

Rendezvous—through synchronization, this is a situation in which one process (or thread) must wait for another to produce a result.

Repetitive stress injuries—a class of injury that arises from performing a repetitive motion over and over in such a way that it damages some part of the body.

Resident monitor—the earliest form of operating system, it remained resident in memory to start new processes and handle user I/O requests.

Resource management—a task of the operating system to allocate and deallocate system resources from program requests.

Resources—components of a computer system including disk access, file access, network access, and memory access.

Restriction—a type of database operation to retrieve records that match given criteria.

Ring network—a computer network topology in which all computers are connected to their nearest two neighbors so that messages must be transmitted from component to component around the ring until the destination is reached; this is a cheap but not very efficient topology.

Risk assessment—a step in risk management where the organization specifically focuses on the risks and vulnerabilities of their assets.

Risk management—a process that most organizations will undergo to determine their threats and vulnerabilities, critical to ensuring information assurance and security.

Root—the name given to the Linux/Unix system administrator(s); also the name given to the top-most point in the Unix/Linux file system (/).

Rotational delay—the time it takes for the proper disk sector to rotate underneath the read/write head as the disk spins (also known as rotational latency).

Round-robin scheduling—the scheduling algorithm used in multitasking whereby the processes in the ready queue are rotated through one at a time, moving to the front of the queue after visiting the last process in the queue.

Router—a network broadcast device that takes the destination address of a message to select a network to pass the message on to.

Routing—the process of a router moving a message onto the appropriate network.

Runlevel (Linux)—the start level used in Linux to select which services should be started or stopped at system initialization time.

Run-time error—an error in a computer program that arises during program execution, typically causing the program to terminate abnormally.

Scheduling—an operating system task that places waiting processes in the order that the processor will execute them.

Script—a small, interpreted program often used to automate simple tasks.

Scripting language—a class of interpreted programming language used to write script programs.

Sector—a region on a disk surface used for addressing.

Security—an operating system task that extends resource permission across networks to ensure that people who remotely access the computer are authorized to do so.

Seek time—the time it takes to relocate the read/write head to the proper track of a hard disk.

Selection statement—a type of programming language instruction that, based on a condition, takes one of two (or more) paths through the code.

SELinux—security enhanced Linux offers a different mechanism for permissions by using access control lists.

Semantic web—extending the World Wide Web to include automated reasonsers so that the knowledge can be more readily processed by computer rather than human.

Server—a stand-alone computer (and the software it runs) that provides a service for remote users; examples include file server, web server, and e-mail server.

Service—a program that runs in the background of an operating system, responding to requests from users and software alike.

Shareware—a category of free software with restrictions placed on the usage of the software such as a limited number of uses or a time limit that, once exceeded, causes the software to be unusable; typically made available to entice the user to buy the full version.

Shell—an environment in an operating system like Linux/Unix, controlled by an interpreter, in which the user has a session.

Shortcut icon—a soft link in Windows to permit easy access to starting programs.

Shut down—the process of stopping the operating system and shutting down the hardware. The process is useful in that it ensures that all files are closed to avoid file corruption problems.

Single tasking—a form of process management in which the operating system only allows one program to execute at a time from start to completion.

Sleeping—when a process (or thread) moves itself to a waiting queue to await some situation such as a rendezvous with another process (or thread) or for a specified time interval.

Social engineering—the effort of a person to obtain confidential information from a user by posing as a person in authority, such as calling someone at home, claiming to be from IT and requesting the person's password.

Soft Link—a pointer used in a file system to point from a filename to another filename; in Linux, these are called symbolic links, and in Windows, these are called shortcut icons.

Soft skills—skills that employers often seek in employees that are not technical in nature. The most useful forms of soft skills revolve around the ability to communicate (speaking, listening and writing skills).

Software—the name we give to computer programs to differentiate them from hardware.

Software engineering—the field of software development, often made up of people who have computer science degrees.

Software exploit—a known security flaw in some software used to attack a computer system or network.

Software release—a major distribution of a piece of software that has new features (perhaps a new look) and has resolved problems of older distributions of the software.

Source code—a program written in a high-level language; the program must be translated into machine language before it can be executed on a computer.

Spaghetti code—the result of using unconditional branches such as GO TO statements in a program so that tracing through the program begins to look like a pile of spaghetti.

Spawn—an event where one process starts another process; the original process is often called the parent and the spawned process is called the child.

Spindle—a motorized cylinder that rotates all hard disk platters at the same time.

SQL injection—an attack on a website where the attacker enters an SQL command as part of a URL (or part of a web form) in an attempt to access the backend database.

SRAM—static RAM, a fast form of memory used to build registers and cache.

Star network—a computer network topology in which all resources in the network connect to a central server. Today, that server is usually a network hub or switch.

Startup level—the level that a service should be started as in Windows, which provides the service with a set of access rights.

Status flags (SF)—a register in the CPU that stores results of the most recent arithmetic or logic operation such as "positive", "negative", "zero", "carry out", "overflow", and "interrupt".

Storage—long-term memory used for permanence, unlike main memory, which is volatile; storage devices include the hard disk, flash memory, and optical disk.

Storage capacity—the amount that can be stored on the given device.

Store—a CPU operation of taking a result stored in a register and copying it into memory.

Strong password—the requirement that passwords meet certain restrictions such as at least 8 characters in length and a combination of upper- and lower-case letters and non-letters; these restrictions make passwords much harder to crack.

Subroutine—a piece of a larger program that solves a specific subproblem, invoked from the main program or another subroutine via a subroutine call; subroutines are often named functions, procedures, or methods depending on the language.

Suspended—a process that has voluntarily stopped processing for the time being; also, in Linux, a process is suspended when the user types control+z in the shell running the process.

Swap space—an area of the hard disk reserved to store program pages as an extension to main memory; swap space supports virtual memory.

Swapping—the process whereby the operating system moves pages between swap space and memory.

Symmetric key encryption—a form of encryption that uses a single, private key for both encryption and decryption.

Synchronization—a requirement in operating systems whereby two or more processes that share some common resource must access that resource one at a time.

Syntax error—an error in a program that is a misuse of the syntax of the language; syntax errors are usually caught at compile time by the compiler.

System administrator—a person in charge of administering a computer system or a collection of computer systems.

System software—software that supports the use of the computer system; in other words, the operating system and other supporting software.

System unit—the main cabinet in which the primary components of a desktop computer are housed including the motherboard, the hard disk and optical disk, a power supply unit, and one (or more) fan(s).

Tab completion—an editing feature of the Bash shell in Linux in which the user can type part of a file name and press the <tab> key and the interpreter will attempt to complete the file's name.

Task manager—a Windows utility that allows the user to control running applications, processes, and services.

TCP/IP—a pair of protocols (TCP and IP) used by all computers on the Internet.

Then clause—the portion of an if–then or if–then–else statement that executes when the condition is true.

Thread—an instance of a process that shares the same process code with other running instances; each thread has its own data.

Timer—a hardware device in a computer used to count clock cycles so that the CPU can be interrupted to force a context switch to another process; used in multitasking/multithreading.

Track—a subdivision of a disk surface for addressing; essentially, tracks are concentric rings.

Transfer time—the time it takes for the operating system to move a block of information between memory and a storage device.

Transmission media—the physical connection between components in a network; the most common forms are twisted wire pair, coaxial cable, and fiber optic although wireless uses radio signals.

Tree topology—a nearest-neighbor form of computer network organization where resources are arranged hierarchically.

Twisted wire pair—a form of transmission media used in computer networks and telecommunications; it is cheap but has low bandwidth.

Two's complement—a binary representation that can store positive and negative integers.

Type mismatch—a syntax error in which a program instruction attempts to place the wrong type of value in a variable, such as attempting to put a real number into an integer.

Ubuntu Linux—a popular distribution platform of Linux based on Debian Linux.

UDP—a protocol used in place of TCP when packet delivery is less important than speed of delivery.

Unconditional branch—a form of program instruction that branches to another location in a program; in FORTRAN, these types of instructions are called GO TO statements.

Unicode—an extension of the ASCII character representation set from 7 bits to 16 bits.

Unix—a popular operating system for mainframe computers developed starting in the late 1960s, it eventually became a model for Linux. Unix and Linux developers are often part of the open source community.

Upgrade—a minor release in software to fix known problems including security holes.

User—a person who uses a computer. Typically, a user is given a user account.

User account—a collection of access rights and login information for a user.

User name—the name associated with a user account.

User interface—the means by which a user interacts with the computer hardware and software. The common two forms of interface are the command line and the graphical user interface (GUI).

Utility program—a piece of system software that the user might decide to run from time to time to help manage some aspect of the system.

Vacuum tube—a hardware component used in first-generation computers used as part of the computer's computational unit; vacuum tubes had short shelf lives such that first generation computers were not very reliable.

Variable declaration—specifying the name and type of a variable in a program; many programming languages require a variable declaration before the variable can be used.

Vertical software—software used within a specific division of an organization or by a single profession.

Virtual machine—an illusionary computer that is stored in computer memory and made available by software in order to permit a computer user to use multiple computers and multiple platforms without the expense of purchasing multiple computers.

Virtual memory—the extension of main memory onto swap space so that the user can run more programs than would normally fit in memory.

Virtual private network—a local area network that permits secure remote access through an authentication mechanism.

Virtual reality—a combination of software and I/O devices that give a user the illusion of being in a world created by the software; often used to control devices remotely and to explore locations that might be too remote or dangerous to reach.

Visual programming—a form of programming in which GUI components are added to software through a "drag-and-drop" approach.

Volatile memory—memory that requires electrical power to retain its contents; if power is turned off, the contents disappear from memory.

Waiting—a process that is not in the ready queue because it is either waiting on I/O to complete, suspended, or waiting for a rendezvous.

Waiting queue—a waiting line for processes in a waiting state.

Wearable—a newer form of I/O device in which the interface device is something that a person would wear on their body.

Web developer—a person hired to build web pages and web sites through the use of html, css, and a variety of scripting languages.

Web server—a hardware device that stores an organization's web site, or the software that provides the service of responding to HTTP requests from clients.

Website administrator—a person in charge of administering, configuring, installing, monitoring, and troubleshooting the hardware and software of a web server.

White space—spaces, tab keys, and return keys in a program that are placed there to make the program more readable, but are ignored by the compiler.

Wide area network—a type of computer network that is distributed across a large geographical area.

Windows—a common platform of operating system, introduced in the mid-1980s.

Windows installer—a program that allows a user in the Windows operating system to easily install new software.

Word—the storage size of the common datum in a computer. Today, typical word sizes are 32 or 64 bits.

Write—the ability to store something to a file.

Appendix B: Linux and DOS Information

LINUX COMMANDS

&—when used after a command, it launches the command in the background.

! (bang)—used to retrieve a command from the history list. !! retrieves the last instruction, !#, where # is a number that retrieves the instruction # from the list.

~—used to denote a user's home directory. When used as ~username, it denotes username's home directory.

.—used to indicate the current directory, as in cp ~foxr/* . (copy all files in foxr's home directory to this directory).

..—used to indicate the parent directory, as in cd ../.. (move up two levels in the file system).

apropos *string*—display all commands whose description field contain the given *string*; used to identify a command when the name of the command is not known.

bash—start a new bash session in the current window.

bg *number*—where number is an integer number; it moves the job corresponding to number to the background.

cat *item(s)*—concatenate the *items* specified (items are typically file names), default output is sent to the screen.

cd *path*—change directory to the directory given in *path*. If path is an absolute path, then you are changed to that directory. Otherwise, you are moved relative from where you currently are (pwd) to the new location. The use of .. moves you up one level in the file system.

chgrp *group item(s)*—change item(s)' group ownership to the given group.

chmod modification item(s)—change the item(s)' permissions to the given modification. Can be specified in one of three ways:

1. Using +/– with ugo and rwx
2. Using = with ugo and rwx
3. Using three digits as in 755 to denote owner, group, world

chown *owner item(s)*—change item(s)' owner to given owner.

configure—execute the configure script to create a makefile, issued as ./configure followed by any desired parameters. Parameters are software specific. See make and makefile.

cp ***source target***—copy file(s) listed as source to target. Target is typically a directory although if source is a single file, target can be a new file name. Two options of note are –i for interactive mode (prompt the user before overwriting a file that may already exist) and –r for a recursive copy (if source is a directory, it not only copies the contents of the directory, but all subdirectories).

echo ***string***—output string to screen. The string can comprise literal values, variables (with a $ before each variable name) and Linux commands. When placed in '', output all items literally. When command is placed in ` `, execute command and use its output as part of the string.

exit—leave the current bash session. If another session exists in the same window, you will resume it, otherwise the window will close.

fg ***number***—where number is an integer number, it moves the job corresponding to number to the foreground.

find ***path*** **options**—the find command searches for files that fill some criteria, the *path* is the starting point of the search, you might use your home directory (~) or you might start at root (/). Keep in mind that if you run this program without being root, some directories and files may not be accessible by you. The options that you most likely will use are –name *string* to specify that you are searching for all files with *string* in their name followed by –print to output the list of files found.

fsck—file system check, used to check the integrity of the file system.

jobs—list all active processes in the given terminal window, often used along with fg and bg.

gcc—the GNUs C compiler used to compile programs. g++ is the GNU's C++ compiler.

grep expression file(s)—match the regular expression to each line of the file, returning any matches. egrep uses the extended regular expression set. Common options include –e (same as using egrep), –c to output just the number of matches, –i to ignore the case of the letters, and –n to output line numbers for each match.

gzip/gunzip ***filename***—zips/unzips filename.

history—display the history list.

ifconfig—may require /sbin/ifconfig—display network interface information for computer including IP address.

kill signal pid—kill the process of the given process ID using the given signal, signal is often –9.

less ***filename***—like cat, displays *filename* to the screen, but pauses at each screen. You can also use the arrow keys to move up and down in the file. The command more is similar but cannot step backward.

ln [–s] ***filename linkname***—create a link (symbolic link if –s is used) from linkname to filename.

ls ***filenames***—list the filenames provided. ls by itself lists all items in the current directory. ls permits wildcards for filename expansion such as ls *.txt. Common options for ls are given:

ls –l—long listing

ls –a—include "hidden" files (filenames that start with a dot)

make—run the makefile script to compile software, options include make all, make install, make tar.

man *command*—provide the manual page for the given command.

mkdir *dirname*—create the directory named *dirname* in the current directory.

mount *physical logical*—a command to mount a partition indicated by *physical* (the actual name of the file system) to the file system at the location *logical*. The command umount is used to unmounts a partition. mount –a mounts all partitions found in the file /etc/fstab.

mv *source destination*—move the item *source* to *destination*. If *destination* is a filename, then this is a rename operation; otherwise, the file is moved to the new directory.

nice *value command*—execute command with the niceness value of *value*. Can also adjust a running process using nice –n *value PID*.

nslookup *alias*—query the local DNS for the IP address of the given alias. Can also be called as nslookup *alias server* if you want to specify a different DNS.

passwd *username*—prompt the user to change username's password. If username is omitted, then change the current user's password. Unless you are root, you will be required to enter username's password before changing it. Options include:
-l—lock the account so that only root can access it
-u—unlock the account
-d—delete the password for the account
-x—establish an expiration date for the password (a date by which the password must be changed)
-w—establish the number of days in advance that the user will be warned to change passwords

ps—process status information, that is, output the running processes. There are many options available. The ps command by itself only prints processes running in the current terminal window owned by the current user. Other options of note include:
a—print processes of all users of the current window
f—denote parent-child relationships using ASCII "art" output
u—user "user-oriented format" for output (see below)
x—print processes no matter which window or console it originated from
Using option u outputs processes by user name, PID, %CPU usage (the percentage of time it has been running versus waiting), %memory usage (as a percentage of the computer's main memory capacity), total amount of memory usage, virtual memory utilization, the console that started the process, the process' execution status, the date that the process started, the amount of CPU time it has used, and the process' name.

pwd—print working directory.

Redirection—although this is not a Linux command, you can use redirection in your Linux commands in order to alter the standard input or output of a command. The redirection operators are
>—redirect output to the given file as in cat foo1.txt foot2.txt > foo3.txt.
>>—redirect output to append to the given file.
<—redirect input from keyboard to the given file.

<<—redirect input from file to keyboard, end input after given string is reached, as in cat << quit > foo.txt, which will allow the user to enter items until the string "quit" is entered.

|—a pipe, take the output of one Linux command and use it as input to another, such as ls –al | egrep "rwx", which takes the output of the ls –al command and uses it as input to the egrep command.

rm item(s)—remove (delete) the item(s). Options include –r for a recursive delete (this deletes subdirectories and their contents recursively), –f to delete without permission, and –i to delete interactively—that is, to ask the user before deleting each item.

rmdir item—remove the given directory. This is only available if the directory is empty.

service *servicename* command—command to change the service *servicename*. The command is one of start, stop, status, and restart.

ssh host—open a secure shell communication with host using encryption. This has replaced telnet, which opens communication with a host but without encryption.

su [username]—switch to username's account. If username is omitted, switch to root. The instruction requires that you provide the account's password (unless you are currently root).

sudo *username command*—literally, this command executes *command* as *username*, that is, it executes the specified command as if it was issued by the given username. Commonly, this command is used without *username* so that the *command* executes as if issued by root. An entry must be placed in the file /etc/sudoers that specifies that the given user has access to the specified command. For instance, if zappaf were permitted to use useradd, the following line would be added to the suoders file:

```
zappaf localhost=/usr/sbin/useradd
```

Before the command executes, the user is required to enter their password.

tar *filename* [*source*]—the tape archive utility bundles one or more files and directories into a single file. When the options –xf are used, the files/directories are extracted from the bundle and copied into the current directory. When used with –cf, a new tar file is created from the entry(ies) listed under source. For instance, tar –cf foo.tar /myfiles would take all of the files in the subdirectory myfiles along with the myfiles directory and place them into foo.tar. An added option, z, will gzip or gunzip the files (gzip when used with –c and gunzip when used with –x). –c stands for create, –x for extract.

useradd *options username*—create a new account with the name username. This instruction can only be executed by root. The instruction adds associated entries in /etc/passwd, /etc/shadow, and /etc/group. The command has many options including:
-m—create a home directory
-d *dir*—use *dir* the in place of the default directory
-s *shell*—use *shell* in place of the default log in shell

-G *groups*—add user to the list of *groups* (the list is separated by commas but no spaces)

-c *comment*—add *comment* to the /etc/passwd file, usually used to specify the user's full name

-p *password*—use the encrypted *password* for an initial password

-u *uid*—give the user the specified *uid* (user ID)

userdel ***username***—delete username's account. The option –f forces the deletion even if the user is still logged in and –r removes the user's home directory and e-mail files along with the user account.

usermod ***options username***—modifies username's account based on the options specified. These options are the same as in useradd.

vi—launch the vi text editor (also vim). Emacs launches the Emacs text editor.

wget ***URL***—download the file specified by the URL. Similar to entering an http request in a web browser except that the file is stored to disk.

who—lists all usernames of currently logged in users.

whoami—outputs the current user's username. This is useful if you cannot remember who you are if you have used su.

yum command package—the Yellowdog Updater, Modifier program is used to install, remove, and update software packages. Command is usually one of install, update, check-update, upgrade, remove, list, info, clean, or reinstall. The yum program uses the rpm program.

LINUX FILES AND DIRECTORIES OF NOTE

/etc/group—file that stores all group information.

/etc/init.d—a Linux directory where services are stored.

/etc/inittab—the initialization script first executed after the Linux operating system boots, it is responsible for setting the runlevel.

/etc/passwd—file that stores all user account information (excluding user passwords).

/etc/rc#.d—a set of directories in Linux which stores symbolic links to services that should be started or stopped at system initialization time. The # is the run-level, one of 0–6.

/etc/rc.sysinit—a script executed during system initialization that starts and stops services.

/etc/shadow—file that stores all user passwords in an encrypted form.

/etc/syslog.conf—the configuration file for the syslogd service that logs application software and non-kernel operating system messages.

DOS COMMANDS OF NOTE

cd or chdir—change directories, used with a path as in cd ..\foo\bar. See also C:, D:, etc.

chkdsk—checks the specified disk for file system integrity.

cls—clear the screen.

copy—copy a file from one location to another in the file system. Wildcards are permissible.

C:—change to the C: partition. D:—change to the D: partition. This also applies for any other partitions.

del (also erase)—delete the given file(s). Wildcards are permissible.
dir—list the contents of the current directory or the given directory if a path is supplied.
echo—output the given string and/or variables to the screen.
edit—launch the DOS text editor program.
exit—can be used to exit the DOS window if running DOS within windows.
format—to format a disk. WARNING: formatting a disk erases its contents.
help—gives general help on DOS. help *command* lists *command*'s help page.
md (also mkdir)—create a new directory.
mem—display memory usage information.
more—displays contents of a file, one screen at a time.
move—move a file to a new location. If the new location is the same directory, this performs a rename. The ren instruction is a rename instruction.
path—display the contents of the path variable. Can also be used to set new paths in the variable.
print—sends a specified file to a printer.
rd (also rmdir)—remove an empty directory.
undel—undelete a previously deleted file. This works on files that are "recoverable" only. A deleted file may or may not be recoverable based on the amount of time that has elapsed since the deletion.

Index

E

F

G

H

I

N

Q

R

S